Global Marine Science and Carlsberg

The Golden Connections of Johannes Schmidt (1877–1933)

Bo Poulsen

BRILL

LEIDEN | BOSTON

Cover illustration: Ove Paulsen, Johannes Schmidt and J. N. Nielsen on board the *Thor*. *From left to right:* Ove Paulsen (plankton), Johannes Schmidt (fisheries) and J. N. Nielsen (hydrography) are posing on board the S/S *Thor* during the 1903 expedition (Carlsberg Archives).

Library of Congress Control Number: 2016947571

ISBN 978-90-04- 31636-2 (hardback)
ISBN 978-90-04- 31639-3 (e-book)

Want or need Open Access? Brill Open offers you the choice to make your research freely accessible online in exchange for a publication charge. Review your various options on brill.com/brill-open.

Typeface for the Latin, Greek, and Cyrillic scripts: "Brill". See and download: brill.com/brill-typeface.

This book is printed on acid-free paper and produced in a sustainable manner.

Contents

Summary *459*

Archival sources *469*

Published sources *497*

Index *515*

List of figures, tables and illustrations

Acknowledgements

For me, this book began in early 2011, when I received an email from Canadian marine scientist Mike Sinclair, who then served as President of the International Council for the Exploration of the Sea (ICES). Mike invited me to the ICES headquarters in Copenhagen about a project on the World famous marine scientist, Johannes Schmidt. Mike thought it was intriguing that Schmidt was once employed by the owners of Carlsberg, one of the World's leading breweries. Yet, to submit a research proposal to the Carlsberg Foundation the applicant needs to be a Dane or based in Denmark. I then scouted the potential to do an archival based study and I realized that the potential was huge. Eventually, the Carlsberg Foundation granted me a generous two years of research time to write this book, and I am very thankful for this chance to do what at least for me at has been a really gratifying piece of work.

As I began working on Schmidt, Flemming Besenbacher became Chairman of the Carlsberg Foundation, and I am very thankful to Flemming for having taken time to follow this project with great enthusiasm and encouraging remarks along the way. This book became twice as long as first planned, simply because there was so much more to Johannes Schmidt. Thus in 2014, the foundation granted me funds for another two months of time to proceed with the project, while my head of department, Marianne Rostgaard at Aalborg University reduced my teaching load for another two months. Thank you both!

Jesse Ausubel and Brian R. McKenzie provided crucial support for the initial project. So did Mogens Rüdiger, Michael Wagner and Benni Winding Hansen. From start to finish I have benefitted tremendously from an unofficial backing group of friendly ears. René Taudal Poulsen, Anne Husum Marboe, Michael Boas Pedersen, Verena Winiwarter, Poul Holm and Anne Lif Lund Jacobsen have all frequently lent an ear to fuzzy and bright ideas as well as my many worries.

Mike Sinclair and Vera Schwach both read and critiqued most of the manuscript, sharing with me their vast knowledge of marine science history. I am also very grateful for discussions with Peder Olesen Larsen, who knows all about biochemistry. He also kindly shared with me his own manuscript, which is now published as: *Iværksætter og forvaltere. Carlsberg Laboratorium 1876–2014*. Copenhagen: Carlsbergfondet; 2015.

Archives and libraries are wonderful places, but I have received an extraordinary kind and efficient help from Ulla Nymand at the Carlsberg Archives, Esther Jarløv at the Carlsberg Foundation, and Henning Knudsen at the Museum of Natural History, Copenhagen.

When this project commenced I had just started working at Aalborg University, and this project has benefitted enormously from frequent conversations with so many of my fellow historians there that it would be unfair to mention some and not others. You know, who you are! Stephanie Erlenbach, however, deserves special mentioning for having spent eight weeks digitizing important parts of the archival material for this manuscript. Also at Aalborg University, Helle Weiergang and Julie Skibsted Larsen have been the most efficient and helpful administrators of this project, one can think of.

Conferences, workshops and networks are wonderful opportunities to discuss ideas and present new material. I am grateful for being given this opportunity at the Oceans Past Conferences in Tallin (2015) and in Fremantle (2012), the 7th Biennial conference of the European Society for Environmental History in Munich (2013), The Dansk Historikermøde in Odense (2014), the 17th Dansk Havforskermøde in Roskilde (2013), the Havet som Ressource conference in Esbjerg (2013) the Centre for Science Studies Colloquium, University of Aarhus (2013) and the 100th Statutory Meeting of the International Council for the Exploration of the Sea on Johannes Schmidt's home turf at the Carlsberg Academy (2012). Finally, this book has benefitted from the COST Action on Oceans Past Platform (lS1403), supported by COST (European Cooperation in Science and Technology).

When I ventured into finding an international publisher I was greatly aided by advice, chiefly from John R. McNeill, but also from Jeff Bolster and Jake Hamblin. Kira Pravato was most helpful and quick with many of the illustrations in the book, and Kenn Tarbensen provided valuable advice. With remarkable attention to detail, my eagle-eyed copy-editor Claire Neesham made countless improvements to each and every page of the manuscript. At Brill Publishers, I have been in the best of hands with Michiel Thijssen and Wendel Scholma as well as Frits Fritschy, whom I am grateful for gently turning my rough manuscript into a fine hardback. Thanks also to two anonymous peer reviewers who provided insightful comments to improve the final manuscript. Any mistakes still in the manuscript however, are mine, not theirs.

Thanks to a generous donation from the Carlsberg Foundation, it was possible to include a large amount of illustrations, while at the same time making the book much more affordable.

The utmost gratitude is reserved for my family. My wife, Anne Dorthe is herself a very busy historian, and my debt to her is immeasurable, as is my love. Our three children, Astrid, Niels and Laura are constant sources of joy, inspiration and the most pleasant distractions, and when I have been absent for archival studies and conferences, my parents-in-law, Anne Grethe and Preben have so often driven from afar to help with babysitting.

The book is dedicated to my father, Jan, who has taught me so many things in life.

Introduction

This manuscript tells the surprising story of how the world-famous brewery Carlsberg – by accident – came to support international marine science and in doing so made Denmark a truly global player in marine science in the 1920s. The story is told through the life and work of marine scientist, research entrepreneur, and expedition leader, Johannes Schmidt (1877–1933).

Schmidt was in charge of a staggering 26 oceangoing expeditions, resulting in landmark achievements such as the discovery of the Sargasso Sea between Bermuda and the Caribbean as the breeding ground for the eel, which had been a mystery since the time of Aristotle 2400 years earlier. In the first decades of the 20th century the study of the eel was subject to intense scientific rivalry, chiefly between Italy and Denmark. The first nation to find their breeding ground might get the upper hand in terms of catching eel in great numbers, not to mention the associated scientific glory.

As it turned out it was Schmidt who gradually became one the most decorated scientists across continents and disciplines, receiving honours such as the Alexander Agassiz Medal of the US National Academy of Sciences, the Darwin Medal from the Royal Society of London, The Weldon Memorial Prize, the French Médaille Geoffrey Saint-Hilaire, and the Danish Galathea Medal. He was bestowed with membership of 22 different learned societies from around the world, while he received no less than 15 official decorations from Denmark, Iceland, France, Italy, Norway, Spain, Tunisia, Morocco, Turkey, Greece, The Netherlands, and what was then the Kingdom of Siam.

A significant part of Schmidt's fame came about as a result of his great pioneering effort in popularizing his research through the new mass media such as radio and cinemas. In this, he was a forerunner to the likes of Jacques Cousteau or Hans Haas, yet Schmidt was first and foremost a field scientist in the mould of 19th century giants such as Charles Darwin with his voyage on the HMS *Beagle* or Wyville Thompson and colleagues on the 1872–1876 Challenger Expedition around the

World, not to mention his contemporary, Norwegian Arctic explorer and marine scientist, Fridtjof Nansen.

The main argument centres on the abovementioned accident, that Carlsberg, a name associated with first class beer making, became a force of international status and fame within marine science during the first three decades of the 20th century. The contingency of a core group of world class scientists at the heart of the confluence of two internationally oriented science institutions is the most important reason for this 'happy' accident.

These institutions included the Board of Directors of the Carlsberg Foundation, which is Denmark's oldest and most important private research endowment, and also the legally controlling body of the brewery bearing the same name. To this day the board is run by a small group of university professors, and before WWII they clearly had the desire and the means to support marine science explorations, chiefly in collaboration with the Danish government and the most important private Danish corporation of the time, the East Asiatic Company. Around the same time the inter-governmental science organization, the International Council for the Exploration of the Sea (ICES) was established. From 1902 onwards ICES has had its headquarters in Copenhagen.

A small group of scientists and research managers, chiefly Johannes Schmidt (1877–1933), found a way to profit from having two internationally oriented science organizations on their doorstep, and in so doing made Denmark a truly global player within marine science. This was possible, in part, because Schmidt proved to be an adept organizer and thinker, but it would not have turned out so successful had Schmidt not been made director (1910–1933) of the Physiological Department of the Carlsberg Laboratory. Gradually, and with the tacit approval of his board, Schmidt turned his lavish laboratory from a centre of beer related research into a veritable marine science laboratory and zoological collection. Thus in 1930 at the pinnacle of Schmidt's career, Carlsberg's name was associated with landmark marine science achievements around the world, just as much as with the production of beer, and the laboratory was a marine science embassy for the world.

In this process spanning three decades, the huge hurdle of the cost of sending research vessels to sea was overcome through ingenious fundraising, while the scientific frontier of the deep sea was pushed literally kilometres down into the ocean and wide across the Atlantic at first, then later the Pacific and Indian Oceans using custom made fishing gear and a burgeoning international network of a couple of hundred marine scientists and laymen.[1]

The power of the international network of marine scientists and the strategic asset of having funds from the Carlsberg Foundation are shown to have been vital in a situation, where World War One was threatening the very existence of ICES, and Danish aspirations as a political power of some note in the North Atlantic realm began to wane. Both of these outside challenges were turned into relative advantages for Danish marine science once the world's oceans became navigable again after 1918. This study is important in three different ways:

First of all, from an empirical point of view this study connects a number of spheres, which have not previously been investigated in relation to one another. In the current literature on the history of international marine science there has been little explanation for the involvement of Carlsberg, which is better known as a brewing company. Meanwhile, within the Carlsberg Foundation Schmidt has sometimes been seen as an embarrassment to the foundation, to the company, and to the laboratory. Indeed, prior to the proposal of this project there was a real danger that Schmidt's legacy was set to disappear in infamy. This book therefore attempts to paint a realistic picture of someone who is neither a villain nor a heroic genius separated from the societal context of his scientific achievements. As another separate historiographical sphere, the political history of the Danish North Atlantic colonial presence in Greenland, Iceland, and the West Indies has not previously been considered in connection with dreams of marine science glory in an otherwise tiny and waning seaborne empire.

Secondly, the empirical results should present a clear picture of a scientific world of marine science, deeply rooted in its societal and political context. In fact, it is apparent from Schmidt's story that no matter how exact and accurate science is practiced in ocean-scale marine science there are always outside influences.

Thirdly, from the point of view of today's world of science management, practices, and fundraising the story told shows the importance of sustained long term commitment as a prerequisite for truly ground-breaking research. It also highlights the importance of interdisciplinary ingenuity, inter-institutional collaboration, and international commitment across traditional disciplines. This is an important aspect of why an otherwise small and insignificant country produced a whole array of world class scientists.

1 In this study the term 'marine scientist' is defined as someone undertaking the scientific study of the physical ocean, marine ecosystems, and various marine life forms. At times over the past century this has been synonymous with the term oceanographer, but in this manuscript marine scientist and marine science are used, rather than oceanographer, oceanography, or oceanology.

A symmetrical biography

Across the natural sciences, social sciences, and humanities, a scholarly argument can be presented in a number of ways ranging from a one page letter to for instance the journal *Nature* to lengthy monographs such as this one. One style of presentation is not *per se* better than the other, but choices need to be made depending on the topic and type of argument presented.[2] In this case the main argument, briefly touched upon in the introductory section is best presented in the form of a monograph of the kind, often referred to as a biography. These are the main reasons for writing a biography on this topic:

The cultural dynamics of marine research was embodied by the men (very few women took an active part in the explorations touched upon in this volume) who carried out the expeditions, the laboratory experiments, and the political negotiations associated with their research. In the case of the life and career of Johannes Schmidt there is a clear connection between man and matter. Thomas Edison famously coined the phrase that genius is 1% inspiration and 99% perspiration.[3] This is very true in this case, and to fully present the sweaty parts of Schmidt's genius the aim of this manuscript is to create a so-called symmetrical biography.[4] This means Schmidt's scientific achievements and their paradigmatic context are presented on an equal footing with the main political, societal, and personal achievements and structures influencing his life and career. Looking at a scientist from both within his science and from the context is rarely possible in the brief format of an academic article, whereas the monograph gives the space for consideration of the most striking or important intersections between science and society in a given context.

The genre of historical biography has been intensely debated in recent decades. Lately, a Norwegian dissertation dealing with the life of Fridtjof Nansen, the Norwegian Arctic explorer, political activist, and marine scientist demonstrated a number of different approaches frequently used in such a biography.[5] In a Danish context, several discussions among historians, have circled on the issue of what is a proper biography. A recurrent theme in these discussions is the extent to which it is possible to write an intimate portrait of a deceased person and in more extreme cases debates as to the scholarly merit of writing about just one individual.[6] Thus, in relation to Thomas Söderqvist's biography of Danish Nobel Laureate Niels Kaj Jerne, the explanatory power of an

2 Poulsen B. Marin miljøhistorie – samfundsrelevant tværvidenskab. In: Enevoldsen T, Jelsøe E, eds. *Tværvidenskab i teori og praksis.* Copenhagen: Hans Reitzel; 2012:180–200.

3 Quoted from: Allen RC. *The British industrial revolution in global perspective.* Cambridge University Press Cambridge; 2009.

4 Knudsen H. *Videnskabens mand. Fysiologen, formidleren og forskningsaktivisten Poul Brandt Rehberg.* Aarhus: Aarhus Universitetsforlag; 2010.

5 Jølle HD. *Innledning til en historiefaglig biografi tillegg til Nansen. Oppdageren.* [Ph.D.]. Tromsø: Universitetet i Tromsø; 2013.

6 In the case of Possing B. *Viljens styrke: Nathalie zahle: En biografi om dannelse, køn og magtfuldkommenhed.* Gyldendal; 1992 a lenghty debate ensued as to whether or not a focus on the life of one individual qualified as a relevant topic for a Danish 2nd doctorate, the so-called Doktordisputats or Habilitation as it is known in German.

investigation of the psychology of an important scientist as a way of linking his main theories with his personal life, or describing his whole existence so to speak, became a topic for debate.[7] This is a fascinating topic in its own right, yet by way of disclaimer, the inner life of Johannes Schmidt is not at the forefront in this biography. This author does not claim to know the inner workings of Schmidt's mind, except for what he trusted to the diary he kept from 1929–1933, or from what can be gleaned from his personal correspondence with his close friends Carl H. Ostenfeld and Ove Paulsen. The focus in this book is on Schmidt as a historical agent, as someone who functioned in a societal context. In this way, the approach chosen for this volume is more akin to many biographies within the sub-discipline of political history. One such example is the biography of Danish diplomat, Henrik Kaufmann, which incidentally forms part of the final discussion in this book.[8]

Actors and structures – money and conflict

In this case the choice of a biography serves to underline the importance of actors and networks within a framework of institutions and structures, as well as the background of the natural environment. Biographies come in many shapes and sizes, and there are limits to what is included in the following. Johannes Schmidt is dealt with as an actor in a network, which is framed within certain societal and paradigmatic structures and institutions. As an actor in a network, the role of Schmidt is that of a principal motif in a group portrait, where the other main figures consist of his closely knit network of family and colleagues; principally marine scientist C. G. Joh. Petersen, the ecologist Eugen Warming, botanist Carl H. Ostenfeld, and the administrator and science organizer Christian F. Drechsel.

With a reference to American historian, Donald Worster this is biography as environmental history trying to see '…that individuals as well as societies are shaped by natural as well as cultural influences.' This has the effect on this narrative that there is a frequent shift in scale from the single individual of Johannes Schmidt to the scale of international politics or even the long-term geological processes of continental drift. In the words of Worster; '…any human life is more complex than the sum of its social relations. Human life has a material and a biological foundation and it has an ecological and an economic context; that context can be as small as a farm or city block or as wide as a continent or planet.'[9]

7 Söderqvist T. *Hvilken kamp for at undslippe. En biografi om immunologen og nobelpristageren Niels Kaj Jerne*. Copenhagen: Borgens Forlag; 1998; Pedersen J. Hvilken kamp for at undslippe (kritik mod Thomas Söderqvist). Historisk *Tidsskrift*. 1999(99):480–491; Söderqvist T. Hvilken kamp for at undslippe. svar til jan pedersen. *Historisk Tidsskrift*. 1999;99(2); Kjørup S. Billeder, kilder og gode idéer. *Personalhistorisk Tidsskrift*. 1999(2):259–266; Koch CH. En ny videnskabshistorisk genre? *Personalhistorisk Tidsskrift*. 1999(2):280–286; Kondrup JJ. Biografen og den biograferede. *Personalhistorisk Tidsskrift*. 1999(1999: 2):252–258.

8 Lidegaard B. *I kongens navn. Henrik Kauffmann i dansk diplomati 1919–1958*. Copenhagen: Samlerens Forlag; 1996.

9 Worster D. Living in nature: Biography and Environmental History. In: Myllyntaus T, ed. *Thinking through the Environment. Green Approaches to Global History*. Cambridge: White Horse Press; 2011:28–39.

The image painted however, is not that of a static network or natural context for that matter. Schmidt's network evolved, expanded, and contracted at various points in his life, under the influence of societal change and encounters with a dynamic marine environment. Schmidt's life was filled with spectacular events, and he corresponded with several hundred different people during his career spanning some 35 years. The French historian Fernand Braudel once opened an article with a metaphor on fireflies as nature's equivalent of historical events: 'I remember a night near Bahia, when I was enveloped in a firework display of phosphorescent fireflies; their pale lights glowed, went out, shone again, all without piercing the night with any true illumination. So it is with events; beyond their glow, darkness prevails.'[10] All fleeting apparitions of events and people, are not all equally important, and it is a deliberate idea that the following several hundred pages are not filled with too many pale lights from fireflies, or trivial pieces of information.

Rather the objective is to detect a number of recurrent features in Schmidt's life and then address a set of similar questions to a number of key projects, where Schmidt was involved. The main questions addressed in the following chapters are:

1. What were the sources of funding for Schmidt's research activities, c. 1898–1933?
2. What were the scientific, institutional, national and socio-economic justifications for the awarding of the diverse grants?
3. Did Schmidt's research impact the image of his private and public sponsors?
4. What was the role of institutions such as ICES in the coordination of Schmidt's research?

Then in the second to last main chapter (Chapter 16 'Measures of impact and legacy'), the results are summarized in relation to Schmidt's over all merit and legacy as a scientist. Here, the focus is more attuned to addressing these latter two overarching questions:

5. What have been the scientific impacts of the results of the research funded by the Carlsberg Foundation and conducted by Schmidt, both in relation to oceanography and beer production?
6. Were the Carlsberg Foundation's subsequent funding strategies negatively impacted by the funding of Schmidt's research?

10 Braudel F. *On history*. University of Chicago Press; 1982. pp. 10–11.

Working with these six questions a number of recurrent features emerged from a dialogue with Schmidt's own publications, a substantial bulk of archival material, newspapers, and other scholarly publications from the past century on diverse topics such as the history of Danish foreign policy, the history of genetics, the history of beer brewing, fisheries investigations past and present, zoology, geology, and hydrographic tables on nutrient flows at great depth. The smallest plankton from Lake Gatun and the drift of the largest continents are equally at risk of being fleeting apparitions on a tropical night, but the questions above add structure to this book.

When looked at from a distance, Schmidt's c. 35 years as an active researcher who launched no less than 26 oceangoing expeditions includes both striking consistencies, as well as some major breakpoints, when following the flow of money or the triggers for conflict in relation to his many oceangoing expeditions.

In a fairly recent anthology on *Scientists and Scholars in the Field* the editors stated by way of introduction that: 'The fact that until recently the study of the history of scientific field practices has received comparably sparse attention may well reflect the relatively low academic status usually ascribed to fieldwork. Since the nineteenth century, standards of scientific stringency, neutrality, and accountability have been defined according to ideals manifested in the limited, generalized, and carefully composed framework of the laboratory. Fieldwork on the other hand must by definition be conducted in intimate and often unpredictable interaction with specific geographic localities with specific characteristics that influence, shape, and to some degree even constitute results.'[11]

Now, Johannes Schmidt himself probably never thought of his expeditions as a low status activity, but the rest of this quotation fits well with the underlying premise of this investigation. This study proposes that the sheer scale of oceanic voyages makes them a particularly interesting study object. It is extremely expensive by any standard to equip a research vessel with men and gear, so significant aspects of this kind of research include the funding bodies and their motivations.

This manuscript further argues that the study of fieldwork and expeditions is important because it lays bare a relationship, which in other circumstances might be hidden as scientists are often portrayed as being above the everyday turmoil of politics. In reality this is rarely the case, as we will see in the following chapters when it comes to a whole series of substantial advances in marine science in the first

11 Nielsen, Kristian H., Harbsmeier, Michael and Ries, Christopher J.: ‚Studying Scientists and Scholars in the Field', in: Kristian H. Nielsen, Michael Harbsmeier and Christopher J. Ries (eds.): *Scientists and Scholars in the Field*, (Aarhus: Aarhus University Press, 2012), p. 12–13.

three decades of the 20th century. Here Schmidt's and his colleagues' constant need to go to sea on a well-equipped research vessel are discussed alongside the scientists' requirement for a lot of money. When searching for money, arguments, or justifications need to be made as to why this or that project ought to receive funding. The method of following the money functions as a way of getting close to the arguments presented, both politically and scientifically. Therefore, the following 17 chapters constantly return to the topic of funding, and some may feel that this is at the expense of a few details about other important aspects of Schmidt's life and work.

For the same political and scientific reasons, the following chapters are filled with conflicts. Personal disagreements, scientific feuds, and geopolitical conflicts are used to illustrate the strong wills of the protagonists and the bold arguments that needed justification. In these justifications it is possible to discover layers of opinion, which one would never find by looking at only for instance Schmidt's research publications or from reading anniversary volumes covering the corporate history of a private or public institution. In such publications conflicts tend to be toned down or erased completely.

Some conflicts have a clear material substance, for instance conflicts involving geopolitics. The American environmental historian Brian Payne has introduced the concept of *environmental territorialization* into the history of fisheries in a North American case study. Payne defines territorialization as the underlying process of encroachment of a territory – in his case the fight for securing fishing rights in the American North-Western North Atlantic Ocean, which was not the property of any one specific fishing operation or fishing community.[12] In the case of Johannes Schmidt the concept of environmental territorialization is a useful term for his and his colleagues' desire to secure scientific results for the glory of their own country, often in explicit competition with other countries such as Norway or Italy. However, in more than one case, the territorialization went beyond the realm of science and into the realm of international business and international politics. This book tries to emphasize Schmidt's contantly evolving perception of nature as well as the impact from frequent very concrete encounters with the natural environment. This situates the story within modern marine environmental history.[13]

Admittedly, there is also an entertainment value in the discussion of conflict, and the narrative in the following chapters does exploit the

12 Payne, Brian J. *Fishing a Borderless Sea: Environmental Territorialism in the North Atlantic 1818–1910*, East Lansing: Michigan State University Press. 2010.

13 Schwerdtner-Máñez, K. & Poulsen, B. (eds.) *Perspectives on Oceans Past: A Handbook of Marine Environmental History,* (Springer Science + Business Media B. V., 2016); Schwerdtner-Mánez, Kathleen; Holm, Poul; Blight, Louise; Coll, Marta; MacDiarmid, Alison; Ojaveer, Henn; Poulsen, Bo; Tull, Malcolm. The future of the oceans past: Towards a global marine historical research initiative. I: *P L o S One*, Vol. 9, Nr. 7, e101466, (2014).

tensions and verbal fights as a means to create suspense for the reader, but primarily the exposé of conflicts is a means to tease out information for the sake of the overall argument.

What emerges then from this focus on conflicts and fundraising schemes is a clear pattern at the structural level, where a few institutions were crucial for the success of Schmidt and his colleagues. In terms of institutional and financial support for his many projects seven different types of funding sources are recurrent. These are the Danish Biologisk Station, the Carlsberg Foundation, the Carlsberg Laboratorium, The Danish Commission for Marine Science operating under the auspices of the International Council for the Exploration of the Sea, Various direct and indirect government support, private corporate funding, and as a strategic reserve the fortune of Schmidt's family-in-law.

Main argument

In short, the recurrent features associated with money and conflict add structure to the questions addressed in this book, and this is what brings forward the main thesis, or argument, in this volume, namely the proposition that:

Within a core group of natural scientists and science managers, Johannes Schmidt was the key actor in an accidental institutionalization of marine science in the first three decades of the 20th century, where the various branches of the Carlsberg Foundation in combination with the International Council for the Exploration of the Sea, the Danish state and a number of private companies supported a remarkable international hub for World class research.

This book further argues that to fully understand the scientific accomplishments of Johannes Schmidt, the actual science needs to be situated in the right political and societal contexts, where Schmidt and his core network represent a distinct cosmopolitan conservative world view with a firm belief in the value of upholding a Danish presence on the international scene, whether in connection to science, business or politics. The formal North Atlantic Empire of the small state of Denmark was waning rapidly during the first three decades of the 20th century, but an informal empire of science was erected instead.

This is synthesized in the final main chapter of the book under the headline, 'A fortunate man.'

To sea – a placeless place

A few years ago American historian John Gillis noted that the sea was underrated, and largely unnoticed in society at large, also historically, and that historical scholarship dealing with aspects of the sea was short in supply. While there has in fact been a tremendous increase in writing about the sea in the last couple of decades both within the field of marine environmental history, history of marine science, and maritime social and economic history, Gillis made an important point as to why the sea is often absent from the societal discourses: 'Until recently, seas were seen as placeless places, without topography. In western cultures, the sea was perceived as pure nature, something best left to the natural sciences.'[14] As mentioned above natural sciences do not thrive in a vacuum separate from the rest of the World, and sciences in the field such as much marine science, are always in need of expensive ship time, yet this aspect of research can be forgotten in a placeless place such as the sea.

The issues relating to Arctic science are similar. The vast majority of people on Earth never visit the Arctic region, and until the last few decades, Arctic societies were difficult to access from the rest of the World. The scores of Arctic explorers and scientists roaming Greenland, the North and South Poles, Arctic Canada, Russia, and the Northern Scandinavian Peninsula from the mid-1800s onwards were exceptional. They embodied the connectivity between the Arctic extremes and modern civilization at large. According to Bravo and Sörlin they did so by way of telling stories.[15] Their narratives became crucial placeholders for the gradual public recognition that there was more to these areas than just 'chunks of ice' in the word of a Danish politician, whom we will encounter in Chapter 9 of this volume.

Public recognition and scholarly or scientific recognition went hand in hand, and convincing storytelling was part of the job for many scientists, if they were to secure corporate and state funding. The same goes for marine science. Hardly anyone ever goes to the Mid North Atlantic Ocean if they are not shipping cargo across the sea between Europe and North America. This is a huge challenge for stirring interest about issues of the seas of the World.

The marine scientists portrayed in this book were keenly aware of this problem. Like the Arctic explorers, the realized that they would need to tell stories of what they did on the high seas far away to drum

14 Gillis, John R., 'Filling the Blue Hole in Environmental History,' http://johnrgillis.com/works/bluehole2.pdf The argument is expanded in: Gillis J. *The human shore: Seacoasts in history.* Chicago & London: The University of Chicago Press; 2012

15 Bravo M, Sörlin S. Narrative and practice – an introduction. In: Bravo M, Sörlin S, eds. *Narrating the Arctic: A cultural history of Nordic scientific practices.* Science History Publications/USA; 2002:3–32.

up interest, and so attract funding, for their endeavours. Early twentieth century marine science operated almost by definition in *mare incognitum*, and in this volume there are several instances where the exploration itself becomes a waypoint in the demarcation of the Danish sphere of influence. While smallish Denmark was hardly a big geopolitical player in the Atlantic Ocean, there was great symbolic national value for its research scientists to be seen on the high seas and to find the eel's breeding place etc. In this way the sea was given a mental place as well as a more literal and concretely observable place, when cod for instance were marked with numbered metal pieces to be retrieved by fishermen at a later stage in the fishes' lives.

Schmidt and several of his colleagues were great communicators of science. They were storytellers just like many Arctic explorers. This offered great advantages to the author of this manuscript, but it also raised some significant challenges, particularly when choosing a story line for a biography of one such storyteller. The great opportunity is the fact that the material presented in a book dealing with so many oceangoing expeditions lends itself to a classic narrative plot with a beginning, where the stage is set, and a formulation of a problem to be overcome, the quest to fulfil the mission at hand and, following the format of a narrative climax, the solution of the problem. Each of Schmidt's expeditions can be written as such a classic narrative: Identification of the scientific problem – followed by sustained fundraising – followed by preparatory phase – followed by an ocean cruise – and finishing with the results, and perhaps the raising of a few refined questions about subsequent expeditions.

Nonetheless, and here lies the challenge in this type of material, in the case of Schmidt the narratives were not only structured around his own success, they were also structuring for his success. Therefore, this book could be viewed as a revision of Schmidt's own storytelling project. The winners are the writers of history, and Schmidt was a winner. The fact that Schmidt came to write substantial parts of his own history, while his legacy was maintained by his former assistants and colleagues in the subsequent decades, means that there is good reason to be careful about the validity of Schmidt's meta-narrative. This can give the impression that each and every obstacle was overcome, and that the outcome of each and every expedition was planned as one successful continuation of the previous one. As we shall see, this is true only in hindsight, but it served Schmidt well when attracting new funds, to sustain the story of him self as a hugely successful and flawless expedition leader. The truth was different from the official story in a number of instances, as we shall see from the descriptions in this manuscript. In the following chapters each major expedition is narrated as one story, however, Schmidt's stories are often revised as these new narratives are developed.

Chapter 1 sets the stage, introducing Schmidt's family background and the scientific and corporate context of his upbringing. In so far as narrative goes the next chapters, 2–4 are characterized by two story lines. One story is centred on the search for offspring of the freshwater eel in Europe (*Anguilla anguilla*) in the Mediterranean Sea and along the continental shelf off the West coast of Europe providing the leitmotiv for the story of much of the rest of Schmidt's life. The other main story revolves around the dominant powerplay between the leading Danish marine scientists, caused by different views on politics and research strategies. Chapters 5–6 are different from the rest in the sense that they do no deal primarily with oceangoing expeditions, but rather with Schmidt's laboratory life at the Carlsberg Laboratorium from 1909–1933. Chapters 7–11 are all interlinked as they form one decade-long quest to search for and discover the breeding places of the eel, located in the Sargasso Sea circa halfway between Bermuda and the Eastern Caribbean islands. Chapter 12 deals more intensely with Schmidt's efforts in relation to fisheries investigations of commercially important species in the North Atlantic Ocean, but this chapter also touches upon geopolitical challenges associated with the Danish possessions in the Northern North Atlantic Ocean. Chapters 13–15 and partly Chapter 16 serve to tell the story of how Schmidt first dreamt of launching a circumnavigation, and secondly how he managed to plan and execute this huge project. Finally Chapter 16 summarizes Schmidt's posthumous legacy.

This book is intended to be readable; a piece of storytelling in its own right. It follows from the symmetry mentioned above that the work presented in this book deals with a subject that crosses traditional disciplinary boundaries. Some arguments presented fall more readily within a natural science context, while others are clearly framed in a context of politics and the history of human society. Nonetheless, this border-crossing material is presented in one extended metanarrative framed around the life and career of Johannes Schmidt. Schmidt is interesting, both from the perspective of marine science, and in a more general historical context. These two perspectives are best understood as two sides of the same coin. Therefore, the style of presentation chosen for this book attempts to intertwine science and society. The sections that deal with Danish foreign policy, for instance, are written to be readily accessible to a marine scientist from Japan or Canada. On the other hand, Schmidt's far flung research of Indo-Pacific eels or his genetic studies are hopefully presented in such a way that a Danish historian with no prior knowledge on the subject should be able to grasp the main ideas as they are presented. In other words, this manuscript hopefully appeals to the generally interested reader.

Historiography

In terms of historiography this monograph is situated mainly within the field of history of marine science, but it is clearly also a more general work of history with a keen eye on the societal context of scientific practices and aspirations. In this way, the angle of the argument is akin to books on the international politics surrounding scientific fisheries management such as Carmel Finley's *All the fish in the sea*, or the politics of Cold War marine science practices discussed in Jacob Hamblin's *Oceanographers and the Cold War. Disciples of Marine Science*.[16]

This is also a work that resembles a number of recent books on the history of marine science institutions such Jennifer Hubbard's *A Science on the Scales: The rise of Canadian Atlantic fisheries biology, 1898–1939*, Helen Rozwadowski's *The Sea Knows No Boundaries* on the history of ICES and Vera Schwach's *Havet, fisken og Vitenskapen. Fra fiskeriundersøgelser til havforskningsinstitutt*.[17] In all three of these books, the societal context, or the institutional history as such is prominent. Apart from inspiration at a conceptual level Rozwadowski's volume on ICES has been an indispensable reference work for developments in the first decades of the history of ICES. On the other hand this volume does not include the same degree of information about the cultural circumstances surrounding the historical narrative of the sea as Rozwadowski's, *Fathoming the Ocean. The Discovery and Exploration of the Deep Sea*.[18]

In several chapters in this volume, the actual marine science history is looked upon in its paradigmatic context, but probably less rigorously than in a book such as Tim Smith's *Scaling Fisheries, the Science of measuring the effects of fishing, 1855–1955* or Eric Mills', *Biological*

16 Finley, C., *All the fish in the sea. Maximum Sustainable Yield and the Failure of Fisheries Management,* (The University of Chicago Press, 2011); Hamblin, J. *Oceanographers and the Cold War. Disciples of Marine Science*, (Seattle & London: University of Washington Press, 2005).

17 Hubbard, J., *A Science on the Scales: The rise of Canadian Atlantic fisheries biology, 1898–1939*, (Toronto: University of Toronto Press, 2005), Rozwadowski, H. *The Sea Knows No Boundaries* (Copenhagen: ICES/University of Washington Press, 2002) and Schwach, V. *Havet, fisken og Vitenskapen. Fra fiskeriundersøgelser til havforskningsinstitutt 1860–2000*, (Oslo: Havforskningsinstituttet, 2000).

18 Rozwadowski, H., *Fathoming the Ocean. The Discovery and Exploration of the Deep Sea*, (Cambridge, MA & London: The Belknap Press of the Harvard University Press, 2005).

Oceanography – An Early history, 1870–1960. The approach in both of these volumes is different from this book in the sense that the internal scientific developments, paradigmatic and technical, are analysed at great depth, whereas the societal context and connectivity is less prominent in these volumes.[19] To understand the international context of the scientific problems conducted by Schmidt's mentor C. G. Joh. Petersen, Smith's *Scaling Fisheries* is a fundamental piece of work. With regards the development of scientific paradigms within marine science this volume is also indebted to a number of articles. A very important source of information in this respect is the 2002 theme issue of ICES *Marine Science Symposia*, which dealt with historical perspectives on various issues in marine science such as fish migration theory.[20] It is worth noting, however, that Sinclair and Solemdal addressed several arguments in this 2002 symposia volume in 1988.[21] The early history of professional Norwegian fisheries science is the topic of Vera Schwach's Dr. Philos. Thesis, a collection of articles dealing with internal scientific issues as well as their societal context.[22]

While internationally there is large and rapidly growing literature on these topics not much has been written about Johannes Schmidt or marine science history in general in Denmark in recent decades. One noteworthy exception from recent years is Kristian Hvidtfelt Nielsen's assessment of the science communication on board the Galathea II Expedition 1950–52.[23] The four volumes of *Dansk Naturvidenskabs Historie*, (the history of natural science in Denmark) summarized the existing body of knowledge of marine science expeditions, drawing mainly on information in older works on Danish marine science history.[24] This is the case also in a recent summary of Danish led eel research from the past century,[25] as well as in the Danish zoologist and explorer Torben Wolff's overview of Danish marine science expeditions from the 18th century until after World War Two, which includes an insightful description of Schmidt's main deep sea expeditions as well as his main scientific results in this field.[26] Tom Fort's *The Book of Eels* also follows in this tradition, but builds on interviews with marine scientist Jan Boëtius, who died in 2012. Fort focuses on Schmidt's lifelong quest to uncover the life history of the World's eel populations.[27]

19 Smith TD. *Scaling fisheries: The science of measuring the effects of fishing, 1855–1955*. 2nd revised 2007 ed. New York: Cambridge University Press; 1994; Mills, E., *Biological Oceanography – An Early history, 1870–1960*, (Toronto: University of Toronto Press, (2nd ed.) 2012).

20 Jakobsson J. Ocean travellers. ICES *Marine Science Symposia*. 2002;215:330–342; Secor, D. H., Historical roots of the migration triangle, ICES *Marine Science Symposia*. 2002;215:323–329.

21 Sinclair M, Solemdal P. The development of 'population thinking' in fisheries biology between 1878 and 1930. *Aquatic Living Resources*. 1988;1(3):189–213.

22 Schwach V. *Til havs med vitenskapen. fiskerirettet havforskning 1860–1970*. [Dr. Philos.]. Oslo: Universitetet i Oslo; 2011.

23 Nielsen KH. Postcolonial partnerships: Deep sea research, media coverage and (inter)national narratives on the Galathea deep sea expedition from 1950 to 1952. *The British Journal for the History of Science*. 2010;43(1):75–98.

24 Nielsen, Kristian Hvidtfelt, Ries, Christopher Jacob and Østergaard, Kristian B., 'Danske Videnskabelige ekspeditioner I perioden', in: Hvidtfelt Nielsen, Kristian and Nielsen, Henry (eds.), *Viden uden Grænser 1920–1970, Dansk Naturvidenskabs Historie vol. 4*, Aarhus: Aarhus University Press; 2006.

25 Rindom S, Tomkiewicz J, Munk P, et al. Eels in culture, fisheries and science in Denmark. In: Tsukamoto K, Kuroki M, eds. *Eels and Humans*. Japan: Springer Science+Business Media; 2014:41–60.

26 Wolff T. *Danske ekspeditioner på verdenshavene: Dansk havforskning gennem 200 år*. Copenhagen: Rhodos; 1967

27 Fort T. *The Book of Eels*. Great Britain: Harper Collins Publishers; 2002.

The most careful and insightful description of Schmidt's achievements as a marine scientist is an anthology published in 1947 by some of his former colleagues and friends in marine science. Entitled *Naturforskeren Johannes Schmidt. Hans liv og ekspeditioner,* this is a true gem of information with regards to the breadth and scope of Schmidt's research.[28] The men who worked alongside Schmidt wrote all the chapters, and there is a lot of value in their assessments of what his research was all about. When it comes to their anecdotes about working with Schmidt, there is much merit in several cases, where the testimonies presented are the only surviving record of particular events. Poul Jespersen's article on the 1913 *Margrethe* expedition for instance, has merit as the only eyewitness account of what actually transpired during this dangerous and unfortunate expedition, although his story was published 34 years after the events had taken place.[29] The chapters written by Aage Vedel Taaning however, are clearly based on first hand knowledge of some of the archival material, which forms the basis of this book.[30]

For all these qualities however, the *Naturforskeren Johannes Schmidt* does not cover everything. Schmidt is generally portrayed sympathetically, as one would expect in this type of memorial volume. Also, Schmidt's efforts as a fundraiser, and the many conflicts he was part of, are not revealed to the readers of this 1947 anthology. Finally, there is much value in some of the obituaries written by fellow scientists after Schmidt died in 1933.[31]

In the historical literature Schmidt is mentioned only in passing in connection with his association with the Carlsberg Foundation. Kristof Glamann, Professor in economic history and later Chairman of the Carlsberg Foundation's Board of Directors, was responsible for the official history of the Carlsberg Foundation, and he also wrote about the history of the Carlsberg Brewery. In these volumes Schmidt plays a very modest role.[32] Johannes Schmidt worked for 23 years as Director of the Physiological Department of the Carlsberg Laboratorium, a sub-section of the Carlsberg Foundation. Schmidt's achievements here are covered in a centennial anniversary volume about the laboratory, a couple of articles, and most recently in a critical assessment of the scientific value

28 Winge Ø, Vedel Taaning Å, eds. *Naturforskeren Johannes Schmidt. Hans liv og ekspeditioner*. Copenhagen: Gyldendalske Boghandel Nordisk Forlag; 1947

29 Jespersen P. Danske handelsskibes og skonnerten *Margrethe*'s indsats. In: Winge Ø, Vedel Taaning Å, eds. *Naturforskeren Johannes Schmidt. Hans liv og ekspeditioner*. Copenhagen: Gyldendalske Boghandel Nordisk Forlag; 1947:57–69.

30 Vedel Taaning Å. Aaleundersøgelserne begynder for alvor. In: Winge Ø, Vedel Taaning Å, eds. *Naturforskeren Johannes Schmidt. Hans Liv og Ekspeditioner.* Copenhagen: Gyldendalske Boghandel Nordisk Forlag; 1947:37–44.; Vedel Taaning Å. Fiskeriundersøgelser i nordlige have efter krigen. In: Winge Ø, Vedel Taaning Å, eds. *Naturforskeren Johannes Schmidt. Hans Liv og Ekspeditioner*. Copenhagen: Gyldendalske Boghandel Nordisk Forlag; 1947:113–131; Vedel Taaning Å. Hovedlinierne trækkes op. In: Winge Ø, Vedel Taaning Å, eds. *Naturforskeren Johannes Schmidt. Hans Liv og Ekspeditioner*. Copenhagen: Gyldendalske Boghandel Nordisk Forlag; 1947:174–187.

31 Bruun AF. The Life and Work of Professor Johannes Schmidt. *Estratto dalla Rivista di Biologia*. 1934;16(1):31; Russell ES. Johannes Schmidt (1877–1933). *Journal du Conseil*. 1933;8(2):145–160; Ege, Vilhelm, Johannes Schmidt 1877–1933, In: Holter H, Max Møller K, eds. *The Carlsberg Laboratory 1876–1976*. Copenhagen: Rhodos; 1976:195–209.

32 Glamann K. *Bryggeren I. C. Jacobsen på Carlsberg*. 2011 Special edition ed. Copenhagen: Gyldendal; 1990; Glamann K. *Carlsbergfondet*. Copenhagen: Rhodos; 1976; Glamann, K. *Beer and Brewing in Pre-Industrial Denmark*. Odense: University of Southern Denmark Press; 2005.

and impact of the Carlsberg Laboratorium over the past almost 140 years.[33]

This is however the first in-depth account of Schmidt's effort at the Carlsberg Laboratorium based on the full archives of the Carlsberg Laboratorium, including correspondence and minutes of the board as well the annual research budgets and research plans. Schmidt's research into genetics at the laboratory has not previously been looked at in connection with the research interests and network of the core people around him, principally Eugen Warming, but also Øjvind Winge and from further away, Wilhelm Johannsen.

More over, this book proposes, for the first time, that during the first three decades of the 20th century an important strategic alliance of like-minded Danish marine scientists, business men, and politicians formed to promote Denmark's status as a sovereign state with a small but noteworthy North Atlantic Empire. Therefore, these ideas are not covered in any existing literature. However, with regards to the general framework of thinking about Denmark's role in matters of foreign policy in this period, this book takes a starting point from Danish historian Bo Lidegaard's thesis that in the first half of the 20th century Danish foreign policy formulation was divided between two world views, where one was *cosmopolitan* and the other was *national*. With regards the factual development of Danish politics in relation to the Danish West Indies and Greenland there has been much to gain from modern political history.[34] This book adds to the existing body of knowledge by stipulating that marine science was a kind of soft diplomacy on behalf of the group of cosmopolitans. The role played by the leading Danish business men in relation to supporting marine science before World War Two has also not been a subject of historical investigation before.

Finally, this is the first major attempt to frame the life and work of Johannes Schmidt as a mediary between two of the most important science institutions based in Copenhagen in the 20th century, The Carlsberg Foundation and the International Council for the Exploration of the Sea (ICES). A recent, short article published by Canadian marine scientist and former ICES President Mike Sinclair in 2010, highlighted the peculiar connection between Schmidt and Carlsberg.[35] This is directly linked to this project, as Mike Sinclair subsequently contacted the author of this book with a proposal to uncover the history of Johannes Schmidt's connection to Carlsberg. This book is the culmination of

33 Holter H, Max Møller K, eds. *The Carlsberg Laboratory 1876–1976*. Copenhagen: Rhodos; 1976; Winge Ø. Andre strenge anslaas – paa Carlsberg Laboratoriet. In: Winge Ø, Vedel Taaning Å, eds. Naturforskeren Johannes Schmidt. Hans liv og ekspeditioner. Copenhagen: Gyldendalske Boghandel Nordisk Forlag; 1947:71–72; Olesen Larsen P. Anderledes – og tak for det. *Carlsbergfondet Årsskrift*. 2011;2011:58–77; Olesen Larsen P. Iværksætter og forvaltere. Carlsberg Laboratorium 1876–2014. Copenhagen: Carlsbergfondet; 2015: p. 78–86.

34 Lidegaard, Bo, Overleveren. 1914–1945, Carsten Due-Nielsen, Ole Feldbæk & Nikolaj Petersen (eds.): *Dansk Udenrigspolitiks Historie,* vol. 4, Copenhagen: Gyldendal Leksikon, 2003. Pp. 179–182; Blom I. *Kampen om Eirik Raudes Land. Pressgruppepolitikk i grønlandsspørgsmålet 1921–1931*. Oslo: Gyldendal Norsk Forlag; 1973; Sjøqvist, Viggo. *Erik Scavenius: Danmarks udenrigsminister under to verdenskrige. Statsminister 1942–1945*. Copenhagen: Gyldendal, 1973.

35 Sinclair M. Marine biodiversity: ICES, Johannes Schmidt, and the Carlsberg connection. ICES *Insight*. 2010(47):4–9.

Sinclair's proposal, but to get here, some 50,000 pages of documents and images preserved in 116 archival boxes were also visited.

Archival source material

Johannes Schmidt was not only an unusually productive scientist and science organizer. He was equally productive in terms of preserving his correspondence with several hundreds of people and institutions around the world. Much of Schmidt's correspondence is preserved in the series, 'Papirer' at the Royal Library's (KB) Centre for Manuscripts and Books (CMB) in Copenhagen, which make up more than 30 archival boxes, somewhat chaotically organized.[36] Schmidt's adult life is fairly well recorded here, since in many cases, a copy of outgoing letters is preserved in connection to incoming documents. A lot less material is preserved from his career up to c. 1910. From then onwards, a bounty of material has been kept in his own archives. This seems to be connected to the fact that he began using a typewriter from c. 1910. At present, it is a mystery to the Royal Library who donated the archives, but judging from the contents some of it appears to have been submitted via Schmidt's widow, Ingeborg Schmidt. In light of how much Schmidt was corresponding, it is striking that virtually no correspondence between Mr and Mrs Schmidt is preserved. One gets a clear sense that their private correspondence has been removed from the collection before submission to the Royal Library, although this is speculative of course. Other papers may have come to the library via the Zoological Museum or the Danish Institute for Fisheries Research.

Another important source of primary material is the Rigsarkivet (Public Records Office) in Copenhagen. Here the archival holding, labelled *Schmidt, Ernst Johannes, biology,* was a true gem of information in the form of Schmidt's personal diary, 1929–1933. In the same boxes Ingeborg Schmidt's travel accounts are found covering her trips with Johannes Schmidt to Tahiti (1926) and their crossing of the Atlantic and Pacific Oceans (1928).[37] From Schmidt's student years not much is preserved, but some information could be gleaned from his friend Ove Paulsen's diary, which is in the Public Records Office.[38] In the 1920s, marine scientist, Anton Frederik Bruun became one of Schmidt's closest collaborators, and Bruun's expedition diaries, 1924–1930 yielded important information with regards to Schmidt's expeditions from 1924

36 Royal Library Copenhagen, Centre for Manuscripts and Books, (CMB). *Johannes Schmidt*: 'Papirer', Ut. 374.

37 RA. *Schmidt, Ernst Johannes, biolog:* Materiale vedr. ekspeditioner og rejser (1876–1933) 1: m.m.

38 Rigsarkivet(RA): *Paulsen, Ove botaniker, og hustru*: Dagbog m.m. (1874–1947). No. 5.

onwards. Thereto, Bruun's archive also held important newspaper cuttings for the same period.[39] All these files have been supplemented with documents from more official origins.

Much information on the 1913 *Margrethe* expedition and the surveys in the Danish West Indies was found in the archives of *Dansk Biologisk Station*[40]. The minutes of the Danish *Kommissionen for Havundersøgelser* were particularly important for reconstructing the early years of Denmark's international effort in marine science, and the feud that took place between C. G. Joh. Petersen and Schmidt's generation of scientists.[41] The Public Records Office also holds archival series originating from the *Dana* expeditions, in the early 1920s, such as the minutes and correspondence of the various '*Dana* committees'.[42]

A crucial set of private archives are those held by the Carlsberg Foundation, where all successful applications to the Carlsberg Foundation are kept for posterity.[43] A list of successful applications within the lifespan of Johannes Schmidt is published as *Carlsbergfondets Understøttelser*, which was used as an entry point to locate Johannes Schmidt's successful applications to the Carlsberg Foundation.[44] To a varying degree, the considerations of the Carlsberg Foundation's Board of Directors are also noticeable through the application files, such as in the case of the 1928–1930 Carlsberg Foundation's Oceanographic Expedition Round the World. As the name suggests it was hosted by the foundation itself.[45] The minutes of the meetings of the Foundation's Board of Directors have proved of indispensible value with regards the viewpoints from inside the foundation.[46]

From 1910 until 1933, when Schmidt worked as a Director of the Carlsberg Laboratory, he was reporting to the laboratory's Board of Trustees. The Carlsberg Archives (CA) holds the minutes of the Board of Trustees of the Carlsberg Laboratory, the so-called *Forhandlingsprotokol*.[47] The annual expenditures of the Carlsberg Laboratory, split between separate departments, were reconstructed from the annual budgets with specifications for each department.[48] In addition to keeping expenditures within a fixed annual budget, Schmidt also submitted annual reports to the Board of Trustees on the research and publications coming out of the laboratory.[49] The guestbook of visitors to the Carlsberg Laboratory from c. 1922–40, however was found in Schmidt's own archives at the Public Records Office.[50]

The chapter covering the political aspects of the Danish marine science presence in the Danish West Indies owes a lot to the archives of

39 RA. *Anton Fr. Bruun. Danaekspeditionerne 1924–1930.*

40 RA: *Dansk Biologisk Station, Johannes Schmidts embedsarkiv*: Korrespondance om åleforskning (1908–1933) E28; RA: *Dansk Biologisk Station, Johannes Schmidts embedsarkiv*: Fiskeriundersøgelser og skildpaddemærkning i Dansk Vestindien (1914–1916) E30.

41 RA. *Komm. Danmarks Fiskeri- og Havundersøgelser:* Forhandlingsprotokoller (1900–1960) N1-N4.

42 Rigsarkivet(RA), *Carlsbergfondets Danaekspeditioner*: 'Forhandlingsprotokoller 1917–1973'. 10291.

43 Carlsbergfondet(CFD), *Bevilgede Ansøgninger*.

44 Spang-Hanssen, E. *Carlsbergfondets Understøttelser 1876–1936*. Copenhagen: Bianco Lunos Bogtrykkeri A/S, 1937.

45 CFD. *DANA. BS3 1 & 2. Dana-Expeditionen*; CFD. *Vedr. DANA II, 1828–1930. 103/1928.*

46 Carlsberg Foundation, Board of Directors (CFD): *Carlsbergfondets Direktions Protokol.*

47 CA, CL, E17, Forhandlingsprotokol, Gærfysiologisk Underafdeling, Regnskabsbog, Bestyrelsen, 1876–1953.

48 Carlsberg Archives(CA). *Carlsbergfondet, Carlsberg Laboratorium*, E31. Endelige budgetter og noter 1914–1972.

49 CA. *Carlsbergfondet: Carlsberg Laboratorium (CL):* Korrespondance, regnskaber og beretninger, boxes no. E06-E10 covering 1909–1935.

50 RA. Carlsbergfondets DANA-ekspeditioner. 1917–1973. 1. Gæstebog, 1922–1940.

the *Foreningen De Danske Atlanterhavsøer*, which include the minutes of the board and minutes from the general assembly as well as membership files and letter books.[51] Some material was also sifted from volumes of the society's journal, *Atlanten*.[52] The transcripts of Schmidt's and other expert appearances in front of the government commission on the Danish West Indies in 1916 were published. This is also the case for the government negotiations on the sale of what became the US Virgin Islands.[53]

51 RA.10154. *De Danske Atlanterhavsøer. 1902–1917*, no 1–18.

52 Anonymous, *Atlanten*, vol. 1–4 (1904–1917).

53 Anonymous, ed. Betænkning angiven af den i henhold til lov nr. 294 af 30. september 1916 nedsatte rigsdagskommission angaaende de dansk vestindiske øer. Copenhagen: J. H. Schultz; 1916; Anonymous, ed. Betænkning over forholdene paa de dansk-vestindiske øer afgiven af den allerhøjeste resolution af 18. november 1902 anordnede vestindiske kommission. Copenhagen: J. H. Schultz; 1903; Anonymous. Forhandlingerne om overdragelsen af de dansk-vestindiske øer til de amerikanske forenede stater 1915–1916. Copenhagen: Udenrigsministeriet; 1916.

1 Formative years (1877–1903)

Schmidt's upbringing was unusual, if not entirely happy, then quite ideal for science nonetheless. While the main focus of this book is to explore the causes and consequences of Johannes Schmidt's deeds in his capacity as a scientist, science manager and organizer, as opposed to presenting a more intimate biography of 'a great man of the past,' at least some introduction to the unusual circumstances of his upbringing is vital for fully appreciating his later achievements

Johannes Schmidt's father, Ernst Schmidt was the estate manager at Jægerspris Slot, a royal palace some 50 km to the Northwest of Copenhagen. In 1876 he married Camilla, *née* Kjeldahl, and the following year on 2 January she gave birth to Ernst Johannes Schmidt. When Johannes was born, both parents were fairly mature, 50 and 38 years old respectively, and while Johannes most certainly had a comfortable first few years in life, a major change occurred in 1884 when his father passed away.[1]

1 Wolff, T. Schmidt, Ernst Johannes. In: Cedergreen Bech, S. ed. *Dansk Biografisk Leksikon*, Copenhagen, Gyldendal; 1983:13,148–149; Paulsen O. Slægt og ungdom. In: Winge Ø, Vedel Taaning Å, eds. *Naturforskeren johannes schmidt. hans liv og ekspeditioner.* Copenhagen: Gyldendalske Boghandel Nordisk Forlag; 1947:9–11.

1.1 The estate manager's house at Jægerspris Slot
Until the age of 7 Johannes Schmidt's childhood unfolded in this house next to the Jægerpris Royal Palace, where his father was the estate manager (Poulsen, B).

Johannes then moved with his widowed mother and two younger brothers to Copenhagen, where they rented a house at Vesterbrogade 125, one of the busiest streets in the South western end of Copenhagen, a couple of kilometres from the central train station in the rapidly growing city. According to the 1890 Danish census, the widowed Camilla Schmidt ran a boarding house from this address. She was registered as the head of a household consisting of 11 boarders, one matron and three maids, and not least her three children. Johannes Schmidt was then 13 years old, while his two brothers, Alfred and Jørgen were 10 and 8 years old respectively.[2] Obviously Camilla Schmidt needed to work for a living, but judging from the number of servants she is likely to have been one of the more affluent among the many, women in particular, running boarding houses in Copenhagen around this time.[3] Later in life, Johannes Schmidt earned himself a reputation as a highly efficient manager of scientific expeditions, dealing with staff from many walks of life. Perhaps he was able to draw on his own first-hand experience of growing up in a busy household, where his mother was managing people as well as finances.

The Schmidt family's Copenhagen residence was situated just a few hundred meters from the Carlsberg Brewery and within walking distance from the house of Johannes Schmidt's maternal uncle, Johan Kjeldahl. Kjeldahl was a renowned chemist and the director of the Chemical Department of the Carlsberg Laboratorium, a research laboratory set up in 1875 by the founder and owner of the Carlsberg Brewery, I. C. Jacobsen. Thus, the events in the Schmidt family in 1884 brought the young Johannes Schmidt close to the epicentre of the expanding commercial and scientific empire of the Carlsberg Brewery.

The Carlsberg Brewery

In 1884, the same year that 7-year old Johannes Schmidt arrived in Copenhagen, the perhaps most famous of all scientists in those years, Louis Pasteur also came to town. As the world-renowned French chemist and founder of microbiology, Louis Pasteur was in Copenhagen for the international medical congress. Pasteur had demonstrated that fermentation processes, crucial for wine making and brewing beer, were caused by the growth of microorganisms. If the microorganisms could be controlled, much more reliable products could be made. Pasteur realized

2 RA. Folketælling 1890, Staden København, Udenbys Vester, Vesterbrogade 125 Matrikel 43. E.1. (p. 293).

3 Mortensen MT. *Pensionatets kulturhistorie. fællesskab, flirt og frikadeller*. Aarhus: Aarhus Universitetsforlag; 2015.

that the growth of microorganisms could spoil beverages such as wine, and in 1866 he invented a means to preserve wine over a longer period of time; killing off pathogenic bacteria through a sequence of rapid heating and cooling. This process still bears his name – pasteurization. Quickly breweries around Europe took up pasteurization as a central aspect of beer making. In Denmark I. C. Jacobsen was one of the pioneers implementing pasteurization from 1874. Two years later, Pasteur himself published his *Étude sur la Bière* widely considered a milestone in the history of brewing. I. C. Jacobsen and later his son, Carl Jacobsen would have their portrait painted with Pasteur's book in front of them.[4]

In 1884, the founder and owner of Carlsberg, I. C. Jacobsen put up Pasteur with his wife and daughter in a string of suites at the luxurious Hotel d'Angleterre in Copenhagen. Brewers around the world were greatly indebted to Pasteur for his landmark achievements, and the Carlsberg family now had the means and opportunity to demonstrate this to Pasteur and the world of science. The Hotel d'Angleterre was also the scene of a dinner for all French and Scandinavian delegates to the medical congress. A few days later Pasteur was the guest of honour at a private dinner at Carlsberg. According to family legend, Pasteur remarked that '…the air in the fermentation cellars at Carlsberg was almost as clean as the air on the top of the French Alps'.[5]

This was a great triumph for the owner of the brewery, I. C. Jacobsen. The previous generation had charged him with pioneering a science-based approach to modern industrial production. Jacob Christian ('I. C.') Jacobsen was the only son of a small time Copenhagen brewer, Chresten Jacobsen. When in 1835, he inherited his father's business the brewery was still a small-scale operation in the courtyard of an inner-city townhouse in Copenhagen. The typical Danish brewery in those days was small in scale and mainly catered for a local market producing 'hvidtøl', a Danish household beer with a low content of alcohol and of very varying quality. Having studied chemistry at the University of Copenhagen, the young brewer, I. C. Jacobsen, eagerly followed the emerging debates over the role of proper hygienic precautions in relation to health hazards,

4 Glamann K. *Bryggeren I. C. Jacobsen på Carlsberg*. 2011 Special edition ed. Copenhagen: Gyldendal; 1990: 220–213.

5 Schmidt I. *Blade af Carlsbergs Historie. kaptajn S. A. van der Aa Kühle og hans liv på Carlsberg*. Copenhagen: Bianco Lunos Bogtrykkeri; 1956. p. 26–28.

1.2 The Carlsberg Breweries in 1934 [pages 40–41]

Carlsberg quickly became the dominant brewery in Denmark, and during the first decades of the twentieth century it was one of the largest factories in the country. The Carlsberg Laboratorium (no. 15) is in the top right corner of this drawing from 1934 (Carlsberg Archives).

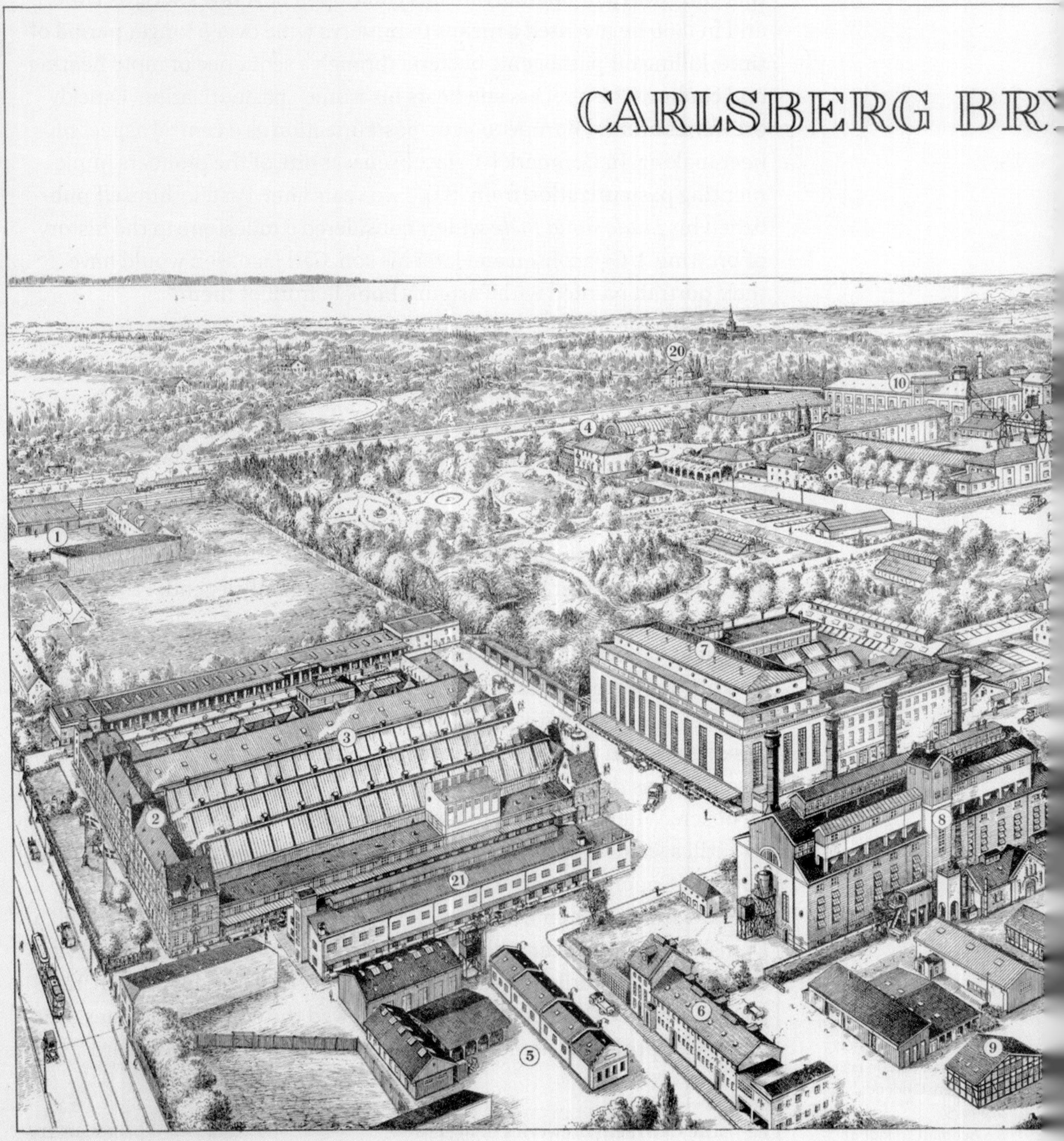

1) Stalde 2) Hovedkontor 3) Aftapningsanstalt 21) Lagerbygning 4) I. C. Jacobsens Villa (Æresboligen) 5) Automobilgarager 6) Stald 7) Mineralvandsfabrik 8) Kedelhus 9) Værksted

11) Maskincentral 12) Mellembryggeriet (Det første Ny Carlsberg) 13) Værksteder 14) Arbejderboliger 15) Carlsberg Laboratorium 16) Ny Carlsberg 17) Jesus Kirken 18) Carl Jacobsens Villa (Direktørbolig) 19) Carlsberg Museum (Det gamle Glyptothek) 20) Funktionærboliger

nutrition and best conditions for beer production. Realizing that he needed to improve the reliability of the final product, Jacobsen in 1847 set up a new industrial plant three kilometres outside the congested inner city of Copenhagen, on the southern slope of the hill, Valby Bakke, in what was then the countryside. The brewery was renamed, Carlsberg, after his only son, Carl.[6]

Two years earlier, in 1845 Jacobsen had visited Munich, from where he brought home different yeast strains. These provided the foundation for the production of Bavarian style beer. This new type of beer was stable and of good quality and consumers soon got a taste for the new beverage. Around the same time the Danish population began to grow rapidly, particularly in urban areas. The population of Copenhagen alone quintupled between 1800 and 1900.[7] This provided a growing customer base for Carlsberg's Bavarian style beer, which was helped by the lack of a sales tax on beer for a number of years. Gradually, I. C. Jacobsen became one of the wealthiest industrialists in Denmark. Over time, other companies started producing Bavarian style beer, and in 1871 I. C. Jacobsen built a new separate brewery next to his original brewery. The new brewery was named Ny Carlsberg, to distinguish it from 'Gamle Carlsberg', the old brewery. I. C. Jacobsen then leased the new brewery to his son, Carl Jacobsen. The idea was that Carl Jacobsen could earn his wings by trying out the brewing business on his own.[8]

The Carlsberg Laboratory

Meanwhile, I. C. Jacobsen sought to step up his already close collaboration with leading Danish natural scientists in order to improve his brew. At Carlsberg, science showed the way to the future. This led to the foundation in 1875 of a completely new structure, The Carlsberg Laboratory, which opened the following year. While the brewery already had a laboratory for more day-to-day analyses, Jacobsen was convinced that he needed a more thorough knowledge of the fundamental processes associated with fermentation, and only scientists working steadily and systematically in an undisturbed way could be expected to come up with real breakthroughs. Focusing on chemistry and biology, The Carlsberg Laboratory became a separate unit detached from the daily running of the business and expectations of the breweries. At the time this separation of research from daily routines was unique by any

6 Glamann K. *Bryggeren I. C. Jacobsen på Carlsberg*. 2011 Special edition ed. Copenhagen: Gyldendal; 1990.

7 Johansen HC. *Danish population history 1600–1939*. University Press of Southern Denmark; 2000

8 Glamann K. *Carlsberg-fondet*. Copenhagen: Rhodos; 1976. p. 29–35. Glamann, K. *Beer and Brewing in Pre-Industrial Denmark*. Odense: University of Southern Denmark Press; 2005. p. 164–180.

international standard. The new laboratory was also unusual in that it had decided not to keep new discoveries a secret, everything would be published. The laboratory was divided into two separate units, the Physiological Department and the Chemical Department, each with an experienced scientist as the leader, who in turn had a laboratory and a few assistants at his disposal.

When in 1884, Pasteur was taken on a tour of Carlsberg he also met with the heads of the Carlsberg Laboratory. According to the memoires of Emil Chr. Hansen, Louis Pasteur kept his tall black hat on throughout his visit to the laboratory, but the famous Frenchman, was still able to look out from underneath it, and able to recognize one of the most interesting industrial laboratories in Europe.[9]

Emil Chr Hansen (1842–1909) headed the Physiological Department, and following in the footsteps of Pasteur, he made an important discovery regarding the brewing process. Pasteur believed the main reason for the frequent occurrence of foul smelling and bitter brews that were unmarketable was due to infections from bacteria. Instead, Hansen identified interference from so-called 'wild yeast' as the source of the problem. Hansen therefore developed a method whereby he could isolate a single yeast cell and propagate it into an absolutely pure culture. Hansen and not least Jacobsen's international stature rose rapidly, as Jacobsen set about spreading news of his discovery to other European brewers. He offered these insights for free to the leading European brewing companies, and soon the use of pure lines of yeast had spread to all industrial breweries.[10]

In the Chemical Department, where Johannes Schmidt's maternal uncle, Johan Kjeldahl, (1849–1900) reigned, Pasteur will have witnessed yet another important advancement in science published only the previous year in 1883. In laboratories around the world, where plant and animal material are analysed it is still common to 'kjeldalize'; this refers to a method that can be used to determine the nitrogen content in organic material. Kjeldahl had initiated studies of protein metabolism during the germination of barley and during alcohol fermentation, both central processes in brewing. Yet, the methods already in existence were too slow and complicated to work with, which made him develop his own.[11]

Less than a decade after its foundation, I. C. Jacobsen could reap the fruits of his innovative institutional setup. All in all, the Carlsberg Laboratory was at the forefront of international research into brewing-related life science. Twenty-five years later, Johannes Schmidt would

9 Glamann K. *Bryggeren I. C. Jacobsen på Carlsberg*. 2011 Special edition ed. Copenhagen: Gyldendal; 1990: 267–270.

10 Klöcker A. Emil Christian Hansen (1842–1909). In: Holter H, Max Møller K, eds. *The carlsberg laboratory 1876/1976*. Copenhagen: Rhodos; 1976: 178–179.

11 Johannsen, W. 'Johan Kjeldahl' 1849–1900. In: Holter, H, Max Møller, K, eds. *The Carlsberg Laboratory 1876–1976*, Copenhagen: Rhodos; 1976. p. 52–62.

1.3 Johan Kjeldahl in the Carlsberg Laboratorium
Johan Kjeldahl, (1849–1900) was Johannes Schmidt's maternal uncle, and the Director of the Chemical Department of the Carlsberg Laboratorium. For some time, Schmidt also lived with his uncle, undoubtedly an inspiring environment for a future scientist. (*Painting from 1897, Carlsberg Archives*)

follow in the footsteps of Hansen and Kjeldahl, as he himself was made director of the Physiological Department. How that experiment played out is unfolded in Chapter 5 and 6.

The Carlsberg Foundation

I. C. Jacobsen was also a great patron of the arts and sciences in general, and in 1876 he established the Carlsberg Foundation, one of the World's oldest private foundations devoted to arts and sciences. The foundation was novel in that alongside its 'Section A' – the laboratory – it also had a

'Section B', which according to the statutes had the remit 'to further the various natural sciences, together with mathematics, philosophy, history and linguistics, in the manner laid down in detail in the statutes of the Foundation'. To promote theses disciplines Jacobsen gave the foundation a gift of 1 million Danish Kroner, which was a huge sum at the time, and came from a mortgage on his Carlsberg Brewery.[12]

Following the death of I. C. Jacobsen in 1887, the assets of his vast empire of businesses and charities came under the control of the Carlsberg Foundation. The Foundation was, and still is, managed by a board of five – all scientists appointed by the members of The Royal Danish Academy of Science and Letters. I. C. Jacobsen believed that the royal academy was the only institution in Denmark which was fortunate enough to be independent of all unscientific considerations and secondary influences. In addition, as a firm believer in science, Jacobsen was particularly drawn to the idea that men with scientific knowledge and abilities were the ones best suited to secure his vision for the future. His vision was that the foundation would operate to the benefit of science and to the honour of Denmark. Quickly, the foundation came to include a third 'Section C', the Frederiksborg Castle with a museum for the national history of Denmark; a further addition, the 'Ny Carlsbergfondet', which supported the arts and arts museum was added to the foundation after Jacobsen had passed away.[13]

Today, the Carlsberg Foundation is the majority shareholder in Carlsberg Breweries, the fourth largest brewery in the World, and this provides the primary source of funding for the foundation. Thus, one might argue that the world of science is in charge of running a business, which in turn generates funds for the arts and sciences. In an international context, The Carlsberg Foundation is probably one of the oldest private funding agencies based on an industrial corporation.[14] Within the borders of the Kingdom of Denmark it is hard to exaggerate the historical importance of the Carlsberg Foundation for the promotion of natural and human sciences. As in most other countries in the period before World War II, in Denmark there was little in the way of national government support for research.

In Denmark there were no public funding opportunities for research outside the few positions at universities, museums and other institutes of higher learning before the establishment of the Rask-Ørsted Foundation in 1919.[15] There was only one university (Copenhagen) in Denmark between 1864 and 1928, and very few positions were available for aspiring scientists who wished to conduct their own research and be paid.

12 Pedersen J. *The Carlsberg Foundation*. Copenhagen: Bianco Luno; 1956.

13 Pedersen J. *The Carlsberg Foundation*. Copenhagen: Bianco Luno; 1956: 12–14.

14 Kjær B. 'A Danish National Institution: The Carlsberg Foundation'. *Scandinavian Review*. 1993;81(2):87–93.

15 Olesen Larsen, P. *Stadier på forskningens vej. Dansk forskningspolitik i går og i dag*. Odense: University of Southern Denmark Press; 2010. p. 13.

This highlights the importance of the Carlsberg Foundation to research in Denmark, more so since, it was the only funding body of this kind during the first 50 years or so of its existence. Early Danish recipients of the Nobel Prize include the likes of Niels Bohr (Physics, 1922), Niels Finsen (Medicine, 1903), August Krogh (Medicine, 1920), Johannes Fibiger (Medicine, 1926) and Henrik Dam (Medicine, 1943).[16] Without exception they all profited tremendously from the Carlsberg Foundation's support for the acquisition of laboratory equipment, salaries for hired assistants, and in the case of Niels Bohr, an entire laboratory.[17]

There is no Nobel Prize for the field of biology, yet as this book will reveal, Schmidt was bestowed with numerous other international recognitions. This book follows how Schmidt as the nexus of a small group of men succeeded within the field of marine science in particular, where a good handful of research positions in government institutions sprung up from the end of the 19th century onwards. In particular though, Johannes Schmidt springs to the forefront as the one who most eminently sought to exploit the possibilities of advancing a research agenda within marine science, through the use of a range of financial sources, but not least the money flowing from the revenue of Carlsberg's steadily growing beer production.

Student years

As a young man and as a natural history student, Johannes Schmidt left little correspondence. The closest testimony of this kind comes from his fellow marine scientist, plankton zoologist, Ove Paulsen (1874–1947). In the mid-1940s Paulsen was asked to reflect on his recollection of Schmidt as a young man, when they were both students of natural history at the University of Copenhagen in the 1890s.[18] Paulsen remained a lifelong friend of Johannes Schmidt, and his portrait of a young man, paints a flowery image of Schmidt as a faithful and caring young man, with simplicity of mind. 'Schmidt had a fondness of plain people and he even loved dogs and other animals'. Whereas Paulsen in some ways may have exaggerated the memory of his then deceased friend it is beyond doubt, that he was on the ball, when he emphasized that Schmidt was a very determined and hardworking student of natural history. Schmidt finished his training for his *Magisterkonferens*, roughly the equivalent of a Master's degree in only four years. His thesis was an analysis of algae,

16 Nielsen, H. & Nielsen, K. Danmark og Nobelprisen gennem 100 år, in: *Aktuel Naturvidenskab*, 2001, 1: 33–35.

17 Spang-Hanssen E. *Carlsbergfondets Understøttelser 1876–1936*. Copenhagen: Bianco Lunos Bogtrykkeri A/S; 1937.

18 Slægt og ungdom. In: Winge Ø, Vedel Taaning Å, eds. *Naturforskeren Johannes Schmidt. Hans Liv og Ekspeditioner.* Copenhagen: Gyldendalske Boghandel Nordisk Forlag; 1947:9–11.

19 KB. Johannes Schmidt 'Papirer'. Ut.374.VI.2. 'Manuskripter'. Eksamensopgave 1898.

and judging from the notes exchanged between his examiners, the three committee members were clearly dazzled by Johannes Schmidt's answers to his exam questions. One external examiner even raised the issue of whether or not his famous uncle, Kjeldahl had helped him, but this was not the case at all, Schmidt's professor, Eugen Warming replied.[19]

Eugen Warming (1841–1924) was one of the brightest stars in Danish natural history. He is widely heralded as the father of the concept of ecology: the interaction of different species in a larger system, where

1.4 Professor Eugen Warming
Eugen Warming (1841–1924) was one of the most influential natural scientists in Denmark in his day. As a professor of natural history at the University of Copenhagen, he spotted the talented student Johannes Schmidt, whom he sought to promote at university, via the Carlsberg Foundation and later within the Carlsberg Laboratory (Carlsberg Archives).

ecological niches are created. As a scientist he was also heavily involved in introducing Darwin's ideas on evolution in various empirical fields. As a devout Christian though, Warming had difficulties accepting natural selection as an evolutionary mechanism.[20]

Warming had spotted a bright talent in Schmidt, and for the rest of Warming's professional life, he would support Schmidt's research plans, no matter what direction they took. Perhaps, Warming recognized a bit of himself in Schmidt as well. Like Schmidt, Warming had lost his father, a Lutheran minister, at the age of two, and Warming's mother had resettled the family to live in the household of her brother in the town of Vejle.[21] The importance of Warming's support from the 1890s onwards is highlighted by this recent statement from the history of Danish natural sciences: 'It was almost impossible to avoid Warming, when submitting a research application'.[22]

According to Paulsen, Schmidt showed no haste in study manners, and was respected among his fellow students for not just memorizing what he had to learn, but for thinking through the underlying problems.[23] Before writing up his account of studying with Schmidt, Paulsen may have consulted his own diary from the 1890s. Today this is archived in a private collection at the Public Records Office in Copenhagen.[24] Here, we get a glimpse of Schmidt's student life, when in the fall of 1897 he and Paulsen were preparing together for three exams. The zoology exam appears to have consisted of a clarification of animal skeletons, whereas the botany test dealt with the fertilization process of flowers. For the geology exam the test revolved around five different rocks from the Danish island of Bornholm. In the entry of the following day, Friday the 18 November, Paulsen has noted that the 'fool's game of examinations' was over, and Paulsen invited his fellow students over for what he calls '…an animated dinner, becoming more animated afterwards!' The animated students that evening are interesting, as Paulsen lists the names of his guests; 'Ostenfeld, Levinsen, Weis, Morten Pedersen and Schmidt'.[25]

Ove Paulsen and Carl Hansen Ostenfeld (1873–1931) would later join Johannes Schmidt on a number of important marine science expeditions, and Ostenfeld in particular came to play a significant role in Schmidt's professional life. In their youth they were close friends already, and in December 1900 Paulsen was the best man at Ostenfeld's wedding.[26] Nonetheless, Schmidt's first large collaboration came about with fellow student 'Weis'. Frederik Anton Weis (1871–1933) would later

20 Sand-Jensen K. Naturhistorie. In: Kjærgaard PC, ed. *Lys over landet 1850–1920. Dansk Naturvidenskabs Historie 3*. Aarhus: Aarhus Universitetsforlag: 164–171.

21 Prytz S. *Warming – botaniker og rejsende*. Lynge: Bogan's Forlag A/S; 1984.

22 Sand-Jensen K. Naturhistorie. In: Kjærgaard PC, ed. *Lys over landet 1850–1920. dansk naturvidenskabs historie 3*. Aarhus: Aarhus Universitetsforlag: 165.

23 Slægt og ungdom. In: Winge Ø, Vedel Taaning Å, eds. *Naturforskeren Johannes Schmidt. Hans Liv og Ekspeditioner.* Copenhagen: Gyldendalske Boghandel Nordisk Forlag; 1947:9–11.

24 Rigsarkivet(RA): *Paulsen, Ove botaniker, og hustru: Dagbog m.m. (1874–1947)*. Booklet for 1896–1901.

25 Rigsarkivet(RA): *Paulsen, Ove botaniker, og hustru: Dagbog m.m. (1874–1947)*. Booklet for 1896–1901. 18 November 1897.

26 Rigsarkivet(RA): *Paulsen, Ove botaniker, og hustru: Dagbog m.m. (1874–1947)*. Booklet for 1896–1901. 8 December 1900.

become professor of plant physiology at the Agricultural University in Copenhagen.[27] However, as early as 1899–1901 he and Johannes Schmidt published a three-part book together on the importance of natural history in the study of bacteriology. Schmidt wrote the first and third part of the 354 pages long study. As a testimony of its usefulness, the book on bacteria was translated and published in Germany as early as 1902.[28] In 1905, it appeared in Polish.[29]

Ingeborg

From his early and prolific scientific writings it is clear that Johannes Schmidt was an efficient man, even in his youth. Also, although efficiency may not be the most appropriate term when it comes to marital matters, Schmidt was equally decisive when it came to choosing a spouse. Johannes Schmidt will have known his future wife ever since moving to Copenhagen through his uncle, Johan Kjeldahl at Carlsberg. In the 1880s the Carlsberg Laboratory still resided in modest buildings. The two laboratory directors, Emil Chr. Hansen and Johan Kjeldahl lived with their families on different floors of the nearby patrician style villa, 'Valbjerg' on the grounds of the Carlsberg Breweries. The young Johannes Schmidt must have encountered his future wife, Ingeborg van der Aa Kühle (1880–1958) on numerous occasions from when she was just a little girl. Ingeborg's parents lived on a separate floor of the same large villa as the Kjeldahl family, until the 1890s.[30] The brewery compound eventually included several villas of this kind.

Ingeborg's father, Søren Anton van der Aa Kühle, also worked for the Carlsberg empire, and following the death of I. C. Jacobsen, Kühle took over the management of the 'Gamle Carlsberg'. When Ingeborg was born in 1880, her father was the CEO of what was the most important brewery in Denmark. Thus several highly educated and resourceful families lived side-by-side on the Carlsberg Brewery's premises in this Copenhagen suburb.

While Johannes Schmidt was an extremely fruitful writer, it is somewhat surprising that hardly any private correspondence remains between him and his wife. There is no reason to doubt the sincere lifelong devotion between Ingeborg and Johannes Schmidt. Yet, the polite memoires of Ingeborg Schmidt are silent as to when she and Johannes first became infatuated with each other. For sure, they were closely connected

27 Christensen, Carl, 'Weis, Frederik Anton', in: Cedergreen Bech, Svend (eds.) *Dansk Biografisk Leksikon,* Copenhagen: Gyldendal, vol. 15: 376.

28 Slægt og ungdom. In: Winge Ø, Vedel Taaning Å, eds. *Naturforskeren Johannes Schmidt. Hans Liv og Ekspeditioner.* Copenhagen: Gyldendalske Boghandel Nordisk Forlag; 1947:11.

29 Bruun AF. The Life and Work of Professor Johannes Schmidt. *Estratto dalla Rivista di Biologia.* 1934;16(1):31.

30 Schmidt I. *Blade af Carlsbergs Historie. kaptajn S. A. van der Aa Kühle og hans liv på Carlsberg.* Copenhagen: Bianco Lunos Bogtrykkeri; 1956. p. 13.

1.5 Ingeborg and Johannes Schmidt
Ingeborg and Johannes Schmidt (Schmidt, I., 1956).

at the time of Johannes Schmidt's first expedition to study mangroves in Siam, today's Thailand, in 1899–1900. Ingeborg had fallen ill and was confined to bed for several months, but she recalled how the hardship of the illness as well as the accompanying harsh winter was softened by the letters she received from Johannes in Siam. In May of 1900, Johannes Schmidt returned from the mangroves, and accompanied by his uncle, Kjeldahl, he went to see Ingeborg's parents and asked for their daughter's hand. The engagement was romantically sealed later that day, when Johannes presented Ingeborg with a ruby ring, he had brought with him from Siam. Three years later, 17 September 1903 they became Mr. and Mrs. Schmidt following a wedding ceremony at the Jacobsen brewing dynasty's very own church, Jesuskirken, a few hundred meters from the Carlsberg Breweries and their former homes. Johan Kjeldahl was no longer alive, as he had died in 1901 following a swimming accident while on holiday in Northern Zealand.[31] Perhaps this is the reason, why Schmidt's best man at his wedding was Rudolph Koefoed, a business partner of his father-in-law at Carlsberg.[32] Schmidt's mother though, lived to see her oldest son getting married, but she eventually died in 1906.[33]

As a married couple Johannes and Ingeborg could now live on their own, and by 1904 they were renting an apartment in Østerbrogade 58, a pleasant neighbourhood in the northern part of central Copenhagen, where they could also afford to have their own maid.[34] In her memoires, Ingeborg Schmidt recalls that the Admiral in the Siamese Navy, Andreas de Richelieu gave a particularly successful speech at their wedding party.[35] By then, Richelieu had already proved helpful for Schmidt's scientific success.

The Siam Expedition, 1899–1900

The idea for Schmidt's first expedition came about as a consequence of Schmidt's encounter with Admiral Richelieu in the fall of 1898.[36] At the turn of the 20th century, Andreas de Richelieu was a wealthy businessman with connections to the Danish Royal family as well the royal house of Siam. Richelieu had risen to become Admiral in the Siamese navy, and he was also an important ally and co-founder with H. N. Andersen of the East Asiatic Company. The East Asiatic Company became the single most important private corporation in Denmark in the first half of the 20th century. The company included a string of global business enterprises, in shipping, forestry and industrial plants controlled in large from the Copenhagen headquarters, and its most important foreign branch was the office in Siam.[37]

According to zoologist, Theodor Mortensen, who was already an established marine scientist at the Zoological Museum, it was the recently graduated Johannes Schmidt, who came to him in the fall of 1898 with a proposition for a joint expedition to Siam. Following the advice of Richelieu, Schmidt had decided to establish a base camp for the expedition on the island of Koh Chang on the eastern banks of the Gulf of Thailand. There was already a military station on Koh Chang, from where the scientific expedition was able to draw man-power and various logistical help.[38]

31 Schmidt I. *Blade af Carlsbergs Historie. kaptajn S. A. van der Aa Kühle og hans liv på Carlsberg*. Copenhagen: Bianco Lunos Bogtrykkeri; 1956. p. 50.

32 RA. Kirkebøger, Københavns Amt, Valby Sogn, 1892 v. – 1905 v. 17. p. 120. 17. September 1903.

33 Wolff, T. ,Schmidt, Ernst Johannes. In: Cedergreen Bech, S. ed. Dansk Biografisk Leksikon, Copenhagen, Gyldendal; 1983:13,148–149.

34 RA. Folketælling 1890, Staden København, Udenbys Klædebo, Østerbrogade 58 Matrikel 136. E. (p. 114).

35 Schmidt I. *Blade af Carlsbergs Historie. kaptajn S. A. van der Aa Kühle og hans liv på Carlsberg*. Copenhagen: Bianco Lunos Bogtrykkeri; 1956. p. 49–52.

36 Mortensen, T. Siam-Ekspeditionen, in: Winge Ø, Vedel Taaning Å, eds. Naturforskeren Johannes Schmidt. Hans Liv og Ekspeditioner. Copenhagen: Gyldendalske Boghandel Nordisk Forlag; 1947: 12–16.

37 Eggers-Lura, Aldo, *Admiralen, Kongen & Kaptajnen. Danskere i elefanternes rige*. Copenhagen: Høst & Søn; 1998. p. 95–102.

38 Mortensen, T. Siam-Ekspeditionen, in: Winge Ø, Vedel Taaning Å, eds. Naturforskeren Johannes Schmidt. Hans Liv og Ekspeditioner. Copenhagen: Gyldendalske Boghandel Nordisk Forlag; 1947: 12–16.

Schmidt signed a joint application for him and Mortensen addressed to the Carlsberg Foundation asking for funds to conduct the research while in Siam as well the costs of living during the expedition. This was Schmidt's first successful application to the Carlsberg Foundation, where he promised to oversee the fruition of three sets of botanical investigations. First of all, they would collect plankton and take water samples during their journey from Europe to East Asia. Secondly, the island of Koh Chang would be subject to a close examination of tropical Siamese vegetation forms, with the intention also to photograph certain characteristic sections of vegetation, as well as charting heights, temperature and humidity on the islands. As a third separate analysis, the expedition intended to scrape the seabed for marine algae, something never done before in the Gulf of Thailand. On his section of the application Mortensen indicated that he as the zoologist would be in charge of collecting plankton as well as scraping the seabed in the Gulf of Thailand with the object of studying animal life there.[39]

At some point, Schmidt must have moved in with his maternal uncle, as the application was signed by 'Cand. Scient Johannes Schmidt, c/o Johan Kjeldahl, Carlsbergvej 10, Valby.' This was hardly an anonymous end for, what was a brief application to that particular foundation. Schmidt's professor from university, Eugen Warming knew all about Schmidt's qualities, and Warming was an important member of the Board of Directors of the Carlsberg Foundation. What the grant application does not reveal is the level of government support, but in a private letter to his son, Warming revealed that the government paid a similar sum of money, 2,500 Kroner.[40]

In addition, the East Asiatic Company supported the expedition with a heavily reduced fee (700 DKK instead of 1,450,–) for travelling from Denmark to East Asia and back, a very substantial part of the total cost of the endeavour.[41] Thus, on the 1 October 1899, the two young scientists embarked on their first trip beyond the borders of Europe, boarding the EAC steamship, the S/S *Siam*. Mainly a freight ship, cargo was loaded onto *Siam* in the ports of Antwerp and Southhampton before it headed through the Suez Canal stopping only in Port Said. During the three-week long voyage across the Indian Ocean Theodor Mortensen fondly remembers how the two men spent their time conducting plankton studies. They gathered their samples by attaching a plankton-net to a water pump on the side of the ship.[42] Later in the 20th century, it became quite common to use whatever was available on a freight ship to

39 CFD. 1898–99. No. 47. Application from J. Schmidt, 24 December 1898.
40 Prytz S. *Warming – botaniker og rejsende*. Lynge: Bogan's Forlag A/S; 1984, p. 121.
41 CFD. 1898–99. No. 47. Application from J. Schmidt, 24 December 1898.
42 Mortensen, T. Siam-Ekspeditionen, in: Winge Ø, Vedel Taaning Å, eds. Naturforskeren Johannes Schmidt. Hans Liv og Ekspeditioner. Copenhagen: Gyldendalske Boghandel Nordisk Forlag; 1947: 12–16.

43 CFD. 1898–99. No. 47. Application from J. Schmidt, 24 December 1898.

44 Mortensen, T. Siam-Ekspeditionen, in: Winge Ø, Vedel Taaning Å, eds. Naturforskeren Johannes Schmidt. Hans Liv og Ekspeditioner. Copenhagen: Gyldendalske Boghandel Nordisk Forlag; 1947: 12–16.

support sample taking, but at the time of Schmidt and Mortensen's voyage, making the best of what was available at sea was considered quite novel. In the grant application Schmidt was able to refer to instances of this method being deployed successfully on steamships that travelled regularly across the North Atlantic between Denmark and the Faroe Islands and Iceland. Following training on how to conduct the sampling, it could be left to the ship's crew to carry out subsequent investigations.[43]

The next stop of the voyage was Singapore, where they both seem to have been infatuated with the flora, fauna and the folklore of the Far East. 'It must be wonderful to be a plant in Singapore', Schmidt remarked.[44] When Schmidt and Mortensen arrived in Siam in January of 1900, they set up quarters on the island of Koh Chang. Courtesy of Admiral Richelieu, all the coastal stations of the Siamese Kingdom as well as the steamships of the East Asiatic Company were placed at the disposal of the two explorers of mangrove forests and inshore waters. This is also known from a contemporary British report. The British trade attaché in Bangkok reported back to London that the two Danish scientists were not really scientists. Instead he thought they were spies sent

1.6 Mangroves in the Gulf of Thailand
Mangroves during low tide, where the roots are clearly visible. Mangroves are salt water tolerant and they thrive in tropical areas, such as the Gulf of Thailand. The formation of shoots in mangroves was the topic for Schmidt's doctoral dissertation (1903). (The Archives of the Natural History Museum, Copenhagen).

out by the Russians and the Danish East Asiatic Company with the intent to get ahead in terms of forestry and mining. In fact, he added, one of them looks more like a navy officer than a scientist.[45] Judging from contemporary images, the looks of both Schmidt and Mortensen could have passed for that of a covert navy officer.

While there is no indication that Schmidt and Mortensen were involved in covert activities as such, they did take an interest in trees. Schmidt studied the so-called, *Dipterocarpaceae*, huge rain forest trees of importance to the timber industry, and a species which was substantially taller than the rest of the rain forest tree cover. Schmidt would either shoot down branches or he would have the soldiers cut down the entire tree. Mortensen remembers how he developed the greatest respect for Schmidt as a scientist, tirelessly collecting samples of a whole range of plant species displaying a keen eye for seeking out new specimen. Schmidt's main research topic at the Koh Chang station was found right on the waterfront, the mangrove plants, *Rhizophoraceae*. Safely back in Denmark again in the summer of 1900, Schmidt would devote much of his spare time working with the material of Koh Chang.[46]

Teaching assistant and doctoral thesis

From 1901–1909, Schmidt worked as a part-time teaching assistant on courses in plant anatomy run by his old professor at the University of Copenhagen, Eugen Warming.[47] Schmidt's job situation is revealed in a letter from Warming dated May 1902, which Schmidt kept in his personal files for posterity. Here, Warming appeared eager to hold on to his bright young protégé, as he emphasized how the job would be fairly easy for him, teaching two hours twice a week. He would earn 800 Danish Kroner per year, or roughly the same as a skilled labourer, and the tasks mainly consisted of making sure that the necessary chemicals and instruments were ready by the start of the exercises, so he would not be required to make a presentation for every lesson. There would be students of natural history as well as from medicine, but for Schmidt's comfort, Warming let him know that it looked like the class of the coming academic year would be fairly small.[48]

Around the same time, Schmidt also worked as a teaching assistant at the technical university, *Polyteknisk Læreanstalt.* Here, Schmidt gave a taught course in 'microscopy and the growth of fermentation

45 Kaarsted T. *Admiralen. Andreas de Richelieu – forretningsmand og politiker i Siam og Danmark*. Odense: Odense Universitetsforlag; 1990. p. 142.

46 Mortensen, T. Siam-Ekspeditionen, in: Winge Ø, Vedel Taaning Å, eds. Naturforskeren Johannes Schmidt. Hans Liv og Ekspeditioner. Copenhagen: Gyldendalske Boghandel Nordisk Forlag; 1947: 15.

47 Strubberg AC. Nyttefiskenes Livshistorie studeres i Nordatlanten og sydpaa til Biscaya. In: *Naturforskeren Johannes Schmidt. Hans Liv og Ekspeditioner*. Copenhagen: Gyldendalske Boghandel Nordisk Forlag; 1947:26.

48 KB. Johannes Schmidt 'Papirer'. Ut.374.V.1. Privat, 'Instrukser og Breve, ca. 1902–1908'. Letter from E. Warming to J. Schmidt, 2 May 1902.

organisms.'[49] All in all, the young aspiring scientist was in demand in several arenas already. As Mortensen remarks about Schmidt's scientific credentials following the Siam Expedition, he was then 'a full-blooded botanist, and as such he would undoubtedly have gone down in history as one of the greatest'.[50]

In 1903, Johannes Schmidt defended his doctoral thesis, which dealt mainly with the formation of shoots on mangrove trees in East Asia. This enabled Schmidt to draw conclusions with regards to the characters of the plant community to which mangroves belong. Schmidt stated explicitly that Eugen Warming, his professor, was his mentor and his inspiration, and Warming was also a member of Schmidt's assessment committee for the doctoral degree.[51]

Schmidt's doctoral thesis however, spanned a number of other conclusions drawn from his earlier work on bacteria as well as two of his then current marine investigations. Thus, in one of the chapters of his thesis Schmidt argued that it is yet to be proven if the Baltic Sea proper has an indigenous stock (bestand in Danish) of plaice (*Platessa platessa*). He also stated that cod (*Gadus morhua*) is a migratory species, and consequently that the Danish Kattegat Sea area did not possess a stock of cod (torskebestand) of its own.[52]

Conclusion

To achieve something with your life at least three different components are significant, namely talent, hard work and good fortune. In the case of Schmidt all three components interacted in a way that gave him a head start as a scientist, while his marriage moved his social status from middle class to the mercantile upper class in Danish society. The loss of his father at the age of seven was tragic, of course, but partly coming of age in the household of Johan Kjeldahl must have been a stimulating experience in more than one way. The uncle's laboratory was in reach for a bright boy coming of age. Furthermore, he was introduced to one of the most international and science oriented business enterprises in Denmark at the time. This was cemented by the decision to marry Ingeborg van der Aa Kühle, the daughter of the CEO of Gamle Carlsberg. While Schmidt's social life was further tied-in to the extended 'Carlsberg Family', his marriage also added an important financial security, which he could rely on for his future enterprises.

49 CA. *Carlsbergfondet, Carlsberg Laboratorium.* E06.'1909'. J. Schmidt to Board of Trustees, 22 September 1909.

50 Mortensen, T. Siam-Ekspeditionen, in: Winge Ø, Vedel Taaning Å, eds. Naturforskeren Johannes Schmidt. Hans Liv og Ekspeditioner. Copenhagen: Gyldendalske Boghandel Nordisk Forlag; 1947: 16.

51 Prytz S. *Warming – botaniker og rejsende.* Lynge: Bogan's Forlag A/S; 1984, p. 127.

52 Schmidt J. *Bidrag til Kundskab om Skuddene hos den gamle Verdens Mangrovetræer.* Copenhagen: Det nordiske Forlag; 1903.

Schmidt would not have made it very far, if he had not been an unusually bright natural history student at Copenhagen University. Writing and publishing books and articles, even as a student certainly made his professors raise an eyebrow or two. By a stroke of good fortune, the professor of natural history happened to be the world class scientist, and founder of ecology, Eugen Warming. Warming in particular, sought to promote Schmidt's chances to excel in botany, hiring him as a teaching assistant, and undoubtedly encouraging him to pursue the chance to go to Siam for what became Schmidt's first of many expeditions.

The funds for the Siam expedition were made up of subsidies from three sources, the Danish state, The East Asiatic Company and the Carlsberg Foundation. This way of pooling resources to meet a declared goal would become another mark of Schmidt's fundraising successes throughout his career, the shape of which he could not have predicted before the turn of the 20th century. Yet, his determination and hard work is manifest from his doctoral thesis on mangrove plants, defended as early as 1903, with Eugen Warming acting as one of the examiners.

By then, as Schmidt entered his mid-twenties, this already accomplished botanist and bacteriologist was headed for his most decisive career move, as he made his way to the sea, and so added another page to his widening research portfolio.

2 The legend of the unintentional eel larvae

All honour to the Danes who solved this age-old mystery,
Who, step by step, and year by year, revealed the history;
Johannes Schmidt the leader, with 'Papa' Petersen behind,
Who made the 'Thor' and 'Dana' ships of fame for all mankind.[1]

The prominent British zoologist Walter Garstang is chiefly remembered for ground breaking studies of the impact of fisheries on North Sea plaice (*Platessa platessa*), but posthumously, a collection of his poems was published, featuring topics such as the discovery of the life history of the freshwater eel (*Anguilla anguilla*), and it is from this volume that the verse cited above is taken. Obviously, Garstang is heralding the effort of Danish scientists Johannes Schmidt and C. G. Joh. Petersen, both of whom Garstang knew personally and had worked with collaboratively. Intentionally or not though, there is a double meaning to the line: 'Johannes Schmidt the leader, with 'Papa' Petersen behind.'

In the most recent literature review of investigations into the life history of the eel, Johannes Schmidt is described as the scientist who identified The Sargasso Sea as the central breeding location of the European eel, based on empirical investigations of the distribution of eel larvae of different sizes.[2] This is correct, but it is equally safe to say that in Schmidt's own papers on his eel investigations, he gradually downplayed just how much he owed to his Danish colleague, and marine science mentor, C. G. Joh. Petersen (1860–1928).

In the many accounts of Schmidt's route to 'fame for all mankind', the pivotal narrative point is the finding in 1904 of the first eel larvae ever caught in the Eastern part of the North Atlantic. Here Schmidt is credited for the discovery, which then set him on track for almost three decades of dedicated pursuit of the life history, not least the breeding places, of the eel worldwide.

1 Garstang W. *Larval forms and zoological verses.*; Basil Blackwell, 1951, 1985 (reprinted 1985, Chicago: University of Chicago Press), p. 60–61.
2 van Ginneken VJT, Maes GE. The European eel (*Anguilla anguilla,* linnaeus), its ifecycle, evolution and reproduction: A literature review. *Rev Fish Biol Fish.* 2005;15(4):367–398.

Nonetheless, in an overview article on eels from 1906, Schmidt fully recognizes the significance of Petersen's achievements on the first page of this lengthy report of his Atlantic investigations.[3] However, when in 1912, Schmidt wrote up an account of the Danish investigations in the Atlantic Ocean and the Mediterranean on the biology of the freshwater eel, Petersen's role had dwindled. Now Petersen was mentioned only as the 'the then leader of the expeditions'. In 1912, Schmidt remembered how finding the first eel larva in the North Atlantic Ocean in 1904 was 'a chance of luck'.[4] Schmidt's own recollection of his achievements continued over the next two decades as his scientific career reached a zenith of international fame and recognition. As the winners are the ones writing history, Schmidt himself may have been the strongest force behind the creation of the myth, or legend, of the sheer luck that led to the discovery of the eel larvae. Several modern chroniclers of marine science have built on and extended this legend, but there is much more to this story than a tale of the luck of a great genius.

In the international literature, the unintentional discovery of the eel larvae is highlighted in Helen Rozwadowski's *The Sea Knows No Boundaries*, and similarly in the most recent contribution in an edited volume on *Eels and Humans*.[5] In Jakob Jakobsson's contribution to the centennial ICES *Marine Science Symposium* from 2002, the Danish fisheries investigations in the Atlantic Ocean are highlighted as having 'particular interest', and Schmidt is once again described as the brilliant young leader, who found an eel larva, 'where none should have been…'[6] Petersen's name is not part of the article. In modern overview works on the history of Danish marine science the view that Schmidt found the eel, unexpectedly, is also the prominent narrative.[7]

One important contemporary exception is Tom Fort's *The Book of Eels* (2002). This book presents a lengthy account of Schmidt's route to fame for his eel research. Fort based his account on interviews with the late Danish marine scientist and famous eel researcher, Jan Boëtius (1918–2012). Fort does mention that Schmidt gradually paid less and less tribute to the fact that he was Petersen's assistant, when finding the eel larvae in 1904, on what Fort erroneously describes as Schmidt's first Atlantic cruise on *Thor*. But, despite this acknowledgement, the full extent of the scholarly credit that Schmidt owed to Petersen is not discussed, as Fort's focus is more on the future rivalry between Schmidt and Italian scientist Grassi.[8]

3 Schmidt J. Contributions to the Life-History of the Eel (*Anguilla vulgaris*, flem.). *Rapport et Proces-Verbaux des Reunions*. 1906;5:137–274.

4 Schmidt J. *Danske Undersøgelser i Atlanterhavet og Middelhavet over Ferskvandsaalens Biologi*. Copenhagen: C. A. Reitzel; 1912, p. 6.

5 Rindom S, Tomkiewicz J, Munk P, et al. Eels in culture, fisheries and science in Denmark. In: Tsukamoto K, Kuroki M, eds. *Eels and Humans*. Japan: Springer Science+Business Media; 2014:41–60; Rozwadowski H. *The sea knows no boundaries. A century of marine science under* ICES. Seattle and London: ICES/University of Washington Press; 2002. p. 71.

6 Jakobsson, Jakob, Ocean travelers, in: ICES *Marine Science Symposia*, 215: 330–342. 2002.

7 Wolff T. *Danske ekspeditioner på verdenshavene: Dansk havforskning gennem 200 år*. Copenhagen: Rhodos; 1967. p. 155–156; Nielsen, Kristian Hvidtfelt, Ries, Christopher Jacob and Østergaard, Kristian B., 'Danske Videnskabelige ekspeditioner I perioden', in: Hvidtfelt Nielsen, Kristian and Nielsen, Henry (eds.), *Viden uden Grænser 1920–1970, Dansk Naturvidenskabs Historie vol. 4*, Aarhus: Aarhus University Press; 2006. p. 285.

8 Fort T. *The Book of Eels*. Great Britain: Harper Collins Publishers; 2002. p. 96–97.

In the commemorative volume from 1947, *Naturforskeren Johannes Schmidt,* it is striking that Petersen's central role is acknowledged in only one out of a total of 18 chapters, namely Aage Vedel Taaning's chapter on the beginning of Schmidt's eel investigations.[9] In the same volume it was A. C. Strubberg's task to explain the finding of the first larvae in the North Atlantic. An obvious choice, since Strubberg worked as Schmidt's assistant on board *Thor* from 1904. Petersen's role, in so far as Strubberg recalled, is limited to a description of how they used a 'Petersen young fish trawl' for their investigations.[10] Petersen receives the same modest credit in Ove Paulsen's account of his Atlantic expeditions with Schmidt on the steamship *Thor*.[11] Aage Vedel Taaning was a good deal younger than Schmidt and Paulsen, a different generation really, and too young to remember what things were like in the early 1900s. Taaning therefore seems to have relied on written records for his account of Schmidt's eel investigations. Therefore, it seems likely that Taaning simply came across Petersen's instructions for *Thor*'s cruise in the same way as this author. These instructions paint a quite different picture.

So why has Petersen all but disappeared from the historiography of eel investigations? And is it really possible to attribute Schmidt's 1904 catch of a single eel larva to 'a chance of luck'?

In reality, there was much more to Schmidt's 1904 'career trawl'. In this chapter, the story of Schmidt's life changing path to become a researcher of the life history of the European freshwater eel is told in its context of the existing world of marine science and the Danish institutions he became part of from the beginning of the 20th century. This is indeed a story of bright ideas and the scientists' and bureaucrats' will to pursue them, but it is equally a tale of scientific and political fights on a personal level, as well as on inter-institutional and international levels. It all began a generation earlier, in the 1880s, when Danish marine science was just about to be professionalized.

The state of fisheries in Danish society

Over the latter half of the 19th century, fisheries all around Europe experienced a tremendous growth. The growing population world wide and the ever-expanding network of railroads criss-crossing entire continents facilitated an unprecedented demand for fish, especially fresh fish and

9 Vedel Taaning Å. Aaleundersøgelserne begynder for alvor. In: Winge Ø, Vedel Taaning Å, eds. Naturforskeren Johannes Schmidt. Hans Liv og Ekspeditioner. Copenhagen: Gyldendalske Boghandel Nordisk Forlag; 1947:39–40.

10 Strubberg AC. Nyttefiskenes Livshistorie studeres i Nordatlanten og sydpaa til Biscaya. In: *Naturforskeren Johannes Schmidt. Hans Liv og Ekspeditioner*. Copenhagen: Gyldendalske Boghandel Nordisk Forlag; 1947:26–36.

11 Paulsen O. Havundersøgelserne i de nordlige Have begynder. In: Winge Ø, Vedel Taaning Å, eds. Naturforskeren *Johannes Schmidt. Hans Liv og Ekspeditioner.* Copenhagen: Gyldendalske Boghandel Nordisk Forlag; 1947:17–25.

canned fish, supplementing the more ancient traditions of consuming salted, smoked and pickled marine fare. The fishing industries in countries such as Great Britain, Canada and the United States were able to meet the improved market situation through the implementation of new vessels and new types of fishing gear and fishing methods. Cotton nets replaced nets made from hemp, steam propulsion and from around the turn of the 20th century also motor propulsion replaced the use of wind, which greatly multiplied the fishing power of each fishing vessel. Towards the end of the 19th century other countries invested heavily in their fishing sectors to further their share of the harvest of the seas. Gradually countries like Japan, Germany, Norway, France, The Netherlands, and to some extent Denmark joined this leap into modern industrial fishing practices and industries.[12]

Denmark was a relative late-comer. Only in the 1880s did larger decked vessels for offshore fishing enter the fleet, firstly from Frederikshavn, later from many fishing ports around the country. The extensive Danish coastline drawing a zig-zagged line for more than 7000 kilometers in and out of fiords and estuaries provided the dwellings for more than 10,000 fishermen before the turn of the 20th century. During the following decades until the outbreak of WWII this number almost doubled.[13] The most important catches for Danish fishermen in the decades around the turn of the 20th century were those that included cod, plaice, herring and eel. In the largest Danish estuary, the Limfjord for instance, eel accounted for roughly 80 percent of the total value of all fisheries.[14] Eel was so prolific in the Limfiord that it was the pride of every housewife to be able to cook eel in as many ways as possible to ensure a variety of dishes. These dishes included eel in curry, eel in jelly and of course smoked, fried and salted eel.

The so-called 'Danish Seine' or 'snurrevoddet' as it is known in Danish was invented in the Danish estuary, Limfjorden in 1848. With this simple method of catching high quality fish, mainly eel and plaice, with a small boat, a net attached to an anchor and a buoy, relatively small fishing units with little capital available could flourish. When the motor engine was installed in most Danish fishing vessels in the 1910s and 1920s, there was seemingly no end to the rapid rise of the fishing industry. At the turn of the 20th century, the first wave of modernization had resulted in a fishing fleet reaping c. 7.8 mio. Danish Kroner for having landed 27,000 tonnes of fish. In 1940, when the war came to Denmark, the fishing industry earned almost 7 times more, 56.7 mio Kroner for a

12 Poulsen, B., 'Fisheries', In: L. Kotzé & S. Morse (eds.) *The Berkshire Encyclopedia of Sustainability vol. 9. Afro-Eurasia: Assessing Sustainability*, Great Barrington, MA: Berkshire Publishing Group, (2012), 136–140.

13 Blegvad H. Fiskeriet i Danmark. København : Selskabet til udgivelse af kulturskrifter; 1946. vol. 1, p. 29–33.

14 Anonymous, *Fiskeriberetning* (for the years 1888–1910).

catch of 122,000 metric tonnes. In other words, the landings had grown by more than 400 percent in four decades, while the turnover had become eight times larger in the same period.[15]

The fisheries received a good deal of public support and attention as well. In 1912 for instance, the fishermen's organization and the fish exporters' organization joined together to create a large fair in Copenhagen, the *Fiskeri- og Motorudstillingen*. The newly appointed Danish king, Christian X, opened the show in the middle of the Port of Copenhagen, and during less than two months, more than 100,000 visitors came to witness the largest, at that time, manifestation of the modern Danish fishing industry.[16]

However, one major threat to continued growth was lurking in the background. Already by the turn of the 20th century it was noticeable for fishermen and marine scientists alike that the catch rates in the fiords and in more closed areas such as Kattegat and the Western Baltic Sea were dwindling. To keep up high catch rates, a necessity when there was high investment in ever more expensive seagoing vessels, fishermen had to venture further and further away from the home shores. Thus by 1940, the North Sea had become the most important fishing grounds in terms of revenue and weight.[17]

The desire to improve the fishing industry caused managers and fishermen to turn to marine science for help. To know how to manage limited stocks of various commercial fish, much more knowledge was needed with regards to how and why fishing stocks fluctuated, how much fishing pressure they could take, while there was also a lurking understanding that the fish were merely one aspect of a larger system of marine species in distinct hydrographical conditions. Here, like in the beer brewing business, the order of the day was to turn to science for help.

The Danish Biological Station and C. G. J. Petersen

The first institution in Denmark dedicated specifically to marine science was the so-called *Biologisk Station*. Government inspection of fisheries was already in place from around 1870, and during the course of the 1880s, the gunboat *Hauch*, doubled as the vessel used for fisheries inspection and as a research vessel with an on-board laboratory. The captain of the vessel was Christian F. Drechsel (1854–1927), who came

15 Blegvad H. Fiskeriet i Danmark. København : Selskabet til udgivelse af kulturskrifter; 1946. vol. 1, p. 29–33.

16 Tarbensen K. *For alle danmarks fiskere – Danmarks Fiskeriforening 125 år, 1887–2012*. Taulov: Danmarks fiskeriforening; 2012, p. 76–79.

17 Blegvad H. *Fiskeriet i Danmark*. København: Selskabet til udgivelse af kulturskrifter; 1946. vol. 1, p. 31–33.

2.1 Carl Georg Johannes (C. G. Joh.) Petersen on board S/S *Sallingsund*

Carl Georg Johannes Petersen (1860–1928) was one of the most influential marine scientists in Europe in the decades before and after 1900. He was a leading intellectual as well as an ingenius scientist in the field constructing several devices and sampling methods. The Petersen grab, the Petersen young fish trawl and the Petersen disc are inventions still bearing his name. Here Petersen is accompanied by his dog on board the S/S Sallingsund.

from a naval background.[19] Somewhat later in life Drechsel became the Secretary General of the International Council for the Exploration of the Sea (hereafter ICES) between 1908–1927, and a pivotal figure for the advancement of Danish and international marine science at large.[20]

From 1883, Drechsel worked closely with the marine scientist chosen to conduct zoological studies from *Hauch*, Carl Georg Johannes (C. G. Joh.) Petersen (1860–1928). In 1883, Petersen was still a student of natural history at the University of Copenhagen, but the years on *Hauch*

18 Blegvad, H., Christian Frederik Drechsel, *Den Store Danske Encyclopædi,* Copenhagen: Gyldendal, http://www.denstoredanske.dk/Dansk_Biografisk_Leksikon/Forsvar_og_politi/S%C3%B8officer/C._F._Drechsel (accessed, 2 January 2014).

19 The acronym ICES is widely used today, therefore this acronym will be used in the following. Knudsen M. De internationale Havundersøgelser. In: Winge Ø, Vedel Taaning Å, eds. *Naturforskeren Johannes Schmidt. Hans Liv og Ekspeditioner*. Copenhagen: Gyldendalske Boghandel Nordisk Forlag; 1947:109.

resulted in a series of publications including Petersen's 1888 doctoral dissertation on the biogeography of Danish molluscs.[20] Based on a combination of trawling and scraping the bottom on more than 500 different sites, the so-called 'stations', scattered evenly over the Kattegat Sea Petersen was able to detect the presence and absence of a multitude of molluscs. Using this dense spatial sampling method Petersen dared to conclude also negatively, that the absence of a certain species in his sample indicated the absence of the species as such in that particular area of the seabed.[21]

Following the 1886 season, the ship's captain Drechsel moved on to a new position as a government fisheries consultant, which for Drechsel marked the beginning of forty years of government and international service for the advancement of marine science and marine fisheries. In 1890, Drechsel published a large volume on the state of Danish fisheries, mapping the size and geographic spread of more than a dozen of the most important commercial fish species.[22] Then, in 1893, Petersen published a lengthy volume on the scientific results obtained on board *Hauch*.[23]

20 Petersen CGJ. *Om Biologisk Stations Virksomhed fra 1889–1914. Efterskrift 1918.* Copenhagen: Centraltrykkeriet; 1919.

21 Petersen CGJ. *Om de skalbærende Molluskers Udbredningsforhold i De danske Have indenfor Skagen.* Copenhagen: Andr. Fred. Høst & Søns Forlag, 1888.

22 Drechsel CF. *Oversigt over vore Saltvandsfiskerier i Nordsøen og Farvandene indenfor Skagen.* 1890. Dansk Fiskerimuseum; 1988, first pub. 1890.

23 Petersen CGJ. *Det videnskabelige udbytte af Kanonbaaden Hauchs' Togter i de danske Farvande indenfor Skagen, i Aarene 1883–86.* Copenhagen: Høst & Søns Forlag; 1893.

2.2 Christian Frederik Drechsel in 1890
Christian Frederik Drechsel (1854–1927) portrayed in 1890. Drechsel had a diverse maritime background. He was a navy captain before working as a government fisheries consultant alongside a job as the Port Director of Copenhagen. He had a successful partnership with C. G. Joh. Petersen for more than 20 years until 1908 (The Royal Library of Copenhagen).

Thus, when the *Biologisk Station*, which had been initiated by Drechsel and Professor Lütken of the University of Copenhagen, was formally established in 1889 it was not a surprise when C. G. Joh. Petersen was appointed as the lead investigator. Biologisk Station then acquired its own vessel, a modified transport ship, which was towed to positions in various Danish fiords each summer. This marked the beginning of a formidable expansion of Danish marine science with the explicit purpose of promoting fisheries and optimizing fisheries legislations.[24] Under Petersen's leadership several important results were obtained, and some of the scientific methods and gear that he and his assistants developed are still in use today.

During the 1890s, Petersen became internationally recognized as one of the first marine scientists to conduct marking experiments with fish. In a marking experiment, fish, such as plaice, are caught and a numbered piece of metal is then attached to each specimen. Then the fish is released again. If the same specimen happens to be caught again by a fisherman or fellow scientist, it is possible to estimate the movement and growth of this particular fish during the period, since it was marked.[25]

Another invention of this era was a grab used in shallow waters for collecting macroscopic fauna from the sea-bed, the so-called 'Petersen bottom sampler'. This grab contains a specific amount of material each time it is hoisted from the bottom of the sea, which means there is some consistency between samples if the same kind of instrument is used for collection. Like the marking technique, the grab is still used today.[26]

The methodological and technical advances of the Biologisk Station immediately yielded a series of important results including some, which would later benefit Johannes Schmidt in his own investigations. The very first publication of the Biologisk Station that dealt with the biological condition of the fish in the Holbæk Fjord presents such an example.[27] Here, Petersen presented the method for measuring the lengths of a number of individual fish from different species, in order to estimate the age and relative fecundity of different year classes of a distinct population. The idea was that the length of a great number of young fish would cluster around distinct length sizes. These length classes, Petersen rightly assumed to be different age groups.[28] This method spread internationally and was used by Johan Hjort in Norwegian cod investigations just after the turn of the 20th century.[29]

24 Blegvad H. *Dansk biologisk station gennem 50 år, 1889–1939*. Copenhagen: C. A. Reitzels Forlag; 1943.

25 Smith TD. *Scaling fisheries: The science of measuring the effects of fishing, 1855–1955*. 2nd revised 2007 ed. New York: Cambridge University Press; 1994

26 Fenchel T, Bendtsen J, Sand-Jensen K. *Naturen i Danmark: Havet*. Copenhagen: Gyldendal; 2006.

27 Petersen CGJ. Fiskenes biologiske Forhold i Holbæk Fjord. *Beretning fra Den Danske Biologiske Station*. 1890–91;1:153.

28 Blegvad H. *Dansk biologisk station gennem 50 år, 1889–1939*. Copenhagen: C. A. Reitzels Forlag; 1943. p. 5–9.

29 Schwach, Vera, *Havet, fisken og Vitenskapen. Fra fiskeriundersøgelser til havforskningsinstitutt 1860–2000*, Oslo: Havforskningsinstituttet; 2000. p. 153.

2.3 The Petersen grab used by R/V *Dana*
The Petersen mud grab used by the R/V Dana. With this grab a standardized amount of the sea bed was collected from each haul. Slightly modified versions are still in use (RA. Carlsbergfondets DANA-ekspeditioner. 1917–1973. 5).

Petersen's eel investigations

It is worth remembering that, at the time when Petersen began his experiments; the general perception was that Danish commercial species bred in near shore areas, which is true in many cases, such as for plaice (*Platessa platessa*). However, when Petersen investigated the freshwater eels (*Anguilla anguilla*) in the Danish fiords, there was a marked absence of small eel. From this, Petersen suggested that it was likely that eel bred in open waters, since only adults were caught.[30] In his first paper on the subject, Petersen put the eel in its own category due to the mysterious

30 Blegvad H. *Dansk biologisk station gennem 50 år, 1889–1939*. Copenhagen: C. A. Reitzels Forlag; 1943. p. 9.

circumstances of its breeding habits.[31] Petersen's eel investigations continued in the following years. Fisheries for eel were particularly important in the many shallow fiords and sounds in Denmark, where small holders supplemented their income with eel fisheries. Relatively little however, was known with regards to the reproduction of the eel.

As early as the fourth century B. C. E., Aristotle was puzzled by the mysteries of the common freshwater eel, which he stated must migrate to sea to breed. In 1684, Francesco Redi from Italy found that adult eel swim to sea to breed and that the glass eel migrate in the other direction from the sea to fresh water systems in the months between January and April. In this way, Redi was fighting the commonly held belief that eel gave birth to live youngsters, a belief that Linnaeus preferred for his classification a century later.[32] In 1874 however, the Austrian scientist, Styrski was the first to describe the male organ of the eel, whereas the female organ had already been known for about a hundred years. No one, however, had ever encountered a spawning eel, or an egg for that matter – quite unlike any other common fish species.[33]

Eel was commonly caught as a three-year-old glass eel, and then it metamorphosed into the so-called yellow eel, which gets its name from the yellow colours on its head. In the 1890s Petersen undertook an experiment in which yellow eel were kept in tanks at the Biologisk Station. Hereby, he was able to assert that after a few weeks some of them began to acquire a silvery tan. Previously it was believed by some that the so-called silver eel was a different species than the yellow eel. Petersen, however, established that the metamorphosis from yellow to silver eel marks a transformation from a process of growth, where the yellow eel has well developed digestive organs towards a phase, where the eel 'puts on' its *breeding dress*. At this point the digestive tract shrinks, so the silver eel has difficulty eating, but its reproductive organs become more pronounced. These observations rendered obsolete the previously held theory that the yellow eel was a separate species from the silver eel.[34]

Petersen's contribution marked a leap forward, but still the rest of the life history of the freshwater eel remained as mysterious as the darkest depths of the ocean. Where did eel go to breed, once it had vanished from the near shore of Danish waters?

31 Petersen CGJ. Fiskenes biologiske Forhold i Holbæk Fjord. *Beretning fra Den Danske Biologiske Station*. 1890–91;1:123–183.

32 Vedel Taaning Å. Aaleundersøgelserne begynder for alvor. In: Winge Ø, Vedel Taaning Å, eds. Naturforskeren Johannes Schmidt. Hans Liv og Ekspeditioner. Copenhagen: Gyldendalske Boghandel Nordisk Forlag; 1947:38.

33 Schmidt J. Contributions to the Life-History of the Eel (*Anguilla vulgaris*, flem.). *Rapport et Proces-Verbaux des Reunions*. 1906; 5:139–140.

34 Petersen CGJ. The Common Eel (*Anguilla vulgaris* T.) gets a particular Breeding-dress before its Migration to the Sea. *Report from the Dansk Biologisk Station*. 1896;5.

Leptocephalus brevirostris

This question also puzzled a research team in Italy, where another great advance was made with regards to the life history of the common freshwater eel. Also in 1896, the Italian scientist, Giovanni Battista Grassi and his former student, Dr. Calandruccio published a study in the prestigious *Transactions of the Royal Society of London*, where they put forward the theory that a strangely looking, lancet shaped and transparent little fish known as the Leptocephalus brevirostris was really the larval stage of the common freshwater eel. The two men had compared the morphology of the Leptocephalus with that of the eel. They had noticed at an aquarium how the larvae of moray like fish such as European conger eel (*Conger conger*) gradually transformed into elvers or glass eel, the stage before reaching adulthood. The larvae were often found in digested form in the stomachs of ocean sunfish (*Mola mola*).

The two Italians summarized their findings on the life history of the eel by stating that it matured in the depths of the sea, where it acquires larger eyes than individuals caught near shore. The abysses of the sea are the spawning places of the eel, where the eggs then float in open water as opposed to being attached to the sea-bed. Another important clue at the time, was that one, larvae, the *Leptocephalus* had been found in the Strait of Messina in Italy, suggesting that perhaps the Mediterranean was the nursery for eel.[35] Grassi did not say explicitly that *all* eel in Europe bred in the deep waters west of Italy, but there is no doubt that he believed the Italian eels to do so. From October to January, the common eel migrates to sea. He noted that currents such as those in the Strait of Messina throw up, from the abysses of the sea, specimens which are more advanced than other silver eel without yet having reached state of 'total maturity'. Then Grassi stated that 'eggs, which according to every probability belong to the Common eel, are found in the sea from the month of August to that of January inclusive; fourth, the *Leptocephalus brevirostris* abounds from February to September'.[36]

Grassi's 1896 paper refers a few times to the advances made by Petersen in Denmark on the transformation of yellow eel into silver eel, while Petersen in his paper from the same year recognizes Grassi's findings on the Leptocephalus as the eel's larva, as well as Grassi's idea of the deep sea as the eel's spawning ground. Petersen though, reveals that he is anxious to hear more about the Italians' explanations.[37]

35 Grassi GB. The Reproduction and Metamorphosis of the Common Eel (*Anguilla vulgaris*). *Proceedings of the Royal Society of London*. 1896;60(359–367):260–271.
36 Grassi GB. The Reproduction and Metamorphosis of the Common Eel (*Anguilla vulgaris*). *Proceedings of the Royal Society of London*. 1896;60(359–367):270–271.
37 Blegvad H. *Dansk biologisk station gennem 50 år, 1889–1939*. Copenhagen: C. A. Reitzels Forlag; 1943. p. 13.

Unfortunately, Grassi appears to have been content with his contributions to the life history of the eel, and during the next couple of decades Grassi devoted his time to terrestrial research topics. In fact, when Grassi is remembered today as a great scientist it is mainly to due to his studies on termites, which won him the 1896 Darwin Medal, and, even more so for his contributions to parasitology. Grassi along with British Nobel laureate Ronald Ross is now chiefly remembered for his discovery of how malaria is transmitted to humans via certain species of mosquitos. The two research teams, one in England and one in Italy competed ferociously to be first with this important discovery, and Ross went so far as to accuse Grassi of scientific fraud, by presenting false evidence as argument. Grassi was furious, but the intervention of German bacteriologist Robert Koch on Ross' side, is today viewed upon as the tipping point, which gave Ross the sole Nobel credits for the advances in malariology. By modern day standards the two scientists ought to have shared the glory of their important discoveries. Italy was plagued by malaria at the turn of the 20th century, but after the cause of the disease was known, politicians and planners launched a series of successful campaigns to reclaim the marshlands that hosted the infectious mosquitos, as well as to introduce more direct biological and chemical control of the mosquitos.[38]

Grassi's successful investigations in parasitology kept him busy, and distanced from contemporary studies of the eel. In a lengthy obituary following his death in 1925, Grassi's eel investigations are not even mentioned.[39] Then again, by 1925, Grassi's theory that eel bred in the Mediterranean Sea had been overthrown by one of Petersen's pupils, Johannes Schmidt.

In 1899 however, when Petersen continued his own eel investigations, this was hardly foreseeable. It was during this year that the Danish state granted Biologisk Station its own steamship *Sallingsund*. This proved of huge benefit to marine investigations. Now, the vessel could go whereever, Petersen saw fit, whereas the transport boat had needed towing each time it was to be moved. One of the first tasks for the new steamship was to count all the pots used for eel in Denmark in the autumn months. Since silver eel were observed to migrate away from the Danish shores in autumn, Petersen speculated that they might turn towards the Baltic Sea for spawning, but his eel pot study indicated otherwise. The proliferation of the Danish autumn eel pot fisheries did not increase towards the Baltic as would be expected if Petersen's theory was correct. Petersen, therefore, proposed that the silver eel caught in Denmark during the autumn must be from Danish waters.[40]

38 Capanna E. Grassi versus Ross: Who solved the riddle of malaria? *International Microbiology*. 2006;9(1):69–74.
39 Janicki C. Giovanni Battista Grassi. *Naturwissenschaften*. 1926;14(13):261–269.
40 Blegvad H. *Dansk biologisk station gennem 50 år, 1889–1939*. Copenhagen: C. A. Reitzels Forlag; 1943. p. 18.

An international research station

41 Blegvad H. *Dansk biologisk station gennem 50 år, 1889–1939*. Copenhagen: C. A. Reitzels Forlag; 1943. p. 13–20.

Petersen's international reputation grew steadily as his investigations progressed, and throughout the 1890s, there was a constant flow of visiting marine scientists on board his floating marine station. In 1897, for instance, the biologists Apstein from Kiel, Johan Hjort and Dr. Gran of Norway and Professor Knipowitsch from Russia, all paid visits to the boat. Petersen in turn travelled to Sweden and Norway the same year and met with the famous hydrographer Otto Pettersson in Holma, Sweden. The following year, it was Otto Pettersson's turn to join Petersen on board his vessel. Professor Vinciguerra from Rome, Dr. Arnold from St. Petersburg and Dr. Malm from Gothenburg all came from abroad to see Petersen's work this year. They were joined by young and senior Danish marine scientists who would later attain significant international stature such as hydrographer Martin Knudsen, who worked for decades within ICES; plankton zoologist and future professor of botany, Carl Hansen Ostenfeld and zoologist Carl Wesenberg-Lund, who later founded the Danish research station for freshwater studies.[41] Johan Hjort came back several times, and he and Petersen seem to have struck their lifelong friendship during these energetic years around the turn of the century.

2.4 The Biologisk Station's barge
One of the research station, Biologisk Station's main assets in the 1890s was this barge, which was towed to where C. G. Joh. Petersen wished to carry out investigations. Petersen's many international visitors included Johan Hjort from Norway, Otto Petterson from Sweden, Knipowitsch and Arnold from Russia and Professor Vinciguerra from Rome (Carlsberg Archives).

Schmidt enters marine science

In short, the Biologisk Station earned a reputation as an attractive training ground for Danish students and international scientists alike. From 1901 until 1908 Johannes Schmidt came to profit from this scientific environment as a promising young scientist eager to learn the ropes at the Biologisk Station.

Schmidt learned to fish on board *Sallingsund*. He learned about salinity, temperature, the different layers of water in the water column and the ways of various currents. He got to know the various Danish fisheries and talked to the fishermen, all with Petersen by his side, ready with his state-of-the-art tools and techniques. According to his friend from university, Ove Paulsen, Schmidt once made a remark to Petersen on a particular fish, which Schmidt thought was much more common, than what he had previously thought. Petersen replied that 'everything is bloody common, when you know all the right places to fish'.[42]

Nonetheless, Schmidt proved to be a quick learner. Thus in 1902, Schmidt was the designated rapporteur of one section in the annual report from Biologisk Station. Petersen himself was frequently listed as the sole author of the entire annual volume of the report from Biologisk Station, but in subheadings there was room for Schmidt. The 1902 study in question dealt with the spatial abundance of the fry of plaice along the Danish inshore waters. The findings indicated that the southernmost spawning areas for plaice were to be found along a line going through the Danish Straits. Petersen was well aware that the study needed repetition and was affected by temperature as well as local fecundity of the resident plaice.[43] A number of different types of gear were used, including the 'Petersen-trawl', indicating that Petersen was well aware of his own stature as a scientist. Even today, the same type of gear is still referred to as the 'Petersen young-fish trawl'.[44]

It appears like a token of appreciation from Petersen that Johannes Schmidt's meticulous work sifting the evidence from hundreds of samples was replicated in the report from Biologisk Station. Actually, several pages are verbatim reproductions of Schmidt's logbook from *Sallingsund*. Six days a week all summer, the vessel was traversing Danish waters, and almost every day, sometimes several times a day, new 'stations' or sampling sites were taken.[45] Petersen must have been utterly content with his young assistant, so for the following summer season, in 1903, Petersen asked the 26 year-old Johannes Schmidt to

42 Paulsen O. Havundersøgelserne i de nordlige have begynder. In: Winge Ø, Vedel Taaning Å, eds. *Naturforskeren Johannes Schmidt. Hans Liv og Ekspeditioner*. Copenhagen: Gyldendalske Boghandel Nordisk Forlag; 1947:17.

43 Petersen CGJ. Hvor og under hvilke Forhold kunne Rødspættens Æg udvikle sig til Unger indenfor Skagen? *Fra Den danske biologiske Station*. 1904;12 (1902–03):3–23.

44 Smith TD. *Scaling Fisheries: The science of measuring the effects of fishing, 1855–1955*. 2nd revised 2007 ed. New York: Cambridge University Press; 1994, p. 147.

45 Petersen CGJ. Hvor og under hvilke Forhold kunne Rødspættens Æg udvikle sig til Unger indenfor Skagen? *Fra Den danske biologiske Station*. 1904;12 (1902–03):8–13.

2.5 The S/S *Sallingsund*
In 1899 the Biologisk Station purchased a former steam ferry, the S/S Sallingsund. In the summer season of 1901, C. G. Joh. Petersen introduced Johannes Schmidt to marine science on board the Sallingsund (*Carlsberg Archives*).

head the scientific investigation on a ship of his own, the steamship *Thor*, scheduled to investigate the North Atlantic waters around Iceland and the Faroe Islands. This had everything to do with Petersen's international stature and the creation of ICES.

Formalizing international marine science

In general, there was an international vitality surrounding marine science in the years around 1900. The Danish Biological Station and Petersen's work flourished through informal international networks, and the same was true for various marine research stations in Europe and North America. However, during the 1890s there was a growing concern, chiefly in the countries of Northern Europe, that more formal structures surrounding marine science were necessary.

Especially in the countries around the North Sea, there was a growing concern in the last decades of the 19th century that the commercial fisheries in that area were not entirely inexhaustible. Fishermen complained about dwindling catch rates and it had been observed that individual specimen of fish became smaller on average. Studies by British zoologist, Walter Garstang showed that in spite of an increase in total catches in the North Sea plaice fishery the catch per vessel did not increase. When accounting for the increment caused by motorization, the catch rates were in serious decline. Meanwhile, C. G. J. Petersen pondered whether it would be possible to address the concept of overfishing, which he also discovered as a local phenomenon in the plaice fishery, by transplanting fish from an abundant area to a less densely populated. Simultaneously and in collaboration with colleagues in other nations, Johan Hjort in Norway was pioneering the study of age composition in fish stocks as a means to determine the potential dynamics of stock fluctuations. Thus, scientific advances were occurring in many national settings. Nevertheless, and given the international character of the fishing grounds around Europe, there was a consensus in the group that it was necessary to embark on a more focused international cooperation to solve the problem of overfishing.[46]

To achieve this, several important personalities joined forces to create the so-called Stockholm conference in 1899. The chief engineer of the conference was the Swedish hydrographer Otto Pettersson. In hydrography, one of the main objectives in the late 1800s was the task of charting the flow of ocean currents and to distinguish these from the influence of meteorological variations in wind directions and air temperature. Whereas the problems associated with fisheries were of economic concern to the governments in Western Europe, the challenges associated with hydrography also caught the interest of national security. Information regarding the weather and currents in high seas was of strategic interest to many countries. Otto Pettersson sought to combine biological and physical oceanography in the interest of improving our understanding of important fisheries' problems.[47] One such classic problem in his own proximity revolved around understanding the dynamics of the sudden appearances of large shoals of herring along the Bohuslen Coast in Western Sweden, roughly once every century since the Viking Age.[48] The possible environmental connection had been proposed to a Scandinavian audience as early as 1879, when another Swedish scientist. Axel Ljungman hypothesized that sunspot eruptions were linked to the herring outbursts.[49]

46 Smith TD. *Scaling Fisheries: The science of measuring the effects of fishing, 1855–1955*. 2nd revised 2007 ed. New York: Cambridge University Press; 1994, p. 96–109.

47 Rozwadowski H. *The sea knows no boundaries. A century of marine science under* ICES. Seattle and London: ICES/University of Washington Press; 2002. p. 21.

48 Smith TD. *Scaling Fisheries: The science of measuring the effects of fishing, 1855–1955*. 2nd revised 2007 ed. New York: Cambridge University Press; 1994, p. 96–109.

49 Ljungman, A. V., 1879. Bidrag till lösningen af Frågan om de stora sillfiskenas sekulära periodicitet. *Nordisk Tidsskrift for Fiskeri* 5, 257–268.

The works of Otto Pettersson, as well as the Arctic explorations conducted by Norwegian explorer, Fridtjof Nansen caught the attention of the then king of Sweden and Norway, Oscar II, and Pettersson persuaded the Swedish king to sponsor the so-called Stockholm conference in 1899. Petersen represented the Danish government at this meeting, which in many ways laid the scientific foundations and organizational principles for what became the International Council for the Exploration of the Sea (ICES). Within hydrography, a serious of simultaneous voyages by different steamship vessels from each of the participating countries was to be undertaken. The participating countries, Great Britain, Denmark, the Netherlands, Finland, Germany, Norway, Russia and Sweden all committed themselves to make hydrographical surveys within their own territorial waters. For Denmark, this meant that surveys were to be made not only in the waters around mainland Denmark, but also in the North East Atlantic ocean off the Faroe Islands and around Iceland.[50]

Next to fisheries and hydrography, a certain spirit of internationalism favoured the collaborative efforts taking place. Several founding members of ICES were firm believers in the idea that international collaboration in the sciences would promote and reinforce economic and social ties between nations, which in turn would enrich the intellectual and material welfare in the developed world.[51] The Stockholm conference of 1899 was followed by a series of intergovernmental negotiations among the eight countries interested in establishing the intergovernmental structure. It was at these meetings that the international spirit gave way to classic diplomacy, where national interests meant the participants wrestled over issues such as where to place the institution's secretariat, the so-called Central Bureau, and where to place the planned international marine laboratory. Neither of the Great Powers would allow the other to host these facilities, and following another conference in 1901 in Christiania (Oslo), Norway, a compromise was reached, where Norway would host the Central Laboratory and Denmark would house the Central Bureau in Copenhagen.[52]

50 Rozwadowski H. *The sea knows no boundaries. A century of marine science under* ICES. Seattle and London: ICES/University of Washington Press; 2002. p. 19–35.
51 Rozwadowski H. *The sea knows no boundaries. A century of marine science under* ICES. Seattle and London: ICES/University of Washington Press; 2002. p. 11.
52 Rozwadowski H. *The sea knows no boundaries. A century of marine science under* ICES. Seattle and London: ICES/University of Washington Press; 2002. p. 38–39.

Central Bureau in Copenhagen

It is safe to say that Danish marine science came to benefit tremendously from ICES. Each member country had delegates on the council, which met each year, and following the preparatory conferences, the Danish delegates C. G. J. Petersen, Christian F. Drechsel and Martin Knudsen had central roles to play in the formative years of ICES' existence. The fisheries' interests were funnelled into a number of committees and sub-committees. Committee A dealt with the migration of fish species, while Petersen's focus was on Committee B, which became commonly known as the 'overfishing committee'. It was to the members of this committee, in particular, that Petersen and Garstang addressed the issue of plaice in the North Sea, and it is likely that these sessions influenced Petersen when he wrote up his thoughts on the concept of overfishing, published in 1903. Here, Petersen presented different types of overfishing depending on whether the size of specimen caught or the size of the fish stock in absolute numbers had declined, while Petersen also emphasized the need to distinguish between natural and fishery induced fluctuations.[53]

Within the hydrographical program, the other Danish delegate, physicist, Martin Knudsen placed Denmark firmly on the scene of hydrography. When ICES was set up, Martin Knudsen (1871–1949) was a lecturer in physics at the University of Copenhagen, and a pioneer in constructing instruments suitable for the challenges of marine science. Knudsen had several years of working experience with Otto Pettersson, and Knudsen became head of the Central Bureau's hydrographical section. Here, Knudsen's work was an instant success, as he prepared batches of so-called standardized seawater.[54] Standard seawater held a certain salinity, which then became the international yardstick for comparing and measuring the quality of water samples taken around the world. As a testimony of Knudsen's success, marine scientists continued to use samples derived from Knudsen's original samples until the 1970s.[55]

Probably as a result of the prominent positions of Knudsen and Petersen, ICES was formally inaugurated at a conference in Copenhagen in 1902. Denmark though, did not occupy all influential positions within the international council. The president of the German Oceanographical Society, Walther Herwig was elected the first president of ICES, with Otto Pettersson as vice president. Each member country was obliged to pay an annual sum to cover the administrative costs of maintaining the Central Bureau, and for the few years in which it was in existence, this money also subsidized the Central Laboratory in Norway. Funding was allocated according to the size of the individual countries, giving a total of c. 1 Million Danish Kroner. The Danish government, however, sponsored the in-

53 Petersen CGJ. What is overfishing? *Journal of the Marine Biological Association.* 1903;6:587–594.
54 Smed, Jens, Martin Knudsen – the oceanographer, in: ICES *Marine Science Symposia*, 2002, 215: 125–127.
55 Rozwadowski H. *The sea knows no boundaries. A century of marine science under* ICES. Seattle and London: ICES/University of Washington Press; 2002. p. 46.

ternational operations substantially with 100,000 Danish Kroner per year, or a tenth of the total contribution. As acting head of the Central Bureau, the Dutch government's fisheries consultant, P. P. C. Hoek was elected General Secretary, while Arctic explorer and scientist, Fridtjof Nansen from Norway was bestowed the task of heading the Central Laboratory.[56]

Implementing ICES agreements in Danish marine science

Meanwhile, the centrally placed Danish delegates at the preparatory conferences took a lead in Denmark with regards planning what tasks lay ahead for Danish marine science. Following the 1899 Stockholm conference a government commission was formed, and by 1902 it transformed into the *Kommissionen for Havundersøgelser*. There were a few changes in the composition of the commission, which initially had included two naval officers, Rear Admiral Wandel (as chairman) and Captain Trolle, but from the start, C. G. J. Petersen was the strong arm in a core group, which also included Martin Knudsen, who represented the interests of hydrography, which is what would be called physical oceanography in today's terminology. C. F. Drechsel was appointed on behalf of the Ministry of Agriculture and Carl Hansen Ostenfeld, represented the interests in plankton research.[57]

Two new laboratories were set up for plankton investigations and for hydrographical research. Within fisheries research, the Biologisk Station headed by C. G. Joh. Petersen continued as a separate entity, even after July 1902, when the Ministry of Agriculture recommended that Petersen become chairman of the commission, following the resignation of Rear Admiral Wandel and Captain Trolle.[58] The commission typically met in Petersen's private apartment. Reading the minutes from the meetings of the first years, there is a clear sense that Petersen is directing the topics discussed and liaising with the international council, when it comes to scientific matters. Drechsel acted more as an organizational manager, liaising with the ministry and the private sector. This also fitted well with Drechsel's recent day job as Director of the Port of Copenhagen.[59] All four marine scientists were former visitors to Petersen's floating research station, *Biologisk Station*, so Garstang's depiction of 'Papa Petersen behind' is accurate in the sense that Petersen was *primus inter pares*, pulling the strings of the show in the early years.

56 Drechsel CF, Knudsen M, Ostenfeld CH, Petersen CGJ. *De Internationale Havundersøgelser 1902–1907*. Copenhagen: Kommissionen for Havundersøgelser / C. A. Reitzel; 1908. p. 3–4.

57 Poulsen EM. *Kommissionen for Danmarks Fiskeri- og Havundersøgelser 1902–1952*. Copenhagen: Kommissionen for Danmarks Fiskeri- og Havundersøgelser / C. A. Reitzel; 1952. p. 12–16.

58 Knudsen M. De internationale havundersøgelser. In: Winge Ø, Vedel Taaning Å, eds. *Naturforskeren Johannes Schmidt. Hans liv og ekspeditioner*. Copenhagen: Gyldendalske Boghandel Nordisk Forlag; 1947:109; RA. *Komm. Danmarks Fiskeri- og Havundersøgelser*: Forhandlingsprotokoller (1900–1960) N1.: 10th meeting, 5 July 1902.

59 RA. *Komm. Danmarks Fiskeri- og Havundersøgelser*: Forhandlingsprotokoller (1900–1960) N1.

Acquiring *Thor*

Right from the first meetings of the commission in 1900 there was a realization that a new and bigger vessel was needed to conduct future investigations.[60] The small yacht, *Sallingsund*, which Petersen was managing as his floating Biologisk Station was not fit for ocean-going voyages.[61] Various solutions, including renting a cutter fitted for plaice fisheries, were discussed, but in late 1900 the commission decided to recommend that the government buy the commission a new research vessel, the fairly new steamship *S/S Thor*.[62] *Thor* was a modern steam trawler of 115 feet, built in 1899 in England for a Copenhagen based fishing company, the 'Dansk Damp-Trawling Aktie-Selskab Dan.' However, following the liquidation of the fishing company the government purchased the

60 RA. *Komm. Danmarks Fiskeri- og Havundersøgelser*: Forhandlingsprotokoller (1900–1960) N1.: 1st meeting, 16 January 1900.

61 Knudsen M. De internationale havundersøgelser. In: Winge Ø, Vedel Taaning Å, eds. *Naturforskeren Johannes Schmidt. Hans liv og ekspeditioner*. Copenhagen: Gyldendalske Boghandel Nordisk Forlag; 1947:110.

62 RA. *Komm. Danmarks Fiskeri- og Havundersøgelser*: Forhandlingsprotokoller (1900–1960) N1.: 4th meeting, 5 December 1900.

2.6 S/S *Thor* in Iceland

The S/S Thor in Iceland. The Thor was a large trawler with a spacious laboratory on deck. From 1903 until 1912, Thor was deployed in the service of fulfilling Denmark's designated obligations in the International Council for the Exploration of the Sea (Carlsberg Archives).

trawler in 1902 for 108,000 Danish Kroner on behalf of the Danish commission for marine science. The vessel was then refitted for marine science to make available a variety of different gear types, and a 3 × 5 meter laboratory was added to the deck of the trawler.[63]

The acquisition of *Thor* was mediated through Christian F. Drechsel, as he sat on both sides of the table buying as well as selling. In today's world this would not be considered good corporate governance, but times were different then. Drechsel was a board member of the liquidated fishing company for whom he secured a price 85 percent of what they had paid for the newbuilt vessel two years earlier.[64]

1903 Instructions

The *Thor* expedition in 1903 was the first Danish North Atlantic research cruise around Iceland and the Faroe Islands for seven years, since the *Ingolf* expeditions in 1895 and 1896.[65] It was also the first major cruise under the auspice of the International Council. Petersen's instructions for Johannes Schmidt and his fellow travellers were fairly broad, which supports the exploratory character of the expedition as well as Petersen's inclination for promoting the practice of the fisheries. As the responsible ichthyologist, Schmidt was tasked with initiating a three-year programme for investigating fisheries and whaling in Iceland and on the Faroe Islands. First, they were to take stock of how other nations exploited the area, and secondly, they were to investigate how it was possible to expand the local Icelandic and Faroese fisheries. Central to the expansion of fisheries in the North Atlantic possessions was the mapping the different fishing grounds, it was also thought important to gain information about the presence and absence of various fishes and how this was linked to the biological conditions in the sea. The only species mentioned by name in the instructions was plaice, for which the expedition was charged with the task of investigating the locality of different age groups.[66]

Hiring Johannes Schmidt

In the fall and winter of 1902–03, the four man commission was busy preparing next year's cruise with *Thor*, and following mutual consultation, they arrived at a set of instructions for the captain and for the

63 Sandbeck T. *Danske havforskningsskibe gennem 250 år*. Stenstrup: Skib Forlag; 2007. p. 132.

64 Holm, P. (1996). Technology transfer and social setting: The experience of Danish steam trawlers in the North Sea and off Iceland, 1879–1903. In P. Holm, O. Janzen & J. Thór (Eds.), *Northern seas yearbook 1996* (pp. 113–157). Esbjerg: Fiskeri- og Søfartsmuseets Forlag.

65 Wolff T. *Danske ekspeditioner på verdenshavene: Dansk havforskning gennem 200 år*. Copenhagen: Rhodos; 1967, p. 117–118.

66 KB. Johannes Schmidt 'Papirer'. Ut.374.V.1. Privat, 'Instrukser og Breve, ca. 1902–1908', Copy of instructions from C. G. J. Petersen to J. Schmidt, 23 April 1903.

67 RA. *Komm. Danmarks Fiskeri- og Havundersøgelser*: Forhandlingsprotokoller (1900–1960) N1.: 19th meeting, 18 February 1903; 20th meeting 18 April 1903; 21st meeting 24 April 1903.

68 Poulsen EM. *Kommissionen for Danmarks Fiskeri- og Havundersøgelser 1902–1952*. Copenhagen: Kommissionen for Danmarks Fiskeri- og Havundersøgelser / C. A. Reitzel; 1952. p. 17.

scientific leader of *Thor*. Petersen was responsible for drafting these so-called ichtyological or fisheries related instructions.[67] Also in the fall of 1902, Johannes Schmidt was hired as the leader of investigations of the fisheries that would be visited during the proposed North Atlantic cruise. Another young scientist, A. C. Johansen was hired to take care of investigating the fisheries around mainland Denmark.[68]

For the 1903 cruise of *Thor*, Schmidt was appointed principal scientist of a team of three marine scientists. The others were hydrographer J. N. Nielsen who would carry out work on behalf of Martin Knudsen and botanist, Ove Paulsen, whose designated role was to support the plankton zoological studies, according to the interests of Carl Ostenfeld. Ove

2.7 Ove Paulsen, Johannes Schmidt and J. N. Nielsen on board the *Thor*
From left to right Ove Paulsen (plankton), Johannes Schmidt (fisheries) and J. N. Nielsen (hydrography) are posing on board the S/S Thor during the 1903 expedition. The 26 year old Schmidt was the leader of the expedition. Schmidt and Nielsen are carrying their chibouks, a long-stemmed turkish pipe (Carlsberg Archives).

Paulsen noted that Petersen presented Schmidt with his instructions for the cruise, and he then underlined Schmidt's responsibility by telling him that he was 'presented with the most important appointment ever given to any natural historian in Denmark'.[69] This responsibility does not seem to have weighed too heavily on the shoulders of the then 26-year old Johannes Schmidt.

Trawling deep waters

Some of the most important results to come out of the *Thor* expeditions, however, were derived from pelagic trawling in deep waters. Petersen deduced that it would be beneficial to know how deep in the water column the commercially important species could be found. Although, unspecified in the instructions, the various whitefish species such as cod, ling (*Molva molva*), haddock (*Melanogrammus aeglefinus*) and saithe (*Pollachius virens*) would be candidates for such an exercise. Trawls, long lines and various seines should be used for the investigations. Finally, Petersen's own invention, the young fish trawl was to be deployed for deep waters in order to find the eggs and larvae, and thereby the spawning areas of the same commercially important species.[70]

The Petersen young fish trawl or 'yngeltrawl' as it was known in Danish, consisted of a bag made from tightly woven hemp fabric, known in Danish as 'straminposer', which was permeable to water, but not tiny and fragile objects such as fish eggs.[71] In the 1880s, German biologist, Victor Hensen had pioneered the use of this type of fabric for bottom surveys on herring eggs.[72] Yet, Petersen's version was for pelagic trawling, where four panels were sewn into a cone shaped seine. Lateral wings extended from the opening of the seine; where ironclad wooden panels acted as trawl doors keeping the entrance to the net open, when trawled through the water.[73]

Schmidt and his team were the first to probe the deep waters in this way, and the new results came in rapid succession. The early life history of the Icelandic cod for instance, was discovered through *Thor's* biological and hydrographical investigations. In early spring, cod spawns were noted at a depth of c. 50–100 meters near the Southern coast of Iceland. The buoyancy of the eggs makes them drift with the currents at a certain depth. As the eggs hatch into larvae they start to drift with the Irminger Current clockwise in a northern direction around Iceland,

69 Paulsen O. Havundersøgelserne i de nordlige have begynder. In: Winge Ø, Vedel Taaning Å, eds. *Naturforskeren Johannes Schmidt. Hans Liv og Ekspeditioner*. Copenhagen: Gyldendalske Boghandel Nordisk Forlag; 1947:18.

70 KB. Johannes Schmidt 'Papirer'. Ut.374.V.1. Privat, 'Instrukser og Breve, ca. 1902–1908', Copy of instructions from C. G. J. Petersen to J. Schmidt, 23 April 1903.

71 CFD. 1909–10. no. 227d. J. Schmidt: Cruise report from Den Danske oceanografiske Expedition til Middelhavet 1908–1909, p. 5.

72 Smith TD. *Scaling fisheries: The science of measuring the effects of fishing, 1855–1955*. 2nd revised 2007 ed. New York: Cambridge University Press; 1994. p. 83.

73 Strubberg AC. Nyttefiskenes Livshistorie studeres i Nordatlanten og sydpaa til Biscaya. In: *Naturforskeren Johannes Schmidt. Hans Liv og Ekspeditioner*. Copenhagen: Gyldendalske Boghandel Nordisk Forlag; 1947: p. 27

towards the eastern shores of Iceland, where the offspring reached 4–5 centimeters. Then, the coastal waters and fjords of Eastern Iceland serve as a nursery area for the young cod for a few years until the fish reaches maturity and returns to the spawning grounds south of the Icelandic coast. Fishermen already knew the migration patterns of Icelandic cod, but following Schmidt's investigations the natural explanation for this migration was revealed. Schmidt plotted the length of the sampled fish larvae and small fish onto a coordinate system, as he had learnt from Petersen. The emerging patterns would then take the shape of a series of bell shaped heaps, each signalling one year class, whose whereabouts could then be deduced.[74]

Cod, however, was just one of a whole series of Atlantic fish species, whose life history could suddenly be charted in this way. Schmidt and his crew must have worked incessantly on board *Thor* during expeditions, and over the next four years of cruising the North Atlantic they succeeded in making life history descriptions, like the one above, for all 17 species within the fish families of cod, hake (*Merluccius sps.*), ling, tadpole fish(*Raniceps*) and tusk(*Brosme brosme*). The spawning area of all these species was charted on maps spanning the area roughly between the Polar circle and the west coast of Spain. These data could then be compared with the hydrographical measurements for depth, temperature and salinity, and it was thus possible to determine where a given species spawns.[75]

Bjarni Sæmundsson

Johannes Schmidt's trips to the North Atlantic brought him to Iceland, and most importantly it brought him in contact with Bjarni Sæmundsson (1867–1940). As part of the planning of the North Atlantic expedition, Christian Drechsel had corresponded with Sæmundsson, and Sæmundsson had written an account of what species such an expedition could expect to encounter in Icelandic waters. Eels were not among the species, which Bjarni Sæmundsson thought should receive special attention from a Danish fisheries expedition.[76] In 1903, Sæmundsson was already investigating the natural history of Icelandic fish, studies of which would eventually take him towards international fame and recognition, but as his day-time job was as a school-teacher in Reykjavik, he could only pursue his zoological interests in his spare time.[77] Therefore,

74 Paulsen O. Havundersøgelserne i de nordlige have begynder. In: Winge Ø, Vedel Taaning Å, eds. *Naturforskeren Johannes Schmidt. Hans liv og ekspeditioner.* Copenhagen: Gyldendalske Boghandel Nordisk Forlag; 1947:24–25; Strubberg AC. Nyttefiskenes Livshistorie studeres i Nordatlanten og sydpaa til Biscaya. In: *Naturforskeren Johannes Schmidt. Hans Liv og Ekspeditioner*. Copenhagen: Gyldendalske Boghandel Nordisk Forlag; 1947: p. 29.

75 Bruun AF. The Life and Work of Professor Johannes Schmidt. *Estratto dalla Rivista di Biologia*. 1934;16(1):6–7.

76 RA: *Dansk Biologisk Station, Johannes Schmidts embedsarkiv:* Korrespondance om åleforskning (1908–1933) E28: 1908–1913: B. Sæmundsson to C. F. Drechsel, 10 January 1903.

77 Bruun AF. Bjarni Sæmundsson 1867–1940. *Dansk Fiskeritidende*. 1941:139.

2.8 Bjarni Sæmundsson seated on board R/V *Dana*
Bjarni Sæmundsson (to the right) is arguably the most famous Icelandic marine scientist. Throughout Schmidt's career he proved a helpful guide in Icelandic waters.

it was a tempting offer, when Schmidt wrote him a letter in April 1903 and asked if he would like to join *Thor* in the months of June and July in return for making himself useful to the fisheries investigations.[78]

Sæmundsson joined *Thor*, whenever possible, and his first hand practical knowledge of Icelandic fisheries was probably crucial for the success of the expedition.[79] None of the members of the expedition had much prior experience of the area, so one can imagine how important it was to have a local Icelandic speaking liaison to communicate with the local fishermen, guiding them through nautical and political territories alike. During the long and arduous days on board *Thor*, Sæmundsson

78 KB. Johannes Schmidt 'Papirer'. Ut.374.III. 'Forskellige Breve fra Johannes Schmidt til Bjarni Sæmundsson'. Letter from J. Schmidt to B. Sæmundsson, 10 April 1903.

79 Paulsen O. Havundersøgelserne i de nordlige have begynder. In: Winge Ø, Vedel Taaning Å, eds. *Naturforskeren Johannes Schmidt. Hans liv og ekspeditioner.* Copenhagen: Gyldendalske Boghandel Nordisk Forlag; 1947: p. 21.

became an important conversation partner of Johannes Schmidt. This we know, from their frequent correspondence, which continued in the autumn of 1903. Here, Schmidt tells Sæmundsson about his findings in relation to the year classes of Icelandic cod in the sympathetic manner only possible among close collaborators.[80]

Schmidt's gratitude towards Sæmundsson is evident from a discussion in the Danish commission for marine science in April 1904. Schmidt had proposed to the commission that his studies from Icelandic waters should be dedicated to Bjarni Sæmundsson. Nonetheless, for whatever reason, the commission advised him not to do so.[81] The more senior scientists in the commission might have been careful not to smear the glory and honour too thinly on too many faces. Nevertheless, Schmidt and Sæmundsson remained friends for life.

Going over the results of the 1903 expedition on *Thor*, the commission appears to have been pleased with Schmidt's accomplishments during the cruise to the North Atlantic. Upon reviewing the activities in 1903, the commission agreed that Schmidt could have his own assistant on board for next year's cruise; a cruise that would alter the course of Schmidt's entire career.[82]

1904 – Eel in the North Atlantic

The next year, 1904, Schmidt's instructions were far more detailed. This can be seen as reflection of the great leap in knowledge obtained the year before. The marking experiments with cod and with plaice were to be continued, and more knowledge was needed on the Icelandic whereabouts of the herring. The Petersen young fish trawl was to be used in deep waters and in the intermediary zones below the surface. The targeted fish fry were those of halibut (*Hippoglossus hippoglossus*), ling, tusk and eel followed by a question mark. Petersen was anxious to know to what extent these species led an intermediary life in deep oceans.[83] The presence of eel in the instructions is highly interesting. Eel was not among the species analysed in 1903, and nowhere does Schmidt appear to have been interested in the life history of the freshwater eel. The interest also, did not seem to have arisen from Iceland. Sæmundsson published a book in 1908, synthesizing the fishes of Iceland, where he underlined the lack of interest in eel prevailing in Iceland proper. Sæmundsson concluded that eel had almost no economic importance

80 KB. Johannes Schmidt 'Papirer'. Ut.374.III. 'Forskellige Breve fra Johannes Schmidt til Bjarni Sæmundsson'. Letter from J. Schmidt to B. Sæmundsson, 16 October 1903.

81 RA. *Komm. Danmarks Fiskeri- og Havundersøgelser*: Forhandlingsprotokoller (1900–1960) N1.: 27th meeting, 22 April 1904.

82 RA. Komm. Danmarks Fiskeri- og Havundersøgelser: Forhandlingsprotokoller (1900–1960) N1.: 24th meeting 24 November 1903.

83 KB. Johannes Schmidt 'Papirer'. Ut.374.V.1. Privat, 'Instrukser og Breve, ca. 1902–1908'. Instructions from C. G. J. Petersen to J. Schmidt, 9 March 1904.

for Icelanders, who typically detested the species. He also noted that there were few export opportunities for eel from Iceland, and that the fecundity of eel did not provide any promise of a sustainable large-scale fishery.[84]

Petersen, however, was the leading expert on the freshwater eel, rivalled internationally only by Grassi and Calandruccio in Italy. Therefore, there is every reason to believe that Petersen was not only the scribe, but also the brains behind the search for eel in the North Atlantic.

The presence of eel would have made but a fleeting appearance in the copious notes on Danish marine science, had it not been for the fact that the larva of the eel did appear in a Petersen young fish net, as *Thor* was trawling the waters between Iceland and the Faroe Islands in 1904. On the night of the 22 May, Schmidt was smoking his pipe in his cabin, while his new assistant, A. C. Strubberg was on watch, overseeing the young fish trawl being hauled on deck from several hundred meters below. Strubberg later recalled that it was a moving moment for him, as he discovered the zigzagging larva in the bag, 77 millimetres long and translucent with silvery eyes in the light from the outdoor lamp. Strubberg hurried, as fast as his boots would allow, down the staircase and into Schmidt's cabin. The smoky atmosphere, however, did not prevent Schmidt from realizing the importance of the night's discovery.[85]

The telegram announcement

Schmidt immediately realized the potential, and he wrote a letter to Petersen to let him know of their successes and he also included some of the samples for Petersen to look at. In a letter from 28 June, Petersen thanked Schmidt for sending him the samples of larvae from halibut, the first ever caught, while the single eel larva perhaps was too precious to be shipped by mail. Petersen told Schmidt that if he really was certain that he had caught a specimen of the *Leptocephalus brevirostris*, then Schmidt should tell him, so that he could publicize the find in the newspapers via the Danish press agency, Ritzaus Bureau. Schmidt could do that himself as well, if he saw fit. Petersen's letter is contained in an amicable slightly paternal tone, where he informs Schmidt of relevant marine science gossip from Denmark and his own studies from *Sallingsund* operating in the inner Danish waters, while Petersen also advised on how to best use the young fish trawl for bottom trawling, as

84 Sæmundsson B. *Oversigt over Islands Fiske med Oplysninger om deres Forekomst, vigtigste biologiske Forhold og økonomiske Betydning*. Copenhagen: Kommissionen for Havundersøgelser, C. A. Reitzel; 1908. p. 106–107.

85 Strubberg AC. Nyttefiskenes Livshistorie studeres i Nordatlanten og sydpaa til Biscaya. In: *Naturforskeren Johannes Schmidt. Hans Liv og Ekspeditioner*. Copenhagen: Gyldendalske Boghandel Nordisk Forlag; 1947: p. 30.

Thor was not in possession of an otter trawl, which would be suited for just that.[86]

Schmidt, self-confident and proud of their achievements in the North Atlantic, then went along and wired the Ritzau Bureau of their spectacular deep-sea findings. 'An important discovery', ran the headline. 'The offspring of the freshwater eel and that of the halibut had been found by a Danish scientist'. Then a short account followed, where it was revealed that a larva three inches long had been found swimming freely in waters, where the bottom was more than 1000 fathoms below. The success arose from a cruise led by biologist Johannes Schmidt, on board a ship sent out by the Danish government. All true, but Schmidt had left out an important part of the affair – the role played by C. G. J. Petersen.[87]

Moreover, the 27-year-old Schmidt probably did not anticipate that glorious scientific discoveries can provoke mordant scorn from older and less fortunate colleagues, but it did. Danish natural historian, William Sørensen (1848–1916), placed a commentary in the daily newspaper, Politiken, whose readers were exposed to a rant of caustic wit: 'For those of us enchained by having to work for a living, it is always a joy to see young and affluent scientists being able to solve scientific riddles (…) Therefore, I was thrilled to see our young and eager biologist, Dr. Schmidt (…) shine far above the laborious investigations conducted by other biologists in Northern Europe'. He then ridicules the fact that it should be something noteworthy to have found an eel larva, since the natural history museum in Copenhagen already possessed one specimen of the *Leptocephalus brevirostris*.[88]

Provoked by William Sørensen's attempt to ridicule Schmidt as well as his own research topic, Petersen therefore, submitted his own commentary in early August. Presumably motivated by the wealth of questions from the general public, which Petersen was confronted with following Schmidt's press release, he felt compelled to assert that it really was true, and that contrary to public belief the eel did not give birth to live young eel. Furthermore, it reinforced that it was Petersen who sent Schmidt out on his cruise, and that Petersen was the one who had instructed Schmidt to search for eel. Petersen told the story of how Grassi had found larvae in the Mediterranean, and showed them to Petersen in Rome, and he then emphasized how this was the first specimen found outside of the Mediterranean.

Petersen may have been annoyed by the fact that his name was omitted from Schmidt's press release, but one could also see his attempt

86 KB. Johannes Schmidt 'Papirer'. Ut.374.V.1. Privat, 'Instrukser og Breve, ca. 1902–1908'. Letter from C. G. J. Petersen to J. Schmidt, 28 June 1904.

87 KB. Johannes Schmidt 'Papirer'. Ut.374.V.1. Privat, 'Instrukser og Breve, ca. 1902–1908'. News clip from undated newspaper, July 1904.

88 KB. Johannes Schmidt 'Papirer'. Ut.374.V.1. Privat, 'Instrukser og Breve, ca. 1902–1908'. Sørensen in *Politiken*, July 1904.

to place him in the forefront of the eel investigations, as a way of standing up for this young protégé, Schmidt, protecting him from William Sørensen. In Danish natural history Sørensen was widely reputed as the *enfant terrible*, constantly mocking opponents. He had lost his part-time position at the museum of natural history over a polemical book, criticizing the most powerful man in Danish natural history, Japetus Steenstrup.[89] C. G. Joh. Petersen and Eugen Warming had been students of Steenstrup in their university years, and it was Steenstrup who recommended Petersen for his first marine science position on board *Hauch* in 1883.[90] Thus, from Sørensen's viewpoint Petersen and his colleagues were potential enemies.

A few days later, Sørensen indeed treated Petersen to a scornful commentary, in which he claimed that it had been commonly known since the dawn of times that the freshwater eel left near shore waters, when they went spawning, so searching for them in Denmark was utterly silly. Sørensen finished his letter-to-the-editor with a polemical remark directed at the Danish newspapers in general, where he told them to treat future press releases such as the one from Schmidt, with the same caution as other advertisements, hence only accepting them, when paid for.[91] The discussion was continued with at least one more commentary from Petersen, stressing that for those people content with only presuming knowledge, the new discovery was perhaps of little interest, but true science needed to be based on facts.[92] Indeed, there was still much to be discovered, and Petersen also took it upon himself to popularize the Danish eel investigations.

Petersen on the Atlantic larvae

In the fall of 1904, Petersen published a small booklet on the natural history of the eel, which included suggestions on how Danish landings of eel could be improved.[93] Petersen's booklet was written in plain Danish, and it appeared in a book series, which also included such popular topics as egg farming and dairy production, and the usability of wild plants. In other words, Petersen's booklet was intended for practising fishermen. The selling point of Petersen's booklet was that fishermen ought to set up lights at night at the mouth of Danish fiords. The idea was that this would deter the eel from leaving the Danish waters, thus increasing the number of fish being landed. Important parts of his argument

89 Sand-Jensen K. Naturhistorie. In: Kjærgaard PC, ed. *Lys over landet 1850–1920. dansk naturvidenskabs historie 3*. Aarhus: Aarhus Universitetsforlag:164–165.

90 Sandbeck T. *Danske havforskningsskibe gennem 250 år*. Stenstrup: Skib Forlag; 2007. p. 122.

91 KB. Johannes Schmidt 'Papirer'. Ut.374.v.1. Privat, 'Instrukser og Breve, ca. 1902–1908'. W. Sørensen in *Politiken*, 6 August 1904.

92 KB. Johannes Schmidt 'Papirer'. Ut.374.v.1. Privat, 'Instrukser og Breve, ca. 1902–1908'. Petersen in *Politiken*, August 1904.

93 Petersen, C. G. J., *Aalen. Hvad man ved om den samt anvisning paa at forøge udbyttet af Aalesfiskeriet*. Copenhagen: Det Schubotheske Forlag, 1904.

hinged upon ideas associated with Schmidt's recent catch of an eel larva in the North Atlantic.

In the booklet Petersen did not assert much credit to his own part in this discovery. Instead he explicitly noted that 'Dr. Johs. Schmidt from Copenhagen' now has found an eel larvae more than 300 meters below the sea surface off the coast of the Faroe Islands.[94] For Petersen, Schmidt's discovery was a convincing argument in favour of the assertion that the *Leptocephalus brevirostris* was in fact the larval state of the eel and not a separate species.

Petersen explained how we are now coming much closer to finding out where the larvae come from, since Schmidt fished the larva in a place, where one would assume that they found it normal to thrive. Unlike Grassi, Schmidt did not make his discovery under 'certain abnormal' hydrographical conditions. The reason why Schmidt found only one larva was to do with the fact that the expedition was disturbed by stormy weather. Petersen hypothesized that the larvae of the eel is not likely to be found closer to Danish waters than the West coast of Scotland, since the larvae must live close to where eggs and small larvae (<2.5 inches) live. If the breeding grounds were to be found, we would find out where the adult eels went in great numbers after leaving Danish waters, thereby increasing the chance of catching many more eel.

According to Petersen, the findings by Grassi and Schmidt suggested that the eel larvae prefer warm waters at great depths, which is why it is not likely that larvae could be found near Denmark. Therefore, in his view, the glass eel which frequent Danish waters in spring would have come a long way before reaching Denmark's shores. This suggested that the Danish eel population must be a tiny local population supplied with fry from spawning grounds common to all the eel in Europe. Petersen felt that Danish fishermen should not refrain from catching as many eel as possible, as their catches would not impact the size of the Danish share of Europe's eel populations.[95]

The hypothesis of the migratory patterns of the eel in different life stages, summarized above, reveals that Petersen had indeed thought long and hard about the whereabouts of eel larvae, and he definitely seems to have grasped the theory of an oceanic origin of the spawning grounds of the eel at a very early stage. Petersen's theory, however, has been largely overlooked in the literature on the discovery of the breeding grounds of the eel.

94 Petersen, C. G. J., *Aalen. Hvad man ved om den samt anvisning paa at forøge udbyttet af Aalesfiskeriet.* Copenhagen: Det Schubotheske Forlag, 1904. p. 14.

95 Petersen, C. G. J., *Aalen. Hvad man ved om den samt anvisning paa at forøge udbyttet af Aalesfiskeriet.* Copenhagen: Det Schubotheske Forlag, 1904. p. 14–20.

The following year, Petersen also published Schmidt's Atlantic discovery in English, in a short two-page article. He also noted that the Irish fisheries investigations had caught an eel larva the same summer, in August of 1904 northwest of Eagle Point off the Irish coast. In combination with the complete absence of eels smaller than elvers, and spawning eels in Danish waters, Petersen now hypothesized that all the eel in Northern Europe spawn well out into Atlantic Ocean, where they find the right amount of warmth and depth, and where modern fishing technologies have not yet been able to catch a spawning eel. Petersen ends his article with a short postscript, where he reveals that Schmidt succeeded in finding great quantities of *Leptocephalus brevirostris* in the depths of the Atlantic, accentuating evidence in the direction that the common eel does not spawn in shallow water, neither in the Baltic nor in the North Sea.[96] This was the result of another set of specific instructions from Petersen to Schmidt.

Straying off the Danish ICES areas

For the 1905 season, *Thor's* cruise was set to cater mainly for furthering the eel investigations. The hypothesis was that the younger life stage of the eel would be more prolific to the south of where they had been fishing so far. Therefore, the 1905 cruise of *Thor* took a more southerly course. This also facilitated an enlargement of the cod-related researches mentioned above.[97] The new course had also been agreed upon in the Danish commission for marine science, and for the hydrographers this was a chance to extend their investigations into the relative changes in salinity and temperature at various depths and currents in the Atlantic, still fairly uncharted water, when moving away from coastal areas.[98]

Throughout June of 1905, *Thor* cruised the waters west of the British Isles, examining the breeding places of cod, in search of more eel larvae. They were successful in both endeavours. No less than 265 eel larvae were caught between the 7 and 26 June between the Hebrides and Brittany, far out in the Atlantic, where the continental shelf falls from c. 1000 to 2000 meters. Then, following a stint in Icelandic waters, the team collected another eight eel larvae on the return trip in September. This time, off the Hebrides, the larvae were large, c. 7–8 centimeters and on the verge of their transformation into elvers.[99]

96 Petersen CGJ. Larval Eels (leptocephalus brevirostris) of the Atlantic Coast of Europe. *Meddelelser fra Kommissionen for Havundersøgelser: serie Fiskeri.* 1905;1(5):3–5.

97 KB. Johannes Schmidt 'Papirer'. Ut.374.V.1. Privat, 'Instrukser og Breve, ca. 1902–1908', Instructions from C. G. J. Petersen to J. Schmidt, 11 April 1905.

98 RA. *Komm. Danmarks Fiskeri- og Havundersøgelser*: Forhandlingsprotokoller (1900–1960) N1.: 29th meeting, 14 February 1905.

99 Vedel Taaning Å. Aaleundersøgelserne begynder for alvor. In: Winge Ø, Vedel Taaning Å, eds. *Naturforskeren Johannes Schmidt. Hans Liv og Ekspeditioner.* Copenhagen: Gyldendalske Boghandel Nordisk Forlag; 1947:40–41.

Schmidt and Petersen now both thought they had hit the jackpot, in terms of finding the breeding places of the eel. Petersen included a three-line postscript to his 1905 article on the eel, reporting that Schmidt had found more eel, and during the fall of 1905, Schmidt wrote up a lengthy report on the recent 'Contributions to the life-history of the eel'.[100] Spanning almost 140 pages Schmidt reviewed all the previous contributions to the life history of the eel, and then went on to explain the findings of *Thor's* expeditions with charts and x-ray images of the various life stages of the freshwater eel. It seemed likely that the spawning ground was outside the continental shelf of Europe below 1000 meters of depth, and in conditions, where the temperature is above seven degrees Celsius accompanied by salinity above 35.3 parts per thousand. From this, Schmidt concludes 'that the eel is not born nor do the larvae develop nearer the coast of North Europe than in the Atlantic off Great Britain and France and that consequently the whole of the North European stock of eels depends upon the young which migrate inwards from that region'.[101] Grassi's work is duly credited, and there is no hint that Schmidt doubts the Italian proposition that their eel in the Mediterranean breed in the great depths of the Tyrrhenian Sea.

Whereas Schmidt this time is the sole author, Petersen receives special thanks as the inclusion of the investigations in the Danish part of the international council's explorations was the result of his great interest in the matter. Schmidt then explicitly notes the importance of remembering the crucial role played by Petersen's young fish trawl.[102]

With regards the Atlantic eel, Schmidt is quite careful not to be too conclusive on the breeding grounds, as they had still not caught an adult eel, or a spawning one or an unhatched egg for that matter. This became part of the expedition plan for the 1906 campaign. This year more than 600 larvae were found, but according to Schmidt in all the wrong places, while they investigated waters further and further into the Atlantic. Schmidt became troubled by the fact that what he thought were breeding grounds seemed to include a much larger territory than what the previous year's investigations had led him to believe. Yet, towards the end of the trip, Schmidt writes home that as they sought eel eggs from far south, without much luck, all he can conclude is that eel simply did not breed when and where *Thor* had traversed the Eastern North Atlantic.[103]

100 Schmidt J. Contributions to the life-history of the eel (anguilla vulgaris, flem.). *Rapport et Proces-Verbaux des Reunions*. 1906;5:137–274.
101 Schmidt J. Contributions to the life-history of the eel (anguilla vulgaris, flem.). *Rapport et Proces-Verbaux des Reunions*. 1906;5:262.
102 Schmidt J. Contributions to the life-history of the eel (anguilla vulgaris, flem.). *Rapport et Proces-Verbaux des Reunions*. 1906;5:137–138.
103 Vedel Taaning Å. Aaleundersøgelserne begynder for alvor. In: Winge Ø, Vedel Taaning Å, eds. *Naturforskeren Johannes Schmidt. Hans Liv og Ekspeditioner*. Copenhagen: Gyldendalske Boghandel Nordisk Forlag; 1947:42–43.

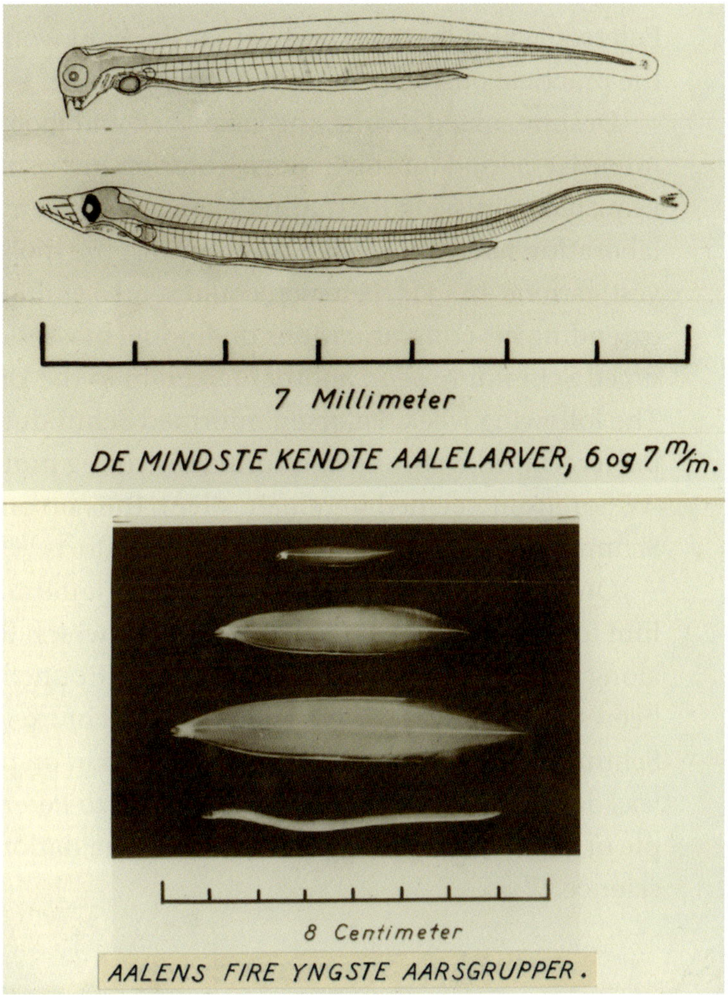

2.9 Eel larvae at different stages *When the S/S Thor first went to the North Atlantic Ocean the metamorphosis of the eel from larval stage (Leptocephalus) to glass eel had already been discovered, but the earliest life stages were not yet known. During the next 20 years, Schmidt led several expeditions that found larvae only 6–7 mm long.*

Petersen's light experiment

In the summers of 1905 and 1906, Petersen also intensified his own eel investigations in Danish waters. Trying out a hypothesis proposed in his booklet from the fall of 1904, Petersen initiated a large-scale experiment, where he tried to prevent the silver eel from leaving Danish territorial waters. In the smaller of the Danish straits, the Lillebælt, Petersen set up a cable running across the sea from Jutland to the Island of Funen. Here, 55 electric lamps were attached to the cable, and the idea was that the lights would deter the home-bound migrant eels from crossing the line of the cable. Instead, they would seek shelter along the shore, where they would be caught in traps. The experiment proved a failure.

Petersen was right in assuming that the light would scare the eel, but the practical obstacles of then catching the eel were insurmountable.[104]

Despite mixed results, the years 1905 and 1906 were the most eel-intensive period for the Biologisk Station just as they were for the Danish commission for marine science. Petersen and Schmidt's collaboration had come to full fruition with the most successful eel investigations. In 1905, Petersen could send a summer letter to Schmidt extending his congratulations on finding masses of eel larvae, and he urged Schmidt to send another telegram to the Danish newspapers.[105] The following week, Petersen informed Schmidt that his telegram had been received well, and comforted his young protégé that the infamous 'W. Sørensen' seemed to remain silent, thus no one interfered with Schmidt and Petersen's public relations efforts.[106]

On the 24 July Petersen wrote another letter to Schmidt to thank him for sending a specimen of eel larvae, which Petersen could take along to the meeting of the International Council, where the news had been successfully received. Petersen is full of excitement as he informs Schmidt of how an international convention on plaice investigations was decided on by the council; according to Petersen yet another example of how high Denmark ranks among the nations conducting marine science.[107]

Conclusion

This chapter began with an account of how Johannes Schmidt is typically portrayed in modern literature on marine science. Schmidt is looked upon as the brilliant young scientist, whose genius permeated the Atlantic waters, whereby he found the first eel larvae of the freshwater eel ever caught in the North Atlantic, which in turn ignited his lifelong quest for investigating the otherwise unknown stages in the life history of the eel. However, only in Walter Garstang's poem, and two other commentaries, does the literature give credit to role played by C. G. J. Petersen. Petersen was not only behind Schmidt he was also left behind by later chroniclers.

In summary, this chapter has tried to set the record straight with regards to the early years of Johannes Schmidt's career. When Schmidt boarded *Thor* for the first time, he carried with him not only expectations and instructions from Petersen, he was also bestowed with the

104 Blegvad H. *Dansk biologisk Station gennem 50 år, 1889–1939*. Copenhagen: C. A. Reitzels Forlag; 1943. p. 23.

105 KB. Johannes Schmidt 'Papirer'. Ut.374.V.1. Privat, 'Instrukser og Breve, ca. 1902–1908'. Letter from C. G. J. Petersen to J. Schmidt, 4 July 1905.

106 KB. Johannes Schmidt 'Papirer'. Ut.374.V.1. Privat, 'Instrukser og Breve, ca. 1902–1908'. Letter from C. G. J. Petersen to J. Schmidt, 9 July 1905.

107 KB. Johannes Schmidt 'Papirer'. Ut.374.V.1. Privat, 'Instrukser og Breve, ca. 1902–1908'. Letter from C. G. J. Petersen to J. Schmidt, 24 July 1905.

experience and know-how associated with state of the art fisheries science, which he had learned during the previous years working as an assistant at the Biologisk Station. Biologisk Station with Petersen at the helm was amongst the most important centres for the advancement of marine science anywhere during this period, and internationally renowned C. G. Joh. Petersen mentored Schmidt.

However, while the personal relationships and networks were of prime importance, the above chapter also serves to show how the landmark achievements in Danish and international marine science in the decades around 1900 owe just as much to distinct institutional and technological developments. After all, Schmidt was still a junior member of the learned republic of marine science, when he began charting the waters on board *Thor*.

Institutionally, it was Schmidt's blessing that he was the young aspiring research leader at a time, when ocean sciences were coming together internationally. The latter half of the 19th century had seen a gradual increase in international collaboration in matters of fisheries and hydrography, and the formation of the intra-governmental organization of the International Council for the Exploration of the Sea from 1902, prompted member countries such as Denmark to invest substantially in their effort. In the Danish case this meant not only charting the near shore waters around mainland Denmark, but also, when seen from Copenhagen, the rather remote deep waters around Iceland and the Faroe Islands.

Technologically however, the advancements made by Johannes Schmidt and his team on-board *Thor* could scarcely have been accomplished in such rapid succession, had it not been for the technological advances made in the previous decades. One thing is the deployment of *Thor* itself. For its day, *Thor* was a powerful state of the art converted fishing trawler, and she was ideally suited for deep-sea investigations in the changeable weather of the North Atlantic. The steam propulsion facilitated a usage scarcely possible just a few decades earlier, when fishing vessels were still powered by the wind alone. With steam propulsion at the disposal of Schmidt and his team, he could negotiate with the ship's captain to improvise the course followed in order to capitalize on new findings. The way in which the life history of Icelandic cod was charted in just one summer, is one example thereof. Therefore, the success in these years can also be attributed to the fact that *Thor* could go to places, where the team was comfortable that they were the first to probe the waters with Knudsen's fine-tuned hydrographic instruments, and Petersen's clever designs, not least with regards to the young fish trawl.

In 1906 therefore, the future of Danish marine science was promising, and the deep-sea adventures seemed to be waiting in line for the highly gifted talents assembled in the Danish *Kommissionen for Havundersøgelser*. It seems likely that no one anticipated the tumultuous twists and turns that were to follow in the years to come, when Schmidt's generation of marine scientists would take control of both the Danish commission, and Denmark's general involvement with ICES at the expense of Petersen, who in turn suffered political humiliation.

3 Petersen's attempted coup d'état

Johannes Schmidt's position as an assistant to Petersen came to an abrupt end in 1909. In the literature on Danish hydrography this is mentioned in passing only, and the single commentary on what actually happened was recorded by hydrographer Martin Knudsen, who was himself a member of the Danish commission for marine science both before and after Petersen ceased to collaborate with the rest of the commission. According to Martin Knudsen, Petersen had already begun to look at alternative projects a year before he left the commission.

Knudsen recorded that Petersen had had a close working relationship with Johan Hjort, the most important figure in Norwegian marine science for half a century. Norway was also a founding member of ICES, and from the start, the Norwegian research related to the ICES program was in the hands of a commission, just as it was in Denmark. Soon, however, the Norwegian commission had been dissolved and all oceanographic research came under the auspice of the Norwegian state, which gave Johan Hjort a significant role in its control. Petersen wanted a similar arrangement for Danish marine science with himself at the helm. Petersen put forward this proposal in spite of reservations from the other members of the Danish commission. In the end, Petersen's proposal was not feasible at government level. Petersen then resigned from the commission but maintained his position as Director of Dansk Biologisk Station. Knudsen further speculated that Petersen's misfortune was linked to the fact that there was a change in government in July 1908. Following a huge fraud scandal involving the minister of justice, P. A. Alberti, the whole cabinet resigned including the minister of agriculture, Ole Hansen, who was responsible for the fisheries investigations.[1] Martin Knudsen wrote this account in 1947, almost 40 years after events took place. Had he been summoned to give a contemporary account, he probably would have been more vocal about how serious this situation was for Danish marine science.

1 Knudsen M. De internationale havundersøgelser. In: Winge Ø, Vedel Taaning Å, eds. *Naturforskeren Johannes Schmidt. Hans liv og ekspeditioner*. Copenhagen: Gyldendalske Boghandel Nordisk Forlag; 1947:110–111.

Downsizing the Danish commission

It is worth remembering that the Danish commission for marine science, like the International Council, was a temporary organizational structure during the first decades of its existence.[2] Thus, the Danish government budget for the commission needed frequent renegotiation. At the commission's meeting in November of 1905, the members learned from Petersen that the Head of the Department of Agriculture, Wulff, had informed him about the future of the commission. They were to be prepared for all investigations to cease from August 1907 onwards.[3] This prompted the members of the commission to think of how to continue the investigations beyond the summer of 1907,[4] and a knock on effect would have been discussions on how to get the most out of the 1906 programme. During the spring of 1906 there seems to have been a consensus with regards changing the course of the *Thor's* upcoming North Atlantic expedition. Fellow marine scientists, Johan Hjort of Norway and D'Arcy W. Thompson in Great Britain were consulted, and neither objected to the idea that the *Thor* would not go to Iceland as she had in the previous years. Instead, Schmidt was to continue his successful and spectacular eel investigations West of Ireland and down to Spain. For Knudsen it appears to have been a welcome shift as well, in terms of charting the hydrographical profile in those waters.[5] They were to harvest the low hanging fruits.

Nonetheless, while the entire commission agreed on this, what happened after the 1906 expedition became a topic of dispute, the first of several over the next couple of years. In April of 1906, Petersen presented the commission with an outline of how to distribute expenses among the interests of the commission, under a future scenario, where the overall budget was diminished. On an international level there was substantial interest in continuing the joint investigations over forthcoming years, yet Petersen, perhaps somewhat surprisingly, argued that Denmark's effort should be only so large that it was possible to claim that the country was continuing the explorations. Petersen suggested that the existing program could continue, but with half the previous budget, or c. 60,000 DKK. This budget cut meant that hydrography, which is highly dependent on ample ship time to take samples, suffered from Petersen's proposals. Knudsen deemed it very difficult to continue hydrographical investigations, but apparently, the majority backed Petersen's overall plan. This included firing Knudsen's assistant,

2 Rozwadowski H. *The sea knows no boundaries. A century of marine science under* ICES. Seattle and London: ICES/University of Washington Press; 2002. p. 52.

3 RA. *Komm. Danmarks Fiskeri- og Havundersøgelser*: Forhandlingsprotokoller (1900–1960) N1.: 30th meeting, 17 November 1905.

4 RA. *Komm. Danmarks Fiskeri- og Havundersøgelser*: Forhandlingsprotokoller (1900–1960) N1.: 31st meeting, 4 January 1906.

5 RA. *Komm. Danmarks Fiskeri- og Havundersøgelser*: Forhandlingsprotokoller (1900–1960) N1.: 32nd meeting, 26 January 1906.

J. N. Nielsen from the autumn of 1906.[6] Once the downsizing of the commission came into effect, Knudsen had good reason to be dissatisfied with Petersen's leadership. Also Schmidt was dissatisfied with the sacking of Nielsen – his close collaborator from the trips to Iceland. 'A mean act' Schmidt labelled it, when he informed Bjarni Sæmundsson about the latest developments in a private letter.[7] As for Petersen and Schmidt's working relationship, some friction arose from 1906.

A strained relationship

When in 1906 Schmidt reported from his summer trip with the *Thor*, Petersen felt obliged to inform Schmidt of his duties. While acknowledging that he had found out 'interesting things' in the Skagerrak area, Petersen noted that this area fell outside of the territory Schmidt was supposed to investigate according to the program. Petersen, however, told Schmidt that he would not press charges against him, if he was delayed by a week or so returning from the investigations in Ireland. On a more serious note, Petersen did not like the fact that he learned of a visit by Schmidt to Norway from Johan Hjort. Petersen, as the head of investigations should have known about it first. Likewise, Petersen was also disappointed when he found out about Schmidt's trip to Frederikshavn from geologist, Steenstrup.[8]

Quite possibly, Schmidt was acting more and more independently as his experience and interests matured over the years with the *Thor* expeditions. Schmidt's confidence grew in tune with his international publications and growing international reputation. Thus, in 1907, Schmidt proposed that he should attend the upcoming meeting of the International Council in London. On being alerted to this trip, Petersen replied somewhat condescendingly that he did not mind Schmidt attending, but that he would have to arrange it himself with the general secretary, P. P. C. Hoek. Petersen suggested that Schmidt could label himself Petersen's secretary, but warned him that he probably would not be eligible to join the official receptions and parties.[9]

These two letters indicate that the working relationship between Petersen and Schmidt was not always a match made in heaven. However, clearly something serious went wrong between the two men around 1906. In the spring of 1907, Schmidt made a couple of unsuccessful job applications, which suggests he sought to move on from his

6 RA. *Komm. Danmarks Fiskeri- og Havundersøgelser*: Forhandlingsprotokoller (1900–1960) N1.: 33rd meeting, 18 April 1906.

7 KB. Johannes Schmidt 'Papirer'. Ut.374.III. 'Forskellige Breve fra Johannes Schmidt til Bjarni Sæmundsson'. Letter from J. Schmidt to B. Sæmundsson, no date, but from context it is surely spring of 1906.

8 KB. Johannes Schmidt 'Papirer'. Ut.374.V.1. Privat, 'Instrukser og Breve, ca. 1902–1908'. Letter from C. G. J. Petersen to J. Schmidt, 4 August 1906.

9 KB. Johannes Schmidt 'Papirer'. Ut.374.V.1. Privat, 'Instrukser og Breve, ca. 1902–1908'. Letter from C. G. J. Petersen to J. Schmidt, 29 May 1907.

position working for Petersen. One of these applications was for a job as the Danish government's fisheries inspector. Schmidt kept in his files a very positive reference, which he obtained from Johan Hjort in support of his application. 'You have shown a rare ability to set yourself practical and scientific goals, as well as to solve these in a practical fashion. No one else could fill this position better than you.'[10]

Another vacant position in the spring of 1907 was the job of curator for the paleontological collection of the Natural History Museum. Evidence for this application comes from a letter written by the Professor of zoology at the Agricultural University in Copenhagen, J. E. V. Boas. Boas and Schmidt appear to have discussed Schmidt's chances of landing the job, and Boas regrets to tell Schmidt that he is unlikely to win the backing of the review committee, who Boas believes will support the other candidate, 'Ravn', (geologist, J. P. J. Ravn, 1866–1951).[11]

Danish commission taking stock of investigations in 1907

One issue that definitely bothered Schmidt in 1907 was the question of whether or not he had been granted sufficient recognition from Petersen. Schmidt now frequently published results from the path breaking studies conducted on board the *Thor,* and he had emerged as a rising star in international marine science. Thus, when in November 1907, the commission was discussing how to report the results of the previous years' research to the Ministry of Agriculture it seemed obvious to include Schmidt among the group of rapporteurs to the government report. Today, this would have been a natural procedure and this was also Drechsel's take on the matter. Petersen had written his own draft report, in which he summarized the many results obtained not only by himself, but also by A. C. Johansen and Schmidt. Petersen was doubtful that Schmidt would offer his results for inclusion if his name was not on the report. Since Schmidt was not a member of the commission, but merely the commission's assistant, Petersen did not see this as an option. Drechsel then proposed that Schmidt could become a full member of the commission, but Petersen did not support this suggestion at all, and nothing came of it.[12]

In the end, it was Petersen's draft, which was later published in 1908 as a summary of the Danish commission's achievements in the first six

10 KB. Johannes Schmidt 'Papirer'. Ut.374.V.1. Privat, 'Instrukser og Breve, ca. 1902–1908'. Letter from J. Hjort to J. Schmidt, 21 May 1907.

11 KB. Johannes Schmidt 'Papirer'. Ut.374.V.1. Privat, 'Correspondance 1980–1916'. Letter from J. E. V. Boas to J. Schmidt, 1 May 1907.

12 RA. Komm. Danmarks Fiskeri- og Havundersøgelser: Forhandlingsprotokoller (1900–1960) N2: 38th meeting, 11 November 1907.

years of its existence. Among the numerous results of investigations that had involved the commission, the eel studies ranked high on the list of achievements. Petersen was even bold enough to write that in the main, the riddle concerning the life history of the freshwater eel in Europe was now solved, thanks to the work undertaken by the commission. The report went on to state that it should be taken for granted that the spawning grounds could be found at great depth in the Atlantic Ocean. Furthermore, the studies by Schmidt could now form the basis for a well-informed regulation of the eel fisheries. The discovery of enormous amounts of the offspring of eel near the shores of Southwest England had the potential to supply the various European waters with cheap young eel and so increase the eel fisheries.[13]

While Petersen was still calling the shots, Schmidt had other sympathizers in the commission such as Carl Ostenfeld. At the first meeting in 1908, the upcoming meeting of the Central Bureau of the International Council was on the agenda. Petersen was dissatisfied with the General Secretary at the time, P. C. Hoek, on grounds of Hoek's (Dutch) interests in the North Sea legislation of plaice fisheries, which ran counter to Denmark's interests and Petersen proposed to have him replaced. Ostenfeld then suggested promoting Schmidt as his replacement, and Drechsel took it upon himself to pursue this idea at the upcoming meeting.[14]

As it turned out, the Berlin meeting was indecisive on this point.[15] However, at a later meeting it was decided that Drechsel himself became the new general secretary of the International Council from 1909 onwards. Also in 1909, the Central Bureau of the International Council was strengthened at the expense of the annual council. The continuation of the international cooperation was also agreed upon with a renewal of the plaice investigations.[16] Looking forward, at this point it seemed that Danish marine science was centrally positioned.

Petersen's memorandum to the Ministry of Agriculture

Following the cut in government support for the work of the commission, enforced from 1907 onwards, it was now, two years later, time to reconsider the strategy of the commission again. In his proposal to the ministry, Petersen favored further prioritizing of fisheries biology at the expense of hydrography and plankton surveys.[17] According to Petersen,

13 Drechsel CF, Knudsen M, Ostenfeld CH, Petersen CGJ. *De Internationale Havundersøgelser 1902–1907*. Copenhagen: Kommissionen for Havundersøgelser / C. A. Reitzel; 1908. p. 16.

14 RA. *Komm. Danmarks Fiskeri- og Havundersøgelser*: Forhandlingsprotokoller (1900–1960) N2: 39th meeting, 21 January 1908.

15 RA. *Komm. Danmarks Fiskeri- og Havundersøgelser*: Forhandlingsprotokoller (1900–1960) N2: 40th meeting, 5 February 1908.

16 Rozwadowski H. *The sea knows no boundaries. A century of marine science under* ICES. Seattle and London: ICES/University of Washington Press; 2002. p. 52–54.

17 KB. Johannes Schmidt 'Papirer'. Ut.374.V.1. Privat, 'April 1908'. Copy of Letter from C. G. J. Petersen to Ministry of Agriculture, received by M. Knudsen 25 July 1908.

the fisheries' investigations in what he calls 'Denmark proper' could be conducted in a sufficient manner without extending investigations to the deep sea. What Petersen means when he wrote 'Denmark proper,' is the territory of present day Denmark surrounded by the North Sea, Skagerack, Kattegat and the western Baltic Sea, whereas the then Danish North Atlantic possessions, the Faroes, Iceland and Greenland, were of little interest. In terms of topical focus, research of direct economic benefits for the Danish fisheries should be highlighted at the cost of other types of marine science. This, he stated was also the impression he received from the other countries involved with ICES, when he had attended the 1908 council meeting in Copenhagen.

Petersen lamented the current situation of the commission, where the work was divided between three independently working units, each with an independent leader and independent laboratories, each operating with their own man on ocean voyages. Petersen felt that the men in question should be credited for their achievements working under such unfortunate circumstances. However, he also noted that it would have been much better if one man, a fisheries scientist, had led all of the work of the ship. Only then, in Petersen's opinion, would it possible for all the work carried out to serve the interest of promoting fisheries. This leader had to be someone seaworthy and accustomed to fishermen and fisheries. There is no doubt Petersen thought of himself as the new leader, and his model of organization was that of Norway, where his close friend Johan Hjort was in charge.

Petersen then referred to, what he found a tragic situation, namely that the delegates at the international council meeting of ICES, had agreed that the international investigations were particularly beneficial in the sense that they supported a development, where the different national marine science institutions obtained a more permanent status in their own countries.

From a contemporary point of view, it sounds odd that Petersen would be against such a development. In 1908, it also provoked animosity from fellow Danish marine scientists, and Petersen's underlying reasons for this drastic move shine through his formal arguments, as he continued to present the alternative he foresaw for Danish marine science institutions:

Petersen wanted to lay up the *Thor* for the time being. He saw no reason, why the investigations in Icelandic waters should be continued, and under the proposed new regime, the *Thor* could assist fishermen for their experiments with transplanting fish fry to under-populated areas, and take part in the internationally required investigations of the flat fish fisheries in the North Sea.[18] In other words, it is quite clear that Petersen wanted the *Thor* to assist his own favorite projects at the time. The plaice fisheries investigations in the North Sea and the Limfjord had been central to Petersen's research for decades, and by 1905, Walter Garstang had launched a tagging experiment from the English side of the North Sea.[19] This alone would give a self-conscious scientist like Petersen motivation to intensify his own studies. Indeed he went on to propose that the government buy a smaller motor cutter instead of the large steamship the *Thor*. Such a motor cutter would be a more a flexible vessel suited for investigations in the shallow waters of Danish fjords and estuaries, as it could easily dock in the various fishing ports.[20]

Another line of argument running through Petersen's proposal is tied in with his notion of Denmark's international position as a fisheries nation, although he presented this in a somewhat defensive and idiosyncratic fashion. On the one hand, Petersen was well aware of what was at stake internationally with regards overfishing. His own research had contributed greatly to the international discussion and awareness of the problem, and his proposal to the Ministry of Agriculture included the sound suggestion that future regulation of the flatfish fisheries was desirable. On the other hand he was very pessimistic as to Denmark's role in the initiation of such regulations. Laconically, Petersen stated that whether or not Denmark had a large research vessel or a small research vessel, The Great Powers of Europe with a vested interest in fisheries would end up dictating international fisheries regulations. Therefore, large-scale research should be left to those powerful countries.[21]

If taken at face value, Petersen's plan may have been right for the practical fisheries in Denmark, the *Thor* continuing to traverse the waters on the edge of the continental shelf probably would not have provided any additional valuable information. Even, if the *Thor* found more hotspots for fishing of gadoids or eel for that matter to the West of the British Isles, West of France or off the Spanish Atlantic coastline, this would have benefitted the fisheries of these large and influential European countries much more than it would have benefitted the small state of Denmark.

What Petersen was telling the Ministry, is that the government could get away with spending less on funding his area of research, i.e. marine science. This would seem a counterintuitive move from Denmark's most

18 KB. Johannes Schmidt 'Papirer'. Ut.374.V.1. Privat, 'April 1908'. Copy of Letter from C. G. J. Petersen to Ministry of Agriculture, received by M. Knudsen 25 July 1908.
19 Smith TD. *Scaling fisheries: The science of measuring the effects of fishing, 1855–1955*. 2nd revised 2007 ed. New York: Cambridge University Press; 1994. p. 143–146.
20 KB. Johannes Schmidt 'Papirer'. Ut.374.V.1. Privat, 'April 1908'. Copy of Letter from C. G. J. Petersen to Ministry of Agriculture, received by M. Knudsen 25 July 1908.
21 KB. Johannes Schmidt 'Papirer'. Ut.374.V.1. Privat, 'April 1908'. Copy of Letter from C. G. J. Petersen to Ministry of Agriculture, received by M. Knudsen 25 July 1908.

renowned marine scientist, representing the interest of Danish science – and Danish scientists.

Unless that is, if what Petersen really wanted, was to prevent his Danish colleagues from sharing the scientific stardom, he had already obtained. What remains from Petersen's proposal is a clear sense that he was not painting a full picture of his own motivations for wanting to change the organizational set up of Danish marine science. Petersen's proposal only makes sense, if he is also considered to be a jealous, senior scientist afraid to lose control over Danish marine science, embodied in his senior and even junior colleagues such as Johannes Schmidt.

Petersen's job offer

In early 1908, Petersen informed Schmidt that he wished to have the commission abolished by the government, whereupon the international fisheries investigations should fall under the auspice of the Biologisk Station. Petersen must have sought to bypass the influence of his peers in this way, and he clearly did not count Schmidt among them. This was apparent from Petersen's offer to Schmidt of the position of first assistant in the new organization. Petersen even had a clear research agenda laid out for Schmidt. In his new capacity Schmidt would be in charge of conducting experiments in aquariums that used chemical, bacteriological and biological research methods to investigate nutritional relationships in plankton, as well the hatching of fish eggs and the ensuing larval stages.[22] Although Schmidt's answer to Petersen's request is only preserved in the form a draft letter, it paints a clear picture of two men, no longer singing to the same tune, neither on a scientific level nor on a personal level.

On a personal level Schmidt bluntly stated that as a mature researcher he had no desire to become 'Hr. Doctorens Assistent' at the Biologisk Station. The reason why he had remained in his current job for so long was the fact he had been so excited about solving the riddles of marine science, he dealt with in the past years. Schmidt also expressed his distrust of Petersen, as he hinted that the young biologist, Harald Blegvad might already have been lined up by Petersen to replace Schmidt.

Schmidt then moved on to what he calls 'factual reasons'. Schmidt was clearly disappointed about the downsizing of the investigations

22 KB. Johannes Schmidt 'Papirer'. Ut.374.V.1. Privat, 'Instrukser og Breve, ca. 1902–1908'. Memo from J. Schmidt undated; KB. Johannes Schmidt 'Papirer'. Ut.374.V.1. Privat, 'Instrukser og Breve, ca. 1902–1908'. Letter from C. G. J. Petersen to J. Schmidt, 9 March 1908.

around Iceland. According to Schmidt it was always desirable to go to Iceland, because of the chance to compare a relatively lightly exploited ecosystem there, with the much more heavy exploitation patterns encountered in the North Sea. In addition, Schmidt repeatedly emphasized the patriotic duty to promote Icelandic fisheries, since the Icelanders were Danish citizens, and since it was the Danish navy patrolling the territorial waters of Iceland.

Schmidt therefore, boldly concludes that for his own part, he would withdraw from his current position in the event that the commission for marine science came under the control of Biologisk Station. Instead he would continue his biological investigations under private circumstances to the extent that time would allow him to do so.[23]

Upon receiving Schmidt's dismissive letter, Petersen took a paternal and rather amicable approach in his reply. He regretted that Schmidt was not trustful of his intentions, and Petersen assured him that he would also find a great degree of freedom in his new position. Furthermore, he had no intention to replace him with Blegvad. The issue of Iceland was clearly not high on Petersen's agenda, as he pointed out how it should be left to the government to decide upon matters regarding the North Atlantic possessions. Finally, he urged Schmidt to reconsider his standpoint, as they were busy and had to act swiftly, if they were to move the government and the Danish parliament in favor of merging the two institutions.[24]

Schmidt however, seemed to have already made up his mind with regards to Petersen's attempted coup. In his second reply, Schmidt emphasized that he already had made plans for how to spend the coming years of his life as a scientist: 'For me it is of utmost importance to make the most of my life in the coming ten years, since these are usually the ones, where a scientist has the greatest capacity for labor, and can work in the most satisfactory way; and I dare not involve myself in the mentioned (institutional) experiment, which I do not think will be successful.' Furthermore, Schmidt emphasized that he was not happy in his current position. Although he claimed to appreciate the degree of freedom he was bestowed in his current job, he stated bluntly that under a new organizational regime, he would not be interested in a similar job.[25] From Petersen's second reply, we can deduce that Schmidt's final letter to Petersen also included a complaint that he felt he should be allowed to become a full member of the Danish commission alongside Petersen and the other research directors. Yet, Petersen dismissed ever to have made such a promise.[26]

23 KB. Johannes Schmidt 'Papirer'. Ut.374.V.1. Privat, 'Instrukser og Breve, ca. 1902–1908'. Draft letter from J. Schmidt to C. G. J. Petersen, 9 March 1908. (letterhead says '1907', but it is clear from the rest of the correspondence that the correct year is 1908).

24 KB. Johannes Schmidt 'Papirer'. Ut.374.V.1. Privat, 'Instrukser og Breve, ca. 1902–1908'. Letter from C. G. J. Petersen to J. Schmidt, 11 March 1908.

25 KB. Johannes Schmidt 'Papirer'. Ut.374.V.1. Privat, 'Instrukser og Breve, ca. 1902–1908'. Draft letter from J. Schmidt to C. G. J. Petersen, 12 March 1908.

26 KB. Johannes Schmidt 'Papirer'. Ut.374.V.1. Privat, 'Instrukser og Breve, ca. 1902–1908'. Letter from C. G. J. Petersen to J. Schmidt, 13 March 1908.

The tone in Petersen's letters did not suggest that he was jealous of Schmidt's Atlantic successes, but as he was effectively shutting down the North Atlantic investigations, he did not have to be, and surely, Schmidt's perception of the whole situation was a feeling of being constrained by Petersen's regime. Meanwhile Schmidt continued to correspond with Petersen on matters in relation to the *Thor's* summer season cruise in the North Atlantic for 1908. These letters deal with the manufacture of various gears for the trip and with the distribution of salary to the various assistants along the route such as Bjarni Sæmundsson in Iceland as well as Schmidt's own salary.[27] Financially, there was a considerable sum of money at stake. Schmidt's salary for four months of work on the *Thor* in 1908 amounted to 833 Kroner plus an additional 400 Kroner as a *per diem* for being stationed abroad.[28] This was equivalent to Schmidt's salary as a university teacher during an academic year, so he was gambling with an amount of money, most of his colleagues could not afford to lose.

Following the rejection of Petersen's job offer, it was apparent for Johannes Schmidt that his days working for Petersen were numbered, but apart from siding clearly with the alternative faction in the commission he would have to wait and see what the outcome would be.

Rebel assistants

By the end of April 1908, Martin Knudsen had received a commentary from his hydrographer colleague, J. N. Nielsen, which read like a character assassination of Petersen. The document bears a signature of authenticity from a secretary, adding reason to believe that the documents served the purpose of scandalizing Petersen in the eyes of readers other than the members of the commission.[29]

In his commentary, Nielsen bluntly stated that it would be difficult to find one person, who would have the scientific overview necessary to effectively lead all of Danish marine science, and in the case of Petersen, that his appointment would be most unfortunate. Nielsen put down the success of the North Atlantic investigations to the fact that Petersen had left the leadership thereof entirely to Johannes Schmidt. Furthermore, if Johannes Schmidt and A. C. Johansen were to retire from their current positions as Petersen's assistants, he doubted that Petersen would be capable of training a new team of investigators.

27 KB. Johannes Schmidt 'Papirer'. Ut.374.v.1. Privat, 'Instrukser og Breve, ca. 1902–1908'. Letters from C. G. J. Petersen to J. Schmidt, 27 March and 24 April 1908.

28 KB. Johannes Schmidt 'Papirer'. Ut.374.v.1. Privat, 'Instrukser og Breve, ca. 1902–1908'. Note by J. Schmidt, spring 1908.

29 KB. Johannes Schmidt 'Papirer'. Ut.374.v.1. Privat, 'April 1908'. Copy of Letter from J. N. Nielsen to M. Knudsen 28 April 1908.

Nielsen was equally critical of Petersen's leadership of the Biologisk Station, arguing that he treated the station's research vessel, the *Sallingsund*, as his own private yacht. The previous year, Nielsen had assisted Petersen on board the *Sallingsund*, where he found it extremely difficult to get any real work done, as the time schedule of the summer season's cruise was dictated by Petersen's desire to take part in various hunting parties around the country. Thus, Nielsen found hydrographical work difficult, as they were expected to go ashore to take Petersen to and collect him from his hunting excursions. In two instances, Petersen was away hunting for several days at a time. Finally, Nielsen also accused Petersen of overlooking important findings about the proliferation of oysters in the Limfjord, when these findings did not match Petersen's own conclusion. Nielsen finished his fierce critique with a reference to Johannes Schmidt and his splendid North Atlantic investigations, and in particular the clever combination of fisheries biology and hydrography.[30]

If there was any doubt as to which basket Schmidt would place his eggs, it certainly vaporized as Schmidt wrote up his comments to Petersen's proposal of merging Biologisk Station with the Kommissionen for Havundersøgelser. While Nielsen had been deeply personal, bitter even, in his critique, Schmidt's critique was scientific in its character, focusing on the potentials of North Atlantic investigations, which would not be fulfilled by Danish scientists and Danish fishermen, were it not for the presence of the *Thor* in the North Atlantic. In fact, in many ways, Schmidt's lengthy account of the state of affairs can be read as a proposal for how to best execute Danish marine science in the North Atlantic in the years to come. While critiquing Petersen's proposal it also reads like a job application for a future leader of the proposed investigations in the Danish North Atlantic.[31]

First up, Schmidt tackled the technological and financial issues at stake, centred on the usefulness of the *Thor*. Schmidt agreed that the 100,000–117,000 Danish Kroner, which had been spent annually on keeping *Thor* in service was a lot of money, but a cut back to c. 30,000 Kroner, by laying up the *Thor*, should not be the only alternative. *Thor* could be kept in service maintained for c. 70,000 Kroner per year. In this way, the vessel could be operational in the North Atlantic for c. 4 months, which was deemed sufficient by Schmidt. He felt that spending half a year or more on the *Thor*, which had been the case in previous years, was in fact too much. He and the other scientists on board the *Thor* had

30 KB. Johannes Schmidt 'Papirer'. Ut.374.V.1. Privat, 'April 1908'. Copy of Letter from J. N. Nielsen to M. Knudsen 28 April 1908.

31 KB. Johannes Schmidt 'Papirer'. Ut.374.V.1. Privat, 'April 1908'. Note by J. Schmidt, undated.

collected so much data that it became difficult to find time for desk-top analyses during time ashore.

Schmidt rejected the idea of replacing the *Thor* with a cheaper motor cutter. While a motor cutter might be able to assist with transplanting plaice in Danish water, it would not be useful for any deep-sea investigations. Also, since a motor cutter had to be acquired and equipped before it could be used for scientific purposes it would prove to be more costly than suggested by Petersen.

Upon discarding Petersen's proposed solution, Schmidt painted an image of what might become of marine science in the area, if the current intensive effort was continued. The central international importance of Iceland was evident to Schmidt, when it came to the then current issue of depletion of the stocks of plaice and haddock in the North Sea. For haddock the individual year classes, 1903–1905, had recently been observed to vary greatly in the North Sea from one year to the next.[32] This is probably why Schmidt pondered to what extent fishing affected the size of the individual year classes of haddock. In addition, Schmidt mentioned that there was the possibility of trawling in the more pristine Icelandic waters to give an indication of the scope of natural fluctuations of these species, which in turn could be compared with the much more intensively exploited fishing ground in the North Sea and Norwegian Seas.[33]

The thought that Denmark should voluntarily give up a research area, which had proved so successful, was a real shame to Schmidt. He pointed out that whereas the eel investigations had led the *Thor* into a more southern route in the cruises of 1905–07, there was still a boundless potential to investigate the Danish parts of the North Atlantic for new banks for the commercially important fish stocks. Schmidt mentioned the halibut (*Hippoglossus hippoglossus*) as one example of a highly priced fish species, whose Icelandic whereabouts where virtually unknown prior to the investigations with the *Thor*. Still it remained to be studied how fast the halibut grows, a pre-requisite for assessing whether or not the growth rates of the fishes can keep up with intensified fishing. Cod is another such species, where at the time, Icelanders in the North and East waited anxiously for the cod to arrive at their doorstep. If further marking experiments were conducted, the migratory patterns of the cod around Iceland would be revealed, thereby facilitating the pursuit of more intensive fishing. Schmidt envisioned the same for the herring, which appeared in terrific numbers for a few summer months, yet 'where does it migrate to for the rest of the year?'[34]

32 Smith TD. *Scaling fisheries: The science of measuring the effects of fishing, 1855–1955*. 2nd revised 2007 ed. New York: Cambridge University Press; 1994. p. 126.
33 KB. Johannes Schmidt 'Papirer'. Ut.374.V.1. Privat, 'April 1908'. Report by J. Schmidt, undated.
34 KB. Johannes Schmidt 'Papirer'. Ut.374.V.1. Privat, 'April 1908'. Report by J. Schmidt, undated.

Schmidt also commented on the importance of tying-in hydrography with fisheries' related questions, when addressing the significance of Iceland. He stated that Icelandic waters were the one area within international waters, where the greatest clarity had been reached with regards to knowledge of the pelagic drift of young fish, and this had been linked to the prevailing drift of ocean currents. As Schmidt asks rhetorically; 'why wouldn't the fishermen use a thermometer for finding fish just as they lower a lead into the sea for depth measurement?'[35]

Towards the end of his report, Schmidt turned to the issue of Denmark's national interest, a political card he would eventually be playing multiple times over throughout his career. Schmidt found that currently, Denmark was in the front line of nations engulfed in marine science. Agreeing to Petersen's proposal would reduce Denmark to a third or fourth rate player in marine science. Hereby, Schmidt echoed the way in which Denmark's political power was ranked as a small state. This was so much more regrettable, Schmidt continued, as the southern provinces of Denmark would be cut off from the Faroe Islands and Iceland, raising the prospect of no other nation picking up the torch and carrying out scientific investigations. Elsewhere in the text, he literally underlined how the Faroes had the status of a Danish county, and similarly Schmidt recalls how the water surrounding Iceland was in fact part of Denmark's territorial waters.[36]

Schmidt summed up his report by stating how it was extremely difficult to understand, why Petersen would want to assume responsibility for stopping, or at any rate inhibiting Danish participation in the international investigations.[37]

While, Schmidt had been in charge of the *Thor* in the North Atlantic, A. C. Johansen acted as Petersen's first assistant on the *Sallingsund*. Still, Johansen did not agree with the chairman's proposed cuts or the merger. Johansen delivered a slightly more condensed report compared to Schmidt's, but the two men might very well have coordinated their efforts, as their arguments are similar. Johansen did not emphasize the patriotic deeds associated with marine science, as Schmidt had done, but the scientific, technological and financial arguments remained the same. Johansen further added that selling the *Thor* was the same as writing off all the expenditure that had gone into equipping her for scientific service, as the equipment would be more or less useless in any other setting.[38]

35 KB. Johannes Schmidt 'Papirer'. Ut.374.V.1. Privat, 'April 1908'. Report by J. Schmidt, undated.
36 KB. Johannes Schmidt 'Papirer'. Ut.374.V.1. Privat, 'April 1908'. Report by J. Schmidt, undated.
37 KB. Johannes Schmidt 'Papirer'. Ut.374.V.1. Privat, 'April 1908'. Report by J. Schmidt, undated.
38 KB. Johannes Schmidt 'Papirer'. Ut.374.V.1. Privat, 'April 1908'. Report by A. C. Johansen, 18 May 1908.

Christian Drechsel's opinion was clear from the context, although he did not deliver the same kind of written report to Knudsen, as did the younger scientists. In 1906, however, Drechsel in his capacity as government fisheries consultant had published a report on the state of fisheries in Iceland. Here, Drechsel explained how Icelandic waters had the potential to be much better exploited through Danish – Icelandic collaboration.[39] Thus we can safely deduce that Drechsel's longstanding interest in exploiting Icelandic fishing banks from Denmark was still very much alive.[40] So, whereas Petersen's practical interest in fisheries lay mainly within the territory of modern day Denmark, i.e. the North Sea, Skagerrak and the Western parts of the Baltic, Drechsel saw the potential of exploiting what was also a part of Denmark, namely Iceland and Icelandic waters. The former compatriots from the days of traversing the Kattegat on the gunboat *Hauch*, now headed for an open confrontation over mutually conflicting interests.

Petersen's last bout in the commission

Once, Martin Knudsen had collected the reports from Nielsen, Schmidt and Johansen, he and Drechsel and Ostenfeld were ready to face Petersen at the next ordinary meeting of the commission. The 43rd meeting of the Danish commission turned out to witness the last bout between Petersen and the other three members of the committee. The four men were summoned to Drechsel's apartment on a Friday afternoon in September, and according to the minutes, Petersen went straight into an attack on the future existence of the committee.[41] Petersen reported that he maintained his earlier position in recommending that the commission should be dissolved as of 22 July in the following year, 1909. He therefore did not see any reason why the commission should proceed with normal procedures and suggestions.

Drechsel replied that he was saddened by the situation at hand, but since he felt he had exhausted all possibilities of settling disagreements, Drechsel now stated that he was strongly against Petersen's proposal to terminate the commission. Knudsen and Ostenfeld both agreed to oppose Petersen's suggestion, and along with Drechsel they argued that disagreements were centred on administrative matters.

39 Drechsel, Christian F., *Island og Fiskeriet*, Copenhagen: J. H. Schultz' Bogtrykkeri, 1906.

40 Holm, P. (1996). Technology transfer and social setting: The experience of Danish steam trawlers in the North Sea and off Iceland, 1879–1903. In P. Holm, O. Janzen & J. Thór (Eds.), *Northern seas yearbook 1996* (pp. 113–157). Esbjerg: Fiskeri- og Søfartsmuseets Forlag.

41 RA. Komm. Danmarks Fiskeri- og Havundersøgelser: Forhandlingsprotokoller (1900–1960) N2: 43rd meeting, 16 September 1908.

3.1 Physicist Martin Knudsen
Phycisist Martin Knudsen (1871–1949) was a pioneer in physical oceanography. He became Head of the Central Bureau for hydrography when the International Council was founded in 1902, and here he prepared batches of standardized seawater. He also crafted the design of water samplers for deep sea investigations (Carlsberg Archives).

As a consequence, the commission agreed to inform the ministry of a joint statement, which reads as an expression of Petersen's minority opinion. In the statement the commission reported that while there was full agreement that Denmark's part of the international oceanographic investigations should be carried out for as long as the other countries also participate, no agreement was reached with regards to administrative matters. Here, it was noted that Petersen maintained the belief that the commission should be dissolved, and future Danish obligations to the international investigations should be handled by Dansk Biologisk Station. Therefore, Petersen did not want to propose a budget for the commission. Finally, it was added that Petersen found it necessary that the leader of the Danish fisheries investigations was also Denmark's voting delegate to ICES.

Knudsen remarked that since Petersen did not wish to propose a budget, the three other committee members would have to formulate and submit such a proposal without Petersen. Knudsen polemically asked Petersen if he thought he could still remain as chairman of the commission and the commission's fisheries investigations. Petersen declined to answer that question, and so ended the meeting after an hour's discussion.[42] The closure of the meeting, therefore, signalled the end of Petersen's active involvement with the Danish commission for marine science.

Restructuring the Danish commission – enter Johannes Schmidt

Following the stalemate of the September meeting, the majority of the members of the commission drew up their own proposal to the ministry. Drawing heavily and explicitly on the reports obtained previously from Schmidt and Johansen, they were able to portray an image to the ministry favoring their own position.[43]

Drechsel, Knudsen and Ostenfeld proposed that the Danish part of the international oceanographic investigations continued to be led by a commission of its own, and they emphasized that contrary to Petersen's position, the commission should be in charge of appointing the voting representative of Denmark at the international council's meetings.

The level of Denmark's engagement in the international investigations was proposed to remain on the current level, where the government would grant the commission 62,000 Kroner á year. They modestly proposed that this would be the case for one more budget year (July 1909 – July 1910). This signalled the end of the affair with regards to uniting the two institutions. The government decided to follow the recommendations of the majority faction.

When the commission was summoned again in January 1909, Petersen was gone from the commission, and furthermore he had suffered the humiliation that the Ministry of Agriculture had relieved him of his duties as Denmark's non-voting representative in ICES.[44] Nonetheless, he maintained leadership over the Biologisk Station. Throughout Petersen's reign until the mid 1920s collaboration between the two Danish marine science institutions was kept to a minimum.

42 RA. Komm. Danmarks Fiskeri- og Havundersøgelser: Forhandlingsprotokoller (1900–1960) N2: 43rd meeting, 16 September 1908.

43 KB. Johannes Schmidt 'Papirer'. Ut.374.V.1. Privat, 'April 1908'. Copy of letter from C. F. Drechsel, M. Knudsen and C. H. O Stenfeld to the Minister of Agriculture, 5 October 1908. One of the documents is dated '5 September', but it is clear from the context that it should more likely be dated '5 October' just as the previous document in the same bundle of papers.

44 RA. *Komm. Danmarks Fiskeri- og Havundersøgelser*: Forhandlingsprotokoller (1900–1960) N2: 46th meeting, 23 March 1909.

Different personalities

Some of the differences between Petersen and the younger generation may be related to different ways of working, and different ways of operating a scientific cruise. Zoologist Harald Blegvad (1886–1951) recalled what it was like to work under the directions of Johannes Schmidt and A. C. Johansen on the *Thor*, and from 1908 he worked for C. G. J. Petersen.[45] Blegvad was Petersen's loyal assistant at the Biologisk Station for 20 years, and later he became head of the Biologisk Station. In Blegvad's portrait of the three men, it is Petersen who stands out.

Schmidt was characterized as a meticulous and intensive worker, who followed a preconceived plan. This made for very efficient use of the short time spans available for research on the *Thor* in the North Atlantic. Johansen was considered to be very hard working, knowledgeable and a curious researcher. In comparison with the two up and coming scientists, Petersen comes across as a renaissance like figure in Blegvad's depiction: Blegvad was awestruck when he first met Petersen, the leading man in Danish marine science at the time, but he quickly adjusted, and he described the hard working environment on Petersen's watch. Petersen had maintained an air of navy officer about him, and his work at sea in some ways resembled a military exercise. The crew wore navy blue uniforms; while Petersen's standard attire was a double-breasted blue jacket and a cap from the Royal Danish Yacht Club. After hours though, Petersen could be a splendid, cheerful and entertaining host, a connoisseur of good food and wine, and according to Blegvad, the *Sallingsund* always had a first rate chef on-board. Petersen would rarely go into port for the night, instead he preferred to anchor just outside the port and have the crew stay on the vessel. In terms of planning, Petersen rarely had a full program for a day's work ready in the morning, and he would issue his orders as the investigations progressed. Always getting new ideas, good ones and bad ones, he wished to try them out at once. In this way, according to Blegvad, many topics would be touched upon, leading to a very heterogeneous set of investigations.[46]

45 Blegvad H. *Dansk biologisk Station gennem 50 år, 1889–1939*. Copenhagen: C. A. Reitzels Forlag; 1943. p. 26.

46 Blegvad H. *Dansk biologisk Station gennem 50 år, 1889–1939*. Copenhagen: C. A. Reitzels Forlag; 1943. p. 26–27.

Biologisk Station after 1908

In 1914, several years after the arguments over the future of the commission, Petersen presented an account of results obtained by the Biologisk Station that related to practical implications of his eel research, of which he was particularly proud The discovery that the eel migrates to sea meant that legislators ought not to consider whether or not the fry of the eel was harmed by seines being dragged along the bottom of the shallow Danish waters, and indeed this had had the implication that the Danish law of fisheries was changed accordingly. The minimum sizes for eel had also been adjusted downwards, so smaller eel could be caught legally, following the scientifically based argument that the male eel were on average smaller than the female. Since the female in most cases is considered the more important asset in the reproduction of fishes, it came to be considered less damaging to catch the smaller yellow eels. With a subtle bitterness Petersen went on to explain that 'Later, investigations in the Atlantic Ocean have greatly advanced the question of the breeding places of the eel; the fact that the eel does not breed in Denmark has been known for a long time'.[47] Johannes Schmidt's name was not to be found in conjunction with Petersen's account. The intermezzo of trying to merge the two institutions, Biologisk Station with the Kommissionen for Havundersøgelser was passed over unnoticed, and Schmidt's name only occurs in a brief listing of Petersen's former assistants at the station.[48]

A few years before the publication of this account in 1909 Petersen had had the pleasure of observing how other researchers were using one of his clever fish survey instruments. The Petersen young fish trawl, famous for capturing the eel larvae of the *Thor* expeditions, was now in use at the English fisheries laboratory at Lowestoft, where H. J. Buchanan-Wollaston successfully conducted a large-scale survey of plaice eggs in the North Sea.[49] Petersen's international reputation also reached Johannes Schmidt's desk. The famous Italian scientist, Grassi, perhaps not aware of the internal Danish skirmish wrote to Schmidt twice in the winter of 1909–1910, asking him for the address of the Danish company manufacturing the Petersen young fish trawl. Grassi eventually purchased the trawl from the N. P. Utzon fish net factory in Copenhagen for the sum of c. 300 Francs.[50] Demand for the trawl continued to grow. E. J. Allen, the Director of the Marine Biological Association's laboratory in Plymouth England also asked Schmidt where he could buy the trawling device.[51] Schmidt, who still used the trawl himself, guided Allen to the Utzon factory.[52]

47 Petersen CGJ. *Om Biologisk Stations Virksomhed fra 1889–1914. Efterskrift 1918.* Copenhagen: Centraltrykkeriet; 1919. p. 4.

48 Petersen CGJ. *Om Biologisk Stations Virksomhed fra 1889–1914. Efterskrift 1918.* Copenhagen: Centraltrykkeriet; 1919. p. 2.

49 Smith TD. *Scaling fisheries: The science of measuring the effects of fishing, 1855–1955.* 2nd revised 2007 ed. New York: Cambridge University Press; 1994, p. 147.

50 RA: Dansk Biologisk Station, Johannes Schmidts embedsarkiv: Korrespondance om åleforskning (1908–1933) E28: 1908–1913: G. B. Grassi to J. Schmidt, 3 November 1909 & 10 January 1910.

51 RA: Dansk Biologisk Station, Johannes Schmidts embedsarkiv: Korrespondance om åleforskning (1908–1933) E28: 1908–1913: E. J. Allen to J. Schmidt, 21 February 1910.

52 RA: Dansk Biologisk Station, Johannes Schmidts embedsarkiv: Korrespondance om åleforskning (1908–1933) E28: 1908–1913: J. Schmidt to E. J. Allen, 24 February 1910.

Conflict resurfaced in 1916

An episode in 1916, testified that the conflict between the Biologisk Station and Kommissionen for Havundersøgelser was still bubbling under the surface. The newly established fisheries advisory council to the Danish government, the so-called Fiskeriraadet, addressed the issue of merging the two institutions as their first agenda item at their very first meeting. Petersen's attitude to these plans are not known, but a letter from Drechsel to Schmidt reveals that he clearly had the eye on the ball and tried to prevent any such thing as long as Petersen was to be in charge of a merged institution. In a memorandum drafted by Drechsel and included in his letter to Schmidt, Drechsel summarizes the developments in 1908, and argues that the duties of the commission for marine science is much wider than the ones entrusted to the Biologisk Station, and Drechsel further underlined that the commission is responsible for upholding international laws and regulations. He summed up his memorandum by stating that the issue at stake was the issue of Petersen and nothing else.[53]

Schmidt becomes a member of the Kommissionen for Havundersøgelser

Many years later, when Martin Knudsen recalled Schmidt's involvement with ICES and the Danish commission for marine science, he also touched upon Petersen's attempt to reorganize the commission with himself as the sole leader. Knudsen however, failed to mention his own central role in the abovementioned events in 1908. Looking back, Knudsen described it as an issue playing out between Petersen and the Ministry of Agriculture. When Petersen presented his plan in the spring, the Minister of Agriculture was Ole Hansen. As noted earlier in this chapter, there was a change of government following the largest and most infamous fraud scandal in Denmark, involving the then Minister of Taxation, P. Alberti. Knudsen speculated that the change in government, where a whole new cabinet was appointed in the early autumn, is what blew Petersen's plan off course.[54] In fact, as this chapter has shown, Knudsen himself played a central and pro-active role in ensuring that nothing came of Petersen's plans.

53 KB. Johannes Schmidt 'Papirer'. Ut.374.V.1. Privat, 'April 1908'. Letter from C. F. Drechsel to J. Schmidt, 19 June 1916 & draft memorandum by C. F. Drechsel, June 1916.

54 Knudsen M. De internationale Havundersøgelser. In: Winge Ø, Vedel Taaning Å, eds. *Naturforskeren Johannes Schmidt. Hans liv og ekspeditioner*. Copenhagen: Gyldendalske Boghandel Nordisk Forlag; 1947: 110.

In any case, Drechsel took over as acting chairman of the Danish commission, having claimed the commission's account books, archive and money from Petersen. The venue was the office of ICES in Jens Koefoedsgade 2, where the new commission decided to rent their own office space. Ostenfeld and Knudsen joined Drechsel and A. C. Johansen was also summoned to the meeting. According to the minutes, they also discussed the future of the commission's work.[55] At the subsequent meetings in the spring of 1909, important decisions were made, such as the establishment of a more fine-grained statistical collection of information from Danish fishing ports.[56]

Importantly, three new members were welcomed to the commission and their workload designated.[57] Inspector of Danish fisheries, F. V. Mortensen was assigned with the task of taking care of the administration of the *Thor*. This had been Drechsel's task, but the new position as General Secretary of ICES came to occupy Drechsel's time. Then, Petersen's former assistants, A. C. Johansen and Johannes Schmidt were given shared responsibility for the fisheries investigations. Johansen became responsible for the investigations in the waters around Denmark, and Schmidt was placed in charge of the North Atlantic investigations.

Schmidt, who had just turned 32, had managed to secure himself a position, where he could not only make important decisions concerning his own research and that of his collaborators, but also had his own budget to allocate. Thus for 1909, he disposed of 6,350 Kroner, of which his own salary amounted to 3,500 Kroner.[58] For comparison, a skilled craftsman in Denmark earned c. 1,000 Kroner in 1909.[59] Schmidt now had a more than decent salary, thereby becoming less dependable on his and his wife's private money. In reality however, Schmidt could dispose of a substantially larger sum as a member of the commission, since the costs associated with running the *Thor* amounted to 25,000 Kroner. In addition to fisheries investigations, the *Thor* still would serve the plankton research led by Ostenfeld, and hydrographical surveys led by Knudsen.[60]

55 RA. *Komm. Danmarks Fiskeri- og Havundersøgelser*: Forhandlingsprotokoller (1900–1960) N2: 44th meeting, 5 January 1909.

56 RA. *Komm. Danmarks Fiskeri- og Havundersøgelser*: Forhandlingsprotokoller (1900–1960) N2: 45th meeting, 27 February 1909 & 46th meeting, 23 March 1909.

57 RA. *Komm. Danmarks Fiskeri- og Havundersøgelser*: Forhandlingsprotokoller (1900–1960) N2: 46th meeting, 23 March 1909.

58 RA. *Komm. Danmarks Fiskeri- og Havundersøgelser*: Forhandlingsprotokoller (1900–1960) N2: 46th meeting, 23 March 1909.

59 Anonymous. *Statistiske meddelelser*. 1909;4. series, vol. 40, issue 1–3. p. 10.

60 RA. *Komm. Danmarks Fiskeri- og Havundersøgelser*: Forhandlingsprotokoller (1900–1960) N2: 46th meeting, 23 March 1909.

Conclusion

This chapter has sought to show what was at stake in Danish marine science, as the first five year term of the international council came to a close, and decisions were to be made, as to whether or not the Danish commission's work was to be continued under a renewed international scheme. While a great many results had been reaped during the first years, not least through the *Thor's* North Atlantic expeditions, the whole organizational structure behind the commission was fragile. This makes the fight for control over the commission interesting, not just in its own right, but also as an example of the vulnerability of both the international structure surrounding marine science, as well as the fragility present in a temporary structure like the Danish commission for marine science.

In the face of what was perhaps a necessary cut in total government support for the Danish commission planned from 1907 onwards, a degree of diplomacy was needed in order to oil the wheels of Danish marine science. It seems that C. G. J. Petersen, as leader, managed to alienate Knudsen by targeting the hydrographical section for the biggest cuts. It also seems that Petersen wanted to curb the independence of the Danish commission for marine science and shut down central parts of the investigations. These suggestions placed Petersen in direct conflict with the other members of the commission, Drechsel, Knudsen and Ostenfeld, as well as the young generation of assistants, principally, Schmidt, Nielsen and Johansen. Petersen's idea seems to have formed sometime during 1907, whereas the situation became explosive during 1908, when Petersen on one side and the rest of the members on the other side, each submitted their diverging work proposals to the Ministry of Agriculture. This conflict erupted over a combination of disagreements, which were scientific, ideological and personal.

With regards to the scientific aspects of the feud, there was a central aspect in what could be labelled the spatial extent of Danish marine science discourse. Everyone agreed that the central purpose of the investigations was to promote practical fisheries. Yet according to Petersen, Denmark's fisheries' interests extended to the North Sea and the Western Baltic, whereas the wider North Atlantic possessions of Iceland and the Faroe Islands were of little interest to Petersen.

This perception of, what was Denmark's sphere of interest, was directly opposed to not least Schmidt and Drechsel's more imperialistic notion of Denmark as a kingdom with important North Atlantic possessions. It so happened that on one hand, these possessions formed valuable testing ground for sampling in a lightly exploited ecosystem, at least compared to the North Sea, where existing fisheries were already thought to have had a negative impact on fish stocks. In addition, the lightly exploited and little known marine ecosystem was ripe for more intensive fishing from Danes and Icelanders alike. In some ways, this forms a link between scientific and personal convictions. Safe to say, Drechsel's ideals were fuelled by his hapless ten years trying to create a successful Danish steam-trawling enterprise around Iceland.

On a more strict personal level, it is hard to overlook the peculiar way in which Petersen handled his leadership as chairman of the Kommissionen for Havundersøgelser and the leadership of Dansk Biologisk Station. From the point of view of securing the future of Danish marine science, Petersen's moves to diminish the commission were counterintuitive. From the correspondence preserved, an image emerges of an old school scientist, afraid to lose control, as the younger generation, not least Johannes Schmidt, came of age and moved towards the scientific limelight.

In any case, Petersen's attempted coup d'état proved unwise, as he failed. This failure in turn provided Johannes Schmidt with a fast track to launch new expeditions furthering his own marine science interests. The younger generation in the commission eventually came to shine in their own ways, while Petersen's legacy gradually fell under a shadow. Petersen's efforts have been partly neglected by later chroniclers of Danish marine science, including stories told by his foremost former assistant, Johannes Schmidt. The same goes for Schmidt's peers and not least an entire generation of Danish marine scientists, maturing under Schmidt's leadership, and who at least partly became dependant for their career on taking part in Schmidt's many expeditions from 1908–1933. Thus, the chapter also serves to show why Schmidt and comrades in the post-1909 commission chose to omit from some of their future scholarship, the extent of 'Papa Petersen's' reach and importance as a founding father of modern Danish marine science.

With regards to Johannes Schmidt, the conflict with Petersen also made him show his colors in terms of his personal character. When he turned 31 in 1908, Schmidt was already a very ambitious and successful scientist. He was fully aware of his own goals for his scientific career, as he flatly turned down Petersen's offer to be part of a renewed institutional setup for Danish marine science. His determination was admirable and no doubt it was supported by a solid private financial situation, where he dared to do without his salary from the Kommissionen for Havundersøgelser. But Schmidt's defiant attitude in the spring of 1908 may also have been shored-up by the knowledge that he had a new and very ambitious project lined up on the side.

As the next chapter will show, Schmidt was able to persuade the Carlsberg Foundation to fund his very own independent oceanographic expedition to the Mediterranean, scheduled for the autumn and winter months of 1908–1909. The frustration about the uncertain prospects for the Kommissionen for Havundersøgelser as well as the personal ambition of Schmidt to solve the outstanding riddles of the Atlantic eel were both good reasons for Schmidt to seek financial support elsewhere. In other words, there was a significant push factor for seeking funds for a Mediterranean expedition as well as the obvious pull factor of investigating, what was largely, from the point of view of modern marine science, a *mare incognitum*. Whereas the commission for marine science was a temporary institution, there was one Danish source of funding for research, which was permanent, the Carlsberg Foundation. In his funding application Schmidt argued that the mystery of the eel could now be solved with a winter expedition into the Mediterranean, explicitly stating how wonderful it would be to claim this for Denmark, before other nations would reap the benefits of the hard Danish labour.

4 The Mediterranean expeditions (1908–1910)

In the fall of 1907, Johannes Schmidt approached the Board of Directors of the Carlsberg Foundation with a proposal for a three months long winter cruise in the Mediterranean with a focus on marine biology. The main objective was to trace and discover the eggs and early larval stages of the freshwater eel as a way of furthering insights into the early life history of the eel.[1]

Schmidt referred to his own previous expeditions in the North Atlantic Ocean, where the larval state had been tracked on the entire stretch of water from Scotland to Spain. The key to find the eel had been to sample in deep waters and in the right season of the year. Schmidt deduced how that in spring-time, there were plenty of eel larvae, while in the fall they seemed to begin metamorphosis towards the glass eel stage. Therefore, winter expeditions should be successful in terms of finding even smaller Leptocephalus, and even eel eggs. However, since the weather in winter was too rough for this type of expedition in the Atlantic Ocean, the Mediterranean was the ideal alternative. Schmidt then refers to how the Mediterranean is a classic site for eel investigations, since it was there that Italian scientists Calandruccio and Grassi first found the larvae of the eel in the 1890s.

The promise of success was backed by reference to technology, as Schmidt argued that the superior modern fishing gear developed during the years of studying the North Atlantic would be a key element in unlocking the deep sea secrets of the eel. In addition, there was reason to believe that the weather and the distances from the shores to the deep waters would be beneficial for such an undertaking. In terms of distance, it is worth recalling how the geographical range of the *Thor* was limited by the fact that she needed coaling every 12 days.

Then, the urgency of getting there first was underlined. Schmidt refreshed the memory of the board of directors that it was the Danish advances in the North Atlantic in the preceding few years that had charted

1 The original application was not available, but it had been reprinted by Ove Paulsen, who will have had access to the original: Paulsen O. Middelhavstogterne med 'Thor'. In: Winge Ø, Vedel Taaning Å, eds. *Naturforskeren Johannes Schmidt. Hans Liv og Ekspeditioner.* Copenhagen: Gyldendalske Boghandel Nordisk Forlag; 1947: 45–47.

the course for hitting the eel larvae jackpot in a near future. Whoever got to the Mediterranean first and with the right gear and know-how, would reap all the glory. If the Danes were not first, then much costly and arduous work would have been virtually fruitless. In fact the 'Danishness' of the proposed operation was paraphrased throughout the application.

Schmidt was aware that he might not achieve his goal in terms of finding tiny eel larvae in the Mediterranean, so in the application he reassured the Board of Directors that the eel investigations would only form part of the full investigations to be carried out. Well aware, of how the North Atlantic Oceans had given much more insight than what was planned in advance, Schmidt banked on collecting a similar wealth of new information about marine species as well as hydrographical conditions in the Mediterranean. Even if the eel investigations fail, the expedition as such cannot fail, Schmidt stated.[2]

In the end, the application was successful. From the Carlsberg Foundation alone, Schmidt successfully raised a total of 20,300 Danish Kroner for his Mediterranean expedition.[3] In addition Schmidt himself as well as his family chipped in with another 'some thousands of Kroner'. Having secured the fiscal needs for the trip in the winter of 1907–1908, Schmidt then needed to ensure that he could borrow the *Thor* in the upcoming winter for his Mediterranean project. In April of 1908, this issue was handled by the Kommissionen for Havundersøgelser, and it appears to have been a fairly smooth decision to grant Schmidt the right to borrow *Thor* for a period of three months following the normal summer sailing season, in order for him to commence his trip to warmer waters in the winter months of 1908–1909. The commission emphasized that for the finances of the commission, there should be no additional costs associated with the trip, and the additional insurance premium would have to be covered by Johannes Schmidt's project funding.[4]

As the previous chapter revealed the future prospects of the commission looked dire in the spring of 1908, so it must have been of tremendous importance, both symbolically and in terms of the actual progress for Danish marine science, that Schmidt was able to carry on the scientific torch, which had been burning so bright during the first five years of the International Council's existence. During the following months, formalities with the Ministry of Agriculture, who was the actual owner of the *Thor* were taken care of, and Schmidt was then ready to pursue, what he seems to have thought would be the last piece of the mosaic portraying the life history of the eel.

2 Paulsen O. Middelhavstogterne med 'Thor'. In: Winge Ø, Vedel Taaning Å, eds. *Naturforskeren Johannes Schmidt. Hans Liv og Ekspeditioner*. Copenhagen: Gyldendalske Boghandel Nordisk Forlag; 1947:47.

3 Ove Paulsen remembers it as a total of 16,300 Kroner, but according to the Carlsberg foundation, the figure is 20,300. Spang-Hanssen E. *Carlsbergfondets Understøttelser 1876–1936*. Copenhagen: Bianco Lunos Bogtrykkeri A/S; 1937. p. 136. Paulsen O. Middelhavstogterne med 'Thor'. In: Winge Ø, Vedel Taaning Å, eds. *Naturforskeren Johannes Schmidt. Hans Liv og Ekspeditioner*. Copenhagen: Gyldendalske Boghandel Nordisk Forlag; 1947:47.

4 RA. Komm. Danmarks Fiskeri- og Havundersøgelser: Forhandlingsprotokoller (1900–1960) N2: 41st meeting, 22 April 1908.

Stormy welcome to the Mediterranean winter

Compared to the previous cruises to the North Atlantic, Schmidt and his team of researchers were able to arrive at the scene of investigations in style and comfort. The *Thor* left northern waters in the late autumn of 1908, and by December she was already in the port of Messina in Sicily. Schmidt and his fellow travellers, medical doctor, Lerche and botanist Ove Paulsen arrived by train. The presence of Ove Paulsen was made possible through an additional grant of 1,000 Danish Kroner from the Carlsberg Foundation, and Paulsen was given the special task of conducing zooplankton studies.[5] Probably in close connection with the fact that Schmidt's family-in-law were co-sponsors of the trip, his wife Ingeborg and his mother-in-law, Mrs. Johanne Kühle were also part of the overland journey. The two ladies however, continued to nearby Taormina, where they could enjoy the touristic sceneries between the Mediterranean and Mount Etna.[6]

In sharp contrast to these pleasant winter quarters, Schmidt and his team now embarked on a winter cruise. According to Ove Paulsen, the first expedition in the Mediterranean was more troublesome than anything they had previously experienced around Iceland or in the Bay of Biscay. The *Thor* first steamed east to the Ionian Sea, but continuously they had to seek shelter from winter storms with rough seas in places such as the Italian Gulf of Taranto, while the entire of Christmas week was spent anchoring in the Corinthian Gulf in Greece. In this instance the waiting time was spent sightseeing in Athens and making diplomatic calls to the royal palace of King George I (1845–1913) as well as on board the Danish naval cruiseship, *Heimdal*, which was on an intelligence gathering training cruise in the Mediterranean. Danish connections were plenty in Greece, as the Greek king George I had been born Prince Vilhelm of Denmark, son of the Danish king Christian IX (1818–1906). King George was then also the older brother of Prince Valdemar (1858–1939), a Danish naval officer whose political support Schmidt would later profit from. According to Ove Paulsen, Schmidt was bored with the social calls, and impatient with regards to continuing his expedition.[7] Nonetheless, Schmidt surely realized the importance of utilizing his royal credentials.

On 28 December 1908, while Schmidt was making social calls and admiring Greek antiquity news arrived of a massive earthquake centred on Messina, which had been followed by a devastating tsunami.

5 CFD. 1908–09. no. 17. Application from J. Schmidt, 20 October 1908.

6 Paulsen O. Middelhavstogterne med 'Thor'. In: Winge Ø, Vedel Taaning Å, eds. *Naturforskeren Johannes Schmidt. Hans Liv og Ekspeditioner*. Copenhagen: Gyldendalske Boghandel Nordisk Forlag; 1947:47.

7 Paulsen O. Middelhavstogterne med 'Thor'. In: Winge Ø, Vedel Taaning Å, eds. *Naturforskeren Johannes Schmidt. Hans Liv og Ekspeditioner*. Copenhagen: Gyldendalske Boghandel Nordisk Forlag; 1947:48.

Initially, Schmidt worried about the fate of his family at Taormina, but by the 1 January 1909 he had learned that they were safe and well. Nonetheless, the *Thor* as well as the *Heimdal* hurried to the rescue of not just the Danish ladies vacationing in Taormina, but also substantial numbers of local inhabitants. It was Ove Paulsen's impression that the Danish vessels were of no real use, upon arrival to the scene of the earthquake at Messina, which saddened them all. At least, the only active help mentioned by Paulsen, is the rescue of Schmidt's relatives, who were ferried to Naples on board the *Thor*.

No doubt, some stories become more dramatic when they are told at some distance from the events unfolding, but there is no reason to believe that the decision to join the rescue operations at Messina was in any way exaggerating the severity of what had taken place. In fact, even today the double catastrophe of the 1908 Messina earthquake and tsunami is remembered as one of the most devastating natural disasters in European history. The earthquake hit 7.5 on the Richther scale, while the tsunami brought along waves 13 metres high, and the combined impacts killed more than 80,000 people.[8] A few years later, Schmidt and the other members of the *Thor's* crew received a medal of honour from the Italian state for their aid in relation to the earthquake, so it is likely that they helped in other ways than just shipping out Schmidt's family.[9]

4.1 Silver Medal commemorating the 1908 earthquake and tsunami in Southern Italy
Schmidt received an offical Italian silver medal in recognition of the Thor's effort to rescue people from Sicily during the Messina earthquake 28 December 1908 (Carlsberg Archives).

In Naples, yet another week was spent away from the sea, as *Thor's* boiler now needed cleaning. Thus it was mid-January, before the pur-

8 Anonymous. Messina earthquake and tsunami of 1908. In: *Encyclopaedia britannica*. Encyclopaedia Britannica Online Academic Edition ed. Encyclopædia Britannica Inc.; 2013; accessed 21 November. http://www.britannica.com/EBchecked/topic/1483443/Messina-earthquake-and-tsunami-of-1908.

9 Paulsen O. Middelhavstogterne med 'Thor'. In: Winge Ø, Vedel Taaning Å, eds. *Naturforskeren Johannes Schmidt. Hans Liv og Ekspeditioner*. Copenhagen: Gyldendalske Boghandel Nordisk Forlag; 1947:48.

pose of the expedition could be pursued again. A few good working days now followed, as the expedition moved North through the Tyrrhenian Sea, but as they entered the Ligurian Sea and approached the French Riviera, stormy weather once again forced the *Thor* to seek refuge, this time in Menton. As in Greece, the involuntary confinement to port was spent wisely, socializing. This time, Schmidt and his fellow scientists paid a visit to the grand new museum of marine science in Monaco, where they met with its director, M. Ricard.[10] The Prince of Monaco, Albert I (1848–1922) was arguably the most important patron of marine science in the Mediterranean region in the decades around the turn of the 20th century. Prince Albert had sponsored, as well as participated in as many as 28 oceanographic surveys in the Mediterranean, the Atlantic and as far away as Svalbard in the Arctic, and his museum contained an aquarium, a library and a natural history collection.[12] This combination of knowledge resources was to inspire Schmidt and his colleagues later in their careers.

Upon leaving the French Riviera, the stormy weather continued. Initially the plan was to head for the Balearic Islands, but rough weather forced the crew to change course, and instead they now tried to escape the strong westerlies by heading south along the east coast of Sardinia before ending up near Tunisia, where they had to anchor on the leeside of the small rock island of Zalita. Now the coals were nearly spent, and as soon as the weather improved, the Danish expedition headed west along the African shoreline to Algier, where they could buy more coal. Again the winter weather confined the *Thor* to port for an entire week. Apart from diplomatic calls, Schmidt spent the days in Algier trying to receive information, on where in the Western Mediterranean; it would be possible to conduct oceanographic research in relatively calm weather. In fact it was not until the 16 February that the *Thor* was able to steam out of the port of Algier, destined for the Alboran Sea between Southern Spain and Morocco, and not least the Strait of Gibraltar. This was the last stretch in the Mediterranean before heading home.[13]

Schmidt's big plan for charting the early life stages of the freshwater eel had been swept away with the torrents and storms of the Mediterranean winter, and during the little time, when they had indeed been able to fish for the fry of the eel, the results had turned out negative. While it may have prompted Schmidt to speculate as to whether or not the eel spawned in the Mediterranean at all in early 1909, it was too early to draw any conclusions based on the absence of tiny eel larvae at the few

10 Paulsen O. Middelhavstogterne med 'Thor'. In: Winge Ø, Vedel Taaning Å, eds. *Naturforskeren Johannes Schmidt. Hans Liv og Ekspeditioner*. Copenhagen: Gyldendalske Boghandel Nordisk Forlag; 1947:48.

11 Carpine-Lancre J. Oceanographic Sovereigns: Prince Albert I of Monaco and King Carlos I of Portugal. In: Deacon M, Rice T, Summerhayes C, eds. *Understanding the Oceans. A century of exploration*. London and New York: UCL Press; 2001: 57–59.

12 Paulsen O. Middelhavstogterne med 'Thor'. In: Winge Ø, Vedel Taaning Å, eds. *Naturforskeren Johannes Schmidt. Hans Liv og Ekspeditioner*. Copenhagen: Gyldendalske Boghandel Nordisk Forlag; 1947:48–49.

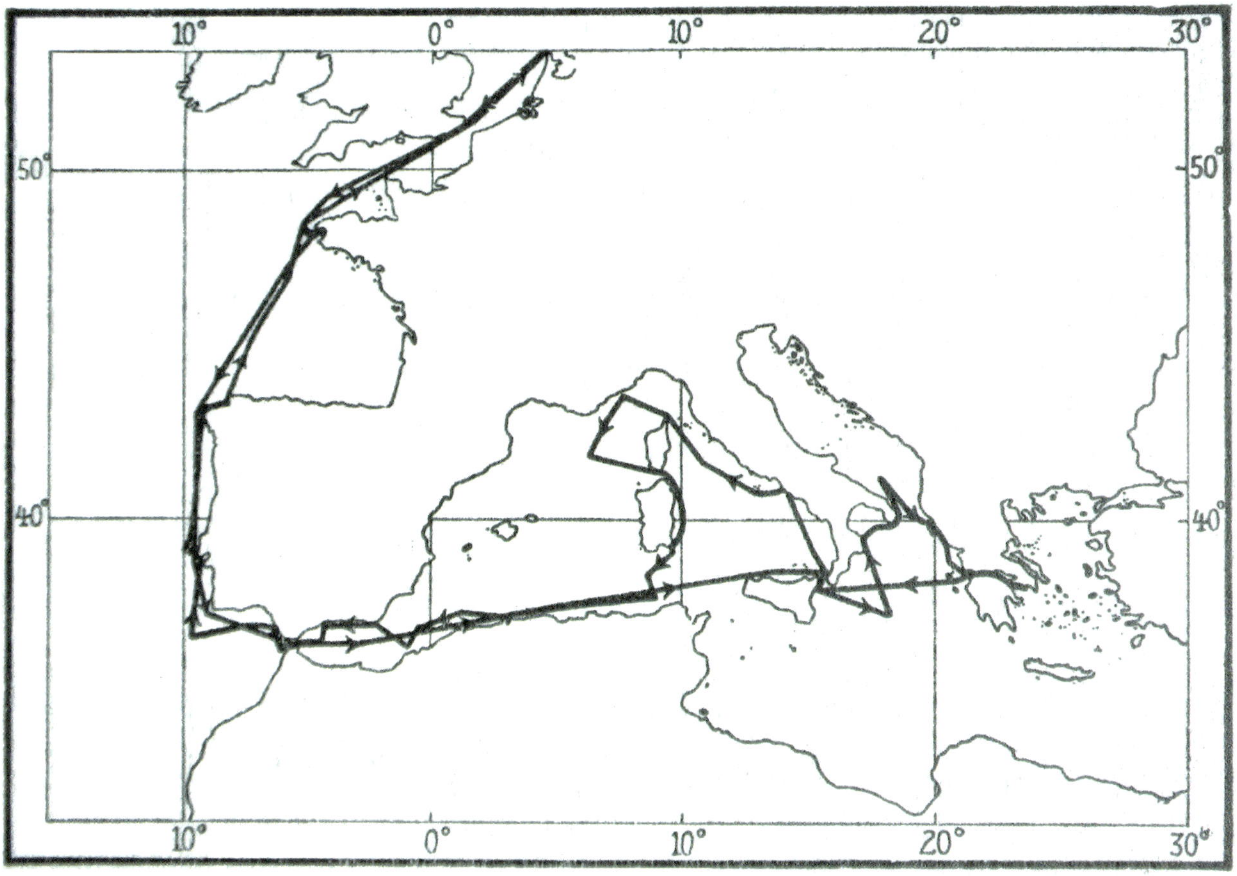

4.2 the first Mediterranean expedition in 1908–1909

The route of the first Mediterranean expedition in 1908–1909 (*Winge Ø. & Vedel Taaning Å. 1947*).

sites tested by a crew with no prior first-hand experience with fishing the Mediterranean. In short, Schmidt had to return, results to show back home were in short supply.

Time was running out, as they approached the last legs of the Mediterranean adventure, before heading home along the Atlantic coastline. The financial situation surrounding the future of the Danish commission for marine science was still not entirely settled, and all that Schmidt knew from the commission with regards to the future of the *Thor*, was that she was to be taken out of service, as soon as Schmidt returned from the Mediterranean.[13] Thus, this could have been Schmidt's last big cruise with the *Thor*, the vessel he had helped to make famous. But just as hope was running out, two good weeks with calmer weather changed the luck of those on board the *Thor*.

13 RA. *Komm. Danmarks Fiskeri- og Havundersøgelser*: Forhandlingsprotokoller (1900–1960) N2: 42nd meeting, 23 July 1908.

Probing a mare incognitum

In calm waters in and outside of the Strait of Gibraltar, the nifty devices for collecting seawater samples came to the rescue in terms of reaping new results. As there was no specialist hydrographer on board, Schmidt himself was in charge of investigating each station, and he worked incessantly, gradually building an image of how the temperature shifted at various depths in the Mediterranean, and how this differed substantially from the Atlantic Ocean. Immediately upon Schmidt's return to Copenhagen, hydrographer J. N. Nielsen was hired to conduct a full analysis of the seawater samples and the records obtained by Schmidt.[14]

Analysing the Ionian Sea, the Tyrrhenian Sea and the Alboran Sea separately, Schmidt noticed how the temperature at great depths such as 1500–2000 meters below sea level, was much higher (13–14 degrees Celsius) in the Tyrrhenian Sea and the Ionian Sea than what previous expeditions had found to the case in the open Atlantic Ocean (2–3 degrees Celsius). Schmidt interpreted this as attributable to the fact that for instance the Tyrrhenian Sea largely was a body of water of its own, where the change of water occurred only at the relatively shallow thresholds around Sicily, Sardinia and Corsica, which separate the Tyrrhenian Sea from the Western Mediterranean Basin.[15]

Furthermore, the surface water temperature in the Alboran Sea between Spain and North Africa was found to be highly dependent on the influx of water from the adjacent parts of the Atlantic Ocean into the Mediterranean. According to Ove Paulsen, Schmidt developed a bit of a reputation for sorcery on board the *Thor*; after he realized this sea surface connection, he began predicting the temperature at a new station, before the water sampler had been hauled on board.[16]

Following Nielsen's analysis, it was now possible to propose a full theory of how the hydrographical features of the Mediterranean, were highly dependent on the Atlantic Ocean. The Strait of Gibraltar forms a barrier, not just because of the narrowness of the strait, but also because of the limited depth in the strait itself, where the maximum depth is c. 400 meters as opposed to up more than 1500 metres immediately inside and outside of the strait. In the surface layers, the water has the same temperature and salinity on both sides of the Strait of Gibraltar, indicating that the water from the Atlantic flows far inside the narrow sections of the Strait without mixing with the waters of the Mediterranean. Having passed the narrowest points the inflowing water spreads

14 CFD. 1909–10. no. 227d. J. Schmidt: Cruise report from Den Danske oceanografiske Expedition til Middelhavet 1908–1909, p. 8–10.

15 CFD. 1909–10. no. 227d. J. Schmidt: Cruise report from Den Danske oceanografiske Expedition til Middelhavet 1908–1909, p. 7–10; Bruun AF. Middelhavstogternes resultater. In: Winge Ø, Vedel Taaning Å, eds. *Naturforskeren Johannes Schmidt. Hans Liv og Ekspeditioner*. Copenhagen: Gyldendalske Boghandel Nordisk Forlag; 1947:52–56

16 Paulsen O. Middelhavstogterne med 'Thor'. In: Winge Ø, Vedel Taaning Å, eds. *Naturforskeren Johannes Schmidt. Hans Liv og Ekspeditioner*. Copenhagen: Gyldendalske Boghandel Nordisk Forlag; 1947:49.

out into the sides of the Alboran Sea, and the density of the inflowing water is diminished. Underneath the inflowing layers of water, along the bottom of the Strait of Gibraltar, there is a current of heavier Mediterranean water flowing in the opposite direction from East to West. The water coming out from the Mediterranean is heavier than Atlantic water at the same depth, due to a higher salinity in the Mediterranean. Therefore, once the Mediterranean water has passed the strait, it descends along the bottom of the continental slope, until this water meets Atlantic water with the same density at a depth of c. 1000 meters. Then the Mediterranean waters continue a horizontal drift into the eastern North Atlantic Ocean. The presence of such an undercurrent, explains why relatively high temperatures were found at these intermediate depths on the Atlantic side of the Strait. The results of the *Thor* thus substantiate a claim for the existence of an undercurrent of Mediterranean origin, which J. N. Nielsen had hypothesized in a 1907 publication from the Kommissionen for Havundersøgelser, while the idea as such first had arisen as a consequence of findings from the British Challenger Expedition in 1877.[17]

The discovery of a Mediterranean undercurrent stretching out into the Atlantic Ocean, and the discovery of the relatively warm intermediate layers in the deep basins of the Mediterranean Sea all justified the time and money dedicated to Schmidt's pioneering first trip to the Mediterranean. Furthermore, the salinity and nutrient levels were also analysed from more or less the same station as the ones facilitating the temperature analyses, and similarly, important new discoveries were made, indicating intermediate layers in the water column, where the salinity and the nutrient levels were fairly high. The overall impression of Mediterranean hydrography though, was that of a sea, with much lower nitrogen levels than what had been found for the Atlantic Ocean.[18]

As the Mediterranean expedition was coming to an end in February of 1909, Schmidt and his colleagues could go ashore in Spain and join the festivities of the on-going carnival, knowing that, while no solid conclusions could be made on the eel's life history, they could be content with their hydrographical results. In addition the storage rooms of the *Thor* were packed with important collections of marine species waiting to be analysed on calm desk-tops back in Copenhagen in years to come. Yet upon the immediate arrival back in Copenhagen there was little to show except for the hydrographic results.

17 CFD. 1909–10. no. 227d. J. Schmidt: Cruise report from Den Danske oceanografiske Expedition til Middelhavet 1908–1909, p. 10–11.

18 CFD. 1909–10. no. 227d. J. Schmidt: Cruise report from Den Danske oceanografiske Expedition til Middelhavet 1908–1909, p. 10–11. Bruun AF. Middelhavstogternes resultater. In: Winge Ø, Vedel Taaning Å, eds. *Naturforskeren Johannes Schmidt. Hans Liv og Ekspeditioner*. Copenhagen: Gyldendalske Boghandel Nordisk Forlag; 1947:52–56.

For almost the first time in his career, it seemed as if Schmidt had failed to return from a trip with the results that he expected, and in this instance the results he had promised to those who gave him financial support including the Danish commission for marine science as well as the Carlsberg Foundation, and his own family. On top of this, Schmidt had spent more money on the cruise than in the initial plan. According to the expedition plan, he was scheduled to be back on the 12 February, but the *S/S Thor* was not in Copenhagen again before the 18 March. Two calamities were to blame. First of all, he assumed responsibility him self for having stayed out longer than agreed upon, but the last month happened to the most successful. Secondly, the return trip was delayed with 10 days, as the *Thor* collided with another vessel in the Kiel Canal on the way back, and some of the blame – and costs landed on the deck of the *Thor*. All in all, Schmidt's expenses had exceeded the budget with 6,834.74 Kroner, as he wrote in an account to the Board of Directors of the Carlsberg Foundation. This was to be of no burden for the foundation however, as Schmidt's mother-in-law Mrs Johanne Kühle now kindly offered to cover this amount so as to lift this burden caused by bad weather.[19] The Board of Directors reviewed this offer from Mrs Kühle at the next board meeting, where it was decided that Schmidt should approach his mother-in-law once more and inquire as to whether or not this was something she would be really happy to do, or whether she felt it more as an obligation.[20] In the end, this sizeable amount never entered the list of expenses at the foundation.[21] Thus, there is good reason to believe that Mrs. Kühle ended up paying for her son-in-law's delay and collision.

Schmidt could be hopeful of his own future as a marine scientist, as well as the future of Danish marine science. Nonetheless, Schmidt does not seem to have placed any emphasis on exploiting *Thor* for his own research during the summer and autumn season of 1909, when *Thor* was allocated entirely to serve in the waters around Denmark.[22] His time would come, and he had had enough of the erratic winter weather in the Mediterranean. Meanwhile he landed another job alongside the one at the commission.

19 RA. *Schmidt, Ernst Johannes, biolog*: Materiale vedr. ekspeditioner og rejser (1876–1933) 1: Letter from J. Schmidt to the Board of Directors of the Carlsberg Foundation, 6 April 1909.

20 CFD. *Carlsbergfondets Direktions Protokol*, vol. 1–3, 7 April 1909.

21 Spang-Hanssen E. *Carlsbergfondets Understøttelser 1876–1936*. Copenhagen: Bianco Lunos Bogtrykkeri A/S; 1937.

22 RA. *Komm. Danmarks Fiskeri- og Havundersøgelser*: Forhandlingsprotokoller (1900–1960) N2: 49th–60th meetings 1909.

National fame and well-paid jobs

As Schmidt returned to Denmark from the gruelling winter expedition to the Mediterranean, he reacquainted himself with the unruly politics of Danish marine science. However, as discussed in the previous chapter, Schmidt could now take a seat of his own as a full member of the Kommissionen for Havundersøgelser.

Then, in late August of 1909, the long-time head of the Physiological Department of the Carlsberg Laboratory, Emil Chr. Hansen passed away following a stroke.[23] One month later Schmidt was appointed the new Director of the Physiological Department of the Carlsberg Laboratory. The rising star of international marine science was becoming the head of a research enterprise devoted to applied science for the promotion of beer brewing. Seemingly, this was a U-turn in Schmidt's career, and his research at the laboratory is dealt with separately in the following chapter. However, his appointment immediately caught the eye of the public.

In October, Schmidt's Carlsberg job was commented upon in leading Danish newspapers. Schmidt is portrayed as a 'natural born scientist', and Eugen Warming was cited for having labelled Schmidt as the most promising of all botanists. Nonetheless, Schmidt had succumbed to his interests in marine science and pursued the breeding of the eel. The article then touches upon his other claim to fame: the fact that he acted heroically during the earthquake and tsunami, which had ravaged the coastline of Sicily and Southern Italy the previous winter. Schmidt had immediately come to the rescue of his wife and his mother-in-law, Mrs. van der Aa Kühle. The article then explains how Schmidt was headed for the *Institut Pasteur* in Paris in the coming winter to frequent himself with the latest in yeast physiology. Then in spring and summer of 1910, he would head out on his very last oceanographic expedition, travelling to Greenland with the *Thor*.[24] Little could they know that Schmidt had a voracious appetite for marine science, which would extend beyond 1 October 1910, when he was scheduled to embark on a terrestrial voyage at the helm of the Carlsberg Laboratory.

At least from 1909, Schmidt was aided in his research by ships of opportunity in the Mediterranean. The Danish East Asiatic Company (EAC) operated a number of shipping lines between Europe and Asia, and the various steamships traversed the Mediterranean Sea and the Suez Canal en route to their destinations. Schmidt had collaborated

23 CA. *Bladudklip, Direktøren*, 1 June 1909–13 September 1910, *Dannebrog*, 28 August 1909.

24 CA. *Bladudklip, Direktøren*, 1 June 1909–13 September 1910, 'Carlsbergs ny Gærfysiolog', *Politiken*, 8 October 1909. 'Carlsberg-Laboratoriet', *Børsen*, 10 October 1910.

with the East Asiatic Company during his Siamese expedition a decade earlier, and now the same company helped him again. Following Schmidt's instructions, the ship captains took care to collect water samples in agreement with the newly appointed Executive Director of the East Asiatic Company, Chr. Schmiegelow.[25] Captain H. Jakobsen of the *S/S Pangan,* for instance, took 40 water samples in the Mediterranean in October 1909 on his way to Bangkok.[26] The samples were then packed and shipped to Copenhagen, as the ship called in Port Said in Egypt.[27]

Simultaneously, Schmidt was planning his own trip, a return to the Mediterranean, but this time in the much more pleasant summer season, and with a schedule for Schmidt to return to Copenhagen in time for taking up his new position at the Carlsberg Laboratory.

Proposing a summer cruise to the Mediterranean

In later commentary on Schmidt's Mediterranean expeditions, the second expedition is usually mentioned as a 'natural extension of the first expedition'.[28] Admittedly, it looks like this when seen from afar, but in the winter of 1909–1910, Schmidt had great difficulties convincing the Carlsberg Foundation that they should back the second trip.

In late November 1909, the Board of Directors of the Carlsberg Foundation received an application from Johannes Schmidt, in which he asked for support for a new expedition to the Mediterranean.[29] Here, they could read that Schmidt had discussed matters with the Danish marine science commission, and the commission had agreed that Schmidt could borrow the *Thor* on the same conditions as he had done once already, namely that he could borrow the vessel for free, but he had to fund the insurance and the expenses of the actual cruise from other sources.

Schmidt wished to borrow the *Thor* for the summer months of 1910, once the *Thor* had finished the tasks associated with the North Atlantic surveys of the Danish commission for marine science. For the planned 1910 expedition Schmidt highlighted two distinct reasons for why it was important to embark on such a huge undertaking once more. These served to show how Schmidt was now pursuing basic research questions of general natural history interest, rather than the just the practical work outlined by the commission.

25 Jansen, Chr. R., 'Schmiegelow, Chr.' in: Cedergreen Bech, S. ed. *Dansk Biografisk Leksikon*, Copenhagen, Gyldendal; 1983:13, 160–161.

26 RA: Dansk Biologisk Station, Johannes Schmidts embedsarkiv: Korrespondance om åleforskning (1908–1933) E28: 1908–1913: H. Jakobsen to J. Schmidt, 1 October 1909.

27 RA: Dansk Biologisk Station, Johannes Schmidts embedsarkiv: Korrespondance om åleforskning (1908–1933) E28: 1908–1913: Schmiegelow to J. Schmidt, 7 September 1909 & J. Schmidt to Schmiegelow, 15 November 1909.

28 Paulsen O. Middelhavstogterne med 'Thor'. In: Winge Ø, Vedel Taaning Å, eds. *Naturforskeren Johannes Schmidt. Hans Liv og Ekspeditioner*. Copenhagen: Gyldendalske Boghandel Nordisk Forlag; 1947:50. Paulsen O. Middelhavstogterne med 'Thor'. In: Winge Ø, Vedel Taaning Å, eds. *Naturforskeren*

29 CFD. 1909–10. no. 227d. Application from J. Schmidt, 23 November 1909.

First of all, Schmidt underlined how most Danish marine science had so far gained a wealth of information and a substantial collection of species test results as a side product of the practical fisheries investigations. However, for reasons of financial restraint it had never been possible to place these auxiliary research results in their right context. Yet, if the previous collections were complemented through a similar effort in a distinctly different ocean such as the Mediterranean Sea, then more general questions could be answered. What these general questions are, however, was not revealed to the Carlsberg Foundation. This may seem somewhat surprising for the modern scientist, but Schmidt probably assumed the Directors of the Board to have a very good knowledge about his work already. Moreover, the application was backed by a report from the first trip, and there is no doubt that the hydrographical results from the winter expedition proved to be Schmidt's linchpin in the story he told in his bid to win funding for a second Mediterranean cruise.

Secondly, Schmidt turned to the specifics of the Mediterranean, where he now argued that in this particular ocean there is such a vast difference between summer and winter that a second trip is really necessary, if they were supposed to paint a full picture of the characteristics of this ocean. In addition, the previous expedition was hampered by poor weather conditions, especially with regards the planned, but unsuccessful research into the Eastern Mediterranean and the Black Sea. In short, the first expedition should be considered a trip of reconnaissance.

Then, Schmidt proceeded to what he labelled as two specific reasons why the trip should be launched as soon as possible in the summer of 1910. The urgency of travelling to the Mediterranean before a planned Norwegian expedition got there first was one good reason. Schmidt referred to correspondence he had had with the Swedish marine scientist, Otto Petterson, who had informed Schmidt that the leading Norwegian marine scientist, Johan Hjort was planning a cruise to the Eastern North Atlantic scheduled to begin also in 1910. Hjort was going to deploy the Norwegian marine research vessel, the steamship, *Michael Sars*, for his voyage, which would be financed by the Scottish philanthropist, Sir John Murray, and like Schmidt, the Scottish-Norwegian expedition would investigate both hydrography and marine biology. The race to beat the Norwegians no doubt fuelled the insatiably ambitious Schmidt. Presenting his ambition to the Directors of the Carlsberg Foundation, Schmidt drew on the card of patriotism. Hjort's expedition was not a national enterprise such as the one Schmidt proposed; yet '...if we are not successful in launching such an expedition with a broad scientific focus, the Norwegians will reap the main part of the glory, while we can only claim the success of the preliminary sketches'.[30]

Finally, Schmidt emphasized that the availability of *Thor* in the summer period, when she was otherwise in service, was a unique chance to conduct research at a very low price compared to the situation when a ship needed to be chartered from scratch. With regards to the proposed budget, the expense levels were similar to what he applied for in connection with the first trip, so evidently the already experienced research organizer Schmidt had been less surprised by the cost of running the expedition, than by the unpredictability of the winter weather. The estimate of total costs amounted to 24,000 Kroner, where the running costs of the actual cruise made up the bulk of the expenditures. From this Schmidt declared that he could provide 10,000 Kroner himself, so the sum he was asking for from the Carlsberg Foundation was the remaining 14,000 Danish Kroner.[31]

Schmidt's willingness to pay a substantial part of the expedition out of his own pocket was a strong indication that he was really serious about this undertaking. He outlined how it was going to be 'one final expedition' solving outstanding issues. No doubt the life history of the eel was the one exceptionally outstanding issue Schmidt wished to pursue, although the eel receives hardly any mentions in the application, even though finding the eggs of the eel could be considered a basic research question, as well as having practical applications for the World's eel fisheries. It is tempting to assume that the lack of results relating to the eel from the first trip was the reason why Schmidt downplayed the quest for the eggs and larvae of the eel in the 1909 application.

Then, at the very end of the application, Schmidt somewhat arrogantly apologized that he was submitting his application more than one week after the deadline had expired for the 1909 round of applications.[32] Modesty did not weigh heavy on the self-confident character of Schmidt's personality.

Network mobilization

The Carlsberg Foundation rejected Schmidt's 1909 application for funding. It is hard to know if the Board of Directors of the Carlsberg Foundation were offended by Schmidt's late application for a second

30 CFD. 1909–10. no. 227d. Application from J. Schmidt, 23 November 1909. p. 4.
31 CFD. 1909–10. no. 227d. Application from J. Schmidt, 23 November 1909. p. 5.
32 CFD. 1909–10. no. 227d. Application from J. Schmidt, 23 November 1909. p. 6.

expedition to the Mediterranean. Three out of five members of the Board of Directors also served on the Board of Trustees of the Carlsberg Laboratory, where they had just hired him to take a lead in beer related research, quite different from the abyss of Mediterranean basins.

In any case, the letter of rejection was anything but a humbling experience for the 32-year old incoming research director. Johannes Schmidt was not going to rest his case, feeling close to solving the great riddle of the freshwater eel. Instead he now tried to convince the Board of Directors to change their mind and revoke the rejection of the expedition. As a member of the Board of Directors, Eugen Warming played a supportive role, when Schmidt chose this rather confrontational style towards the one non-governmental funding agency, which had been most helpful to him in terms of providing financial support since the first application from 1898 to go to Thailand.[33]

In a letter from 28 December 1909 Schmidt explained to Warming that he had just the previous day paid a visit to the prominent Danish chemist, S. M. Jørgensen.[34] Jørgensen was Chairman of the Board of Trustees of the Carlsberg Laboratory, and since January of 1909 he had also served as the Chairman of the Board of Directors of the Carlsberg Foundation.[35] Warming was already on Schmidt's side, and the two professors representing the humanities, might follow the direction of the natural scientists in a matter belonging to their turf. The two representatives from the humanities were the influential professor of history, Kristian Erslev and the renowned professor of indo-european languages, Vilhelm Thomasen.[36]

Hence, Jørgensen and the other natural scientist on the board, Physicist, C. Christiansen were the ones to convince. From the letter to Warming, however, it appears that Schmidt had told Jørgensen that his training in preparation of taking over the position at the laboratory would not be jeopardized by the new project, and he had also let him know that he was already going on a research cruise on behalf of the Kommissionen for Havundersøgelser in spring, which had been arranged long in advance. Jørgensen allegedly replied that '…it was very unlikely that the board would change their position, but of course there was the possibility Warming could push through a different outcome'.[37] This is exactly what Schmidt was banking on, as he proposed to Warming that the physicist, Christiansen, might reconsider if it was emphasized to him, just how much physics there was in the proposed project, with regards to the proposed hydrographical measurements.[38]

33 CFD. 1898–99. No. 47. Application from J. Schmidt, 24 December 1898.
34 CFD. 1909–10. no. 227d. J. Schmidt to E. Warming, 28 December 1909.
35 Pedersen J. *The Carlsberg Foundation*. Copenhagen: Bianco Luno; 1956. 91–92.
36 Glamann K. *Carlsbergfondet*. Copenhagen: Rhodos; 1976, p. 179–181.
37 CFD. 1909–10. no. 227d. J. Schmidt to E. Warming, 28 December 1909.
38 CFD. 1909–10. no. 227d. J. Schmidt to E. Warming, 28 December 1909.

In addition, Warming now called upon his peers to have them write letters of support for Schmidt's research proposal. Shortly after New Year, the prominent professor of zoology and member of the Royal Academy of Science and Letters, Hector Jungersen expressed his support for Schmidt in a letter to Warming.[39] According to Jungersen, Schmidt is an unusually talented young scientist, and it was mainly Schmidt's merit that Denmark had obtained such a prominent position in international marine science. In fact, Schmidt was the only one, who could be trusted to carry out an expedition of this size and scientific scope. Jungersen ends his letter with praise for Schmidt's results with regards to charting the life history of the eel, and Jungersen expresses his wish that the final solution to the question of the eel would be solved by a Danish scientist.[40]

Physicist and leading international hydrographer, Martin Knudsen was next in the line of support. As a fellow member of the Kommissionen for Havundersøgelser, Knudsen had a vested interest in Schmidt's project becoming successful. Knudsen likewise emphasized that the proposed project were almost a guaranteed success with Schmidt at the helm. The hydrographical investigations would be at least as ground-breaking as they had been on the first expedition in the previous winter. Thereto, Knudsen added that there was probably only this one chance for such an expedition to borrow the *Thor* in the summer season. Under different circumstances, when a ship would have to be chartered, an expedition this size would cost at least 100,000 Kroner. Thus, Schmidt's proposal was a real bargain at 24,000 Kroner.[41]

Schmidt and Warming continued to coordinate their effort in January, when correspondence shows that Warming circulated Schmidt's cruise report from the first expedition to the various members of the Board of Directors of the Carlsberg Foundation,[42] and on the 30 January Schmidt personally submitted some of his writings to the historian, Kristian Erslev.[43]

The plan succeeded and the Carlsberg Foundation decided to grant the 14,000 Kroner to Schmidt's second Mediterranean expedition. Thus, already on the 15 February of 1910 Schmidt was ready to officially approach the commission with a wish to borrow *Thor* again. Naturally, the commission was very supportive of launching yet another oceanographic endeavor.[44] The conditions for borrowing the *Thor* were the same as concerning the first trip only the date was changed in the wording of the contract.[45]

39 Wolff, T. ,Jungersen, Hector F. E. in: Cedergreen Bech, S. ed. *Dansk Biografisk Leksikon*, Copenhagen, Gyldendal; 1981:7, 487–488.

40 CFD. 1909–10. no. 227d. H. Jungersen to Board of Directors, 3 January 1910.

41 CFD. 1909–10. no. 227d. M. Knudsen to Board of Directors, 31 December 1909.

42 CFD. 1909–10. no. 227d. J. Schmidt to E. Warming, 4 & E. Warming to Board of Directors, 4 January 1910.

43 CFD. 1909–10. no. 227d. J. Schmidt to K. Erslev, 30 January 1910.

44 RA. *Komm. Danmarks Fiskeri- og Havundersøgelser*: Forhandlingsprotokoller (1900–1960) N2: 61st meeting, 15 February 1910.

45 RA: Dansk Biologisk Station, Johannes Schmidts embedsarkiv: Korrespondance om åleforskning (1908–1933) E28: 1908–1913: Agreement between J. Schmidt and Kommissionen for Havundersøgelser.

Return to the Mediterranean in the summer of 1910

The 2nd expedition to the Mediterranean commenced on the 10 June 1910, as the *Thor* steamed out of the port of Fallmouth near Land's End in Cornwall, UK. The ship's captain Hansen had arrived straight from the Faroe Islands, where *Thor* had been in service of the Kommissionen for Havundersøgelser. Schmidt was on board again, and the natural leader of the expedition. This time he was accompanied by his fellow member of the commission and then curator at the Botanisk Museum, Carl H. Ostenfeld, who was responsible for plankton research. Hydrographer and assistant at the meteorological institute in Copenhagen, J. N. Nielsen was able to continue his analyses of Mediterranean hydrography based on the first expedition, and this time he had the chance to take his own samples and continue charting the *mare incognitum*. Also in the tally of the group of scientists tallied was the chemist, Sven Palitzsch.[46] The 28-year old Palitzsch was on loan from the Chemical Department of the Carlsberg Laboratory, where from 1905–1919 he was an assistant of the director of the department, S. P. L. Sørensen. Sørensen and Palitzsch became known as leading specialists on examining hydrogen ion concentrations in seawater.[47] On-board the *Thor*, this was exactly what he was tasked to do, as Palitzsch became responsible for examining the seawater samples for concentrations of both oxygen and hydrogen levels.[48]

In this way, we can safely deduce that both sections of the Carlsberg Laboratory had a stake in the Mediterranean summer cruise. Schmidt was the designated leader of the Physiological Department, while the head of the Chemical Department, S. P. L. Sørensen supported a fundamental basic research issue through Palitzsch' work, which was not exactly related to the advancement of beer brewing processes. Since Sørensen was an old student of the chairman of the board, S. M. Jørgensen, it may be that Palitzsch' presence was a fait accompli between the interests of Jørgensen and Warming, although this is speculative.

Cruising the Mediterranean in summer time proved to be much more pleasant than in the winter, with many more opportunities to probe the *mare incognitum*. This time all major seas in the Mediterranean were charted from Gibraltar in the west via Algier and across the Balaerean Sea to Genova. Then the route went south, past the eastern shores of Corsica and Sardinia destined for Tunis. Afterwards Malta was passed en route to the Gulf of Sidra off the coastline of Syria. From here, they

46 Paulsen O. Middelhavstogterne med 'Thor'. In: Winge Ø, Vedel Taaning Å, eds. *Naturforskeren Johannes Schmidt. Hans Liv og Ekspeditioner*. Copenhagen: Gyldendalske Boghandel Nordisk Forlag; 1947:51. Paulsen O. Middelhavstogterne med 'Thor'. In: Winge Ø, Vedel Taaning Å, eds. *Naturforskeren*

47 Holter H. An outline of the history of the two departments. In: Holter H, Max Møller K, eds. *The Carlsberg Laboratory 1876–1976*. Copenhagen: Rhodos; 1976: 33. Mayor AG. Detecting ocean currents by observing their hydrogen-ion concentration. *Proc Am Philos Soc*. 1919;58(2):150–160.

48 Schmidt, J. (ed), *Report on the Danish oceanographical expeditions 1908—1910 to the Mediterranean and adjacent seas — vol I: Introduction, Hydrography, Deposits of the seabottom*. Copenhagen: 1912.

49 Paulsen O. Middelhavstogterne med 'Thor'. In: Winge Ø, Vedel Taaning Å, eds. *Naturforskeren Johannes Schmidt. Hans Liv og Ekspeditioner*. Copenhagen: Gyldendalske Boghandel Nordisk Forlag; 1947:51.

sailed northeast and probed the waters of the Cretian Sea and the Aegean Sea. The Bosporus was traversed, and then some of the first modern investigations of the Black Sea hydrography were carried out, before the expedition headed back. The return trip went through the Corinth Canal in Greece, through the Ionian Sea to Messina, whereupon Naples was revisited. Finally, the *Thor* steamed south of Sardinia into the Balaerean Sea and docked in Barcelona, before the last stretches of investigations were carried out, as the expedition went back to Gibraltar.[49]

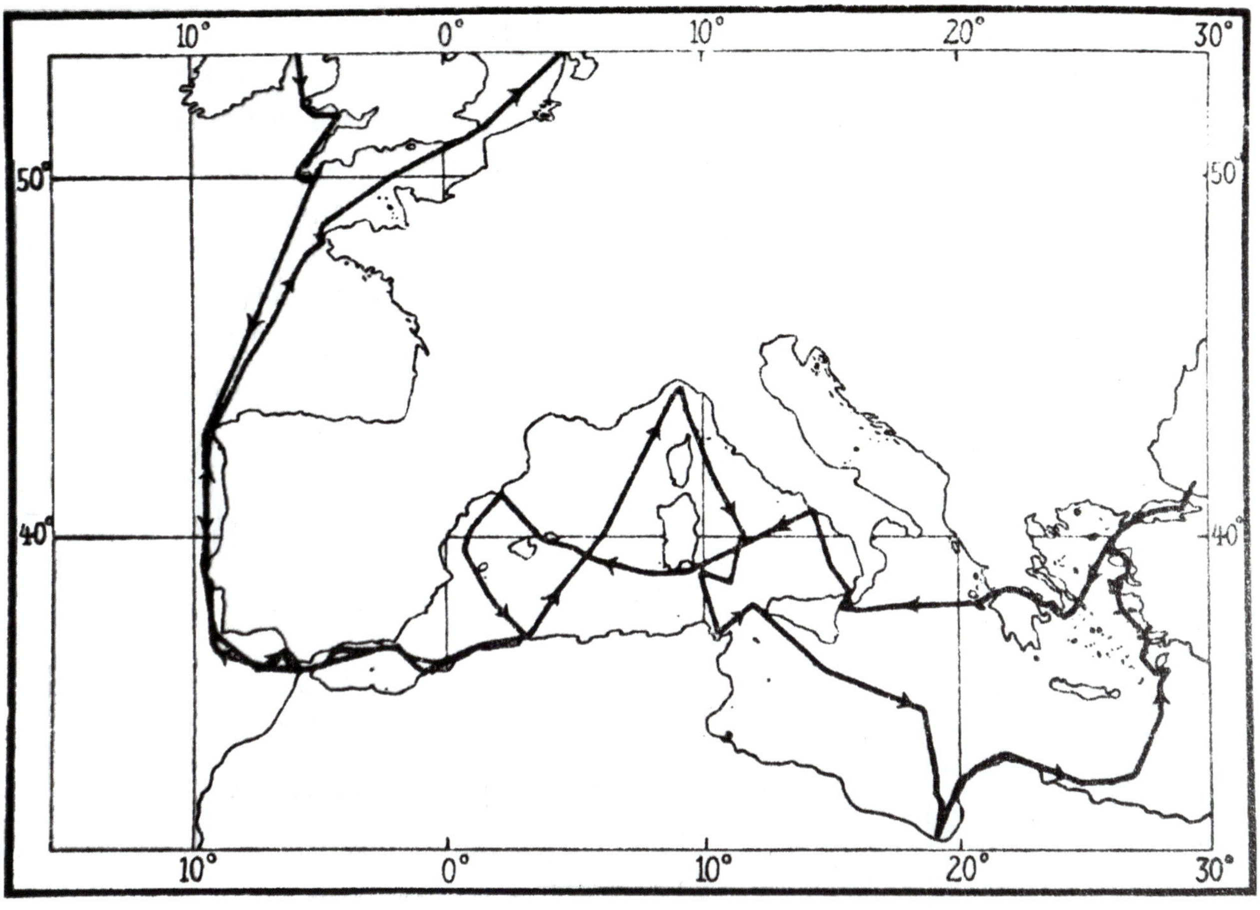

4.3 The second Mediterranean expedition with S/S *Thor*, 1910
The route of the second Mediterranean expedition in the summer of 1910 (*Winge Ø, Vedel Taaning Å., 1947*).

As the weather was favourable, the crew could fish in as many as 185 stations along the route. On 323 occasions they succeeded in fishing with the pelagic trawls, mainly for fish larvae, and 268 times they used the silk like stramin net for catching fish eggs and smaller organisms deployed at various depths in the water column. In addition no less than 1421 water samples were hauled on board for analysis, as well as a large number of bottom samples.[50] Once again, the many clever devices invented by C. G. Joh. Petersen were key elements in uncovering the secrets of the deep seas. Roughly ten times as much research was carried out during the pleasant summer cruise compared with the stormy winter cruise. So, did Schmidt succeed with what he had promised would be his last oceanographic expedition?

50 Bruun AF. Middelhavstogternes resultater. In: Winge Ø, Vedel Taaning Å, eds. *Naturforskeren Johannes Schmidt. Hans Liv og Ekspeditioner.* Copenhagen: Gyldendalske Boghandel Nordisk Forlag; 1947:52.

51 Bruun AF. Middelhavstogternes resultater. In: Winge Ø, Vedel Taaning Å, eds. *Naturforskeren Johannes Schmidt. Hans Liv og Ekspeditioner.* Copenhagen: Gyldendalske Boghandel Nordisk Forlag; 1947:52–56

52 CFD.1909–10.no.227d. Letter from Johannes Schmidt to Eugen Warming, 28 December 1909.

Scientific results from the Mediterranean

The main results from the two expeditions to the Mediterranean fell within the area of hydrography.[51] In order to secure funding for the 2nd trip to the Mediterranean with *Thor*, it had been crucial to persuade the Carlsberg Foundation board member, professor Christiansen that physical marine science was highlighted in the outline of the expedition.[52]

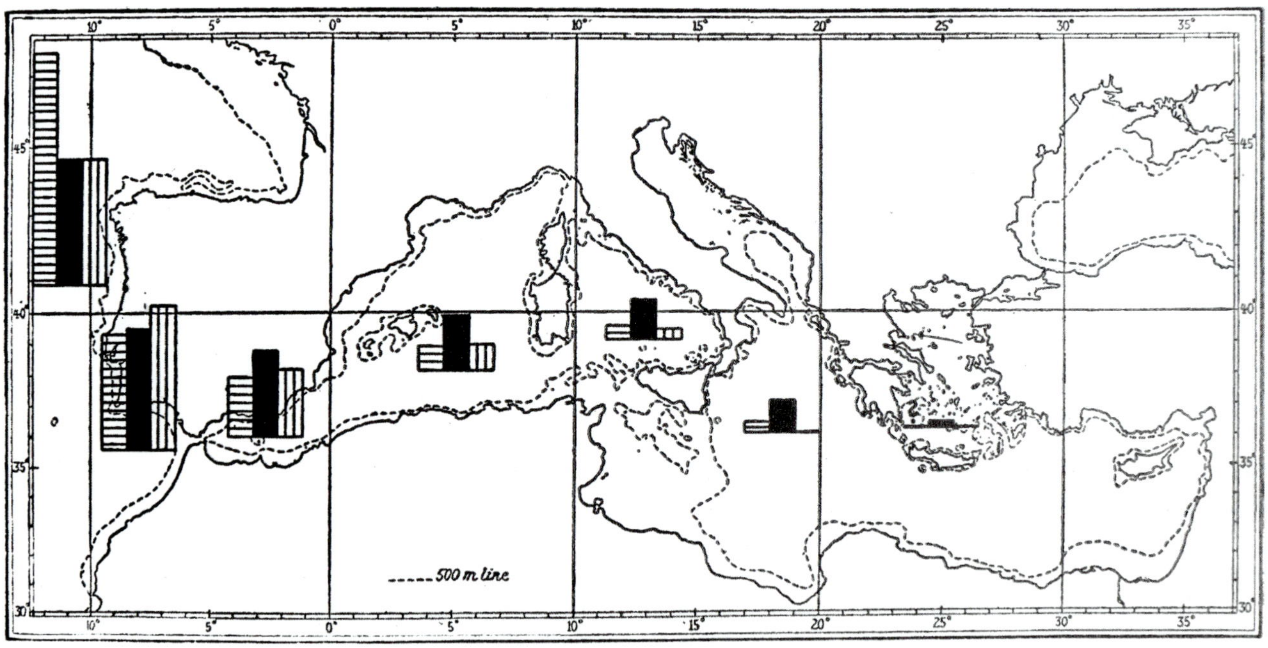

4.4 Nutrient gradient in the Mediterranean Sea

Expeditions with the Thor and later the Dana vessels documented very low levels of nutrients (plankton organisms, nitrate and phosphate) in The Mediterranean Sea compared with the North Atlantic Ocean (Winge Ø, Vedel Taaning Å., 1947).

It fell in the hands of J. N. Nielsen to conduct the hydrographical analyses aboard *Thor*. In Greek waters for instance, in the eastern Aegean from Rhodes to the Dardanelles, Nielsen's analyses of the hydrographic data collected is still considered as pioneering for the region and his results are still referred to.[53]

As a token of appreciation of the Danish effort in the Mediterranean, the scientific staff on board the *Thor*, were all bestowed with a Tunisian order. Thus, Schmidt, Ostenfeld and Nielsen now had the title of Commander of the Order of Nichan-Iftikhar, while Captain Hansen earned the Officer's Cross of the same order.[54] With regards to the life history of the freshwater eel, the results from the two Mediterranean expeditions were negative in the sense that the solution to the problem was not found here.

'Sickening to think that the Norwegians will come first'

In virtually all previous accounts of Johannes Schmidt's Mediterranean eel searches, the story ends with the return of *Thor* from the second trip in the summer of 1910. On the 25 September, *Thor* docked in Copenhagen, and attention was directed elsewhere.[55] As the expedition did not find evidence for the eel to be spawning in the Mediterranean, Schmidt directed his attention to the Atlantic Ocean. This was not entirely the case though. Only Anton Bruun mentions briefly that Captain Hansen stayed behind to continue investigations in the Mediterranean.[56] As late as the spring of 1911, Schmidt still tried to prove that Grassi was right in his assumption that the Strait of Messina between Sicily and the Italian mainland was a breeding ground for the eel. Captain Hansen was himself a diligent investigator of species showing up in the hauls of the *Thor*, and during the winter of 1910–11, Hansen was operating a one-man investigation from the port of Messina in Sicily. This is evident from letters, which Hansen sent to Schmidt in Copenhagen in March 1911.[57]

Hansen gives thorough accounts of how he tried to find eel larvae from great depths in the turbulent waters off the Strait of Messina and to the North of there, off the coast of the town of Faro. Hansen asked local fishermen to help him out setting a bounty for every eel larva they would catch for him. Hansen in turn supplied them with glasses with a solution of formaldehyde in water to preserve the findings, as well as

53 Boussoulengas A. 80 years of oceanographic research in The Aegean Sea. In: Lenz W, Deacon M, eds. *Ocean sciences: Their history and relation to man. Proceedings of the 4th international congress on the history of oceanography*. Hamburg: Deutsches Hydrographisches Institut; 1990:245–249.

54 RA: Dansk Biologisk Station, Johannes Schmidts embedsarkiv: Korrespondance om åleforskning (1908–1933) E28: 1908–1913: 5 November 1910.

55 Paulsen O. Middelhavstogterne med 'Thor'. In: Winge Ø, Vedel Taaning Å, eds. *Naturforskeren Johannes Schmidt. Hans Liv og Ekspeditioner*. Copenhagen: Gyldendalske Boghandel Nordisk Forlag; 1947: p. 51.

56 Bruun AF. Middelhavstogternes resultater. In: Winge Ø, Vedel Taaning Å, eds. *Naturforskeren Johannes Schmidt. Hans Liv og Ekspeditioner.* Copenhagen: Gyldendalske Boghandel Nordisk Forlag; 1947:52.

57 KB. Johannes Schmidt 'Papirer'. Ut.374.V.1. Privat Korrespondance 1902–1916. Letters from G. Hansen to J. Schmidt, 26 & 29 March 1911.

4.5 Captain Georg Hansen on board the R/V *Dana*
Captain Hansen commanded Schmidt's vessels for almost 25 years. Hansen possessed several valuable skills for deep sea investigations. First of all, he knew how to handle a ship at high seas when it was probing deep waters with custom made gear that extended several kilometers into the sea. Secondly, Hansen learned how to sort the harvested specimen be it larvae or fully grown fish. Finally, he was some one Schmidt trusted and confided in regarding the management of the expedition (RA. *Carlsbergfondets* DANA*-ekspeditioner. 1917–1973. 5*).

small sample glasses and wooden holsters for safe shipment. Adult eel was caught commercially in the area, but Hansen's frustration is evident from his letters, where he asks polemically, what a puzzle it is that the Italian scientists had previously succeeded in finding an eel larva. Hansen only found glass eel, 183 specimens at the time of writing, and he had a hard time getting his hands on them. He also searched in vain for ocean sunfish, (*Mola mola*), where the principal idea was to investigate its stomach content for traces of eel. According to Hansen however, the ocean sunfish was a rare catch in general and therefore expensive to set a bounty on, and the season for catching them was still two months away.

Hansen concludes his first letter by telling Schmidt that he will embark on the return voyage to Denmark on the 7 April first by sea and then over the Alps via Munich to Copenhagen. His final words to Schmidt were: 'It is sickening to think that the Norwegians might get there first. Greetings respectfully, G. Hansen'.[58] By this, he undoubtedly hinted at the fact that the Norwegians could surpass Schmidt in solving the ever-growing riddle of the breeding places of the Atlantic eel. The 'Norwegians' in this context, referred to the prominent Norwegian marine research under the directorship of Johan Hjort. Fortunately for both Schmidt and Hansen, Schmidt now finally gave up searching in the Mediterranean and directed attention fully to the Atlantic Ocean.

When reading the internal correspondence between the members of the Kommissionen for Havundersøgelser or the Schmidt funding applications to the Carlsberg Foundation for that matter, one clearly gets a sense of competition, or even rivalry between Danish marine scientists, and marine scientists from other countries such as Norway or Italy. Undoubtedly, a sportive attitude was integral to much of Johannes Schmidt's success, but this healthy competition, however, did not prevent international cooperation. Schmidt and Johan Hjort of Norway for instance were also helpful to each other in terms of sharing technological advances and data, but when Hjort informed Schmidt about his upcoming trip with *Michael Sars* in the spring of 1910, Hjort undoubtedly kept many details to himself.[59] Schmidt would have done the same.

Conclusion

The above investigation into Johannes Schmidt's first two independent expeditions has revealed a number of previously unknown issues, mainly with regards to the organization and execution of the two expeditions.

First of all, this chapter has shown that Schmidt was interested in the eels in the Mediterranean for much longer than previously believed in the literature. Whereas previous commentary has portrayed the eel investigations in the Mediterranean coming to an end after the second expedition, here I have discussed evidence for explorations up to the spring of 1911, when Schmidt requested Captain Hansen to return from his stay in Sicily, where he had continued, unsuccessfully, to sample for eel larvae for several months after the *Thor* had left the Mediterranean.

58 KB. Johannes Schmidt 'Papirer'. Ut.374.V.1. Privat Korrespondance 1902–1916. Letters from G. Hansen to J. Schmidt, 23 March 1911.

59 RA: *Dansk Biologisk Station, Johannes Schmidts embedsarkiv*: Korrespondance om åleforskning (1908–1933) E28: 1908–1913: J. Hjort to J. Schmidt, 31 January 1910.

Nonetheless, Schmidt's two expeditions were both successful, but in ways that he had not predicted, and along with the publication of his North Atlantic investigations, the Mediteranean *Thor* expeditions, firmly established Schmidt as the leading man in Danish deep sea marine science. He had already become a sort of research celebrity in Denmark, talked about daily newspapers, and he was also highly respected by his peers. Whereas the expeditions did not provide much of a solution to the eel problem, considerable and important advances were made within hydrography. The inclusion of hydrography and plankton research in his expeditions increased the gratitude and respect in the eyes of his peers. The clever technological devices developed in Northern waters, such as Knudsen's pipette and the young fish trawl were now for the first time deployed in the Mediterranean. The measurement of for instance the flow of water in and out of the Strait of Gibraltar and the mapping of the different temperature gradients in the various major basins in both the Western and Eastern Mediterranean were immediate successes.

With regards to the funding and organization of the *Thor* expeditions the above chapter has uncovered a number of hitherto unknown aspects and shed more light on others. While Schmidt's private means have been reported to aid him, we can now see that he actively used his private money as an important strategic asset promoting his scientific dreams, illustrated by the fact that his mother-in-law proposed to the Carlsberg Foundation to cover the sum Schmidt had overspent in relation to the budget he had drawn on from the Carlsberg Foundation. In the literature, the two Mediterranean expeditions are often treated as one, but it took a lot of effort to find support for the second trip, when Schmidt had to call on help from several Danish professors, with Eugen Warming as his foremost champion. In addition to the direct support of Schmidt's expeditions the Carlsberg Foundation subsequently chose to fund several projects arising from the collections on board the Mediterranean cruises.

Three and a half months passed from the departure from Falmouth before the *Thor* landed in Copenhagen on the 25 September, six days before Schmidt was scheduled to begin his new career as a head of Carlsberg Laboratory's Physiological Department. In hindsight this was another good career move by Schmidt, which provided him with a terrific research platform for the next 23 years, but probably not in the way intended by most people, when he set out on this new journey in the fall of 1910.

5 Director at the Carlsberg Laboratorium

In 1910 Johannes Schmidt replaced Emil Chr. Hansen as director of the Carlsberg Laboratory. (...) This was an act of treason against The Brewer, what they let happen, with regards to management, was a total failure of leadership.[1]

Laboratory legacy

'Treason' was the verdict in 2011, when the outgoing chairman of the Board of Directors of the Carlsberg Foundation, chairman of the Carlsberg Laboratory, and professor of medicinal chemistry, Povl Krogsgaard-Larsen commented on his predecessors' decision to hire Johannes Schmidt. Krogsgaard-Larsen expressed this view in an autobiography, looking back and taking stock of his nine years (2003–2011) at the helm of the Carlsberg conglomerate of science funds and global businesses, established by 'The Brewer', I. C. Jacobsen. In addition, he had been a member of the Board of Trustees of the Carlsberg Laboratory for close to two decades from 1993–2011.

Within the Carlsberg Foundation Schmidt's legacy has been complicated. Johannes Schmidt is a celebrated and famous scientist within marine science circles. Yet now, more than 100 years after he was first appointed director of the Physiological Department of the Carlsberg Laboratory, Johannes Schmidt is nothing short of infamous in Carlsberg circles. Povl Krogsgaard-Larsen conducted his own inquiry into this topic, as he found it odd that an accomplished marine scientist such as Schmidt would suddenly switch to beer related research. Krogsgaard-Larsen cites a critical commentary from Professor P. Boysen Jensen (1883–1959), who in 1933 complained to the Board of Trustees of the Carlsberg Laboratory that they had appointed Professor Øjvind Winge, of the Veterinary and Agricultural University as Schmidt's successor.

1 Bjerager A. *Bryggerens Lærling. Povl Krogsgaard-Larsens erindringer.* Gylling, Denmark: Gyldendal; 2011: 235–236.

Here, according to Krogsgaard-Larsen, Boysen Jensen laments the decision to hire Schmidt back in 1910, as Schmidt had possessed very few qualifications for the job. Subsequently, Schmidt had turned the Carlsberg Laboratory into a zoological museum, as Krogsgaard-Larsen put it.

Krogsgaard-Larsen further referred to the 1976 anniversary volume on the Carlsberg Laboratory, where the editor, Heinz Holter cautiously remarked that the steady inflow of money from the brewery had allowed the laboratory researchers to assume a 'very liberal interpretation' of their raison d'être. In Krogsgaard-Larsen's opinion, the founding brewer, I. C. Jacobsen would never have accepted that employees of the Carlsberg Laboratory would 'study the wanderings of the eel in remote oceans,' when in charge of research at a laboratory, where the focus should adhere to the guiding principles of the laboratory statutes, namely the promotion of research to enhance the quality of beer.[2]

Trying to explain the appointment of Schmidt, Krogsgaard-Larsen noted that Johannes Schmidt had married the daughter of the first CEO of the Gammel Carlsberg Brewery, Søren A. van der Aa Kühle. 'One gets an ugly sense of nepotism,' was Krogsgaard-Larsen's conclusion.[3]

Such a harsh and direct critique is quite rare, when it comes to events taking place close to a hundred years ago, and it is worth mentioning that Krogsgaard-Larsen openly connected his commentary on Schmidt to his own attempt to steer the research of the Carlsberg Laboratory in a more focused direction that concentrated on beer relevant issues. Krogsgaard-Larsen had fought for this – and succeeded, during a process lasting a decade, where clashes between a majority on the Board of Trustees and the researchers in the laboratory had been, in his own words; 'very intense.'[4] No doubt the seemingly blatant delinquency of Johannes Schmidt was used as a symbol of a direction the laboratory should steer clear of. Nonetheless, he could be right. Was Schmidt's more than 22 years long term at the Carlsberg Laboratory a mere scam for the pursuit of marine science?

Previously, other commentators on Schmidt's career have mentioned that a whiff of nepotism had tainted the sweet scents of beer brewing on the Carlsberg premises. Among chroniclers of Schmidt's accomplishments as a marine scientist, his connections with the Carlsberg social circles are mentioned in passing as well. Tom Fort presents an implicit causation when stating in one sentence: 'In 1903 he had married Ingeborg Kühle, daughter of the chief director of the Carlsberg Brewery, and seven years later he was appointed director of the prestigious physio-

2 Bjerager A. *Bryggerens Lærling. Povl Krogsgaard-Larsens erindringer.* Gylling, Denmark: Gyldendal; 2011: 235–234–237.

3 Bjerager A. *Bryggerens Lærling. Povl Krogsgaard-Larsens erindringer.* Gylling, Denmark: Gyldendal; 2011: 235–236.

4 Bjerager A. *Bryggerens Lærling. Povl Krogsgaard-Larsens erindringer.* Gylling, Denmark: Gyldendal; 2011: 237–241.

logical institute funded by the brewery'.[5] The same view is echoed in a footnote in Rozwadowski's volume on the history of the International Council for the Exploration of the Sea: 'Schmidt's wife came from the Carlsberg family, famous Danish brewers, which certainly helped him gain access to funding.'[6] This is contrasted by the discreteness on this point in an article by Peder Olesen Larsen.[7] Recently though, Olesen Larsen published a book on the institutional history of the Carlsberg Laboratorium (1876–2014), where Schmidt's old professor Eugen Warming is credited with having promoted Schmidt to his job.[8]

In the 1947 commemorative volume on Schmidt's life and deeds, Schmidt's former collaborators stayed clear of this feature of his professional life.[9] Only Schmidt's assistant from the 1920s expeditions, Aage Vedel Taaning remarked that it had been beneficial for Schmidt to land his prominent position at the Carlsberg Laboratory at an early age. Vedel Taaning also underlined the support from his wife Ingeborg as a vital asset for his scientific aspirations in the sense that she was a splendid hostess for parties in Schmidt's home. Vedel Taaning, however, remarked that it was Schmidt's professor from university Eugen Warming whose influence secured him the position as head of the laboratory.[10]

The following investigation of Schmidt's career as the Director of the Physiological Department of the Carlsberg Laboratory will shed additional light on this issue. While most of this book is occupied with Schmidt's work as a marine scientist, this chapter deals with his other efforts as a researcher and research director in what was effectively Denmark's most prominent privately owned research laboratory before World War II. The chapter presents an account of the main non-marine science areas of research pursued by Schmidt during his long tenure, 1910–1933 at the Carlsberg Laboratory. This is followed by an analysis of how Schmidt's research related to the desires of his superiors, the overseeing Board of Trustees operating under the auspice of the Board of Directors of the Carlsberg Foundation, as well as the demands from the management of the Carlsberg Brewery.

Terms of employment

When Schmidt landed the job as head of the physiological department, he was promoted not primarily by the Carlsberg family clan, but by his old professor from university, Eugen Warming. In early September,

5 Fort T. *The book of eels*. Great Britain: Harper Collins Publishers; 2002, p. 97.

6 Rozwadowski H. *The sea knows no boundaries. A century of marine science under* ICES. Seattle and London: ICES/University of Washington Press; 2002: 303. Rozwadowski cites Went AEJ. *Seventy years agrowing. A history of the international council for the exploration of the sea, 1902–1972*. Copenhagen: International Council for the Exploration of the Sea; 1972. p. 76. for this piece of information. However, Went's volume on the history of ICES is silent on this issue throughout the book.

7 Olesen Larsen P. Anderledes – og tak for det. *Carlsbergfondet Årsskrift*. 2011;2011:58–77.

8 Olesen Larsen P. Iværksætter og forvaltere. Carlsberg Laboratorium 1876–2014. Copenhagen: Carlsbergfondet; 2015: p. 78–86.

9 Winge Ø, Vedel Taaning Å, eds. *Naturforskeren Johannes Schmidt. Hans Liv og Ekspeditioner*. Copenhagen: Gyldendalske Boghandel Nordisk Forlag; 1947.

10 Vedel Taaning Å. Hovedlinierne trækkes op. In: Winge Ø, Vedel Taaning Å, eds. *Naturforskeren Johannes Schmidt. Hans Liv og Ekspeditioner*. Copenhagen: Gyldendalske Boghandel Nordisk Forlag; 1947:176.

Three weeks after Emil Chr. Hansen had passed away, the managing Board of Trustees of the Carlsberg Laboratory adjourned for a board meeting, where the most important topic on the agenda was the task of finding a new head of the physiological section of the laboratory. According to the minutes, the board quickly moved to ask Johannes Schmidt to write an application for this vacancy, and Schmidt was the only candidate mentioned in the minutes. Eugen Warming also had a seat on the board of the laboratory, and he was the one who moved to promote his former student, Johannes Schmidt. It was blatantly clear to the board members that Schmidt's career as a marine scientist was in full swing, even though the main work at the laboratory was supposed to support the improvement of beer products.[11]

It took until the 22 September 1909 before Schmidt submitted a formal application, which was a one and a half page document where he briefly stated his resume. The only item on his list of accomplishments, which had a direct bearing on the brewing of beer, was his former job as a teaching assistant in 'microscopy and growth of fermentation organisms' at the technical university, *Polyteknisk Læreanstalt*.[12] Schmidt's job application bore no hint as to what he would like to work on as the director. All in all, it was a *fait accompli*. Schmidt was headhunted for the job.

From the discussion recorded in the minutes of the Board of Trustees, it seems that Schmidt was allowed to continue his work at the marine science commission. Also, as Schmidt was already planning the work with the *Thor* for the spring of 1910, it was agreed that Schmidt should not commence his new work at the laboratory until the 1 October 1910.[13] This means that although Schmidt was headhunted very quickly, his position would remain vacant for more than a year after the death of his predecessor.

The terms of employment, which were drawn up by the chairman of the Board of Trustees, S. M. Jørgensen, took these time constraints into account. Nonetheless, Jørgensen explicitly pointed to the statutes of the foundations, where the objective of the laboratory work is mentioned, as they had been laid down by the founder of the laboratory, I. C. Jacobsen.[14] This objective was and still is, although the wording and content of the foundation has been altered since, to initiate and carry out basic research in support of beer brewing.[15]

However, it is notable that the Board of Trustees violated the statutes themselves, since in the second to last paragraph of the terms of the

11 CA. *Carlsbergfondet, Carlsberg Laboratorium*, E17. Forhandlingsprotokol 1, p. 248–255.
12 CA. *Carlsbergfondet, Carlsberg Laboratorium*. E06.'1909'. J. Schmidt to Board of Trustees, 22 September 1909.
13 CA. *Carlsbergfondet, Carlsberg Laboratorium*, E17. Forhandlingsprotokol 1, p. 248–255.
14 CA. *Carlsbergfondet, Carlsberg Laboratorium*. E06.'1909'. S. M. Jørgensen to J. Schmidt, 2 October 1909.
15 Anon. *Carlsbergfondet Fundats*, 2013. http://www.carlsbergfondet.dk/~/media/pdf-filer/Fundats/CF%20Fundats%202013.ashx. The original statutes are printed as an appendix in: Olesen Larsen P. *Iværksætter og forvaltere. Carlsberg Laboratorium 1876–2014*. Copenhagen: Carlsbergfondet; 2015: Appendix 3, p. 267–271.

16 Olesen Larsen P. Anderledes – og tak for det. *Carlsbergfondet Årsskrift*. 2011;2011:58–77.

17 Olesen Larsen P. *Iværksætter og forvaltere. Carlsberg Laboratorium 1876–2014*. Copenhagen: Carlsbergfondet; 2015: Appendix 3, p. 267–271.

employment, it is stated that Schmidt would be allowed to retain his seat as member of the Danish commission for marine science. According to the statutes, the director should be fully devoted to the work of the laboratory, and it was entirely up to the board to decide whether or not the director could take up a membership of a board or another honorary position outside of the Carlsberg Laboratory. It does seem that precedence for such dispensations existed. The director of the Chemical Department, S. P. L. Sørensen for instance had a seat on a number of advisory boards.[16] The statutes, however, do not suggest that it is possible for the board to issue such a dispensation in a case where the director wanted to take up a *paid* job outside the Carlsberg Laboratory.[17]

5.1 Carlsberg Laboratorium c. 1900

The Carlsberg Laboratory is an impressive building. From 1910–1933 Schmidt disposed of the entire first floor of the main building, while his private apartment was located in the adjoined building to the right in the picture (Carlsberg Archives).

Schmidt's membership of the marine science commission was not an honorary title; it was a very well paid job with substantial responsibilities earning him 3,600 Danish Kroner a year. Such a sum of money was almost equivalent to his new salary as a research director. The laboratory salary was settled so that the director would keep an income equivalent to the earnings of a full professor at a university with equivalent seniority, plus an additional 15%. At the age of 32 years old, Schmidt's starting salary was 3,600 Danish Kroner plus 15% or 520 Kroner totalling 4,140 Kroner. In addition to his fiscal revenue, the position included other benefits such as free housing in a spacious apartment in the northern end of the laboratory building. The laboratory also paid for heating and lighting, as well as gardening.[18] 'Beer etc.' was delivered by the Brewery free of charge.[19]

An inside job

From the doorstep of his ground floor apartment, Schmidt could look south along the stately front of the main building towards the southern end of the complex, where his maternal uncle, Johan Kjeldahl had moved in, when the building had been erected in 1897.[20] Schmidt had now returned home, or at any rate come back to where he lived for a couple of years as a young graduate student in botany. Since 1901 though, S. P. L. Sørensen had presided over the Chemical Department following the death of Kjeldahl. Nonetheless, in several ways Schmidt was still well connected to people within the Carlsberg conglomerate.

First of all, Eugen Warming was immensely important. Whether or not he had fought 'like a lion' for Schmidt's candidacy as one newspaper later reported, is hard to know. Yet, Warming clearly thought very highly of his former student, whom he had followed and supported at university, and in the Carlsberg Foundation on several occasions already. The affection went both ways. For Warming's 70th Birthday in 1911, Schmidt presented him with a silver cup, a gift which Warming's wife noted in a private letter shortly after.[21] When Warming turned 80, he received another cup from Schmidt, '...as a deeply felt token of appreciation for everything you have been for me, since I came to you as a young student.[22]

While Warming was the actual promoter of Schmidt taking up the director's position, he will have acted in concordance with the sympathies of the board of the laboratory at large. Here, Schmidt was unusually

18 CA. *Carlsbergfondet, Carlsberg Laboratorium.* E06.'1909'. S. M. Jørgensen to J. Schmidt, 2 October 1909.
19 CA. *Carlsbergfondet, Carlsberg Laboratorium.* E06.'1912'. S. M. Jørgensen, 'Administrative bestemmelser vedrørende Carlsberg Laboratoriet,' 1912: 8
20 Tovborg Jensen, A., The men behind the scene (The Board of Trustees), in: Holter H, Max Møller K, eds. *The Carlsberg Laboratory 1876–1976*. Copenhagen: Rhodos; 1976: 259–281.
21 Prytz S. *Warming – botaniker og rejsende*. Lynge: Bogan's Forlag A/S; 1984, p. 152.
22 Prytz S. *Warming – botaniker og rejsende*. Lynge: Bogan's Forlag A/S; 1984, p. 151.

well positioned through his own upbringing as well as his marriage. In 1909 the Board of Trustees consisted of the following five members: The chemist, S. M. Jørgensen was the chairman, and along with Warming and physicist, C. Christiansen, he was appointed by the Board of Directors of the Carlsberg Foundation, where they all had seats too. Then, there were two adjunct board members, the managing director and owner of the Ny Carlsberg Brewery, Carl Jacobsen, and Rudolph Koefoed who was the former co-director of Gl. Carlsberg Brewery. Koefoed had also worked as a research assistant to Johan Kjeldahl around 1890.[23]

One could argue that since S. A. van der Aa Kühle, Schmidt's father-in-law and the first CEO of the Carlsberg Breweries had passed away in 1906, he would have no longer been in a position to pave the way for his son-in-law, and the same goes for Schmidt's maternal uncle, Johan Kjeldahl who drew his last breath in the summer of 1900.

Nonetheless, Koefoed had been headhunted as a talented brewer by Kühle, when he was working for Kjeldahl.[24] Hence, Koefoed's career had depended on some of Schmidt's close relatives, and he surely will have known Johannes Schmidt personally, when Schmidt was just a bright young school boy living in the neighbourhood.

Also, one should also not overlook the impact of being on friendly terms with the managing director and owner of Ny Carlsberg Brewery, Carl Jacobsen. According to Ingeborg Schmidt's memoirs, her mother's closest friend was the wife of Carl Jacobsen, Ottilia, and the social lives of the two families had been entangled for decades. In fact, in the ensuing conflict between I. C. Jacobsen and Carl Jacobsen, where Kühle was in charge of the old brewery and thereby siding with I. C. Jacobsen, the women in the families upheld close connections across the divide of the separate businesses and family feud in the male line. As a young school girl, Ingeborg had also shared a private teacher with the children of Carl and Ottilia Jacobsen.[25] This means that when the board of trustees convened in 1910 to the find the successor of Emil Chr. Hansen, Carl Jacobsen supported hiring the son-in-law of his wife's best friend.

When Johannes Schmidt entered the Carlsberg Laboratory as the designated leader, his family ties surely worked in his favour from at least two directions. Although this does not necessarily support Krogsgaard-Larsen's 'ugly sense of nepotism' it is certainly worth noting, along with Vedel Taaning's no less important observation of Eugen Warming's warm support for Schmidt's candidacy.

23 Tovborg Jensen, A., The men behind the scene (The Board of Trustees), in: Holter H, Max Møller K, eds. *The Carlsberg Laboratory 1876–1976*. Copenhagen: Rhodos; 1976: 259–281.
24 Rancke-Madsen E. Koefoed, Rudolph Georg. In: Cedergreen Bech S, ed. *Dansk Biografisk Leksikon*, vol. 8. 3rd ed. Copenhagen: Gyldendal; 1981:146–147.
25 Schmidt I. *Blade af Carlsbergs Historie. Kaptajn S. A. van der Aa Kühle og hans liv på Carlsberg*. Copenhagen: Bianco Lunos Bogtrykkeri; 1956. p. 20–24.

At any rate, good connections can only take you so far. Schmidt now had to deliver the research results to match his double professor's salary. Already an accomplished marine scientist of international repute, Schmidt had several landmark publications under the belt to ensure his legacy. However, on taking up the post he, had to resume his botanical studies, which he had so successfully studied ten years earlier, in what must have felt almost in a different lifetime for the now 33-year old Schmidt.

Research topics at the laboratory

So did Johannes Schmidt fulfil his duties as the Director of the Carlsberg Laboratory? The most direct way to assess this is to look at his publication record during the years when he presided over the Physiological Department. Following Schmidt's death in early February 1933, his younger marine science colleague, zoologist Anton Fr. Bruun (1901–1961) compiled a list of Schmidt's scientific publications.[26] Towards the end of an extensive necrology, Bruun listed 131 publications authored by Johannes Schmidt. Only on four occasions, mainly during the early part of his career, did Schmidt co-author his papers. As a rule, he was the sole author. Therefore, the sheer number of publications is ample testimony of his large productivity as a scientist. Yet, it is equally interesting to see how Schmidt devoted his efforts across the various topics he pursued.

Sorting Schmidt's publications by year, it becomes clear that Schmidt was immensely productive from a young age, and furthermore continued to publish a great number of works at a steady pace throughout his life (Figure 5.2). The last four papers of Johannes Schmidt emerged as late as 1932.

During his first five years as an active researcher, seven publications on bacteria saw the light of day, while one of these, co-authored with Ostenfeld, had a marine dimension to it. Schmidt's bio-geographical field studies in Thailand around the turn of the century resulted in no less than 9 research publications in the years to come. From 1904 onwards though, marine science gradually took over as Schmidt's most prolific area of research resulting in 91 publications. Close to half of these, 44 publications, dealt with aspects of the life history of the freshwater eel.

26 Bruun AF. The Life and Work of Professor Johannes Schmidt. *Estratto dalla Rivista di Biologia*. 1934;16(1):3–22.

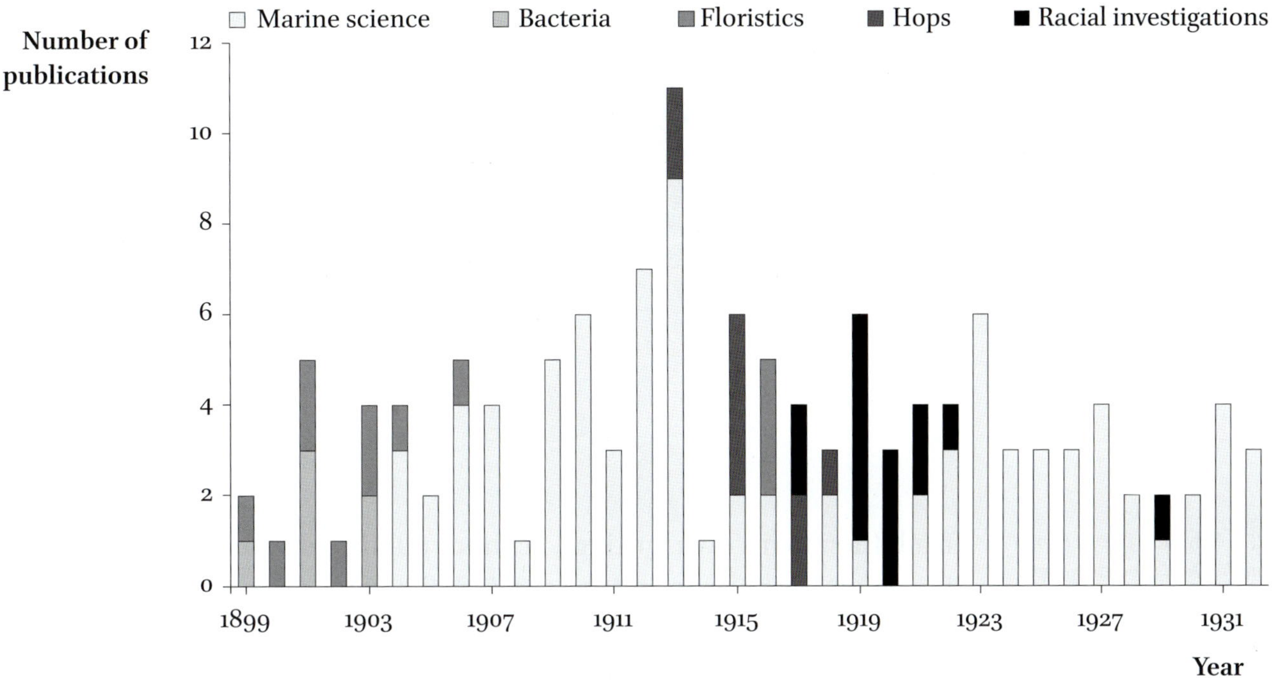

5.2 Johannes Schmidt's publications ($n = 131$) by year and topic, 1899–1932

At the Carlsberg Laboratory, where Schmidt commenced his directorship 1 October 1910, his range of publications attained another two sets of colours, as his research came to include studies of hops and a series of hereditary studies, labelled by Schmidt as his 'Racial investigations.' The latter two areas, of which the racial investigations were also linked with his marine science interests, received Schmidt's attention during the first ten years at the laboratory, but after 1921 they seem to have received very little consideration in the Carlsberg Laboratory. Hereditary studies – or genetics, as we would label it today, received the attention of fifteen publications in total.

It would be more than stretching the argument to claim that Schmidt devoted his entire workload to issues of relevance for brewing from 1910 onwards. Figure 5.3 shows Schmidt's publications during his laboratory years, 1911–1932, and it is evident that 75% of his publications still focused on marine science issues. In addition, the 'racial investigations' were mainly aquatic in character, to some degree boosting his marine science output by a further 15%. Although the methodological and empirical advances made on hereditary qualities had a more general scientific relevance, they were mainly obtained through studies on various fish species, and in some cases specifically linked to the issue of races in fish populations.

5.3 Schmidt's publications by research topic, 1911–1932

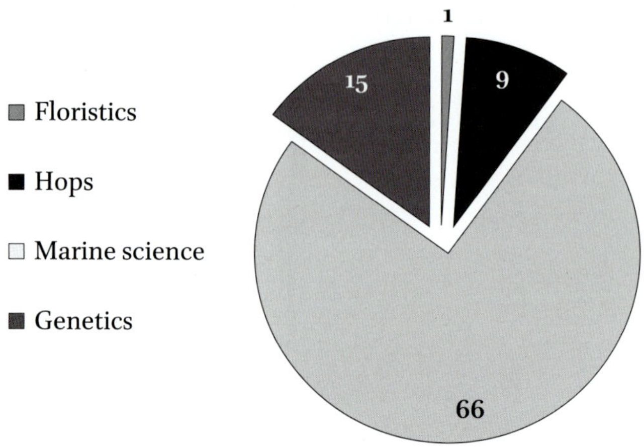

In addition to his own writings, Schmidt published seven different scientific report series as an editor. Schmidt's first large scale job as an editor was the publication of the *Report on the Danish oceanographical expeditions 1908—1910 to the Mediterranean and adjacent seas*.[27] In 1916, Schmidt edited a volume of reprints of studies from the Danish journal, *Botanisk Tidsskrift* on the vegetation in the Gulf of Siam, focusing on the island of Koh Chang, from where he and Theodor Mortensen had returned in 1900.[28] Schmidt's *Investigations on Hops* as well as his *Racial Investigations* each formed part of special series of the so-called *Comptes-rendus des travaux du Laboratoire Carlsberg*, which was published by the Carlsberg Laboratory. Both series were reprinted, once they were completed in 1918 and 1931 respectively.[29] Meanwhile, Schmidt edited and co-edited several volumes covering results from the marine science expeditions in the Atlantic Ocean from 1920–22.[30] Finally, Schmidt orchestrated the publication of a series with results from the circumnavigation, a series which outlived Schmidt by more than 50 years.[31] Finally, Schmidt also edited three marine science volumes of the *Rapport et Proces-Verbaux des Reunions*, when he was chairing the North Western Area Committee of ICES in the 1920s.[32] In some cases these series spanned more than two decades, underlining Schmidt's long-term pursuit of his ideas as well as the scale of his enterprise.

With this breakdown of Schmidt's publication interests in mind, it is of interest to know to what extent it was possible for Schmidt to serve two masters at the same time; the Board of Trustees of the Carls-

27 Schmidt, J. (ed), *Report on the Danish oceanographical expeditions 1908—1910 to the Mediterranean and adjacent seas*. No. 1–10. Copenhagen: 1912-.

28 Schmidt, J., (ed.) *Flora of Koh Chang, Contributions to the knowledge of the vegetations in the Gulf of Siam, 1900–1916*, (Reprint from *Botanisk Tidsskrift*). Copenhagen: Bianco Luno, 1916.

29 Schmidt, Johannes, (ed.) *Investigations on Hops (Humulus lupulus L.) 1913–1918*, i-xi, (reprint from *Comptes-rendus des travaux du Laboratoire Carlsberg* vol. 10–14), Copenhagen, 1918; Schmidt, Johannes, (ed.) Racial Investigations, 1917–1930, i-x, (reprint from *Comptes-rendus des travaux du Laboratoire Carlsberg*, vol. 13–18). Copenhagen, 1931.

30 Schmidt J, ed. *The Danish 'Dana'-Expeditions 1920–22 in the North Atlantic and Gulf of Panama : Leader: Johs. Schmidt : Oceanographical reports ed. by the 'Dana'-committee* : No. 1–8. 2 vol. Copenhagen: ; 1926–31.

31 Schmidt J, ed. *The Carlsberg Foundation's oceanographical Expedition round the World 1928–30*. Report no. 2. Copenhagen: 1932.

32 Schmidt J, ed. *Scientific reports of the north western area committee 1924–29*. 3 reports. ICES, *Rapport et Proces-Verbaux des Reunions*, vol. 39, 57, 72; 1929.

berg Laboratory on the one hand, the marine science commission on the other hand. Schmidt's marine science research while he was in the service of the laboratory forms the backbone for most chapters in this book, but it is also worth considering the other topics mentioned above. As we shall see the hops investigations formed an attempt at fulfilling the obligations of the Carlsberg conglomerate, while the racial investigations aimed at solving more general basic problems in science, even if they appeared fully applicable to his other aquatic interest.

Hereditary qualities of a scent

The Board of Trustees was fully aware that Schmidt was not a new Emil Chr. Hansen, the world famous pioneer of yeast physiology. Therefore, the board appointed Pharmaceutical scientist, Albert Klöcker as *Director Extraordinaire* in 1909. Klöcker had been working as Hansen's assistant since 1892, and the idea was that he should now continue and finish the investigations, which Hansen had staked out before he fell ill and died. This may not have been the easiest task, but Klöcker succeeded in publishing a few articles on this topic in the next few years.[33] With Klöcker working with the legacy from Hansen, Schmidt was free to start afresh on his own research agenda.

Schmidt chose the study of experimental breeding of hops as his first major project at the laboratory. In late January 1911, Schmidt submitted a full research plan to his board, specifically adhering to §8 of the statutes, the ones stipulating that the research was to serve the interest of brewing. Schmidt wanted to introduce studies of hops for two reasons. First of all, the breeding of hops was an understudied topic in the history of the laboratory, where yeast and grain had been studied more thoroughly than hops. Secondly, there was no precedence in Denmark for conducting such studies, in spite of the fact that hops had been grown in Denmark for at least 300 years. At the cost of 5,708 Kroner, Schmidt had a greenhouse erected on the premises of the laboratory, to ensure maturation of more fragile hop stems from southern Europe.[34] The greenhouse was ready in May 1911, and he could start planting hops. A new plot of land was to be farmed with hops as well, even though this area was not operational until 1912.[35]

33 Schiönning, H., Albert Klöcker 1862–1923, in: Holter H, Max Møller K, eds. *The Carlsberg Laboratory 1876–1976*. Copenhagen: Rhodos; 1976:191–194.

34 CA. *Carlsbergfondet, Carlsberg Laboratorium.* E06.'1911'. J. Schmidt to Board of Trustees, 23 January 1911.

35 *CA. Carlsbergfondet, Carlsberg Laboratorium.* E06.'1912'. J. Schmidt to Board of Trustees, 27 September 1911 & 30 September 1912.

5.4 The harvest of hops at the Carlsberg Laboratorium
The harvest of hops at the Carlsberg Laboratorium in 1912. Schmidt had the green house erected for his first project at the laboratory, researching into the aromatic qualities of hops (Carlsberg Archives).

Meanwhile, Schmidt went on his first excursion in his new capacity as director. Schmidt was required to ask permission from the board to be absent from the laboratory outside the holiday season. Therefore, it is possible to follow many of his journeys through his applications. Schmidt had arranged with producers of hops in Bavaria to visit their production. Then, while he was in the neighbourhood, more or less, he wished to make a detour to Rome, where the Danish Ministry of Agriculture had asked him to represent Denmark at the 5th International Congress on Fisheries and Fisheries investigations.[36] Marine science was present throughout Schmidt's tenure at the laboratory and right from the start.

Next up, Schmidt began assembling his own team at the laboratory. As his first *assistant extraordinaire* of his own choosing, Schmidt hired Øjvind Winge (1886–1964), who took up this position on 15 August 1911.[37] Like Schmidt, Winge was a candidate (1910) in natural history from the University of Copenhagen and specialized in botany. Unlike Schmidt though, Winge was an expert in mycology, the study of fungi, especially their cell formation and other microscopic characteristics, also of relevance to beer brewing.[38] This would prove a wise choice, as Winge would serve the next 10 years as Schmidt's assistant creating landmark investigations, also in his own name, before moving on to a position as professor at the Veterinary and Agricultural University in the spring of 1921. Winge's freedom to select research topics was possible because Schmidt operated the laboratory in what was later called 'the old European tradition.' This means that Winge was obliged to work according to Schmidt's directions during official office hours. After hours he could use the laboratory for his own studies.[39]

Schmidt's first efforts went into studies of the longitudinal and rotational movement of hop stems.[40] The hop stem rotates its tip, as it grows, and Schmidt wanted to know, which factors influenced the speed of rotation of the hop stem. Schmidt attached a disc to the stem to indicate the degree of rotation in a given time period, and then tested for the influence of light, humidity, and temperature on the degree and speed of rotation. At a fairly high temperature, it was noted that the plant rotated one full circle once every three hours.[41] These studies, along with Schmidt's later studies on hops were published in the Carlsberg Laboratory's own publication series, *Comptes-rendus des Travaux du Laboratoire Carlsberg*.[42]

Then Schmidt embarked on a more ambitious aspect of his research plan. The desirable bitterness in beer comes from the substance named lupulin, which is also the host of a number of essential oils adding

36 CA. *Carlsbergfondet, Carlsberg Laboratorium*. E06. '1911'. J. Schmidt to Board of Trustees, 13 May 1911; CA. Carlsbergfondet, Carlsberg Laboratorium. E06.'1911'. J. Schmidt to Board of Trustees, 10 June 1911.

37 CA. Carlsbergfondet, Carlsberg Laboratorium. E06. '1911'. J. Schmidt to Board of Trustees, 10 June 1911.

38 *CA. Carlsbergfondet, Carlsberg Laboratorium*. E06. '1911'. J. Schmidt to Board of Trustees, 27 September 1911.

39 Westergaard, M., 'Øjvind Winge 1886–1964', in: Holter H, Max Møller K, eds. *The Carlsberg Laboratory 1876–1976*. Copenhagen: Rhodos; 1976: 212–217.

40 Ege, Vilhelm, Johannes Schmidt 1877–1933, In: Holter H, Max Møller K, eds. *The Carlsberg Laboratory 1876–1976*. Copenhagen: Rhodos; 1976:198–200.

41 Winge Ø. Andre strenge anslaas – paa Carlsberg Laboratoriet. In: Winge Ø, Vedel Taaning Å, eds. Naturforskeren Johannes Schmidt. Hans liv og ekspeditioner. Copenhagen: Gyldendalske Boghandel Nordisk Forlag; 1947:71–72.

42 Schmidt J. Can different clones be characterised by the number of marginal teeth in the leaves. *Comptes-rendus des travaux du Laboratoire Carlsberg*. 1918;14(2):1–24; Schmidt J. On the amount of lupulin in plants raised by crossing. *Comptes-rendus des travaux du Laboratoire Carlsberg*. 1915;11:165–183; Schmidt J. On the aroma of hops. *Comptes-rendus des travaux du Laboratoire Carlsberg*. 1915;11:149–163; Schmidt J. On the aroma of in plants raised by crossing. *Comptes-rendus des travaux du Laboratoire Carlsberg*. 1917;11:330–332; Schmidt J. On the flowering time of plants raised by crossing. *Comptes-rendus des travaux du Laboratoire Carlsberg*. 1915;11:188–198; Schmidt J. The growth in length of hop stems and its diurnal periodicity. *Comptes-rendus des travaux du Laboratoire Carlsberg*. 1913;10:233–251; Schmidt J. The occurrence of the wild hop in denmark. *Comptes-rendus des travaux du Laboratoire Carlsberg*. 1917;11:314–329; Schmidt J. The rotational movement of hop stems and its diurnal periodicity. *Comptes-rendus des travaux du Laboratoire Carlsberg*. 1913;10:267–283.

fruitiness to beer. Lupulin is contained in the fragrant flower cones of the female hop plants. Danish hop breeds were adapted to the relative harsh climate in Denmark, but they were inferior to many foreign plants in the sense that their aromatic qualities were weak. Schmidt wished to cross Danish hops with foreign plants with a high content of lupulin, but to do so, several investigations needed to be carried out first.

In the fall of 1913, Johannes Schmidt travelled to the Eastern states of the USA to visit hop plantations. It is well known in the literature that Schmidt visited the Danish West Indies in the winter of 1913/14 in his capacity as a member of the commission for marine science, but this trip had an equally practical prelude, since Schmidt had come to the Caribbean by way of the USA where he met with local hop planters thereby gaining access to American hop plants for cross fertilization.[43]

Schmidt discovered that the aroma remained constant in older clones of hop plants. Then he succeeded in transferring the aroma to the offspring of a cross between a Danish plant and a foreign plant. The male plants were thought to have a lower lupulin content compared to the female plant, but the species could then be improved through nonsexual proliferation. Then the lupulin rich offspring could be retained. An additional improvement, which was investigated by Schmidt and his assistants, was the timing of flowering. In a relatively cold Danish climate, an early flowering season was a prerequisite for a successful harvest.[44]

According to Øjvind Winge later commentary, the lupulin study was most likely the first study anywhere examining the hereditary qualities of a scent. Reflecting on these investigations, Winge remembered how nearby plots of land at Valbygårdsvej and around the Jesuskirken were used to farm hop plants from all over the world. Nonetheless, Schmidt never took the investigations to the next more practical level, where an industrial scale use could have been investigated.[45] In other words, the hop investigations never came to constitute the level of applied science, which the foundation statutes had stipulated for I. C. Jacobsen's research laboratory.

43 CA. Carlsbergfondet, Carlsberg Laboratorium. E06.1913. Letters from J. Schmidt to Board of Trustees 15, 16 and 23 September 1913.

44 Ege, Vilhelm, Johannes Schmidt 1877–1933, In: Holter H, Max Møller K, eds. *The Carlsberg Laboratory 1876–1976*. Copenhagen: Rhodos; 1976:198–200.

45 Winge Ø. Andre strenge anslaas – paa Carlsberg Laboratoriet. In: Winge Ø, Vedel Taaning Å, eds. Naturforskeren Johannes Schmidt. Hans liv og ekspeditioner. Copenhagen: Gyldendalske Boghandel Nordisk Forlag; 1947:71–72.

'Racial Investigations'

For Schmidt, his next effort at the laboratory became much more fruitful. In 1903 the Danish plant physiologist Wilhelm Johannsen had published a book in Danish, *Arvelighedslærens Elementer*, which dealt with the elements of heredity. Here Johannsen introduced for the first time, the concept of *genes*, as they became known world wide from 1909, when his studies were published in English. In Johannsen's interpretation genes were the basic unit of heredity, and he further made the influential distinction between genotype and phenotype. The phenotype was used to describe the outward appearance of a specimen, whereas the genotype is defined as the genetic characteristics. The structure of DNA was not to be discovered for another 40 years, which meant Johannsen's work was significant in its own right. Johannsen's insights were transmitted to the Carlsberg Laboratory via Schmidt's assistant scientist, Øjwind Winge who had been Johannsen's student.

5.5 Øjvind Winge *Mycologist Øjvind Winge was Schmidt's assistant from 1911–21, where they both embarked on landmark studies of animal heredity (The Archives of the Natural History Museum).*

5.6 Wilhelm Johannsen
Plant physiologist Wilhelm Johannsen invented the concept of genes, and he was an ardent advocate of Darwin's principle of natural selection. While Johannsen experimented with beans, Schmidt pioneered the study of hereditary qualities in fish (The Archives of the Natural History Museum).

Certainly the scientific issues of heredity would have been appealing to Eugen Warming as well, as Warming was an ardent lamarckist, and in fact had a feud with Johannsen over Darwin around the same time, when Warming in 1915 published his own book on heredity.[46] While this is speculative, Warming may well have prioritized his personal scientific interests in the case of letting Schmidt – perhaps even encouraging him to take up his own studies in the emerging field of genetics, although nothing seems to have been written down between them on this topic.

46 Prytz S. *Warming – botaniker og rejsende*. Lynge: Bogan's Forlag A/S; 1984, p. 115.

5.7 Eugen Warming
Eugen Warming is widely heralded as the father of the concept of ecology: the interaction of different species in a larger system, where ecological niches are created. He was also heavily involved in introducing Darwin's ideas on evolution in various empirical fields, but as a devout Christian he refused to accept natural selection as an evolutionary mechanism.

For Schmidt personally, the interest in genetics may have sprung out of a desire to know more about different races, or as we would put it today, different populations of fish species. In virtually all fish stocks there are small differences in size, coloration, growth rates, and the number of vertebrae depending on, where the fish lives. The main question then was: To what extent are the observed differences in fish species dictated by differences in their living environment, and to what extent is this due to the effect of their hereditary qualities?[47]

To begin with, Schmidt initiated a number of experiments in fish tanks with two species: the common European trout (*Salmo trutta*) and the guppy (*Lebistes reticulatus*). The guppy is a small tropical fish from South America, mainly known today as a very popular fish in freshwater aquaria. In the case of both species, Schmidt's studies showed that the temperature of the water had an impact on the number of vertebrae. This meant that not every difference in appearance within a fish species was due to hereditary qualities. The environment also played a role in the outcome of the offspring. However, when looking at trout, in particular, it appeared as if the number of vertebrae fluctuated within a certain range, regardless of the temperature in the hatchery.[48] In other words, these investigations seemed to confirm Johannsen's distinction between phenotype and genotype. The genotype responded to the environment, thereby providing a range of phenotype qualities.

The 'diallel crossings'

What Schmidt now added to this understanding of the role of genes, was a novel method to assess the hereditary aspect of quantifiable characteristics such as the number of vertebrae in a fish, in a chicken, or as the laboratory tried to test around 1920, the number of mammary glands in pig sows. Johannsen had already established that there was a range of phenotype variations possible when the seed of beans germinated. The continuous germination from the same gene, created a so-called *pure line*. This was not possible in the case of fish, where a male and a female are needed to create the offspring. When two different specimens are crossed, the mixture of genes corrupts any attempt to maintain a pure line.

Schmidt's idea was to systematically cross the male and the female in a cohort in such a way that all males were crossed with all females. First the eggs of a female would be taken out and divided in a number of portions. Then each portion of eggs would be germinated with the sperm from each of the male participants in the experiment. Thus, four male and four female fish would give sixteen combinations of

47 Winge Ø. Andre strenge anslaas – paa Carlsberg Laboratoriet. In: Winge Ø, Vedel Taaning Å, eds. *Naturforskeren Johannes Schmidt. Hans liv og ekspeditioner*. Copenhagen: Gyldendalske Boghandel Nordisk Forlag; 1947: 72.

48 Winge Ø. Andre strenge anslaas – paa Carlsberg Laboratoriet. In: Winge Ø, Vedel Taaning Å, eds. *Naturforskeren Johannes Schmidt. Hans liv og ekspeditioner*. Copenhagen: Gyldendalske Boghandel Nordisk Forlag; 1947: 72.

offspring. When the young fish had reached a certain age, they were killed and examined, for instance to count the number of vertebrae. As there were plenty of fish with each combination of parents, it was possible to measure the average vertebrae count for each combination of parents. The number of vertebrae in the mother plus the number of vertebrae in the father divided by two would equal the average value of their offspring. As an illustration of this, two specimens of trout had 59 vertebrae each. However, when measuring their genetic disposition the experiments found that one of the two trouts possessed a hereditary average of 59.54 vertebrae. The other trout with a vertebrae count of 59 was genetically disposed for having 58.57 vertebrae, or almost one vertebra less than the other one.[49]

When in 1947, Winge recalled his work on fish genetics at Schmidt's laboratory he stated that the method mentioned above had caused considerable interest among theorists and practitioners of animal breeding. Winge proposed that Schmidt's method for diallel crossings had great potential in areas such as animal husbandry, where it would indeed be possible to breed for the desirable hereditary qualities of a particular animal.[50] According to a more recent article of the legacy of Winge, this was one legacy of the two men working together.[51] While the link may not be direct, modern DNA coding has shown that the breeding of the domestic pig has resulted in ever more ribs to be present in modern pig husbandry.[52]

The hereditary qualities of chromosomes were another novelty of the Carlsberg Laboratory during Schmidt's reign with Winge as his assistant. Schmidt's and not least Winge's work was ground breaking in this important field of genetics as well. The crossing of guppies again proved fruitful. The male of these fish is shiny and speckled with red, yellow, and black spots on the skin, whereas the female is grey. Crossing specimens, where the male was one type, or race, of guppy and the female was from another type, indicated that the mother had no impact on the outer appearance of the male offspring. The father alone was indicative for the looks of the sons, so to speak. The hypothesis was then proposed that some genetic qualities were linked specifically to the gendered Y-chromosome. Following Winge, this was the first experimental proof anywhere in the world that some hereditary qualities were exclusively tied to gender.[53] Subsequently, Schmidt and Winge were both cited and credited with this discovery in an overview paper on Mendelian inheritance in fish.[54] While the first paper appeared with Schmidt as the

49 Winge Ø. Andre strenge anslaas – paa Carlsberg Laboratoriet. In: Winge Ø, Vedel Taaning Å, eds. *Naturforskeren Johannes Schmidt. Hans liv og ekspeditioner.* Copenhagen: Gyldendalske Boghandel Nordisk Forlag; 1947: 73.

50 Winge Ø. Andre strenge anslaas – paa Carlsberg Laboratoriet. In: Winge Ø, Vedel Taaning Å, eds. *Naturforskeren Johannes Schmidt. Hans liv og ekspeditioner.* Copenhagen: Gyldendalske Boghandel Nordisk Forlag; 1947: 73.

51 Szybalski W. My road to Øjvind Winge, the father of yeast genetics, *Genetics* (1 May, 2001) vol. 158 no. 1, 1–6.

52 Rubin C, Megens H, Barrio AM, et al. Strong signatures of selection in the domestic pig genome. *Proceedings of the National Academy of Sciences.* 2012;109(48):19529–19536.

53 Winge Ø. Andre strenge anslaas – paa Carlsberg Laboratoriet. In: Winge Ø, Vedel Taaning Å, eds. *Naturforskeren Johannes Schmidt. Hans liv og ekspeditioner.* Copenhagen: Gyldendalske Boghandel Nordisk Forlag; 1947: 74.

54 Goodrich HB. Mendelian inheritance in fish. *The Quarterly Review of Biology* 1929;4(1):83–99.

sole author, Winge subsequently published several studies in his own name.[55]

55 Schmidt J. Racial studies in fishes. II. Experimental investigations with Lebistes reticulatus (Peters) Regan. *Journal of Genetics*. 1919;8(3):147–153; Winge Ø. The location of eighteen genes in lebistes reticulatus. *Journal of Genetics*. 1927;18(1):1–43; Winge Ø. Crossing-over between the X-and the Y-chromosome in-Lebistes. *Journal of Genetics*. 1923;13(2):201–217; Winge Ø. A peculiar mode of inheritance and its cytological explanation. *Journal of Genetics*. 1922;12(2):137–144; Winge Ø. One-sided masculine and sex-linked inheritance in lebistes reticulatus. *Journal of Genetics*. 1922;12(2):145–162.

Investigations with *Zoarces*

Despite these achievements in the laboratory, Schmidt's main interest still lay in problems of marine science. With a series of studies of the eelpout (*Zoarces viviparous*) and a study of common trout (*Salmo trutta*), Schmidt investigated the problem of why and how most fish species develop local characteristics.[56] For several decades, scientists had observed and wondered why it is that the fish in the oceans are made up of different 'local forms,' 'races,' or 'families,' as Schmidt remarked in brackets in his first of several *Racial Studies* in the *Journal of Genetics*.[57] Here, Schmidt gives an example that provides information about fish fluctuations and fish migrations. Why for instance, were herring in the Baltic Sea and herring in the White Sea slightly different from one another in terms of size and other physical characteristics such as the number of vertebrae and fin rays as well as their behaviour in matters of spawning time and age of maturity? This question was not only of practical interest but also interesting from the point of view of examining fishes and how hereditary differences are influenced by environmental impacts such as differences in salinity and temperature.[58]

In other words, Schmidt was debating Darwin vs. Lamarck in the case of fishes. Someone like Eugen Warming with his lamarckist views would propose that these external conditions had a direct determining impact on the offspring in subsequent generations. On the contrary, Winge's professor from university, Wilhelm Johannsen would be convinced that external conditions had an indirect impact through a process of selection.

With his own recent method of diallel crossing Schmidt was now able to address this issue, at least with species, where it was possible to keep track of the offspring of individual fish. Eelpout is a viviparous fish, which means that it gives birth to live fish, some 400 towards the end of the winter season. Therefore, Schmidt could examine the progeny of a single female and compare the qualities of the 'babies' with those of the mother. From 1914 until 1918 Schmidt was in charge of examining more than 25,000 specimen of eelpout caught in shallow fiords all over Denmark, where they thrived quite prolifically.[59] Until 1917 he did the

56 Schmidt J. Racial studies in fishes. IV. experimental investigations with *zoarces viviparus* L. *Journal of Genetics*. 1920;10(3):179–191; Schmidt J. Racial studies in fishes. III. diallel crossings with trout (*salmo trutta L.*). *Journal of Genetics*. 1919;9(1):61–67; Schmidt J. Racial studies in fishes. I. statistical investigations with *zoarces viviparus* L. *Journal of Genetics*. 1918;7:105–119; Schmidt J. Diallel crossings with the domestic fowl. *Journal of Genetics*. 1922;12(3):241–245.

57 Schmidt J. Racial studies in fishes. I. statistical investigations with *Zoarces viviparus L.* *Journal of Genetics*. 1918;7:105–119.

58 Schmidt J. Racial studies in fishes. I. statistical investigations with *Zoarces viviparus L.* *Journal of Genetics*. 1918;7:105–119.

59 Schmidt J. Racial studies in fishes. I. statistical investigations with *Zoarces viviparus L.* *Journal of Genetics*. 1918;7:105–119.

work himself, while from 1917–1919 marine scientist Vilhelm Ege was in charge of counting.[60]

The eelpout has the additional characteristic of being very localized. Many subpopulations as we would call them today lived separately within fairly short distance from one another. Schmidt and Ege counted the vertebrae of large numbers (>200) of eelpout from each of a number of different localities while at the same time measuring salinity and temperature. Schmidt found that the number of vertebrae could vary from 101 to 126, quite a range for one fish species. However, for any particular population of eelpout the average number of vertebrae remained stable for the three consecutive years, for which the investigation took place. However the average of particular subgroups ranged from 108 up to 119.5 vertebrae. In his concluding remarks, Schmidt explicitly refers to the debate over hereditary qualities: 'My own interpretation of the offspring analyses, which is supported by the results arrived at in 1903 by W. Johannsen in his studies on bean communities, is roughly as follows. It was found possible by population analyses to subdivide *Zoarces viviparous* into numerous local 'races.' The individual offspring analyses showed that the smallest unit hitherto considered, the 'race,' may be resolved into still smaller elements, expressed by the means of the offspring samples. These smaller elements may differ widely one from another, and likewise from the average of the population ('race'), but when added together they reproduce the picture of the race itself as the latter is expressed in the results of the population analyses.'[61]

Schmidt had also investigated the possible link between the average number of vertebrae in different subgroups, where there was a very large difference in salinity, but there was no direct linkage. Nor did he find such a link when comparing two subgroups in Kjelds Nor and Roskilde Fjord with more or less identical salinity and temperature conditions. Nonetheless, these two sub groups exhibited completely different numbers when accounting for vertebrae. While Schmidt was leaning towards the conclusion that 'A fish "race" is largely a statistical conception. It implies a mixing of different genotypes and the average values characterizing the "race" are primarily dependent upon the quantitative proportion between these; only secondarily on the environment.'[62] According to Schmidt a more final conclusion to this problem would be found only through experimental analyses, and soon he would be able to supply just that.

60 Schmidt J. Racial studies in fishes. IV. experimental investigations with *Zoarces viviparus* L. *Journal of Genetics*. 1920;10(3):180.

61 Schmidt J. Racial studies in fishes. I. statistical investigations with *zoarces viviparus L. Journal of Genetics*. 1918;7:115.

62 Schmidt J. Racial studies in fishes. I. statistical investigations with *zoarces viviparus L. Journal of Genetics*. 1918;7:117.

To further investigate the genotypical characteristics of eelpout Schmidt now transplanted a number of eelpout specimen from one part of the Isefjord and Roskilde Fjord system to another, where he would place both the transplanted fish, and the resident control group, in big perforated wooden boxes to keep them alive and thriving. This was a test of the role of internal factors, and the results showed that the difference in vertebrae count of five, which had been found to exist between the two groups under normal conditions, did not disappear, when the transplanted group was moved to a different location. To Schmidt, this suggested that the genotypical characteristics of the transplanted fish outweighed the influence of environmental change.

Another type of investigation was to compare two samples of the same population developed under different environmental conditions. This time, eelpout was moved from the brackish inner parts of Roskilde Fjord to the much more saline conditions towards the mouth of the fjord system near the Kattegat. Here the results suggested that under the influence of a changed environment the average number of vertebrae rose. Schmidt though, was certain that this was a case of phenotypical variation under certain environmental conditions, yet he was not sure if temperature, salinity, or something else was responsible for these observations. The results were clear though. Schmidt still concluded that the gradation of the average number of vertebrae was to be interpreted as the result of a selection process weeding out the misfits – for instance that the larger vertebrae count in bigger specimen was somehow better fitted to a situation, where salinity and temperature were higher. It was not the case that the environment directly impacted the offspring.[63]

Schmidt's first study in the series of so-called Racial Investigations, on *Zoarces* was reviewed in *Nature* in November of 1918, so the international scientific community did pick up these manuscripts from a private laboratory in Copenhagen.[64] The *Comptes-rendus des travaux du Laboratoire Carlsberg* was a fine showcase in its own right not least when bearing in mind the substantial costs associated with publishing in those days. Nonetheless Schmidt took care to also publish his 'racial studies' in the international literature in the *Journal of Genetics*, which ensured him an international reputation.[65]

63 Schmidt J. Racial studies in fishes. IV. experimental investigations with *zoarces viviparus* L. *Journal of Genetics*. 1920;10(3):179–191.

64 KB. Johannes Schmidt 'Papirer'. Ut.374.V.2. 'Privat og Blandet korr. 1912–1927'. *Nature*, 2558, 102, 7 November 1918.

65 Schmidt J. Racial studies in fishes. IV. experimental investigations with *zoarces viviparus* L. *Journal of Genetics*. 1920;10(3):179–191; Schmidt J. Racial studies in fishes. III. diallel crossings with trout (*Salmo trutta* L.). *Journal of Genetics*. 1919;9(1):61–67; Schmidt J. Racial studies in fishes. I. statistical investigations with *zoarces vivparus* L. *Journal of Genetics*. 1918;7:105–119; Schmidt J. Diallel crossings with the domestic fowl. *Journal of Genetics*. 1922;12(3):241–245; Schmidt J. Racial studies in fishes. II. Experimental investigations with *Lebistes reticulatus* (Peters) Regan. *Journal of Genetics*. 1919;8(3):147–153.

A chicken farm

From 1921 until the early death of Johannes Schmidt in 1933, the Carlsberg Laboratory ceased to be an important hub for racial studies. This has a lot to do with the fact that in 1921, Schmidt's brilliant assistant, Øjvind Winge left the laboratory for a newly established chair in genetics at the Veterinary and Agricultural University in Copenhagen.[66] From Winge's own account of his years as an assistant, Schmidt had completely lost interest in the experiments conducted on the small tropical fish, *Lebistes reticulatus.*

By 1921, Schmidt's interests were entirely devoted to his oceangoing expeditions, and Winge was allowed to take all the Lebistes material with him, to his new position at the Agricultural University. Already the following year, 1922, Winge published some of the abovementioned studies in genetics based on his work in Schmidt's laboratory. Through cross breeding experiments Winge was able to present the first positive example of how hereditary dispositions could travel from an X-chromosome of the parent to the Y-chromosome of the offspring. Schmidt had housed, and at a significant rate fathered a scientific breakthrough of great future impact. Winge remembered many years later that he was overly excited by his new discovery. He therefore paid Schmidt a visit at Carlsberg a few kilometres down the road from the Agricultural University, only to find his former mentor completely absorbed in the problems of the great oceans. Perhaps therefore, Schmidt listened with little interest to Winge's story, and Schmidt then made the following comment: 'Well, it may be that you yourself find this result peculiar, but isn't it utterly trivial, how a pair of red dots is transmitted with the *Lebistes*!'[67]

This episode clearly made a deep impression on Øjvind Winge, but looking back Winge recalls the episode as an example of how Schmidt could be totally immersed in one scientific problem, only to become completely indifferent to other problems, even in the case of an issue Schmidt had written about only three years earlier, as was the case with the tiny guppies.[68]

In hindsight, Schmidt's effort in genetics has been overshadowed by the works of Øjvind Winge, his apprentice and assistant, and later his successor as laboratory director. However, according to a later commentary on Winge, Schmidt was deeply interested in genetics, and without a doubt he influenced Winge's thinking and his transformation from a

66 Westergaard, M., 'Øjvind Winge 1886–1964', in: Holter H, Max Møller K, eds. *The Carlsberg Laboratory 1876–1976*. Copenhagen: Rhodos; 1976: 217–218.

67 Winge Ø. Andre strenge anslaas – paa Carlsberg Laboratoriet. In: Winge Ø, Vedel Taaning Å, eds. *Naturforskeren Johannes Schmidt. Hans liv og ekspeditioner*. Copenhagen: Gyldendalske Boghandel Nordisk Forlag; 1947:78.

68 Schmidt J. Racial studies in fishes. II. Experimental investigations with *Lebistes reticulatus* (Peters) Regan. *Journal of Genetics*. 1919;8(3):147–153; Winge Ø. Andre strenge anslaas – paa Carlsberg Laboratoriet. In: Winge Ø, Vedel Taaning Å, eds. *Naturforskeren Johannes Schmidt. Hans liv og ekspeditioner*. Copenhagen: Gyldendalske Boghandel Nordisk Forlag; 1947:78

mycologist and cytologist studying cell forms of fungi, to one of Denmark's leading geneticists.[69]

At any rate, the Carlsberg Laboratory evolved into a sizeable chicken farm around 1920. According to Winge, the diallel crossing with the domestic fowl gradually took up a significant amount of space in the laboratory garden and in the green house. Over a span of several years, no less than 10 electric egg hatching machines were in operation simultaneously, each containing 200 eggs. At a highpoint of these experiments some 3,000 chicken were hatched each month.[70] The hatching of egg provided one instance where the Board of Trustees actually intervened with Schmidt's research, even if it was in the most trivial way. In 1924, one of the laboratory's neighbours filed a complaint to the Board of Trustees, as Dr. Schmidt's cockerels were making too much noise, in particular in the morning. The board's response was to pay for having the henhouse moved further away from that particular neighbour.[71]

Winge was replaced as Schmidt's research assistant by Vilhelm Ege. Ege was a fisheries scientist by training and already working part-time for Schmidt, so this marked a final turn towards focusing on marine science, as we will see in the next chapter.

Conclusion

This chapter opened with the blunt statement from Povl Krogsgaard-Larsen that the Carlsberg Laboratorium's Board of Trustees had committed an act of treason towards the founder of the brewery, I. C. Jacobsen, when they hired Johannes Schmidt to replace Emil Chr. Hansen as head of the Physiological Department of the laboratory. In hindsight, there is no doubt Schmidt's career spanned a vast array of research interests, yet very few of these had any bearing on what might be termed as conducive for the improvement of beer brewing. On the contrary Schmidt became more and more famous over the years – in areas completely separate from what one might think he should have been working on.

Schmidt's old professor from university, Eugen Warming put Schmidt forward for the job at the Carlsberg Laboratory. Warming was also the one who proposed that Schmidt should be allowed to continue his involvement in the Danish commission for marine science. The rest of the Board of Trustees however, approved of this arrangement, which they will have embraced openheartedly at the time. When reviewing

69 Westergaard, M., 'Øjvind Winge 1886–1964', in: Holter H, Max Møller K, eds. *The Carlsberg Laboratory 1876–1976*. Copenhagen: Rhodos; 1976: 213.

70 Winge Ø. Andre strenge anslaas – paa Carlsberg Laboratoriet. In: Winge Ø, Vedel Taaning Å, eds. *Naturforskeren Johannes Schmidt. Hans liv og ekspeditioner*. Copenhagen: Gyldendalske Boghandel Nordisk Forlag; 1947: 73.

71 Olesen Larsen P. Iværksætter og forvaltere. Carlsberg Laboratorium 1876–2014. Copenhagen: Carlsbergfondet; 2015. p. 95.

the composition of the board, it springs to mind that Schmidt had a special personal relationship with a clear majority of members. It will no doubt have been an advantage that Schmidt had grown up on the Carlsberg premises, and that he had married Ingeborg van der Aa Kühle, the daughter of former CEO Søren van der Aa Kühle. In short, Schmidt was a well-known and well-liked character in the 'extended Carlsberg family,' so to speak.

In spite of all this trust, or perhaps because of it, Schmidt's tenure at the Carlsberg Laboratory had a rather slow start. From the time he was hired more than one year lapsed before he took up office, although the position was vacant. Further, it was not until 1911 that he produced his first research plan for his department. Schmidt found a vacant research field through his investigations into the hereditary qualities of scent in hops. However, this first and only attempt at conducting a truly beer related piece of research never made it beyond some fairly trivial results. Schmidt maintained a garden on the laboratory premises where he grew hops throughout his career, but from around 1914 he seems to have lost interest in the prospects of taking this further. The bibliometric review of Schmidt's research publications revealed that during his 23 years at the helm of the Carlsberg Laboratory less than 10 per cent of his publications were related to beer brewing.

In terms of breaking new ground within the sciences Schmidt was much more fortunate with his next adventure. From c. 1914–1920, Schmidt, possibly encouraged by Warming, who was chairman of the board from 1914–1921, was breaking new ground in the nascent field of genetics with his studies of various types of plants and animals, principally fish. This coincides with Schmidt hiring Øjvind Winge as his assistant researcher in 1914. With the acquisition of Winge, Schmidt gained access to state of the art expertise in genetics, which he would benefit from until Winge left the laboratory in 1922 to become a professor at the Agricultural University. During this time Schmidt is widely credited with having developed the method of so-called diallel crossing, which measured the hereditary qualities of animals by looking at the number of vertebrae.

Schmidt and Winge also discovered that some hereditary qualities were gendered, that is that they were tied to the Y-chromosome, something they discovered through diallel crossing of the small guppy, *Lebistes*. Moreoever, through cross breeding experiments at the Carlsberg Laboratorium Winge was also able to present the first positive example of how hereditary dispositions were able to travel from an X-chromosome of a parent to the Y-chromosome of the offspring.

Nonetheless, in spite of working as a laboratory director for years, Schmidt's main interest lay still within problems of marine science. With a series of experimental studies of eelpout Schmidt investigated the problem of why and how most fish species develop local characteristics. This question was not only of practical interest but also interesting from the point of view of examining how hereditary differences in fish are influenced by environmental impacts such as differences in salinity and temperature. Someone like Eugen Warming was an ardent defender of the lamarckist view that the external conditions would have a direct impact on the offspring in subsequent generations. Warming's greatest opponent in Denmark on this issue was Wilhelm Johannsen who was convinced that external conditions had an indirect impact through a process of selection. Eventually, Schmidt placed his weight behind Johannsen following a six-year-long investigation, sampling more than 25,000 specimens of eelpout.

In this light, it makes a lot of sense that Warming would be content with the on-going work at the Carlsberg Laboratorium from 1914–1921, where he was Chairman of the Board of Trustees. As soon as Schmidt began investigating the hereditary qualities of fishes he found himself merging the greatest interest of both of his employers.The marine science commission had a stake in the practical implications of his studies, while Warming had a deeply rooted theoretical foible for the emerging field of genetics. This observation does not render Warming innocent in terms of not complying with the ethos embodied in the statutes of the Carlsberg Laboratorium, but it serves to highlight one very plausible reason, why the Board of Trustees so readily accepted Schmidt's choice of research. The extent of just how much Schmidt's superiors appreciated his work over more than two decades is the topic of investigation in the following chapter.

6 Carlsberg appreciation

Schmidt's laboratory expenses

Marine science, including various hereditary studies took up roughly 90 percent of Schmidt's time at the Carlsberg Laboratory. Hereby, the Carlsberg Laboratory automatically should be ranked among the most significant scientific hubs of marine science in Denmark before World War II. The laboratory was funded through an annual lump sum provided by the Carlsberg Foundation. Thus it follows that the Carlsberg Laboratory could sponsor significant amounts of research in addition to, what was granted directly through the Carlsberg Foundation. Clearly, Schmidt and colleagues, in particular Øjvind Winge, excelled in a number of areas. In addition to Schmidt's famous marine research, he deserves credit for coining and using the concept of *diallel crossings*, while his racial investigations were ground breaking in their own right. The investigations on hops were probably of minor value. Nonetheless, all this excellence came at a cost, which can be calculated from the annual budgets for the Carlsberg Laboratory.

To track the flow of money from the Carlsberg Foundation via the Carlsberg Laboratory and into the wide range of research activities undertaken by Johannes Schmidt during his laboratory years a database was created covering the annual budgets for the period 1908–1935.[1] The Carlsberg Archives are also in possession of the annual accounts from the laboratory, so it would have been possible to undertake this study as an analysis of money *spent* at the laboratory. This would also be interesting, however, for comparison with the assessment of Schmidt's other fundraising activities, where the focus is on what he was *granted* as opposed to what he spent, the following attempt at quantifying the price of Schmidt's laboratory research is based on the budget instead of the turnover. However, the two figures rarely differed much from each other,

1 CA. *Carlsbergfondet, Carlsberg Laboratorium*. E06. '1908–1913'.Korrespondance, regnskaber og beretninger; CA. *Carlsbergfondet, Carlsberg Laboratorium*, E31. Endelige budgetter og noter 1914–1972.

as the annual lump sums flowing in from the Foundation were the only sources of revenue for the Carlsberg Laboratory.

The budget principles for the lump sum for the laboratory were fairly similar throughout the 23 budget years, when Schmidt was in charge of the Department of Physiology. The money flowed from the accounts of the Carlsberg Foundations to four sections of expenditure within the laboratory. The two main departments each received a lump sum, which went mainly into paying the salary of the scientists and the salary of the laboratory technicians. From its inauguration in 1931, the extraordinary Department of Yeast Physiology was bestowed with a separate lump sum. Finally, there were the expenditures entitled 'shared costs'. These included items such as heating and electricity, diverse maintenance, laboratory equipment, acquisition of books, and last but not least the publication of the Laboratory's own journal, the *Comptes-rendus des travaux du Laboratoire Carlsberg*.

Schmidt officially began his term as director on the 1 October 1910. This also coincided with the beginning of a new fiscal year. Throughout Schmidt's directorship the fiscal year ran from 1 October to 30 September in the following calendar year. The database includes the years from 1908–1935 to see if there was much change in expenditure in connection with the change of directorship in 1910 and again in 1933. This was the case for the first year, when Schmidt took office, but not in the following years, and neither did Øjvind Winge's appointment in 1933 put any immediate strain on the budget for 1933–34. As we will see below, Winge's appointment prompted a large renovation of the laboratory rooms and equipment of the Department of Physiology, but the extra costs associated with the change of directorship were covered directly from the Carlsberg Foundation.[2]

Figure 6.1. presents an overview of the scale of the budget of the Carlsberg Laboratory during Schmidt's reign, 23 years from 1910–1933. Until 1917, the total lump sum for the laboratory fluctuated between 78,000–96,000 Danish Kroner annually. For the following five years, the economic situation of the laboratory improved as the subsidy more than doubled, peaking in 1921/22 with close to 200,000 Kroner of funding. Thereafter, a slower decrease followed. This is mainly due to the termination of Albert Klöcker's appointment as Director extraordinaire, which resulted in savings on his and his assistant's salaries.

2 CA. *Carlsbergfondet, Carlsberg Laboratorium.* E10.1933. no. 20. Circular Board of Trustees, 7 June 1933.

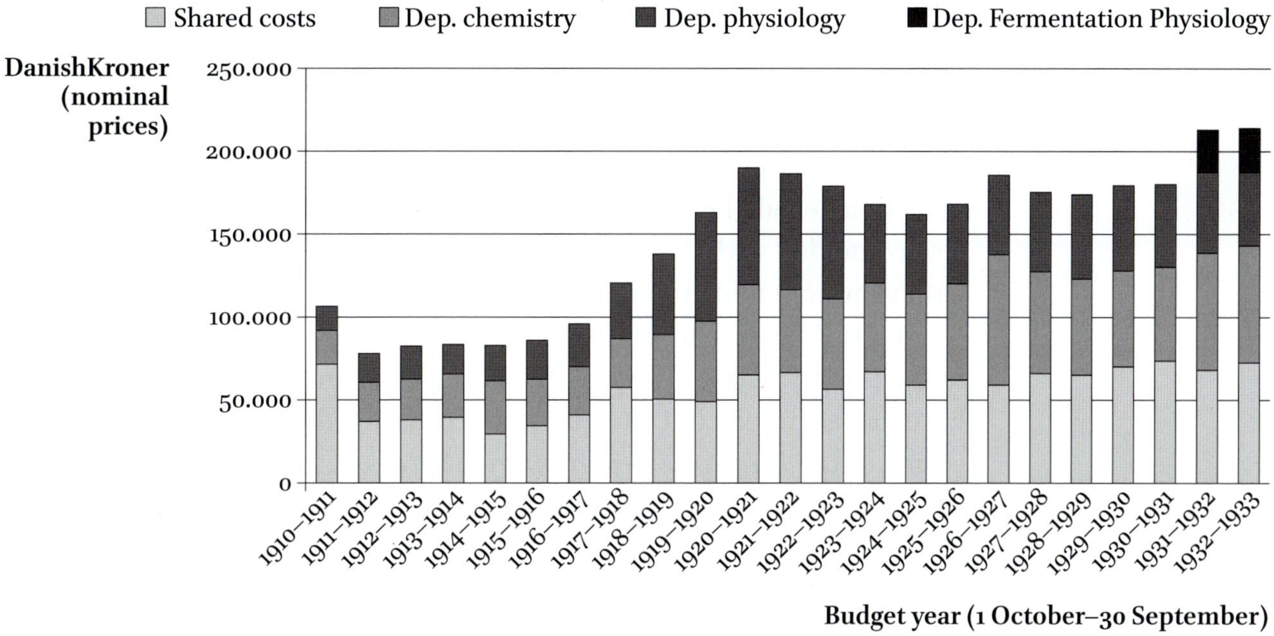

6.1 Sums of annual budgets, Carlsberg Laboratory, 1910–1933

During the last ten years of Schmidt's term in office, the laboratory received between 162,000 and 180,000 Danish Kroner a year, except for the latter two years. The new Department of Fermentation Physiology received an additional c. 25,000 Kroner a year, and Schmidt's Physiology Department received an additional 10,000 Kroner for 'temporary assistants' for the last two years of Schmidt's era. It should be noted though, that the figures are listed as nominal prices. The latter years of World War One were marked by a fair amount of inflation, whereas deflation hit the country in the early 1920s. Taking this into consideration the annual subsidy was larger in the 1920s than what the graph appears to show.

In any case, the economic situation of the laboratory provided some very stable working conditions from one year to the next, and the financial condition was sound throughout the period in question. Looking at the correspondence and accounts from the laboratory one gets a clear picture that lack of money never seemed to be an insurmountable issue.[3] During the time of World War One, the price of heating soared as British coal was in short demand, but at least in the written evidence from the Board of Trustees, there is no indication that the laboratory was seriously hampered by this. The backing of the Carlsberg Foundation was a guarantee for stability.

3 CA. *Carlsbergfondet, Carlsberg Laboratorium.* E06-E10.'1909–1933'.Korrespondance, regnskaber og beretninger.

When looked upon as an entity in its own right, Schmidt's laboratory witnessed a similar upswing in resources allocated during the latter half of the 1910s. As evident from figure 6.2 in 1910 when Schmidt took charge of the budget he was able to spend a little less than 15,000 Kroner, but a decade later he disposed of more than 70,000 Kroner. This also coincided with the most intensive phase of the experimental work with genetics taking up much space and labour.

Albert Klöcker's accounts figured in the overall lump sum given to the Department of Physiology. Schmidt however, does not seem to have had as much control as he would have liked with respect to the efforts of the *Director Extraordinaire*. Therefore, the money for Klöcker's salary and the salary of Klöcker's assistant, Folker might not in reality form part of the funds, which Schmidt was handed, but formally he was Klöcker's boss and thus formally Klöcker's part of the laboratory is included in the Department of Physiology.

Nonetheless, to give a fair picture of the economic assets at Schmidt's disposal, his portion of the shared costs of the laboratory should also be considered part of his resourcefulness. The Department of Chemistry and the Department of Physiology were more or less of equal size in terms of square meters of office space and the equipment needed. Also there is no reason to believe that one department got a significantly larger share of for instance book acquisitions than the other. During the 1920s Schmidt stacked the laboratory library with marine science titles. We cannot assess to what extent Professor Sørensen of the Chemical Department found this a nuisance, but it will not have created much synergy to his own work, except perhaps for Palitzsch's studies of hydrogen concentrations in sea water (see chapter 4).

Therefore, to get a full picture of Schmidt's total resources at the Carlsberg Laboratory, one half of the shared costs have been added to the sum Schmidt was in charge of directly via his department. This brings the total assets for Schmidt's Laboratory research to 40–50,000 during the first seven years of his office, while for almost every year thereafter, Schmidt was responsible for spending in the area of 75–100,000 Kroner a year. This brings the total laboratory revenue for Schmidt's research during the period 1910–1933 to a staggering 1,640,000 Danish Kroner. Hereby, the laboratory was arguably the single largest sponsor of Schmidt's research.

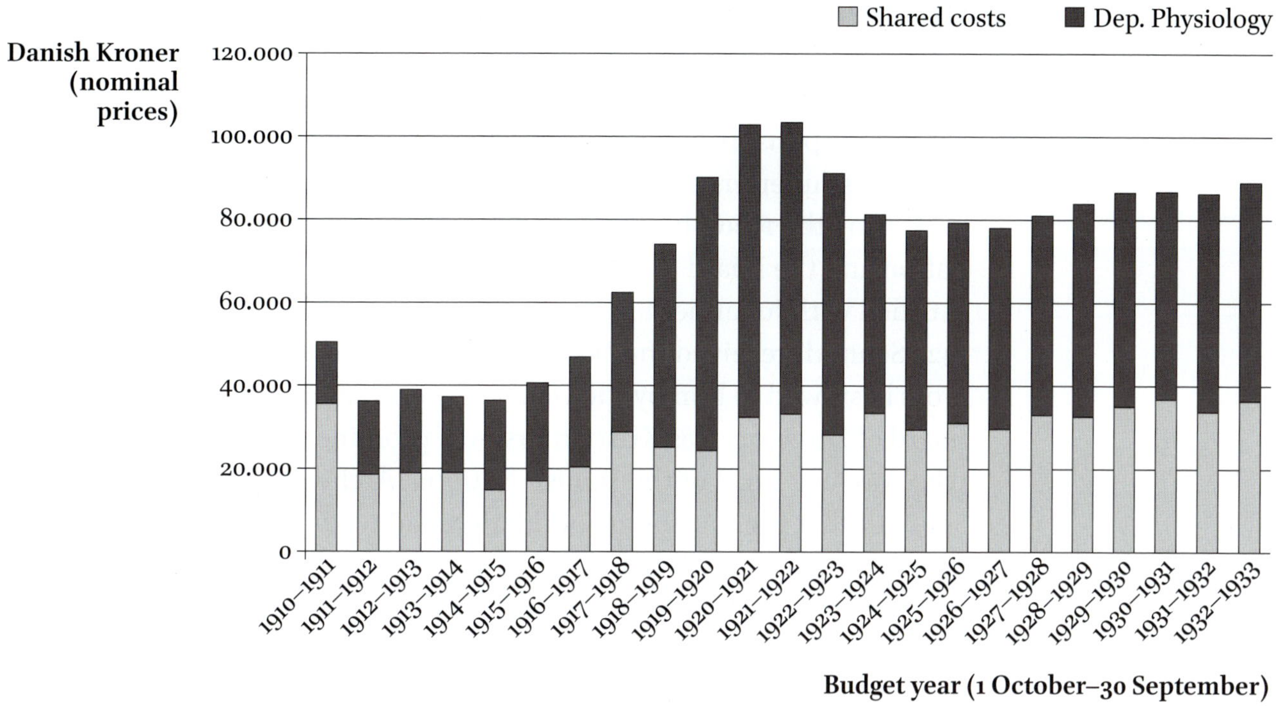

6.2 Johannes Schmidt's annual ressources at Carlsberg Laboratory

Personally, Schmidt also fared well from being a director. From the starting salary of 4,140 Schmidt's earnings rose quickly. In the expensive years in the aftermath of World War One, the gross pay exceeded 12,000 Kroner, but then stayed there for the rest of his employment. Øjvind Winge also earned a decent living from being Schmidt's assistant. He started off with 1,800 a year, but in his last couple of years, when he had emerged as an established Doctor of natural history, he reaped in more than 7,000 Kroner. Finally, Vilhelm Ege was granted 6,170 Kroner for his first year as Schmidt's first assistant in 1921/22, while his salary had risen to 7,800 in his last year working for Schmidt. When Winge returned to the laboratory to replace Schmidt as a senior professor he made 12,000 Kroner per year plus the benefits of the free housing and free beer. In addition, Schmidt employed a number of non-academically trained assistants, such as the gardener, Smith, and indeed with all the aquariums, greenhouses and rearing of chicken, pigs and rabbits there was plenty of work to be carried out. The many thousand specimens of fish for instance, first had to be boiled to remove the flesh so the vertebrates could be examined. Wages always weighed in significantly in the overall budget, on average a little more than 50 percent of the total.

Carlsberg Laboratory appreciation

With the Carlsberg Laboratory emerging as the most prominent of Schmidt's sponsors in quantitative terms it would be only natural that Schmidt's superiors took an interest in how the funds were spent.

During the 1920s, both the Board of Directors of the Carlsberg Foundation and the Board of Trustees of the Carlsberg Laboratory were satisfied with Johannes Schmidt's accomplishments. In the summer of 1922 Schmidt received the *Dannebrogsmændenes Hæderstegn*, a silver cross bestowed to him by the Danish government. The Chairman of the board of Directors was Kristian Erslev, a prominent Danish historian who was also Head of the Danish National Archives. Erslev informed Schmidt that the Carlsberg Foundation had recommended him to the government for this distinction.[4] Schmidt also thanked his immediate employer, the Carlsberg Laboratory for their support in a letter to the board: 'In this I dare to see an appreciation of my work in the service of the laboratory.'[5]

Indeed the laboratory board had initiated the official decoration for Schmidt, something which several board members observed to be a meagre sort of recognition. What the members of the board really wanted, was to have Schmidt bestowed with the honorary title of professor. The laboratory board had encouraged the Board of Directors to submit a request to the Ministry of Education that Schmidt should be honoured with the title of 'Professor'. Officially only employees at a university could be entitled professor, but the Carlsberg Laboratory obviously had a special status.

Schmidt's champion, Warming and the chairman of the Board of Directors, Kristian Erslev actively pursued the idea of securing Schmidt the title of professor as early as 1920. Following conversations with the Prime Minister, Professor Erslev remained rather pessimistic. Then, Warming on behalf of the Board of Directors went to see the Minster of Education, but apart from a brief encounter in a Copenhagen tram, Warming could not access the minister, and so called on the Head of the Department of Education, Weis, who could also not make any promises. In the end, the Carlsberg Foundation was unsuccessful in 1920 with its attempt to obtain official recognition for their laboratory director, who by this point had held his post for almost ten years.[6]

In 1916, the silver cross could not be granted because too little time had passed since Schmidt had been knighted in 1908, then in 1920

4 KB. Johannes Schmidt 'Papirer'. Ut.374.I.4. '1922–1925'. K. Erslev to J. Schmidt, 27 July 1922.

5 KB. Johannes Schmidt 'Papirer'. Ut.374.I.4. '1922–1925'. J. Schmidt to Board of Trustees, 27 July 1922.

6 RA. Komm. Danmarks Fiskeri- og Havundersøgelser: A 23 Dana-ekspeditionerne 1917–1924, E. Warming to C. F. Drechsel, 28 July 1920.

there was another unsuccessful try, but in the spring of 1922 the board submitted yet another thorough recommendation of Schmidt's accomplishments. The board members Hjelmslev and Henriques emphasized how pleased the laboratory was with Schmidt's racial investigations and the studies on the hereditary qualities and natural history of hops. 'His fortunate ability to think through and plan a task, which he sees through with great stamina, has reaped great results, cementing the reputation of the Carlsberg Laboratory, whilst providing Schmidt with a world famous name. The above mentioned investigations bode well for future valuable results.'[7]

Schmidt's work within marine science was mentioned only as a secondary achievement, and Schmidt's eel investigations were passed over unnoticed, something which board member J. Hjelmslev found striking, since this is what Schmidt was famous for in the public arena. Hjelmslev was very supportive, however, and he declared that it would be a poor surrogate for a professorship, if only the silver cross was granted to Schmidt.[8] Ostenfeld added that the other laboratory director, S. P. L. Sørensen had already obtained both a knighthood and a professorship while in the service of the Carlsberg Laboratory. Schmidt on the other hand, had received nothing of this kind.[9] Also, in 1922, Schmidt had to content himself with the silver cross. In 1926, however, there was a more positive exchange with the Minister of Education, Nina Bang who went some way to accommodating the wishes of the Carlsberg Foundation. She noted that while in principle the title of professor should be restricted to positions within the university system, she would not mind if this title was used by Johannes Schmidt, considering the Carlsberg Laboratory's outstanding position in the sciences.[10]

The globetrotter director

Overall, the Carlsberg Breweries, represented by the CEO, Poul C. Poulsen, also seemed to be pleased with Schmidt. In a letter from 1920, Poul C. Poulsen told Schmidt that for principal business reasons the brewery was not in a position to sponsor beer and soft drinks for Schmidt's upcoming expedition, as they could only deliver this to private homes. Poulsen then informed Schmidt that he was sorry to learn that Schmidt's health had forced him to go away again.[11]

7 CA. *Carlsbergfondet, Carlsberg Laboratorium.* E08.1922. No. 5. Letter from V. Henriques to Board of Directors, 5 April 1922.

8 CA. *Carlsbergfondet, Carlsberg Laboratorium.* E08.1922. No. 5. Letter from J. Hjelmslev to V. Henriques, 1 April 1922.

9 CA. *Carlsbergfondet, Carlsberg Laboratorium.* E08.1922. No. 5. Letter from C. Ostenfeld to V. Henriques, April 1922.

10 CA. *Carlsbergfondet, Carlsberg Laboratorium.* E08.1926. no. 3. Letter from Ministry of Education, Nina Bang to the Carlsberg Foundation's Board of Directors, 8 January 1926.

11 KB. Johannes Schmidt 'Papirer'. Ut.374.V.2. 'Privat og Blandet korr. 1912–1927'. P. C. Poulsen to J. Schmidt, 27 January 1920.

One might wonder why health issues would force a laboratory director to go to sea, as opposed to visiting a relaxing sanatorium or a spa in Central Europe, but it was a doctor's recommendation that he spend an extended period of time in a warmer and more pleasant climate than what was on offer in Denmark. When Schmidt visited the doctor in 1920 he was suffering from chronic bronchitis, which had been a problem for Schmidt for some time. For the past few years he had been confined indoors for large parts of the year. However, as Schmidt informed the Board of Trustees he now had the chance to go to sea for 4 months in the spring of 1920. Thus, he could combine recreation and science.[12] What he did not elaborate on here, is that the expedition he was about to embark on was a deep sea investigation to the Sargasso Sea, which he had been planning for several years. In any case, the board approved his leave, and in early July of the same year, Schmidt was bubbling with good news from the Atlantic Ocean. He had greatly advanced the riddle of the eel, and he had no problems at all with bronchitis during his voyage criss-crossing the North Atlantic Ocean.[13]

Schmidt's condition is likely to have been triggered by an allergic reaction to the powdered dried daphnia, which was used to feed the various fish in the laboratory, alternatively it could have been linked to the heart disease (*degeneratio myocardii*), which according to Winge ended up caused Schmidt's death in 1933.[14] In today's language this 'deterioration of the heart' would be labelled as heart failure, a common diagnosis, which includes symptoms such as fluid on the lungs or around the ankles.[15] Schmidt was aware that he had heart disease at least as early as 1922, when he conceded to having this problem in a letter to the French scientist M. Rabot.[16]

Back in Copenhagen at the start of 1923, Poul C. Poulsen congratulated Schmidt on his recent publications on the eel in prestigious international journals: 'Please let me express my admiration for the untiring scientific spirit, which has led to this result – and your equally untiring work.'[17] The only complaint from the brewery, however, also seems to have come from Poul C. Poulsen. Later in 1923 he sent a letter to Schmidt, in Schmidt's capacity as a Member of the Kommissionen for Havundersøgelser, where he regrets to have learned that the *Dana* vessel was not provisioned with products from the Carlsberg Breweries.[18] Apparently the principal reasons for not provisioning the expedition in 1920 had changed in the meantime.

12 CA. Carlsbergfondet, Carlsberg Laboratorium. E08. 1920. No. 6. Letter from J. Schmidt to Board of Trustees, 14 January 1920.
13 CA. Carlsbergfondet, Carlsberg Laboratorium. E08.1920. No. 29. Letter from J. Schmidt to Board of Trustees, 7 July 1920.14 January 1920.
14 Winge Ø. Andre Strenge anslaas – paa Carlsberg Laboratoriet. In: Winge Ø, Vedel Taaning Å, eds. *Naturforskeren Johannes Schmidt. Hans Liv og Ekspeditioner*. Copenhagen: Gyldendalske Boghandel Nordisk Forlag; 1947:79.
15 Marboe, Peter (MD & GP). Personal communication 5 January 2014.
16 KB. Johannes Schmidt 'Papirer'. Ut.374.I.4. '1922–1925'. J. Schmidt to M. Rabot, November 1922.
17 KB. Johannes Schmidt 'Papirer'. Ut.374.V.2. 'Privat og Blandet korr. 1912–1927'.
18 KB. Johannes Schmidt 'Papirer'. Ut.374.I.6. 'korrespondance 1925-(32)'. P. C. Poulsen to J. Schmidt, 30 May 1923. P. C. Poulsen to J. Schmidt, 2 January 1923.

In his annual laboratory report for 1923 Schmidt referred to the last of Winge's publications appearing in the *Comptes-Rendus* along with his own latest publication, a racial investigation on domestic fowl published in the *Journal of Genetics*.[19] Nonetheless, from here on the research directed by Schmidt, as well as his own publications drifted further and further away from what could be interpreted processes linked even remotely with beer brewing. Some hops investigations did continue according to Schmidt's annual reports of the 1920s, but they did not lead to any publications.[20]

In 1924, however, Schmidt's hops investigations became a topic for debate between Schmidt and CEO Poulsen from the brewery. It appears that Poulsen was optimistic about the prospects of the hops investigations and recommended that Schmidt continued along these lines.[21] While Schmidt and most of the staff were busy incorporating eel studies into the realm of the laboratory's racial investigations, it appears that the department's pharmacist, C. Svarrer carried out the investigations into the hops. Svarrer for instance, measured the relative content of different resins in hops.[22]

Meanwhile, Schmidt received a whole series of international decorations, coming to him like pearls on a string. Later chapters will present a more in depth analysis of Schmidt's efforts within marine science, which earned him his international decorations, and from 1923 onwards, his achievements in genetics were also recognized. As he had to report to the Board of Trustees and seek their permission for leave, whenever he intended to travel to another country, Schmidt's letters of permission give a good indication of how his duties became more and more organizational and representational in character. Thus, in 1923, he was away twice for longer trips. In May and June Schmidt spent five weeks in Paris, where Schmidt, accompanied by his wife, Ingeborg, represented Carlsberg at the celebrations in connection with the centennial anniversary of Louis Pasteur. This included a private dinner with relatives of Pasteur, as well as official ceremonies.[23]

While he was in Paris in May and June of 1923, Schmidt gave several public lectures, and he was bestowed with honourable fellowships at the Société Zoologique de France and the Société Centrale d'Aquiculture et de Pêche, while the Institut Océanographique made him a council member.[24] Then in the fall of 1923 the laboratory also sponsored Schmidt's participation in the Annual Meeting of the British Academy of Sciences in Liverpool with 1,000 Kroner.[25] Here, Schmidt presented

19 CA. *Carlsbergfondet, Carlsberg Laboratorium.* E08.1923. No. 36. Letter from J. Schmidt to Board of Trustees, 16 October 1923.

20 CA. *Carlsbergfondet, Carlsberg Laboratorium.* E08.1925. No. 28. Letter from J. Schmidt to Board of Trustees, 30 September 1925.

21 CA. *Carlsbergfondet, Carlsberg Laboratorium.* E08.1924. No. 28. Letters from J. Schmidt to V. Henriques, 24 September 1924.

22 CA. Carlsbergfondet, Carlsberg Laboratorium. E08.1926. no. 28. Letter from J. Schmidt to Board of Trustees, 30 September 1926.

23 CA. *Carlsbergfondet, Carlsberg Laboratorium.* E08. 1923. No. 36. Letters from J. Schmidt to V. Henriques, 18 April, 10 May & 25 May 1923.

24 Vedel Taaning Å. Hovedlinierne trækkes op. In: Winge Ø, Vedel Taaning Å, eds. Naturforskeren Johannes Schmidt. Hans Liv og Ekspeditioner. Copenhagen: Gyldendalske Boghandel Nordisk Forlag; 1947:184–185.

25 CA. *Carlsbergfondet, Carlsberg Laboratorium.* E08.1923. No. 10. Letter from V. Henriques to Board of Trustees, 14 April 1923.

his eel studies, while he was subsequently made honourable doctor of the University of Liverpool.[26] Finally, this trip also provided an occasion for Schmidt to receive the Weldon Memorial Prize at the University of Oxford. This prize is awarded annually 'for a noteworthy contribution to the development of mathematical or statistical methods applied to problems in Biology'.[27] Schmidt's racial investigations fitted this formula well.

The next year the pallet of Schmidt's scientific interests widened even more, as he was able to report to Henriques that his autumn trip to Madrid had been successful. Madrid was the setting for the conference of the International Union for Geodesy and Geophysics.[28] Then on the way home from Madrid, Schmidt stopped in Paris to give a presentation at the Académie des Sciences, of which he was made a corresponding member.[29] Then in 1925, Schmidt went to London, where the University of London hosted a lecture series, which included talks on racial investigations.[30]

Both in 1924 and 1925, Schmidt spent time at sea investigating the waters around Iceland and the Faroes with the government research vessel, the R/V *Dana*. He successfully applied for a three months leave of absence from the laboratory in 1924.[31] The following year he was back in the North Atlantic, this time he spent only his summer holiday on board the *Dana*.[32] Schmidt did not claim any publications in connection to the laboratory work, while he informed the board that racial investigations of the freshwater eel were now part of the racial studies of his laboratory. In this, Schmidt was aided by two marine scientists, Ege and Schmidt's assistant from the days of the *Thor* expeditions 20 years earlier, A. Strubberg.[33]

In any case, the end of 1925 found Schmidt on a trip to Tahiti, as the main stoppage point on a trip around the World. The main reason for such a long trip was stated in his request to the Board of Trustees to have a leave of absence. On November 10th Schmidt informed Henriques that he had now returned home following a 14 days long stay at the hospital, where he was undergoing a cure for bronchitis. Three different doctors had recommended him to rest in a warmer climate. However, as Schmidt had no desire to spend time at the French Riveira, he had set his mind on a longer trip to one of the islands in the South Seas – in the Pacific.[34] While undertaking this trip, he would be able to continue his studies on racial characteristics. He also intended to bring along his wife Ingeborg, although she would pay her own way there. In the end,

26 KB. Johannes Schmidt 'Papirer'. Ut.374.VI.3. Preliminary Programme, BAAS Liverpool meeting, 1923.

27 KB. Johannes Schmidt 'Papirer'. Ut.374.I.4. University of Oxford to J. Schmidt, November 1923.

28 CA. *Carlsbergfondet, Carlsberg Laboratorium.* E08.1924. No. 27. Letter from J. Schmidt to V. Henriques, 21 September 1924.

29 CA. Carlsbergfondet, Carlsberg Laboratorium. E08.1924. No. 33. Letter from J. Schmidt to V. Henriques, 17 October 1924.

30 CA. Carlsbergfondet, Carlsberg Laboratorium. E08.1925. No. 14. Letter from J. Schmidt to Board of Trustees, 18 May 1925.

31 CA. *Carlsbergfondet, Carlsberg Laboratorium.* E08.1924. no. 11. Letter from J. Schmidt to Board of Trustees, 22 April 1924 & No. 21. Letter from J. Schmidt to V. Henriques, 21 August 1924.

32 CA. *Carlsbergfondet, Carlsberg Laboratorium.* E08.1925. Letter from J. Schmidt to V. Henriques, 1 May 1925.

33 CA. *Carlsbergfondet, Carlsberg Laboratorium.* E08.1924. No. 26. Letter from J. Schmidt to Board of Trustees, 24 September 1924.

34 CA. *Carlsbergfondet, Carlsberg Laboratorium.* E08.1926. Letter from J. Schmidt to V. Henriques, 10 November 1926.

this voyage turned out to be a circumnavigation lasting 6 months from mid-December 1925 – mid-June 1926.[35]

Henriques thought it was splendid idea, and added that it was of the utmost importance to the laboratory that Schmidt could recover, while also continuing his research.[36] Perhaps it also helped the application for a leave of absence that Schmidt brought along one Gertrud Ostenfeld on the trip to Tahiti. Gertrud was a young natural scientist, and not least she was the daughter of Carl H. Ostenfeld, Schmidt's friend for more than 25 years. Ostenfeld had replaced Eugen Warming as a board member on both the Board of Directors of the Carlsberg Foundation and the Board of Trustees of the Carlsberg Laboratory. There is no doubt Schmidt benefitted from the fact that Ostenfeld was now his overseer both in the case of the Carlsberg Foundation and the Carlsberg Laboratory. Not only did Ostenfeld not object to the fact that Schmidt was absent for so long from his daytime job at the laboratory, he actively encouraged and supported Schmidt's marine science adventures. Thus in the case of the trip to Tahiti the Carlsberg Foundation paid Ostenfeld an advance on his honorarium in the order of 8,000 Danish Kroner, explicitly to enable him to support his daughter's trip to the South Seas with Schmidt.[37]

During the years that followed this, his first circumnavigation, the results of the investigations into the eels were written-up and were published in the French journal, *La Nature*, as well as in an Australian and a New Zealand based journal.[38] These publications now featured in Schmidt's annual report to the board. Schmidt also revealed that an additional marine scientist, Aage Vedel Taaning was now involved in laboratory investigations.[39] However, neither Vedel Taaning nor Strubberg featured in the laboratory budgets for the relevant years, although it is possible that they were paid from the lump sum set aside for 'temporary assistance', and Schmidt may also have persuaded the Danish marine science commission to contribute to the researchers' salaries. Nonetheless, the annual reports started to read like copy-pasted documents from one year to the next. In 1929 for instance, when Schmidt was absent throughout the year, Vilhelm Ege put together the annual report. The wording of this report was hardly changed from the previous year.[40]

In 1930 though, as Schmidt had returned to the laboratory following his two year voyage round the globe, he was able to present to the board the tenth and final publication of the racial investigations, an article on 'the Atlantic cod, (*Gadus Callarias* L.) and its local races.' The staff now

35 CA. *Carlsbergfondet, Carlsberg Laboratorium.* E08. 1926. Letter from J. Schmidt to V. Henriques, 22 November 1926.

36 CA. *Carlsbergfondet, Carlsberg Laboratorium.* E08. 1926. Letter from V. Henriques to Board of Trustees, 26 November 1926.

37 CFD. *Carlsbergfondets Direktions Protokol*, vol. 5, 25 February 1926.

38 CA. *Carlsbergfondet, Carlsberg Laboratorium.* E09. 1928. Letter from J. Schmidt to Board of Trustees, 30 September 1928.

39 CA. *Carlsbergfondet, Carlsberg Laboratorium.* E09. 1927. Letter from J. Schmidt to Board of Trustees, 30 September 1927.

40 CA. *Carlsbergfondet, Carlsberg Laboratorium.* E09. 1929. Letter from V. Ege to V. Henriques, Annual Report 1929.

6.3 Carl Hansen Ostenfeld in 1930
Botanist Carl Hansen Ostenfeld replaced Warming on the Board of Drectors of the Carlsberg Foundation in 1921, a position he held until his death in 1931, one year after this portrait was painted. Ostenfeld was crucial for the foundation's support for Schmidt's epxeditions in the 1920s. (The Archives of the Natural History Museum, Copenhagen).

included two new additions from the realm of marine science, Poul Jespersen and Anton Bruun both of whom had followed Schmidt around the globe on board the R/V *Dana*.[41]

In other words, during the 1920s Schmidt was housing an entire generation of up and coming Danish marine scientists at the Carlsberg Laboratory. In early June of 1928 Copenhagen once more was the scene of the council meeting of ICES, and the home team will

41 CA. *Carlsbergfondet, Carlsberg Laboratorium*. E09. 1930. Letter from J. Schmidt to V. Henriques, 19 October 1930.

have placed a solid stamp on the academic sessions, where one special session was devoted to 'racial investigations.'[42] From the laboratory guestbook, we can infer that the participants in the meeting visited the Carlsberg Laboratory, where Schmidt hosted an evening event in his stylish surroundings.[43]

The slightest undercurrent of critique

Considering Schmidt's obvious neglect of the duties laid down in the statutes of the laboratory during the 1920s, it is surprising that it was not until 1927 that the board started wondering about the opportuneness of Schmidt's priorities. From one of Schmidt's letters to his then chairman at the laboratory board, Valdemar Henriques, it follows that Schmidt had been asked to clarify how he had spent his time during the spring of 1927. Schmidt's account of his extensive travels to Amsterdam, Ostende, Paris, Madrid, and Stockholm reads like an echo of any modern day top scientist:

> 29/3–13/5 1927: Travelling abroad. Lectures in Amsterdam in early April concerning my eel investigations etc. by invitation from 1) students at the Natural Science Faculty of the university. 2) The Royal Geographical Society in Amsterdam. 3) The Royal Colonial Society in Amsterdam. Danish representative at two committee meetings of the International Council for the Exploration of the Sea: 4–6 April in Ostende on the issue of closing the trawl fisheries in the Scottish Firth of Forth, and 7–9 April in Paris for a meeting on measures to investigate the biology of whales.
> At the end of April and the beginning of May I was invited by the Spanish government to give lectures on my investigations into the freshwater eel, racial investigations etc.
> 25/5–2/6 Danish delegate at the annual council meeting of the International Council for the Exploration of the Sea in Stockholm.[44]

The one and a half months' leave of absence in 1927, was surpassed the next couple of years. For two months in the summer of 1928 Schmidt took part in the first stages of The Carlsberg Foundation's Oceanographic Expedition Round the World.[45] Schmidt was still obliged to apply for a leave of absence from the laboratory, but he appears to have

42 Went AEJ. *Seventy years agrowing. A history of the international council for the exploration of the sea, 1902–1972.* Copenhagen: International Council for the Exploration of the Sea; 1972. P. 74–75.
43 RA. Carlsbergfondets DANA-ekspeditioner. 1917–1973. 1. Gæstebog, 1922–1940.
44 CA. *Carlsbergfondet, Carlsberg Laboratorium.* E09. 1927. Letter from J. Schmidt to V. Henriques, 29 December 1927.
45 CA. Carlsbergfondet, Carlsberg Laboratorium. E09. 1928. Letter from J. Schmidt to Board of Trustees, 30 August 1928.

taken this quite light heartedly. Obviously, it was no secret to anyone that Schmidt was about to embark on an oceangoing voyage, whose main source of funding came from the Carlsberg Foundation, from where three of the Board of Directors also functioned as board members of the Carlsberg Laboratory. Yet, it does appear more than a trifle nonchalant that Schmidt asked for official permission to go abroad via postcard sent from Lisbon two weeks after his expedition had left the Port of Copenhagen.[46] Schmidt himself though, was in Lisbon in connection with a meeting of ICES. Following a couple of months at home in the Carlsberg Laboratory, Schmidt commenced his second circumnavigation around the world on 22 October.[47] Accompanied by Ingeborg, Schmidt embarked on the cruise liner *S/S Rotorua* with the goal of catching up with the expedition again in New Zealand.[48]

As long as Schmidt was present physically in Copenhagen and his laboratory, this highly successful research entrepreneur seemed untouchable. However, the 2nd circumnavigation, this time on board the R/V *Dana*, did provoke critique of his Laboratory work from Henriques. This story however, is unfolded later in the chapters 13–15 on the Carlsberg Foundation's Oceanographic Expedition Round the World.

The unfortunate Klöcker

Still, there is also plenty of evidence to suggest that Johannes Schmidt's work at the Carlsberg Laboratory was highly appreciated by the Board of Trustees. Although, his long tenure was not without difficulties even in the laboratory, as the story of Albert Klöcker illustrates. Albert Klöcker was the Director Extraordinaire appointed to continue Emil Chr. Hansen's work, and his case provides an example of how the Board of Trustees, and Johannes Schmidt, went about handling a colleague of Schmidt's, who was much less successful in several ways. Klöcker was to an extent Schmidt's subordinate, and for five years they worked side by side in the same rooms at the Carlsberg Laboratory.

By 1912, the Board of Trustees were beginning to comment on Klöcker's work ethics. Board member, Rudolph Koefoed wrote an animated complaint to the Board, in which he noted that Klöcker had not advanced any of Emil Hansen's old projects in yeast physiology in a satisfactory manner. Even more personally, Koefoed accused Klöcker of exhibiting a vindictive moral in one of his articles, disrespecting the

46 CA. Carlsbergfondet, Carlsberg Laboratorium. E09. 1928. Postcard from J. Schmidt to V. Henriques, 1 July 1928.

47 CA. *Carlsbergfondet, Carlsberg Laboratorium*. E09. 1928. Letter from J. Schmidt to Board of Trustees, 17 October 1928.

48 RA. *Schmidt, Ernst Johannes, biolog*: Materiale vedr. ekspeditioner og rejser (1876–1933) 2: m.m. Diary of I. Schmidt, 1928.

6.4 Albert Klöcker
Albert Klöcker photographed in 1901, when he was Emil Chr. Hansen's assistant at the Carlsberg Laboratorium. Twenty years later he found himself in dire straits at the same workplace (The Archives of the Natural History Museum).

legacy of Emil Chr. Hansen. Koefoed therefore, proposed that Johannes Schmidt should be tasked with overseeing Klöcker's job. Although, he admitted that Schmidt was still very young, he had already demonstrated a remarkable talent for organization and administration.[49] In the first instance, however, S. M. Jørgensen asked Klöcker to submit a research plan, since three years had passed since he had taken over as

49 CA. *Carlsbergfondet, Carlsberg Laboratorium.* E06.'1912'. R. Koefoed to Board of Trustees, 10 October 1912.

specially appointed director.[50] Klöcker promptly replied at length, apologizing, and he indicated that he was busy with cataloguing Hansen's yeast cultures, which were still preserved at the laboratory. One of his ideas was to test their long-term ability to survive.[51] Three weeks later, Schmidt was able to report to the board that Klöcker had now submitted a manuscript concerning the character of seventeen different yeast cultures.[52]

The following year, Klöcker was ill. This was probably linked to the diabetes, he developed over the years.[53] This prompted Koefoed to once more air his opinion that Schmidt should take control over more of Klöcker's present responsibilities, and judging from Schmidt's involvement with the correspondence, he was in tune with Koefoed. An assistant to Klöcker, v. Topp, had resigned, and now the debate was whether or not Klöcker himself, or Schmidt, was responsible for hiring his replacement.[54] Jørgensen, however, intervened on the side of Klöcker, and in a personal letter to Koefoed, he strongly disagreed with Koefoed's opinions on this matter, and so Klöcker was authorized by Jørgensen to employ his own assistant, pharmacist O. Madsen.[55] Clearly, the working relationship between Schmidt and Klöcker was not a match made in heaven. One reason for this may have been the fact that Schmidt and Klöcker shared the same premises on the 1st floor of the main building in the Carlsberg Laboratory. Klöcker was 15 years older than Schmidt, and for the first four years of Schmidt's directorship Klöcker's salary was higher than Schmidt's; yet, technically Klöcker was the subordinate of Schmidt.

In the summer of 1915, Schmidt saw a solution to the problem of sharing laboratory facilities, when the old laboratory buildings on the nearby brewery premises became vacant. Schmidt proposed to the Board of Trustees that Klöcker could take over the old laboratory buildings, which according to Schmidt would result in only minor extra costs for the overall laboratory. The advantages, as Schmidt saw it, would be that both scientists would have more space. Schmidt could then lay claim to the entire 1st floor of the main building, which was the same space provided for his predecessor Emil Chr. Hansen.[56] As chairman of the board, Warming handled this issue, and recommended to the rest of the board to approve Schmidt's proposition. From Warming's letter however, it is clear that Klöcker felt ousted from the main building by Schmidt. Nonetheless, Warming argued that since Schmidt was the director of the Department of Physiology, he should have the same working conditions as those of his predecessor.[57]

50 CA. *Carlsbergfondet, Carlsberg Laboratorium.* E06.'1912'. S. M. Jørgensen to A. Klöcker, 27 October 1912.
51 CA. *Carlsbergfondet, Carlsberg Laboratorium.* E06.'1912'. A. Klöcker to Board of Trustees, 31 October 1912.
52 *CA. Carlsbergfondet, Carlsberg Laboratorium.* E06.'1912'. J. Schmidt to Board of Trustees, 19 November 1912.
53 Schiönning, H., Albert Klöcker 1862–1923, in: Holter H, Max Møller K, eds. *The Carlsberg Laboratory 1876–1976.* Copenhagen: Rhodos; 1976:192.
54 *CA. Carlsbergfondet, Carlsberg Laboratorium.* E06.'1913'. R. Koefoed to Board of Trustees, 2 June 1913; *CA. Carlsbergfondet, Carlsberg Laboratorium.* E06.'1913'. J. Schmidt to S. M. Jørgensen, 22 May 1913.
55 *CA. Carlsbergfondet, Carlsberg Laboratorium.* E06.'1913'. S. M. Jørgensen to R. Koefoed, 28 May 1913; *CA. Carlsbergfondet, Carlsberg Laboratorium.* E06.'1913'. S. M. Jørgensen to A. Klöcker, 31 May 1913.
56 *CA. Carlsbergfondet, Carlsberg Laboratorium.* E07.'1915'. J. Schmidt to Board of Trustees, 9 August 1915.
57 CA. *Carlsbergfondet, Carlsberg Laboratorium.* E07.'1915'. E. Warming to Board of Trustees, 15 August 1915.

Klöcker's relocation seems to have solved whatever issues there were between him and Schmidt. Later the same year Klöcker received the 'Schibbyeske Præmie,' a scientific prize issued by the Society for Natural History for a paper he had published the same year dealing with the forms of seventeen different types of yeast.[58] Thus, Klöcker's work cannot have been completely without merit. Yet, it appears that the physical split between Schmidt and Klöcker did not enhance the output from Klöcker's share of the department. The Board remained silent on this issue until 1918, when Klöcker troubled the laboratory with an entirely personal problem. He was short of money.

From this year onwards, Klöcker developed a habit of asking the board for an advance payment of his salary to be able to pay his bills.[59] In 1921 he admitted to the board that he would be pawned by his creditors if he did not immediately receive an additional 2,000 Kroner from the laboratory. This caused the board to have a serious debate. In what today appears like an old fashioned patriarchal sense of responsibility, Eugen Warming clearly wished to help Klöcker meet his creditors, although it is apparent from his letters that Warming was appalled at Klöcker's lavish manners.[60] Warming handled this 'hot potato' by asking Schmidt for an opinion.

Schmidt agreed that Klöcker's actions were most unfortunate and inappropriate. However, considering Klöcker's long seniority and his ailing health, Schmidt proposed to help Klöcker rather than push him away.[61] Rudolf Koefoed on the other hand, saw an opportunity to get rid of Klöcker once and for all. A few months later, in late August of 1921, Koefoed proposed to fire Klöcker on grounds that the scientific fruits of his labour had been meagre ever since his employment. Warming though, insisted that no such thing would occur during his last few months in charge of the laboratory. At this point Warming was aproaching his 80th birthday and would soon be retiring. His successor was to be Ostenfeld. Moreover, since Schmidt was out sailing for many months with the *Dana,* any such decision should await Schmidt's return.[62]

By the autumn of 1922 however, it was game over for Klöcker. The new chairman of the Board of Trustees of the Laboratory, Henriques went to see Klöcker in his laboratory and told him he was fired. Henriques also let him know that considering the financial situation, the board had also decided to close the special laboratory, which Klöcker had been in charge of. As Klöcker had been employed for decades, he would have one year's notice, thus his contract of employment would be

58 CA. Carlsbergfondet, Carlsberg Laboratorium. E07. '1915'. J. Schmidt to Board of Trustees, 30 September 1915.

59 CA. *Carlsbergfondet, Carlsberg Laboratorium.* E07. '1919'. A. Klöcker to E. Warming, 9 September 1919; CA. *Carlsbergfondet, Carlsberg Laboratorium.* E07. '1919'. A. Klöcker to E. Warming, 11 October 1919; CA. *Carlsbergfondet, Carlsberg Laboratorium.* E08. 1920. No. 21. Letter from E. Warming to Board of Trustees, 10 June 1920.

60 CA. *Carlsbergfondet, Carlsberg Laboratorium.* E08. 1921. Letter from E. Warming to Board of Trustees, 28 March 1921; CA. *Carlsbergfondet, Carlsberg Laboratorium.* E08. 1921. Letter from E. Warming to Board of Trustees, 2 April 1921.

61 CA. *Carlsbergfondet, Carlsberg Laboratorium.* E08. 1921. Letter from J. Schmidt to E. Warming, 3 April 1921.

62 CA. *Carlsbergfondet, Carlsberg Laboratorium.* E08. 1921. Letter from E. Warming to Board of Trustees, early September 1921.

terminated on 1 November 1923. In addition, Klöcker's assistants, Folker and his other employee, P. Andersen lost their jobs on the same date.[63] The whole board was behind this decision.[64] In the end though, Klöcker never made it to retirement as he passed away on 3 October 1923 after a very short period of illness.[65] Two weeks later Schmidt took over Klöcker's laboratory, where Schmidt's assistant, the marine scientist Vilhelm Ege had begun drawing up an inventory of Klöcker's possessions.[66]

In 1922, Schmidt was approaching his zenith in terms of how much the laboratory appreciated his work, providing some irony to the fact that his onetime officemate Klöcker was laid off the same year. What the story of Klöcker reveals is that in the event that the Board of Trustees was not satisfied with the work of a laboratory director, they were able to act in unison and terminate his contract. They were no lame ducks. Nonetheless, Warming's chairmanship appears to have been marked by a compassionate attitude towards Klöcker, which gave him a longer leash than he otherwise would have had. Koefoed's attitude on the other hand was quite different and he would have been happy to sack Klöcker 10 years earlier.

Personal convictions and sentiments clearly affected decisions made by the Board of Trustees, which for Schmidt, unlike Klöcker, proved very useful.

Taking stock of the lab after Schmidt

When Schmidt died on the 21 February 1933, a sizeable scientific kingdom on the Carlsberg premises had lost its head. The laboratory board was confronted with a number of options for the future direction of the laboratory. Decisions had to be taken regarding to what extent Schmidt's strand of research was to be altered, once a new head was in place. On the 17 March the board convened to commemorate Schmidt's passing away. They were also faced with the question of what to do with the large zoological collection left by Schmidt, as well as the ongoing research projects in the laboratory. The future of Schmidt's assistant, Vilhelm Ege was to be decided by the incoming director. The first step was to appoint S. P. L. Sørensen of the Chemical Department as interim director of the Physiological Department until a new director was found. Then decisions were taken as to the continuation of the different ongoing investigations at the laboratory.[67]

63 CA. *Carlsbergfondet, Carlsberg Laboratorium.* E08.1923. No. 23. Letter from V. Henriques to Board of Trustees, 14 September 1923.
64 CA. *Carlsbergfondet, Carlsberg Laboratorium.* E08.1922. No. 28. Letter from V. Henriques to Board of Trustees, 6 October 1922.
65 Schiönning, H., Albert Klöcker 1862–1923, in: Holter H, Max Møller K, eds. *The Carlsberg Laboratory 1876–1976*. Copenhagen: Rhodos; 1976:192.
66 CA. *Carlsbergfondet, Carlsberg Laboratorium.* E08.1923. No. 23. Letter from J. Schmidt to V. Henriques, 16 October 1923.
67 CA. *Carlsberg Laboratorium*, E17. Forhandlingsprotokol II, 108th meeting, 17 March 1933, p. 29–30.

Professor Sørensen and Vilhelm Ege drew up an inventory of ongoing research activities at the laboratory for the board to act upon. Here, Sørensen proposed that the Carlsberg Laboratory should continue those of Schmidt's investigations that belonged to the racial investigations, principally statistical work with fishes, as well as his 'zoo-geographical examinations', that is the experiments with transplantation of eel pout.[68]

The Board, however, had its own ideas about the future of various projects. For instance, it decided that the results of the *Carlsberg Foundation's Oceanographic Expedition Round the World*, was to be taken care of by the Carlsberg Foundation in the future. Similarly, the ongoing researches into the material collected by the *Dana* expeditions from 1920–22, were to be handed over to Schmidt's other employer, the Danish commission for marine science.

The Laboratory's own investigations included five different topics. First up, Schmidt had recently initiated new racial investigations concerning the influence of temperature on the offspring from trout. These were to be discontinued. Secondly, a series of investigations where eel pout was being transplanted to other fjords were more advanced, and the board recommended that they were transferred to the marine science commission. The same recommendation was made for the statistical examinations of 'eel and cod etc.' As a fourth topic, the genetic studies of chicken were to be summarized and published by Vilhelm Ege, who had already concluded the experiments associated with this topic.

Finally, the laboratory was still conducting investigations into hops with the intention of cross breeding hops that were suited for both quality brewing and the demands of the Danish climate. The hops investigations were presented to the brewery's CEO and adjunct member of the laboratory board, Poul C. Poulsen, who indicated that a continuation of these studies was of little direct interest to the activities of the Carlsberg Brewery. Poulsen, however, was not inclined to close the investigations into the different flavours of hops, and the board decided to postpone the decision on the hops investigations until a new director had been appointed.[69]

What is apparent from this compilation of Schmidt's laboratory research at the time of his death in 1933, is that virtually everything Schmidt had been working on in the laboratory before he died was either discontinued or handed over to other organizations. Sørensen and Ege's plan to keep some of Schmidt's aquatic research interests

68 CA. *Carlsbergfondet, Carlsberg Laboratorium.* E10.'1933'. S. P. L. Sørensen to Board of Trustees, 9 March 1933.

69 CA. *Carlsberg Laboratorium*, E17. Forhandlingsprotokol II, 108th meeting, 17 March 1933, p. 30–31.

within the confinements of the Carlsberg Laboratory was also ultimately rejected. The new director would be able to embark on a whole new research agenda. There were no less than three candidates emerged for this position.

Such a position was not advertised. Traditionally, the board would engage in consultations and then offer the position to a suitable candidate. For Schmidt's successor, three names were brought to the table, namely university Professor Dr. Boysen Jensen, Professor at the Veterinary and Agricultural University, Øjvind Winge, and as the third candidate, Dr. Niels Nielsen (1900–1960).[70] Since 1931, Nielsen had served as head of a newly established Sub-Department of Fermentation Physiology under the auspice of the Carlsberg Foundation.[71] The discussion of successor was continued at the next board meeting on the 24 March 1933, where it was decided that Øjving Winge was to be first to be asked if he was interested in the job, while Niels Nielsen was the designated second choice, if they could not persuade Winge to take up the job.[72] Winge was indeed interested.

Boysen Jensen thus came in last. Was Boysen Jensen aware that he had been a candidate for this job? There is no written evidence to suggest that he was, but it would have been an unusual move by any board member to promote his name at the board meeting, and then have it recorded in the minutes, without the person in question having acknowledged an interest in moving to the Carlsberg Laboratory. In any case, Boysen Jensen's failed candidacy should be taken into consideration, when reading his critique of the Carlsberg Laboratory in general and Johannes Schmidt in particular, as cited by Krogsgaard Larsen in the opening paragraph of this chapter.

On 11 April 1933, Boysen-Jensen addressed professor of physiology, Valdemar Henriques, in his capacity as chairman of the Board of Trustees of the laboratory. Boysen Jensen was a fellow of the Royal Danish Academy of Sciences, the club of scientists, from where the Board of Directors of the Carlsberg Foundation was drawn. Boysen Jensen informed his fellow companion, Henriques that the appointment of Øjvind Winge had been a 'very large disappointment' for him.[73]

There were two aspects of Boysen Jensen's critique; one was looking backwards at Schmidt, while the other confronted Øjvind Winge. In both cases the laboratory statutes were used as leverage for his negative evaluation of the two appointments. Boysen Jensen referred to paragraph 8 stipulating that the research was intended to support

70 CA. *Carlsberg Laboratorium*, E17. Forhandlingsprotokol II, 108th meeting, 17 March 1933, p. 32.

71 Larsen, Poul, 'Niels Nielsen,' in Holter, H, Max Møller, K, eds. *The Carlsberg Laboratory 1876–1976*, Copenhagen: Rhodos; 1976. p. 232.

72 CA. *Carlsberg Laboratorium*, E17. Forhandlingsprotokol II, 109th meeting, 24 March 1933, p. 33.

73 CA. *Carlsbergfondet, Carlsberg Laboratorium*. E10.1933. no. 19. P. Boysen Jensen to Valdemar Henriques, 11 April 1933.

knowledge relevant to the brewing of beer, and he also underlined the relevance of paragraph 9, where it is mentioned that a laboratory board is to be appointed with the obligation to keep oversight of the research being carried out, not least to make sure that the requirements of paragraph 8 are met.

The deceased Schmidt was an easy target. Boysen Jensen wrote in his complaint that the appointment of a new head of the Physiological Department was a chance to start afresh, and undo the wrongs of the past. With this, he hinted at previous decisions of the board to allow the department to develop into a laboratory for fisheries investigations under Schmidt's regime. Following Boysen Jensen, the developments under the reign of Schmidt were so familiar that they hardly needed any commentary. He recalled that when Schmidt was appointed he had possessed very few relevant skills to fulfil the job. Although, Boysen Jensen admitted that at first Schmidt did make an honest attempt to meet the laboratory requirements, he soon drifted fully into the racial investigations. Boysen Jensen further noted that from 1923 onwards the Physiological Department ceased to publish research in the laboratory publication series, the *Comptes-Rendus*.

Then the direction of Boysen Jensen's critique slid towards the role of the board, whose tacit approval of Schmidt's actions had been met with widespread anger and astonishment. It was not specified, however, who was behind these accusations. But Boysen Jensen wrote that he regretted not having highlighted the board's inaction over Schmidt's broadening research portfolio at an earlier stage, but Jensen said that he had been confident that the lack of action was due to the fact that it was difficult to bring Schmidt to order. This belief of his was confirmed, when the board in 1930 took the initiative to set up a new sub-department of fermentation physiology in a villa of its own, and put Niels Nielsen in charge. Boysen Jensen seems to have been spot on here. His critique of Winge, however, seemed a more misguided attack.

Much to Boysen Jensen's disappointment, the recent vacancy of the position as head of Physiological Department had not been filled with a plant physiologist. The appointment of Winge was an act of opportunism, Boysen Jensen declared. The laboratory statutes clearly stipulated that there was a physiological and a chemical department. Hence, Winge's field of research, genetics, did not comply with this format. Only in the case when genetics could be seen as a field within physiology, would this be the case, and Boysen Jensen clearly thought that this was not the case: 'It is irresponsible to hire a researcher specializing in genetics to serve as head of a plant physiological laboratory as long as plant physiology is such a thriving theme.[74]

Contrasting the lengthy complaint, Henriques' reply was as short as possible. In one sentence, Henriques replied that he could not agree with Boysen Jensen's views on the actions of the board.[75] Then, Boysen Jensen's critique was forwarded to Winge, and four days later Winge responded directly to Boysen Jensen, while the board received a copy of the correspondence. Unsurprisingly, Winge argued that genetics indeed was a field of research sitting firmly within the discipline of physiology. Winge declined to comment on Schmidt's role, but provided extended commentary on the laboratory's history. Winge used the iconic scientist and Schmidt's predecessor, Emil Christian Hansen's research as an example of physiological research, where genetics played a significant role. Following Winge's account, Hansen's investigations served to study the systematics and variation in fermentation organisms. The way in which Hansen had proved the existence of constancy in the variation of individual isolated fermentation cells, and his studies of artificially induced hereditary changes, are both examples of what today would be called hereditary examinations. 'And,' Winge wrote, '…it would be slightly bold to claim that Emil Chr. Hansen's investigations were at odds with the laboratory statutes.[76]

The turmoil of replacing Schmidt eventually came to an end, and Winge took up his position as director of the physiological department on the 1 September 1933. At that time he had been a full professor at the Veterinary and Agricultural University since 1921. He was therefore automatically placed on the top pay grade, which at the time was 12,000 Danish Kroner per year, or the same as Schmidt and S. P. L. Sørensen were paid.[77]

As soon as Winge had been appointed to the position as Schmidt's successor, a large renovation project followed, and Winge made sure to emphasize how useless much of the laboratory equipment was for his future work. Winge was already a well-established senior scientist with a proven and relevant track record, so he could put weight behind his words, and he surely knew that he could make larger demands to the Board of Trustees in the transitional phase before he got too settled in the new position. This also means that he had perfectly legitimate reason for highlighting the shortcomings of the research infrastructure he inherited from Johannes Schmidt.

74 CA. *Carlsbergfondet, Carlsberg Laboratorium.* E10.1933. no. 19. P. Boysen Jensen to Valdemar Henriques, 11 April 1933.

75 CA. *Carlsbergfondet, Carlsberg Laboratorium.* E10.1933. no. 20. Valdemar Henriques to P. Boysen Jensen, 12 April 1933.

76 CA. *Carlsbergfondet, Carlsberg Laboratorium.* E10.1933. no. 21. Øjvind Winge to P. Boysen Jensen, 15 April 1933.

77 CA. *Carlsbergfondet, Carlsberg Laboratorium.* E10.1933. no. 30. Board of Trustees to Board of Directors, 15 June 1933.

Winge commenced with a lengthy list of items such as a new much stronger microscope, new electrically controlled thermostats, improved lighting, equipment for microphotography, and much more. There were only two useful microscopes left, but one of them was heavily battered. 'Supposedly it has been used at sea,' Winge remarked. One gets a sense of how laboratories were advancing at a rapid pace in terms of technology, when Winge also asked for such novelties as a refrigerator and a machine for 'inducing x-rays on specimens'. Finally, Winge somewhat ironically wished to bring with him, a number of aquaria, which he was using for investigations of the chromosomes of the *Lebistes reticulatus*, the tiny guppy. This was an ongoing research project he felt unable to stop at the time.[78]

The Board of Trustees reviewed Winge's suggestions without reservations, and he was subsequently granted the full amount of 27,000 Kroner to stock the laboratory with modern equipment, buy a new plot of land for farming, and renovate the two floors of the laboratory housing the Physiological Department.[79]

Winge subsequently went on to lead path breaking studies within the hereditary qualities of yeast. In a recent review of Winge's work, the Polish born professor of oncology, Waclaw Szybalski heralded Winge's work, which appeared in over 150 research publications. Even prior to World War II, Winge's work was recognized in Poland. Szybalski had attended a class at university on the industrial uses of yeast. Winge's efforts in fermentation biology prompted Szybalski's university teacher, Joszt to lecture on how Winge combined desirable fermentation processes into one yeast strain through his use of genetic crosses. Winge's discovery, Joszt prophesised, would revolutionize biotechnology and ignite a whole new subfield of research, namely genetic engineering.[80]

Winge's research had come a long way since the days of diallel crossings of animals in Schmidt's laboratory. Although the tiny guppy, *Lebistes*, which he had worked on as Schmidt's assistant, stayed with Winge throughout his long tenure at the Carlsberg Laboratory. In summer 1933, as Winge had 11,000 Kroner on his budget to spend on 'temporary assistance', he informed the laboratory board that he would spend 2,400 Kroner of this allowance to hire an assistant to conduct examinations of the gender of guppies.[81] Thus Schmidt's aquatic legacy stayed with the laboratory, and when in 1949, Waclaw Szybalski made his last visit to Winge's laboratory, there was still an aquarium installed outside his office – filled with guppies.[82]

78 CA. *Carlsbergfondet, Carlsberg Laboratorium.* E10.1933. no. 26. Ø. Winge to Board of Trustees, 15 May 1933.

79 CA. *Carlsbergfondet, Carlsberg Laboratorium.* E10.1933. no. 27. Circular Board of Trustees, 7 June 1933.

80 Szybalski, Waclaw, My Road to Øjvind Winge, the Father of Yeast Genetics, *Genetics* (1 May, 2001) vol. 158 no. 1, 1–6.

81 CA. *Carlsbergfondet, Carlsberg Laboratorium.* E10.1933. no. 31. Ø. Winge to Board of Trustees, 13 June 1933.

82 Szybalski, Waclaw, My Road to Øjvind Winge, the Father of Yeast Genetics, *Genetics* (1 May, 2001) vol. 158 no. 1, 1–6.

Conclusion

This chapter has sought to establish an overview of the costs associated with Schmidt's somewhat alternative interpretation of what his laboratory was for, and the job of the Board of Trustees. Looking at the long history of the conglomerate of the Carlsberg Breweries, the Carlsberg Foundation, and the Carlsberg Laboratory, it appears strange that one of the laboratory directors excelled in a field of research, as far from beer brewing as marine science. Nonetheless, such was the set up during the more than 22 years from 1910–1933, when Johannes Schmidt was the Director of the Physiological Department of the Carlsberg Laboratory.

Schmidt was allowed to continue his work in marine science, and the fine modern Carlsberg Laboratory was instrumental for much progress, as Schmidt gradually opened the doors for several marine scientists, finding a temporary home at Carlsberg, while paid for by the Danish commission for marine science. Nonetheless, most of the research being carried out at the Carlsberg Laboratory was paid for by the Carlsberg Foundation, overseen by the Board of Trustees, and no less than 1,640,000 Danish Kroner went in the direction of Johannes Schmidt during his time as director. Most of this money went into paying for the staff, the equipment used, and the maintenance and heating of such a large laboratory. Nonetheless, the directorship also earned Schmidt a secure salary equivalent of a university professor, while he enjoyed the comfort of free housing in an appealing area of Copenhagen.

It is safe to say that following Winge's departure from the Physiological Department, Schmidt's research lost any bearing on brewery related research. From c. 1921–1933, Schmidt focused virtually all research in his department towards marine science. Probably this was in part due to the fact that his long term fellow oceanographic traveller and friend, Ostenfeld became trustee of the Board of Trustees in 1921, replacing Schmidt's former champion, Eugen Warming. Schmidt's lack of interest in what was supposed to be his core activity will have been blatantly clear to the board throughout the 1920s and early 1930s. Increasingly Schmidt was not just out of town, but out of the continent of Europe for months and years during the latter decade of his tenure, while his annual reports shrunk to mere one page summaries, often verbatim repetitions of last year's activities.

In spite of his lack of interest in brewery related research, the Board of Trustees, sometimes aided by the Board of Directors of the Carlsberg

Foundation were full of praise for Schmidt and made real efforts to secure official Danish recognition of Schmidt's work. This paid of in the form off a silver cross from the Danish state, and finally in 1926 the right to officially use the title of professor. The actual brewery represented by the CEO Paulsen joined the chorus of appraisal for Schmidt, exhibiting no reservations as to the way he prioritized his time and efforts.

If the board is somehow to be excused, they would be so in light of the fact that the corporate world and the expectations of research leadership might have been different in the 1910s-1930s from today. When Warming was chairman of the Carlsberg Laboratory, the whole set up of the Carlsberg Foundation running a laboratory fairly independent from the daily routines of the brewery was quite unique.

This raises the question of whether or not the Board of Trustees were just lame ducks not really paying attention to what was going on in the laboratory. Not at all, judging from the way they handled the ill fortune of Schmidt's colleague Albert Klöcker. Klöcker is an example of how the Board was able to sack a director, if they were not pleased with his lack of performance. At any rate, Klöcker's research, if not impressive in quantity was actually relevant for the brewery, and he had worked for decades for the Carlsberg Laboratory before he was finally laid off in 1922, but his lack of economic common sense in his private affairs and the fact that he was simply disliked among members of the Board of Trustees turned him into a marked man, eventually dying three weeks before his employment ended. If indeed the board had not approved of what Schmidt was doing, the fate of Klöcker is a testimony of what they were capable of.

In the case of Schmidt the disproval from the board came very late in his life, and as we will see later in this book, much of the critique eventually arose from personal matters, just as much as it was founded on issues of research strategy and research performance. came from outside the Carlsberg organization. Overall it can be said that Schmidt came to use the considerable powers at his disposal for a long and dedicated service to marine science research and science organization. How, Schmidt did this is the main topic for the rest of this volume.

7 Atlantic adventures

Following the two Carlsberg funded expeditions to the Mediterranean and Georg Hansen's shore based investigations out of Messina, it was clear to Johannes Schmidt that the breeding place of the eel was not to be found East of Gibraltar. Wherever the eels went to spawn, it had to be somewhere in the Atlantic outside the perimeter of any previous oceanographic enterprise. But where should Schmidt look? Ideally he wanted to sample as much of the Atlantic Ocean as possible, in order to precisely locate eel larvae, but also to identify those areas where eel larvae would probably not appear.[1]

The task was daunting as the government research vessel the *S/S Thor* could not sail for more than 12 days without coaling. Investigating the deep seas of the Atlantic Ocean required a vessel with a bigger range. Once more, Schmidt's quest to find the cradle of the European eel made it necessary to raise funds for large scale oceanographic enterprises, and once more, Johannes Schmidt would prove himself to be one of the most successful fundraisers and expedition leaders in the history of marine science. In spite of interruptions such as World War One, Schmidt launched four Atlantic expeditions between 1913 and 1922, and following the return of the last of the four he had been catapulted into international stardom as one of the World's most prominent scientists. By 1923, Schmidt's theory of how the Sargasso Sea is the breeding place of all the freshwater eel in the Atlantic Ocean had become widely recognized.[2]

Today, eel research still features prominently in Danish marine science.[3] The scientific results of Schmidt's trips are well known, and they are still cited in the academic literature. Nonetheless, little is known about *why* it came about that a relatively small country such as Denmark would excel in Atlantic marine science.

Several commentaries on the history of Danish marine science and Johannes Schmidt's expeditions have noted that private sponsors made

1 Jespersen P. Danske handelsskibes og skonnerten '*Margrethe*'s indsats. In: Winge Ø, Vedel Taaning Å, eds. *Naturforskeren Johannes Schmidt. Hans liv og ekspeditioner*. Copenhagen: Gyldendalske Boghandel Nordisk Forlag; 1947:57–69.

2 Schmidt J. Breeding places and migrations of the eel. *Nature*. 1923;111:51–54; Schmidt JS. Die laichplätze des Flußaals. *Internationale Revue der gesamten Hydrobiologie und Hydrographie*. 1923;11(1-2):1–40; Schmidt J. The breeding places of the eel. *Philosophical Transactions of the Royal Society of London.Series B, Containing Papers of a Biological Character*. 1923;211:179–208 and Schmidt J. The transatlantic migration of the eel-larvae. *Nature*. 1924;113:12.

3 Møller Hansen M, Sparholt H. Den europæiske ål. http://www.galathea3.dk/dk/Menu/Forskning/Den+europ%C3% A6iske+%C3%A5l/Projektbeskrivelse. Accessed 04/19, 2013.

Schmidt's journeys possible.[4] Nevertheless, little is known about these sponsors, how much they contributed and not least why they chose to support deep-sea endeavors.

The following chapter tells the story of *how* Schmidt succeeded in attracting funds from private, institutional and government sources. Here, for the first time Danish marine science is seen through the lens of Danish and international oceanic trade and geopolitics. Schmidt and his marine science colleagues relied upon funding from various parts of society in order to carry out their research projects. In turn, the marine scientists would offer insights on how to promote and increase Atlantic marine resource exploitation.

The promise of ground breaking scientific output was equally important. At a time when Danish national aspirations were craving heroic deeds, scientific investigations became a more intangible asset, but important nonetheless. While the shrinking Danish political empire offered a story of decline and defeat, the evolving empire of marine science offered stories of clever marine scientists unlocking the secrets of, what many saw as society's last frontier – the World's oceans.

Following the flow of money towards Schmidt's research enterprises, we will observe how imperialism and natural resource exploitation merged to glue together science and business in the waning Danish North Atlantic empire of the first decades of the 20th century. Leading Danish marine scientists, fishermen, ship owners, bankers, politicians and the more national conservative segments in Danish society shared a belief that Denmark had a role to play in the Atlantic Ocean far beyond the borders of present day Denmark.

Therefore, the funding of Schmidt's four expeditions into the Atlantic Ocean, as well as the widespread use of ships of opportunity, is best viewed in relation to the geopolitics surrounding the Danish colonial possessions in the West Indies, nowadays The US Virgin Islands.

The Danish West Indies and marine science

At the dawn of the 20th century, The Great Powers of the Western World were approaching the zenith of imperialist aspiration, colonizing and dominating every last corner of the Earth. Meanwhile, the fairly tiny European power of Denmark was in the process of dismantling its last tropical colony, The Danish West Indies. Part of the Antillean archi-

4 Winge Ø, Vedel Taaning Å, eds. *Naturforskeren johannes schmidt. hans liv og ekspeditioner*. Copenhagen: Gyldendalske Boghandel Nordisk Forlag; 1947; Morcos S, Smed J, Svansson A. The eel and mediterranean oceanography – A danish contribution to ocean science. In: Morcos S, ed. *Ocean sciences bridging the millenia: A spectrum of historical accounts*. Beijing: UNESCO; 2004:197–207; Wolff T. *Danske ekspeditioner på verdenshavene: Dansk havforskning gennem 200 år*. Copenhagen: Rhodos; 1967 and Fort T. *The Book of Eels*. Great Britain: Harper Collins Publishers; 2002.

pelago, The Danish West Indies consisted of the three small islands, St. Croix, St. Thomas and St. John, situated in the Eastern Caribbean. Nowadays, they are known as The US Virgin Islands. Denmark had acquired the three islands one by one from c. 1670–1733, and over the course of the next centuries, sugar production became the mainstay of the Danish West Indian economy.[5] From the latter half of the 19th century however, running the remote tropical colony led to substantial deficits. The poor state of the economy, as well as the geographical and mental distance between Denmark and the Caribbean, favored sentiments geared at selling the islands.[6]

The sale eventually came about in 1917, when the United States paid $ 25 million for the three islands. The decision to sell the colony was reached after a public referendum in 1916, but this is not to say that the decision had come about easily or without a price for certain sectors in Danish society. For almost two decades an alliance formed by some of the leading conservative minded politicians, business men, navy officers and not least marine scientists had come together to keep the Virgin Islands in close contact with the Danish fatherland. This Atlantic alliance of politics, business and science extended back to the turn of the 20th century, where the fate of the Danish West Indies had also been a hot issue in Danish and international politics.

Fearing that the German Empire would conquer the Danish colony, the US presidents McKinley and Theodore Roosevelt both took an interest in the Danish West Indies. The islands were strategically well placed en route between Europe and Central America at a time, when steamships needed frequent bunker stations for fueling, and the construction of the Panama Canal, which eventually opened for shipping in 1913, was on going. In other words, the Caribbean was well inside the perimeter of the political sphere of influence of the United States, the so-called Monroe Doctrine, and the Virgin Islands were but a pawn in the power struggle of the great powers.

Secret negotiations between Denmark and the United States ensued, and in 1901, the US senate approved the ratification of a treaty handing over the islands to the USA. In Denmark, the issue was perhaps more sensitive. The Danish parliament decided not to ask the local population in the colony, if they wanted to be Danish or American citizens. The lower chamber in parliament, *Folketinget*, approved the treaty with a huge majority, but then came the vote of the second chamber, that of *Landstinget*. The seats in the Landstinget were elected by the landed

5 Gøbel, Erik, *A Guide to the Sources for the History of the Danish West Indies (U.S. Virgin Islands), 1671–1917*, (Odense: University of Southern Denmark Press, 2002).

6 Toft P. Kriser og nedgang. den økonomiske udvikling i dansk vestindien 1875–1917. *Erhvervshistorisk Årbog*. 1982; 32:146–191.

elite and the wealthy classes in the cities, which led to a conservative dominance, and Landstinget repudiated the ratification of the treaty. The majority in Landstinget argued that it was unpatriotic to sell off Danish territory, and they further remarked that the economic failures in the Danish West Indies were caused by colonial mismanagement. If only the Danish government and Danish society would renew its investments in the colony and wait until the Panama Canal was finished, the situation of the Danish West Indies would greatly improve.[7]

H.N. Andersen, a wealthy Danish business entrepreneur and founder-owner of the Danish East Asiatic Company, intervened in the situation with a practical solution on behalf of a consortium of leading Danish banks and industrialists and ship owners. On the condition that the islands stayed in Danish hands, Andersen promised to set up an investment scheme, where a monthly steamship service between Copenhagen and St. Thomas would be set up to boost Atlantic trade. The trade was to come from renewed investments in plantations and port facilities. In any case, the plan of selling the islands to the USA was abandoned in 1902. The consortium now had to keep their patriotic promises to the peoples of Denmark and the West Indies.[8]

The scale of investments now flowing into the Danish West Indies was illustrated by the setting up of a new commercial bank, Dansk Vestindisk Bank in St. Thomas. A Danish led consortium consisting of the leaders of the Danish national bank as well as the three largest commercial banks, Landmandsbanken, Privatbanken and Handelsbanken. Soon the Dansk Vestindisk Bank became leverage for Danish investments flowing into plantations and port facilities, and H. N. Andersen's EAC became the main operator at the port of St. Thomas.[9]

Such investment of capital was a positive prospect for the Danish West Indies, and in the event of a successful economic outcome the whole country was likely to gain from the increased level of activity, but Denmark had now become a full democracy in the sense that from 1901 onwards the government was chosen by the majority of seats in the lower house, Folketinget, where all adult men, not yet women, were allowed to vote. So, in order to protect their investments, the wealthy elite became blatantly aware that they needed to promote a wider public interest in the three remotely located islands more than 5,000 kilometers from mainland Denmark.

A lesson or two had been learnt from the trade negotiations, where the fate of the colonies rested in the hands of the publicly elected

7 Nørregaard G. *Vore gamle tropekolonier, vol. 4. dansk vestindien 1880–1917. reformforsøg og forhandlinger.* Copenhagen: Fremad; 1966, p. 56–75.

8 Nørregaard G. *Vore gamle tropekolonier, vol. 4. dansk vestindien 1880–1917. reformforsøg og forhandlinger.* Copenhagen: Fremad; 1966, p. 56–75.

9 Worm Begtrup, Jens, 'Den Dansk-vestindiske Nationalbank 1904–1935', in: *Erhvervshistorisk Årbog*, 59, (2010) p. 183–211; Gøbel, Erik, 'Sankt Thomas havn i det 19. og tidlige 20. århundrede', in: *Handels- og Søfartsmuseets Årbog*, 60, (2001), p. 23–64.

parliamentarians. Public interest would arise from an increase in state revenue on the colonial activities, but rather from more intangible sentimental qualities associated with being a seaborne empire. For this purpose the scientific republic of letters was enrolled to assist the monarchy's ailing empire. The task of agitating the virtues of Danish interests in the Atlantic realm fell within the newly formed society, *Foreningen De Danske Atlanterhavsøer*.

10 RA.10154. *De Danske Atlanterhavsøer*. 1902–1917 Forhandlingsprotokol for foreningen, bind 1. p. 2.

Foreningen De Danske Atlanterhavsøer

On the night of the October, 19 1901, Restaurant Wivel at the Tivoli Gardens in Copenhagen was the venue for the inaugural meeting of what became the society, *Foreningen De Danske Atlanterhavsøer*. The idea of the society was to promote the development of those parts of Denmark, which were situated within the perimeter of the Atlantic Ocean, far away from mainland Denmark. In other words the society would promote the possibilities and virtues of the Faroe Islands, Iceland, Greenland and the Danish West Indies. 28 businessmen, navy officers, scientists and politicians joined forces.[10]

7.1 Cover of the journal *Atlanten*
The cover page of the the journal Atlanten published by the interest group, De Danske Atlanterhavsøer. The map gives a sense of what was at stake: A far flung Atlantic Danish empire stretching from Greenland, Iceland and the Faroes in the North to the Danish West Indies in the South. The map suggests that the West Indies is the natural gateway through the Panama Canal, which was still under construction when this map was drawn in 1904.

The port director of Copenhagen and would-be first general secretary of ICES, C. F. Drechsel was present from the beginning, and it was Drechsel who made sure that the interest of marine science and fisheries was explicit from the start. As the first speaker, when the discussion opened at the meeting, Drechsel spoke about how it would be highly desirable if people in mainland Denmark became more acquainted with fisheries in Iceland. He then promoted the Danish West Indies, focusing on the huge commercial importance of the port of St. Thomas, and the possibilities of emulating the successful banana plantations of the British Caribbean colonies.[11] Another founding member was Arthur Feddersen, a veteran promoter of modern fishing practices. By 1902 Feddersen was the editor-in-chief of the periodical, *Fiskeritidende*, published by the Danish fishermen's association.[12] H. N. Andersen was also among the founding members, and he took up a position as board member in the West Indies section, as an indication of where his motivation and priorities lay. Andersen was also one of several members paying a 200 Kroner contribution to become a permanent member of the society.[13]

The society quickly became a formalized entanglement of patriotism, science communication and business opportunities, where two types of activities dominated. First of all, the society launched a series of public lectures on a range of topics linked to the Danish Atlantic possessions. Secondly, the society decided to disseminate knowledge on the Danish Atlantic possessions through the establishment in 1904 of the journal, *Atlanten*. In an age when printing costs were often the most expensive part of knowledge production, such a journal was an important outlet for information. In the following thirteen years the journal covered themes ranging from tourist reports on Iceland to the possibilities of reindeer farming in Greenland.[14] Drechsel for one, made sure that fishery related topics featured prominently, and as early as 1904, Drechsel wrote an account of the state of fisheries in Iceland with a recommendation that Danish fishermen and capital investments were guided in the direction of the waters around Iceland.[15]

Johannes Schmidt joined De Danske Atlanterhavsøer in August of 1905, upon his return from the trip to the North Atlantic. Schmidt would remain a member in good standing throughout the existence of the society, but according to the minutes of the various committees, the letter book of the society and the journal *Atlanten*, Johannes Schmidt does not seem to have made a big impact on the society during the next ten years apart from paying a 5 Kroner annual subscription. But as we will

11 RA.10154. *De Danske Atlanterhavsøer*. 1902–1917 Forhandlingsprotokol for foreningen, bind 1. p. 5.
12 Tarbensen K. *For alle danmarks fiskere – Danmarks Fiskeriforening 125 år, 1887–2012*. Taulov: Danmarks fiskeriforening; 2012, p. 68.
13 Møller, Frederik, 'Foreningens Stiftelse og første Virksomhed. Uddrag af Foreningens Forhandlingsprotokol': in: *Atlanten*, vol. 1, 1904–1906. p. 27–30.
14 Møller, Frederik, 'Foreningens Stiftelse og første Virksomhed. Uddrag af Foreningens Forhandlingsprotokol': in: *Atlanten*, vol. 1, 1904–1906. p. 23–32.
15 Drechsel, C. F., 'Nogle Bemærkninger om Fiskeriet ved Island og Færøerne og dets Fremtid', in: *Atlanten*, vol. 1, 1904–1906. p. 109–117.

see below, Schmidt's professional network extended well into the heart of *Foreningen De Danske Atlanterhavsøer.*[16]

Marine research initiatives and the Danish West Indies

The marine resources of the Danish West Indies also received their share of attention in the new society as well as from the government and the Carlsberg Foundation. The West Indies had been visited by Danish naturalists several times in the preceding decades. Eugen Warming had been fishing in the Sargasso Sea as early as 1866 on a return trip from an expedition to Brazil, and in 1982, Warming spent four months in the Danish West Indies as part of larger trip to Venezuela.[17] From 1889–1921 Professor Warming was also one of five members of the Board of Directors of the Carlsberg Foundation, and in 1903 Warming became a member of De Danske Atlanterhavsøer.[18] Politically, Warming was a nationalist and conservative, and Warming took great interest in preserving Danish cultural interests in Schleswig-Holstein, an area, which was part of the German Empire until 1920.

Warming's 1892 expedition was sponsored mainly by the University of Copenhagen and by the Danish Ministry of Education, and cost 6,000 Danish Kroner.[19] In addition, two young botanists Børgesen and Paulsen joined Warming on field trips to the Danish West Indies at their own expense. The would-be Danish fisheries consultant, Chr. Løfting also stayed for several months in the Danish West Indies together with Warming, and again in 1895 Løfting was back independently as a zoologist on an expedition with the Frigate *Fyn.*[20]

In addition to the Danish West Indies' expedition, Warming was a veteran Greenland explorer, and Warming's political and scientific inclinations may well have had an influence on the fact that the Carlsberg Foundation became a frequent benefactor for studies on the natural history of the West Indies. In 1897 for instance, botanists Børgesen and Paulsen were awarded 1,200 Kroner by the foundation in support of an illustrated publication on the vegetation in the Danish West Indies.[21]

Thus by the time of the 1902 sales negotiations with the USA, Danish natural historians were among the few people in mainland Denmark who had real life experience of the country's Atlantic possessions. However, the interests of this cohort were not reflected in the Danish

16 RA.10154. *De Danske Atlanterhavsøer.*1902–1919. Medlemsprotokoller, bind 9. p. 15, 28 & 77.

17 Christensen C. Eug. Warming. En Levnedsskildring. In: Kolderup Rosenvinge L, ed. *Eugenius Warming 3. november 1841–2. april 1924.* Copenhagen; 1924: 14 & 23.

18 Anonymous, 'Medlemsfortegnelse', in: *Atlanten*, vol. 1, 1904–1906. p. 56; Glamann K. *Carlsbergfondet.* Copenhagen: Rhodos; 1976, p. 199.

19 Prytz S. *Warming – botaniker og rejsende.* Lynge: Bogan's Forlag A/S; 1984, p. 90–92.

20 Anonymous, ed. *Betænkning angiven af den i henhold til lov nr. 294 af 30. september 1916 nedsatte rigsdagskommission angaaende de dansk vestindiske øer.* Copenhagen: J. H. Schultz; 1916. P. 51–52.

21 Spang-Hanssen E. *Carlsbergfondets Understøttelser 1876–1936.* Copenhagen: Bianco Lunos Bogtrykkeri A/S; 1937. p. 72.

government committee's report of 1902. This report contained just a few recommendations on how to improve the state of fisheries, which seemed to have played a minor role in the local economy. These recommendations included the observation that the 'Mackerel type' fisheries had the potential for greater catches, especially if a modern canning industry was set up. Likewise, the report suggested that the langoustine fisheries and the hunt for sea turtles ought to be carried out more efficiently. In terms of the sustainability of the ocean resources, two concerns emerged. The first problem was that fishermen from the nearby island of Tortola were violating Danish sea territory in their fishing activities. Secondly, the resident population of sea turtles was found to be vulnerable during mating season, when eggs were collected. The report suggested greater control on fisheries in the area to prevent these activities.[22]

In a 1904 article in, *Atlanten,* Danish zoologist, Theodor Mortensen and the botanist, F. Børgesen proposed that a marine biological field station should be established in the Danish West Indies.[23] Nothing came of the plan though due to lack of financial commitment from the Danish government, but Mortensen instead secured funding for a field trip on his own for four months in 1905–1906. This resulted in a topographical overview of the natural history of the Danish West Indies, which he ended with a call for the necessary establishment of a marine biological station on the island of St. Thomas.[24] The Carlsberg Foundation undoubtedly paid for part of this research, since in 1906 Th. Mortensen successfully applied for an annual stipend of 1,000 Kroner per year for six years in a row.[25] Mortensen explicitly stated in his application that he wished to pursue the idea of setting up a biological station in St. Thomas. He argued that following consultations with the Ministry of Finances and the colonial administration in the Danish West Indies, there was a desire to establish such a marine station, and Mortensen felt that he needed to be ready to take over leadership of such a station, once it was set up. However, as the government was not currently willing to fund this station, Mortensen needed a stipend from Carlsberg Foundation to sustain his research in the meantime.[26] For Mortensen this dream lived on. In 1915 he received 2,000 Danish Kroner from the Carlsberg Foundation to conduct deep-sea investigations in the Danish West Indies.[27]

Nonetheless, nothing ever came of the idea of establishing a Danish field station in the West Indies, which undoubtedly would have been an

22 Anonymous, ed. *Betænkning over forholdene paa de dansk-vestindiske øer afgiven af den allerhøjeste resolution af 18. november 1902 anordnede vestindiske kommission.* Copenhagen: J. H. Schultz; 1903, p. 91–92.

23 Mortensen T, Børgesen F. 'En biologisk station i dansk vestindien', in: *Atlanten.* 1904; 1(6):1–15

24 Mortensen T. Fra de dansk-vestindiske øer. *Geografisk Tidsskrift.* 1907; 19:73–94.

25 Spang-Hanssen E. *Carlsbergfondets Understøttelser 1876–1936.* Copenhagen: Bianco Lunos Bogtrykkeri A/S; 1937. p. 117.

26 CFD. 1906–07. no. 28. Application from Th. Mortensen, 1 November 1906.

27 Spang-Hanssen E. *Carlsbergfondets Understøttelser 1876–1936.* Copenhagen: Bianco Lunos Bogtrykkeri A/S; 1937. p. 117.

important piece of research infrastructure for various types of research. The Danish marine scientists thus continuously had to seek specific funding support for any undertaking in the Caribbean.

Ships of opportunity

One of the ways in which the Danish marine scientists reached the *mare incognitum* of the otherwise inaccessible deep seas, was through the aid of ships of opportunity. From 1911–1915, ships from the Danish merchant fleet and ships of the Danish navy collected material on their way across the Atlantic, when servicing Europe, the USA and The Danish West Indies. It is recorded that a vessel made a *station* on 550 occassions, that is fished or in other ways took a sample from the sea.[28]

The East Asiatic Company (EAC) and The United Danish Steam Ship Companies (DFDS) were the most important sources of information in terms of ships of opportunity. The oceanographic survey conducted by the merchant fleet was orchestrated by Johannes Schmidt, although approved by the Danish commission of marine science. In May of 1911 Schmidt reported to the commission that he had already agreed with the East Asiatic Company that three of their steamships would take stations for his research, when crossing the Atlantic Ocean. The commission paid for four young fish nets in order to equip the ships of opportunity.[29]

In a letter from December, 26 1911, Schmidt explained to Captain Kall of the East Asiatic Company, how he would like to have the next round of surveys carried out.[30] Kall was in charge of the steamship *St. Croix* and over Christmas that year he was coaling in the free port of Copenhagen. Schmidt took the opportunity to thank him for the samples he had already collected on a previous journey. Schmidt let him know that out of the four vessels from EAC currently making surveys for the Kommissionen for Havundersøgelser, the *St. Croix* had been the first vessel to collect specimens of the freshwater eel. Schmidt then explained to the captain how he believed, that in accordance with the direction of currents, the younger and recently spawned larvae will be found further to the West and the South, although Schmidt could only be sure of this after some had been caught. Schmidt then emphasised to the captain that he should steam as slowly as possible when pulling the nets, and he urged the captain to do it as close to midnight as possible.

28 Wolff T. *Danske ekspeditioner på verdenshavene: Dansk havforskning gennem 200 år*. Copenhagen: Rhodos; 1967. p. 160

29 RA. *Komm. Danmarks Fiskeri- og Havundersøgelser*: Forhandlingsprotokoller (1900–1960) N3: 72nd meeting, 4 May 1911.

30 KB. Johannes Schmidt 'Papirer'. Ut.374.V.1. Privat Korrespondance 1902–1916. Letter from J. Schmidt to Kaptain Kall, 26 December 1911.

Schmidt also instructed the captain of the *St. Croix* on investigations of the ocean currents themselves. He wanted any sighting of seaweed from the Sargasso Sea to be recorded to measure the extent and density of its proliferation. Finally, Schmidt asked to have champagne bottles thrown overboard with identifying letters; 50 bottles on the 40th Western longitude and 50 bottles on the 50th Western longitude.[31]

The schooner *Ingolf*, owned by the Danish navy was another vessel, which served to help Johannes Schmidt's Atlantic adventures. Since the late 1870s *Ingolf* had served to inspect the waters of the Danish Atlantic possessions not least fisheries inspections, and back in the 1890s *Ingolf* had served Danish marine science on expeditions to Iceland and Greenland. Later *Ingolf* patrolled the shipping routes between Denmark and the Danish West Indies, and as such, Schmidt found it a useful ship of opportunity. In 1911 for instance, marine scientist Harald Blegvad was onboard searching for eel larvae in the Caribbean.[32] On a return trip from the West Indies, the *Ingolf* had taken samples from 20 fishing stations, and eel larvae measuring 2 inches had been found twice in the nets.[33] From late 1912 a report from the ship's Commander Franz Gotschalk reveals that it was quite a service, which the *Ingolf* offered to Johannes Schmidt and the Kommissionen for Havundersøgelser. In the report Gotschalk inquired whether or not it would be enough to fish once every second night for a little longer, rather than fishing every night. Gotschalk explained that it was very time consuming for the ship to have to take down the sails, turn towards the wind and brace the sails aback, which in effect meant to sail backwards. In any case the commander was happy to report that the surveys had been conducted successfully on the westbound journey, and that the samples from that trip had been shipped directly to Copenhagen with the *St. Jan* of the East Asiatic Company.[34]

Funding the trip with *Margrethe* – Vendsyssel Packing Co Ltd

The *Thor* was unfit for deep sea research in the middle of the Atlantic, since she could steam for only 12 days at a time, before she needed coaling. Therefore, the Danish Kommissionen for Havundersøgelser and Johannes Schmidt needed another vessel to undertake deep sea investigations in the North Atlantic.[35] In addition, *Thor* had been operating

31 KB. Johannes Schmidt 'Papirer'. Ut.374.V.1. Privat Korrespondance 1902–1916. Letter from J. Schmidt to Kaptain Kall, 26 December 1911.

32 Sandbeck T. *Danske havforskningsskibe gennem 250 år*. Stenstrup: Skib Forlag; 2007: 97–98.

33 RA. *Komm. Danmarks Fiskeri- og Havundersøgelser*: Forhandlingsprotokoller (1900–1960) N3: 72nd meeting, 4 May 1911.

34 KB. Johannes Schmidt 'Papirer'. Ut.374.V.1. Privat Korrespondance 1902–1916. Letter from Commandør F. Gotschalk to J. Schmidt, 24 November 1912.

35 Paulsen O. Havundersøgelserne i de nordlige have begynder. In: Winge Ø, Vedel Taaning Å, eds. *Naturforskeren Johannes Schmidt. Hans liv og ekspeditioner*. Copenhagen: Gyldendalske Boghandel Nordisk Forlag; 1947:17–25.

in all kinds of weather for almost a decade now and had to undergo major maintenance. The commission decided in early 1912 not to send the *Thor* to the North Sea and the North Atlantic that year due to financial restraints. When news of the withdrawal of a Danish presence hit the International Council, the Swedish representative, Otto Pettersson responded with astonishment. Pettersson was one of the foremost hydrographers of the time, and he was heralded as the initiator of the International Council. Hence Pettersson's disappointment was noteworthy for the Danish marine scientists, who were not able to carry out their share of the scheduled international effort for hydrographic measurements in the North Sea.[36] The decision was not going to be altered by the Danish commission, but one can see Pettersson's interference as a sign of how serious the international collaboration was taken among ICES member states.

Nonetheless, Schmidt's own hobby horse, the eel research received a welcome blue stamp from ICES in 1912. Schmidt successfully proposed to the ICES council that a new special eel-committee was to be set up. He was seconded in this by Swedish fisheries scientist, Philip Trybom. The idea of this working committee was to have two sections, as was often the case in ICES. First of all, it was approved that the committee would work on purely scientific issues such as charting the life history of the species. Secondly, there was a more practical side to it, where green light was given to investigating the potential of sub-planting specimen of glass eel into freshwater streams, which were otherwise inaccessible to the eel. Symbolically, it became of significant value to Schmidt that he could point to international commitments in his further quest to solve the life history riddles of eel.[37]

For the following year, 1913, however, the commission needed to find another vessel to carry out the investigations in Icelandic and Faroese waters. In late November 1912, the fish exporter, K. Mikkelsen Vendsyssel offered to loan the commission the motor schooner, *Margrethe*.[38] *Margrethe* was actually the property of the Ministry of the Navy; but the navy had lent *Margrethe* to the fishermen's association, *Dansk Fiskeriforening*. The fishermen's association however, could not find any use for the schooner, so one of its board members, K. Mikkelsen Vendsyssel rented the vessel in 1906. K. Mikkelsen Vendsyssel was the founder and owner of *Vendsyssel Packing Company Ltd.*, an international wholesaler of fish, which at the time had their headquarters in Copenhagen.[39] While there is no record of how Mikkelsen Vendsyssel made contact

36 RA. *Komm. Danmarks Fiskeri- og Havundersøgelser*: Forhandlingsprotokoller (1900–1960) N3: 77th meeting, 29 January 1912.

37 Vedel Taaning Å. Fiskeriundersøgelser i nordlige have efter krigen. In: Winge Ø, Vedel Taaning Å, eds. *Naturforskeren Johannes Schmidt. Hans Liv og Ekspeditioner*. Copenhagen: Gyldendalske Boghandel Nordisk Forlag; 1947:113.

38 RA. Komm. Danmarks Fiskeri- og Havundersøgelser: Forhandlingsprotokoller (1900–1960) N3: 85th meeting, 26 November 1912.

39 Anonymous. *Vendsyssel packing co. A/S 1898–1948 blade af selskabets historie*. Denmark: Vendsyssel Packing Co. A/S; 1948.

with Johannes Schmidt, it is most likely to have been mediated by Christian F. Drechsel, since they had both been members of the board of directors of the lacklustre Iceland trawling adventure with the Dan company from 1892–1902.[40]

Mikkelsen Vendsyssel's own company was successful though, as they were among the pioneers in salting and freezing Canadian salmon to be shipped and sold to Europe, and Mikkelsen Vendsyssel had freezing technology installed on board the *Margrethe*. Nevertheless, it did not prove profitable for him to ship frozen salmon across the Atlantic on the *Margrethe*, so in 1912 the company decided to lend her to *Kommissionen for Havundersøgelser*, knowing that she was needed in the on-going search for the breeding places of the eel. Fairly little is known about Vendsyssel Packing Co., but in a commemorative volume from 1948 the company boasts of having contributed to the Danish marine science in this way. In a different anniversary publication, for the 40-years jubilee of the Danish fish exporter's organization, Mikkelsen Vendsyssel, remembered how his organization had been a key player in setting up a very large and successful Scandinavian fish fair in Copenhagen in 1912.[41] More than 100,000 spectators visited the Scandinavian fish fair in 1912.[42] Drechsel was also a central figure at this fair, and although speculative, it is highly likely that Drechsel and Mikkelsen Vendsyssel had struck a deal on borrowing the *Margrethe* in the wake of the successful fish fair. Anyway, it seems that Vendsyssel did not miss the *Margrethe* over the summer of 1913, and in November 1913 the commission accepted an offer to borrow the ship for another season at no extra costs.[43]

We can only speculate about the underlying reasons for Vendsyssel Packing Company's generosity towards the marine scientists. Philanthropy should not be overlooked completely, but if indeed the company stood to gain first-hand and inside knowledge on the breeding place of the eel, they would potentially have their hands on a valuable asset. Mindful of C. G. Johannes Petersen's 1904 comments on the breeding place of the eel, that he who got there first would potentially be on top of a veritable hot spot for catching eel. Petersen had reasoned that, since all eel *seemed* to breed in the same area, this would be where the wriggling fish would congregate.[44]

For Schmidt, borrowing the *Margrethe* offered a golden opportunity to pursue his Atlantic aspirations, and having secured the crucial vessel to take him across the Atlantic, he approached Eugen Warming at the Carlsberg Foundation, at least in a tentative way, for further support.

40 Holm, P. (1996). Technology transfer and social setting: The experience of Danish steam trawlers in the North Sea and off Iceland, 1879–1903. In P. Holm, O. Janzen & J. Thór (Eds.), *Northern seas yearbook 1996* (pp. 113–157). Esbjerg: Fiskeri- og Søfartsmuseets Forlag.

41 Anonymous. *Vendsyssel packing co. A/S 1898–1948 blade af selskabets historie*. Denmark: Vendsyssel Packing Co. A/S; 1948.

42 Tarbensen K. *For alle danmarks fiskere – Danmarks Fiskeriforening 125 år, 1887–2012*. Taulov: Danmarks fiskeriforening; 2012. p. 76–79.

43 RA. *Komm. Danmarks Fiskeri- og Havundersøgelser*: Forhandlingsprotokoller (1900–1960) N4: 98th meeting, 20 November 1913.

44 Petersen CGJ. *Aalen. hvad man ved om den samt anvisning paa at forøge udbyttet af aalefiskeriet*. Copenhagen: Det Schubotheske Forlag; 1904. p. 15.

7.2 The schooner M/S *Margrethe*
The motor schooner, M/S Margrethe was chosen for the first deep sea expedition to the West Indies in 1913. Unfortunately Margrethe was an old ship with a hogging keel. The Atlantic cruise with the Margrethe was probably the most dangerous undertaking of all of Schmidt's 26 oceangoing expeditions, and the wreckage of the ship could easily have been fatal for all men on board (*RA. Carlsbergfondets DANA-ekspeditioner. 1917–1973. 7c*).

In a letter from November, 14 1912 Schmidt revealed his plans to head for the Sargasso Sea, where the smallest eel larvae had been found so far. The *Margrethe* is mentioned as a 'splendid vessel', which will cost around 2,000 Kroner in running costs per month at sea, or 12,000 Kroner for the duration of the planned trip. Counting the cost of equipping the motor schooner for a research expedition, Schmidt added another 6,000 Kroner to the budget. This included the installation of a winch for deep sea trawling, which was particularly expensive. Other expenditures cited, included monies to cover the costs of calling at foreign ports including diverse expenses which would amount to 2,000 Kroner. The total cost, Schmidt estimated to be 20,000 Kroner. With regards covering the costs, Schmidt once more drew on his family fortune as a strategic asset. 'I believe, or rather my wife and I believe we could provide 6,000 Kroner for the cause'.[45] The Kommissionen for Havundersøgelser would probably be able to cover some 6,000 Kroner as well, which left an outstanding amount of 8,000 Kroner for which Schmidt did not have funding, so this is the amount of funding he requested from the Carlsberg Foundation.

We can assume that Schmidt's rather informal letter to Professor Warming was a topic for discussion among the Board of Directors, since the letter is archived among the files of the Board of Directors of the Carlsberg Foundation. Nonetheless, there is no indication that Schmidt was asked to write a full proposal for another Carlsberg grant, so the result was likely negative. Then again, Schmidt might have stretched his wishes a bit too far for that year's donations. The printing costs associated with publishing the findings from the Mediterranean expeditions were already drawing substantially on the benevolence of the Carlsberg Foundation. In 1912 alone, this amounted to 4,775 Kroner for the publication of a manuscript on flatfishes (*Pleuronectidae*).[46]

In the end, it was the Kommissionen for Havundersøgelser that chipped in with the bulk of the funds that Schmidt needed for the trip to the Caribbean. The commission had disposed of their income cautiously over their 1912 budget and so were able to carry the balance forward to the 1913 budget. It was decided to invest 19,000 out of the total of 44,000 Kroner in Schmidt's investigations. The commission further agreed that from March-June the *Margrethe* would serve the routines established with the *Thor*. Currents, temperature and salinity would be measured off the Faroe Islands, and Carl Ostenfeld's assistant, Ove Paulsen would be in charge of the zooplankton investigations. With regards to the fisheries investigations, emphasis was placed on studying the age of maturation for cod spawning around The Faeroes, supplemented by herring (*Clupea harengus*), lemon sole (*Microstomus kitt*) and the eel (*Anguilla anguilla*) investigations.

Crowning the planned expedition with the *Margrethe* was the idea to head into the deep sea waters of the Atlantic Ocean to try and locate eel larvae at different stages in their life history in an attempt to localize the breeding grounds of the freshwater eel.[47] As events unfolded in the autumn of 1913, the commission revised the budget for the West Indies leg of the expedition. While they had predicted an extra cost of 7,000 for crossing the Atlantic, the expected costs rose to 10,500 Kroner. The

45 CFD. 1912–13, no. 209. J. Schmidt to E. Warming, 14 November 1912.
46 CFD. 1912–13, no. 209. Application from J. Schmidt, 12 November 1912.
47 RA. *Komm. Danmarks Fiskeri- og Havundersøgelser*: Forhandlingsprotokoller (1900–1960) N3: 86th meeting, 5 December 1912.

commission expected to fund 7,000 from savings made, as the *Thor* was inoperative, while the Ministry of Finances might allocate the remaining 3,500.[48] Following the successful fundraising for the *Margrethe*, it was now time to look to Danish businessmen operating in the Danish West Indies for assistance.

In the summer of 1913, Schmidt wrote a letter to the owner of the East Asiatic Company, H. N. Andersen asking him for support in connection with his upcoming expedition with the *Margrethe*. He noted that Andersen had already been of great assistance to Danish marine science, Schmidt then attached to his letter one of his scientific reports on Mediterranean and Atlantic marine science, and pointed to one of the maps which had used findings from one of the company's ships. What Schmidt now wished to gain from Andersen was: First of all, references to local connections in the Danish West Indies as a way of possibly defraying costs associated with near shore fisheries investigations, once the expedition had reached the island of St. Thomas. Secondly, Schmidt asked more directly if indeed it was possible that the East Asiatic Company would be able to support the expedition with a moderation in the cost of shipping, both when the materials collected by the sea were to be sent home to Copenhagen and with regards to the Atlantic passage back to Europe for Schmidt and one other marine scientist, possibly Carl Ostenfeld, who was scheduled to join him for the Caribbean part of the expedition.[49] While Schmidt's travel funds came from the laboratory, The Carlsberg Foundation financed Ostenfeld's trip to the Danish West Indies with a donation of 2,800 Danish Kroner.[50]

Having found a suitable ship and funding from the commission for investigating the Atlantic Ocean further away from the European mainland than previously, Schmidt still needed to convince his superiors, the Board of Trustees of the Carlsberg Laboratory that they could do without his services for a period of time.

48 RA. *Komm. Danmarks Fiskeri- og Havundersøgelser:* Forhandlingsprotokoller (1900–1960) N4: 97th meeting, 13 November 1913.

49 KB. Johannes Schmidt 'Papirer'. Ut.374.V.1. Privat Korrespondance 1902–1916. 'Privat 1913'. Letter from J. Schmidt to H. N. Andersen, 30 June 1913.

50 Spang-Hanssen E. *Carlsbergfondets Understøttelser 1876–1936*. Copenhagen: Bianco Lunos Bogtrykkeri A/S; 1937. p. 125.

Hops across the Atlantic

In 1913 the Board of Trustees of the Carlsberg Laboratory appears to have expected Johannes Schmidt to be committed to serious work on topics related to the core interests of the charter of the laboratory, namely the improvement of beer production. In September 1913, Schmidt was able to present his idea of a trip to the USA and onwards to the Danish West

Indies to the Board of Trustees of the Carlsberg Foundation. Schmidt did not hide the fact that the expedition about to be undertaken with the *Margrethe* was the main reason for his intended absence. However, he claimed not to have had any real holiday during the previous three years, so he asked for – and received a leave of absence of four months. The idea was to first go to the East coast of the USA to see for himself, what were allegedly the most well run plantations for cultivating hops (*Humulus lupulus*) in the world. In connection with this site visit, Schmidt's intention was to meet up with one of the leading American scientists in the field, Dr. Stockberger. The expenses associated with his work at the Commission for Oceanography did not concern the laboratory board, but Schmidt wished to dispose of 1,500 Danish Kroner for the American part of his trip, and this was granted.[51]

The months that followed the successful fundraising were busy for Schmidt. While he made his way through fields of hops in the Eastern USA, his thoughts may well have wandered from his landlocked duties and made their way to the Atlantic Ocean, where the *Margrethe* was zigzagging across the sea from Northern Europe to the Danish West Indies. By Christmas 1913, Johannes Schmidt had finished his survey of the American production of hops. Having travelled on to Puerto Rico he was set to meet up with the members of the expedition on board the *Margrethe*, and on Christmas Eve he wrote two reports to the Carlsberg Laboratory's Board of Trustees.[52]

First of all he gave an account of his meetings with Dr. Stockberger, which had resulted in a large scale exchange of hops. Circa 100 different types of hops were to be grown simultaneously by the Carlsberg Laboratory in Denmark and the U.S. Department of Agriculture. The intention was to compare the influence of diverse climates on the quality and growth rate of different types of hops. Secondly, Schmidt had proposed an exchange of flower cones from hops, which were to undergo chemical analyses. The purpose of this undertaking was mainly to establish the content of lupulin, which is the substance adding bitter flavor to beer. Finally, Stockberger had proposed that an international network was initiated, whereby it would be possible to standardize methods for scientific and practical assessments of hops.[53]

In a postscript to his first letter Schmidt then proceeded to give a short account of the fate of the *Margrethe*. Following an otherwise very successful research trip, the *Margrethe* had become stranded on the island of Anegada a couple of days before Christmas, and although

51 CA. Carlsbergfondet, Carlsberg Laboratorium. E06. 1913. Letters from J. Schmidt to Board of Trustess 15, 16 and 23 September 1913.

52 CA. Carlsbergfondet, Carlsberg Laboratorium. E07. 1914. Letter from J. Schmidt to Board of Trustees 24 December 1913.

53 CA. Carlsbergfondet, Carlsberg Laboratorium. E07. 1914. Letter from J. Schmidt to Board of Trustees 24 December 1913.

several important collections had been rescued, the ship was now a 'total wreck'. This had thwarted plans for making further surveys of the near shore areas of the Danish West Indies. Therefore, Schmidt told his superiors that he was most likely not going to use the whole of the leave of absence, which he had been granted in the fall. Nonetheless, for the time being Schmidt would stay with his expedition to assure that the young scientific leader of the voyage, Stud. Mag. Poul Jespersen and the crew knew how to proceed.

Troubled testing of *Margrethe*

The *Margrethe*'s voyage across the Atlantic Ocean was indeed marked by a succession of highpoints and finally a submerged low point. By spring of 1913, Schmidt was certain that the eel larvae flowed with the Gulf Stream in a North Eastern direction, and the idea was to conduct surveys with the *Margrethe* in the opposite direction of the Gulf Stream. The *Margrethe* measured only 90 gross register tonnage, and most space on deck had to serve several functions. The laboratory for instance, doubled as the mess for Jespersen and the ship's captain, Hansen.[54] Having sailed the *Thor* for several years, Hansen was well accustomed to the practical crafts of oceanographic field research, which may help to explain why Jespersen, only 22 years old at the time and still a university student, was given command of the scientific part of the voyage.[55]

Before embarking on the Atlantic crossing the *Margrethe*, was still serving the commission's investigations on the fishing banks off the Faroe Islands. Plankton zoologist, Ove Paulsen and fisheries scientist, A. C. Strubberg were on board the *Margrethe* for this part of the trip, and both of them, as well as Captain Hansen sent regular reports to Johannes Schmidt on the state of affairs. For a long time it seemed as if the plan to cross the Atlantic with the *Margrethe* might be cancelled.[56]

The research team left Denmark in April, and soon it became clear that the engine could not be used for more than an hour at a time, before becoming too hot. We know this, because Schmidt complained about the situation to the manufacturer of engines, DAN A/S, in Copenhagen, where the engine of the *Margrethe* had undergone service before leaving Danish waters. Captain Hansen later reported that two crew members had worked for several days, and finally they had succeeded in cleaning the engine, so it would no longer heat up too much.[57]

54 Jespersen P. Danske handelsskibes og skonnerten *Margrethe*'s indsats. In: Winge Ø, Vedel Taaning Å, eds. *Naturforskeren Johannes Schmidt. Hans Liv og Ekspeditioner*. Copenhagen: Gyldendalske Boghandel Nordisk Forlag; 1947: p. 57–69.

55 Spärck, Ragnar, 'Jespersen, P.', in: Povl Engelstoft & Svend Dahl (eds.), *Dansk Biografisk Leksikon*, vol. 11, Copenhagen: J. H. Schultz Forlag; 1937. p. 594–595.

56 KB. Johannes Schmidt 'Papirer'. Ut.374.V.1. Privat Korrespondance 1902–1916. 'Privat kuvert især breve 1912–15'.

57 KB. Johannes Schmidt 'Papirer'. Ut.374.V.1. Privat Korrespondance 1902–1916. 'Privat kuvert især breve 1912–15'. Letter from J. Schmidt to A/S Motorfabrikken 'Dan', 19 May 1913; G. Hansen to J. Schmidt 7 June 1913.

A more serious problem however, seems to have been that basically the *Margrethe* was a weary old vessel. According to a letter written by Ove Paulsen in early June 1913, the *Margrethe* was a 'huge step backwards' compared to the *Thor*. The motor engine had great difficulties sailing against the wind, and when the ship was sailing close to the wind the bowsprit would cause a roaring sound when hitting the sea over and over again. On one of these occasions a wooden beam supporting the front sail, the so-called *flying jib stay* was torn off the ship, but according to Paulsen this was the only way to manoeuver the ship against the wind. When crossing against the wind she was doing fine, and she did not take in too much water. Nonetheless, Paulsen underlined that the *Margrethe* had a tendency to dip the bowsprit, as the keel was hogging. 'It is not amusing that the galley figure has been smashed on one occasion', he said. Paulsen also gave a rare glimpse of the relationship between the scientists on the one hand and the sailors on the other, when he told Schmidt how the crew of sailors stopped playing music and stopped talking in the mess, as soon as he or the other scientists entered the room. The awkward silence was broken only by murmuring and whispering behind their back. 'All in all, the whole situation is sour,' Paulsen concluded.[58]

Captain Hansen though, was reasonably comfortable with the *Margrethe*, but as late as June, 7, a letter from Hansen to Schmidt reveals that four weeks before the Atlantic crossing started, Hansen had still not received a final order from Schmidt as to whether or not the voyage would take place. Also, the final course of the ship's voyage had not been fully decided on at this stage. Referring to the poor state of the *Margrethe*, Hansen urged Schmidt to set the course in a navigable direction to avoid too much crossing against the wind. Hansen suggested that upon leaving the Faeroes, the expedition would first go to the Azores and then towards the central Sargasso Sea. Hansen then asks Schmidt to place an order for ship's provisions in Copenhagen, including maps and guide books of the waters of the Danish West Indies as well as for the Azores.[59]

Final instructions

It was not until June, 23 that Schmidt wrote down the final instructions for the voyage and sent it to Captain Hansen and Poul Jespersen.[60] The

58 KB. Johannes Schmidt 'Papirer'. Ut.374.V.1. Privat Korrespondance 1902–1916. 'Privat kuvert især breve 1912–15'. Letter from O. Paulsen to J. Schmidt, 2 June 1913.

59 KB. Johannes Schmidt 'Papirer'. Ut.374.V.1. Privat Korrespondance 1902–1916. 'Privat kuvert især breve 1912–15'. Letter from G. Hansen to J. Schmidt 7 June 1913.

60 RA: *Dansk Biologisk Station, Johannes Schmidts embedsarkiv*: Korrespondance om åleforskning (1908–1933) E28: 1908–1913. 'Instruks for Undersøgelsern i Atlanterhavet' from J. Schmidt to G. Hansen & P. Jespersen, 23 June 1913.

instruction was to fish with pelagic young fish trawls at various depths. In any one trawl a number of ring nets were to be placed on different sections of the main line at the intervals of 25 m, 65 m., 150 m, 3,000 m, 1,000 m, 1,600 m, 2,500 m and 4,000 meters depth. Such a trawl would comprise a 'full station', where Schmidt emphasized that they should take particular care to aim for samples of young eel larvae. The objective was to sample as wide as possible, to find a reliable picture of the average size of larvae in different areas of the Atlantic Ocean and relate this to the direction of the ocean currents.

In addition to eel larvae, Schmidt also wanted to obtain knowledge on the larvae from swordfish (*Xiphias gladius*), tuna (*Thunnus thynnus*) and ocean sunfish (*Mola mola*), in order to get closer to their breeding grounds as well. Whenever the chance arose, Jespersen and Hansen should also fish with gear such as long lines, herring trawls and gill nets. When sailing in less deep waters they should also scrape the bottom for samples and try to acquire material from resident fishermen. The expedition should cross the Atlantic and take as many stations as possible and aim to arrive in St. Thomas via Bermuda following a fairly rudimentary course. However, the arrival in St. Thomas need not be before November, 1. The Azores could be visited if the team saw fit to do so, but according to Schmidt this was not necessary for the success of the scientific part of the expedition. Upon arrival in St. Thomas the team was to report at the headquarters of the East Asiatic Company, where further instructions would await them. Meanwhile, every opportunity should be used to report back to the commission.[61]

Eel plans stranded in the Caribbean

The Atlantic expedition set off from the Faeroes in July of 1913, and every second night a fishing station was logged. Thus, five weeks passed before reaching the first anchorage on the Azores. The expedition then embarked on a lengthy 66 days' stretch across the Atlantic Ocean, sailing in long legs from North to South and back in search of eel. Consequently, the crew encountered changing weather from chilly Northern winds from Newfoundland, to tropical storms with torrential rains and waterspouts. One such waterspout caused the bowsprit to splinter. In spite of these obstacles eel larvae were continuously found in the trawls, and processed right away in the ship's cabin. The crew now came to real-

61 RA: *Dansk Biologisk Station, Johannes Schmidts embedsarkiv*: Korrespondance om åleforskning (1908–1933) E28: 1908–1913. 'Instruks for Undersøgelsern i Atlanterhavet' from J. Schmidt to G. Hansen & P. Jespersen, 23 June 1913.

ize that the size of the individual eel larvae dwindled the further to the West they were fishing. The other theory was that the eel would become smaller the more to the South the team went fishing. This was proposed by Johan Hjort following his Atlantic trip with the *Michael Sars* in 1910. Nonetheless, as Johannes Schmidt had anticipated in the spring of 1913, Hjort's theory did not hold up against empirical findings.[62]

By late October the expedition had reached Bermuda, and significant catches of eel larvae had been collected. The majority of the eel larvae were between 3 and 4 cm, while a few samples were only 17 mm long, the smallest that anyone had seen. This clustering of sizes in the collection illustrated that the spawning did not occur evenly over the year. In autumn at least, there seemed to have been very little spawning activity. Otherwise, the amount of tiny larvae less than 2 cm would have been greater.[63] Some of the smallest eel larvae were shipped to Washington D. C., where Johannes Schmidt had taken up quarters, so the organizer of the expedition could begin his own inspection of the material as soon as possible.

Hopes were high as the expedition left Bermuda again, charting a course where a big easterly swing would take the *Margrethe* towards St. Thomas in the Danish West Indies. According to Jespersen, the crew fished several hundreds of eel larvae on this stretch of the trip, and by the 21 December they first caught sight of some of the West Indian islands. So close to a long awaited Christmas break in the Caribbean, disaster struck in the middle of the night.

In foggy weather and capricious currents the *Margrethe* ran aground on a near shore coral reef. Jespersen was not considered to be a born sailor, so Captain Hansen secured him to the gunwale, to prevent him from being swept overboard by the waves flooding the vessel. From this position, Jespersen watched on, as the crew struggled to relieve the ship of some of its weight by cutting down the main mast. This did not help the ship to float clear of the reef. As no other vessels were in sight to help the *Margrethe*, he crew fired emergency rockets and flares, but no one appeared to help them. As dawn approached, the ship itself could no longer be saved, and all men boarded the ship's dinghy and rowed ashore with little else in their hands but the ship's papers. From ashore on the tiny island of Anegada, the crew could only watch on, as the ship sunk with the impact of waves hammering against the reef.

The next day was spent trying to salvage as many goods as possible from the *Margrethe*. Then, the exhausted crew paid some local inhabitants

62 Jespersen P. Danske handelsskibes og skonnerten *Margrethe*'s indsats. In: Winge Ø, Vedel Taaning Å, eds. *Naturforskeren Johannes Schmidt. Hans Liv og Ekspeditioner*. Copenhagen: Gyldendalske Boghandel Nordisk Forlag; 1947: p. 57–69.

63 Jespersen P. Danske handelsskibes og skonnerten *Margrethe*'s indsats. In: Winge Ø, Vedel Taaning Å, eds. *Naturforskeren Johannes Schmidt. Hans Liv og Ekspeditioner*. Copenhagen: Gyldendalske Boghandel Nordisk Forlag; 1947: p. 57–69.

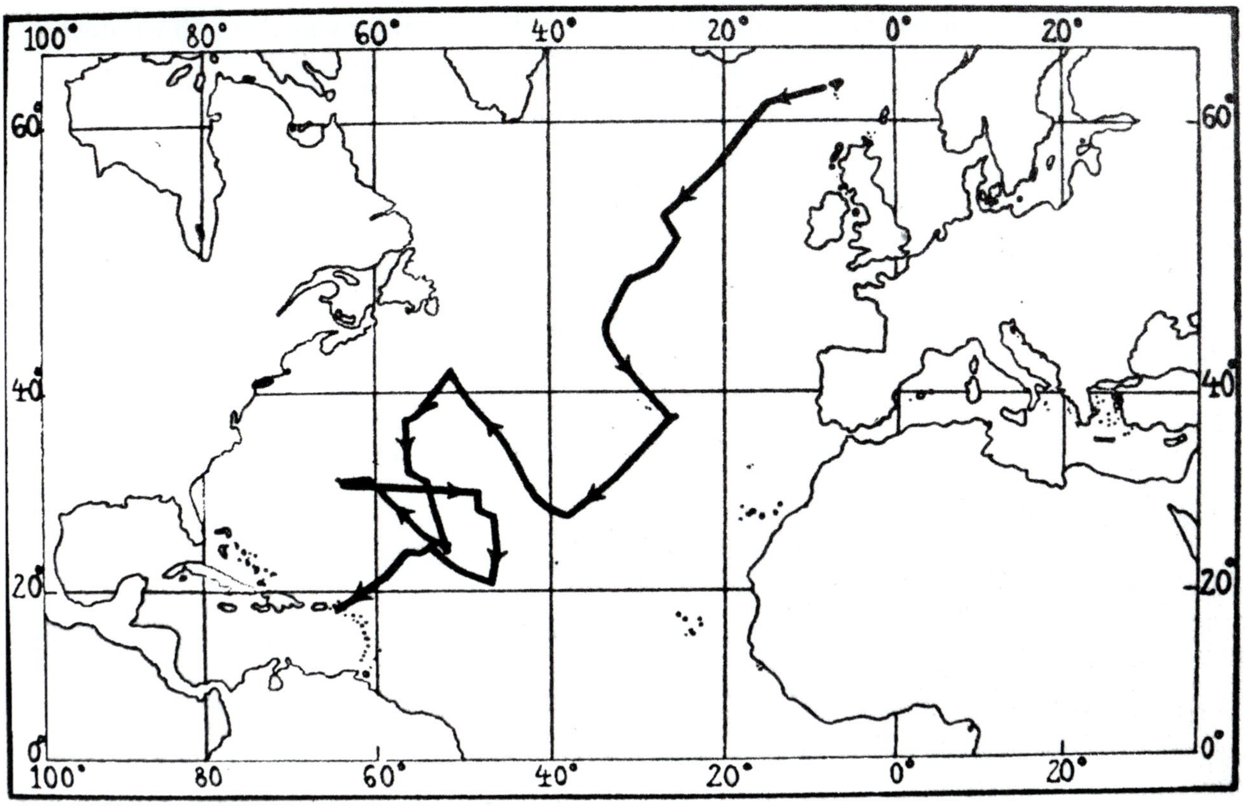

7.3 The *Margrethe* expedition in 1913

The 1913 expedition with the M/S Margrethe, which ended distastrously on a Caribbean reef (Winge Ø. & Vedel Taaning Å. 1947).

to sail them to nearby Tortola in the British West Indies, where authorities took them in for questioning. Thus it was late on Christmas Eve before Captain Hansen, Poul Jespersen and the rest of the crew were able to reach the port town of Charlotte Amalie on St. Thomas. By the time the sun set on Christmas Eve, Hansen and Jespersen appeared in front of the Danish governor's house, and when Jespersen was able to identify himself by presenting a sea-worn business card from his dirty clothes, the two men were allowed to join the festivities of the night.

Having recovered well over Christmas, it was now time to attend to the goods rescued from the *Margrethe*, and before New Year the expedition in St. Thomas was joined by Johannes Schmidt and botanist, Carl Ostenfeld, who both came to St. Thomas via Puerto Rico. Obviously, any plans for collecting material on a return trip had to be abandoned, and an intensive month followed. The Danish marine scientists set up an interim laboratory in the city's jailhouse, the old Fort Christian in

Charlotte Amalie. Here they could work their way through whatever glasses had been recovered from the *Margrethe*.[64]

Exciting days must have followed, and on the January, 24 1914, Schmidt was able to write a triumphant letter home: 'Today, I have had a happy day going through all our collections of leptocephali (eel larvae). By a wonderful stroke of good fortune, all four boxes containing leptocephali are saved, out of the scores of boxes which are all lost. (…) The material is so important, and large, that nothing, not even remotely similar, is in existence. I have today, just by comparing a few sets of data, been able to see that the material allows for extraordinarily significant conclusions regarding the direction and strength of ocean currents.'[65]

64 Jespersen P. Danske handelsskibes og skonnerten *Margrethe*'s indsats. In: Winge Ø, Vedel Taaning Å, eds. *Naturforskeren Johannes Schmidt. Hans Liv og Ekspeditioner*. Copenhagen: Gyldendalske Boghandel Nordisk Forlag; 1947: p. 68–69.

65 Jespersen P. Danske handelsskibes og skonnerten *Margrethe*'s indsats. In: Winge Ø, Vedel Taaning Å, eds. *Naturforskeren Johannes Schmidt. Hans Liv og Ekspeditioner*. Copenhagen: Gyldendalske Boghandel Nordisk Forlag; 1947: p. 68–69.

7.4 Poul Jespersen and Johannes Schmidt in the Danish West Indies

Poul Jespersen (left) and Johannes Schmidt both had good reasons to smile in the Danish West Indies in the early days of January 1914. Jespersen had just survived the wreckage of the M/S Margrethe, and Schmidt had come down from USA *to help sorting out the eel specimen taken on the Margrethe expedition (Carlsberg Archives).*

While, the expedition with the *Margrethe* was successful for as far as she went, the Danish marine scientists had now lost the opportunity to continue their fisheries investigations in open waters around the Danish West Indies. In terms of the return trip however, it appears that H. N. Andersen of the East Asiatic Company had looked favorably at the abovementioned request, which Johannes Schmidt had made in the summer of 1913.[66] For whatever the fare for crossing the Atlantic, it was a ship from the East Asiatic Company which brought back to Denmark, the expedition's collection as well as Johannes Schmidt.[67]

While Schmidt went back to Copenhagen in February, Ostenfeld and Jespersen stayed for another month.[68] This is known in the literature, but neither how they spent their time in the West Indies is revealed, nor the important role of the *Margrethe's* captain, Georg Hansen.

Green turtles and red goatfish

By 1914, the members of the Danish commission for marine science were fully aware that the sea turtles nesting on the beaches of the Danish West Indies were endangered species. Zoologist Theodor Mortensen had written a memorandum on the state of the West Indies sea turtles in 1906, in which he recommended that there should be a total ban on catching turtles on land, when they were nesting and also a ban on collecting the eggs, when these had been laid on the beaches.[69]

In connection with the *Margrethe* expedition in 1913–1914, the commission wanted to investigate the life history of the hawksbill turtle (*Eretmochelys imbricata*) and the green turtle (*Chelonia mydas*) in order to gain more knowledge about how to implement a conservation strategy. Realizing that time was limited, the Danish marine scientists could only make initial examinations of the sea turtles on the islands, and then Schmidt instructed the Colonial administration, the *Gouvernement*, on how to proceed. Following the previous decade of experiences with tagging experiments in North Atlantic waters, Schmidt instructed Georg Hansen on how to set up a similar system for the administrators in St. Croix. The port officers of the Gouvernement would administer a bounty system, whereby fishermen catching sea turtles would be granted a reward for reporting on the turtles. Standardized templates were circulated to the ports of the colony, with different columns to be filled with information on the time and place of the catch,

66 KB. Johannes Schmidt 'Papirer'. Ut.374.V.1. Privat Korrespondance 1902–1916. 'Privat 1913'. Letter from J. Schmidt to H. N. Andersen, 30 June 1913.

67 Jespersen P. Danske handelsskibes og skonnerten *Margrethe*'s indsats. In: Winge Ø, Vedel Taaning Å, eds. *Naturforskeren Johannes Schmidt. Hans Liv og Ekspeditioner*. Copenhagen: Gyldendalske Boghandel Nordisk Forlag; 1947: p. 69.

68 Sandbeck T. *Danske havforskningsskibe gennem 250 år*. Stenstrup: Skib Forlag; 2007: 142.

69 RA. *Dansk Biologisk Station, Johannes Schmidts embedsarkiv*: Fiskeriundersøgelser og skildpaddemærkning i Dansk Vestindien (1914–1916) E30: Memorandum from Th. Mortensen to Gouvernementet for de dansk-vestindiske Øer, 4 April 1906.

the weight, length and gender of the specimen, and additional information on the number of eggs per female.[70] Georg Hansen reported back to Schmidt and complained that the tagging of turtles was slow to take off.[71] However, from later reports from the colony sent to Johannes Schmidt it seems that more than 100 turtles were identified and marked with lead strappings on their shell and released again afterwards during the year of 1914.[72]

The results of these studies were published by Johannes Schmidt in 1916 as explorative paper, printed at cost of the commission for marine science.[73] This explorative paper seems to have been the first to analyze the growth rate of juvenile Green Turtles living in a natural population. This was enabled through studies of the recaptured animals, although the sample size for Green Turtles, (n: 8) seems rather poor by modern standards. Schmidt's study, however, has been explored and referred to several times in modern studies of Caribbean turtle populations.[74] In a recent overview paper on the status of hawksbill turtles in the Caribbean, Schmidt's study serves to underline the fact that human harvests and possible overexploitation of sea turtles has a long history.[75] No doubt the Danish fisheries' scientists brought state of the art marking techniques to the Caribbean, which they had developed over the previous decade of marking experiments with commercial fish in the North Atlantic.

While sea turtles received special attention from the scientists and marine resource managers, Hansen's efforts also included a search for more eel specimens. Carrying a Petersen young fish trawl and a seine nets with him, Hansen travelled to a number of different coastal areas of the Danish West Indies, where he investigated the different fisheries. Hansen managed to catch 28 different species of fish, although he was unhappy not to have found any eel.[76] What remains more impressive is the report Hansen drafted on the state of the fisheries in terms of how many fishermen were active on the islands, what technologies were deployed as well as the various types of fish they caught.[77]

Again, the commission in Copenhagen, in particular Johannes Schmidt, was the chief designer of the fisheries studies in the Danish West Indies. According to the plans presented by Schmidt in October 1913, the intention of the study was deliberately targeted at improving the economic output of the fisheries. Biological studies of the life

70 RA. *Dansk Biologisk Station, Johannes Schmidts embedsarkiv*: Fiskeriundersøgelser og skildpaddemærkning i Dansk Vestindien (1914–1916) E30: Letter from J. Schmidt to Gouvernementet for de dansk-vestindiske Øer, 16 February 1914.

71 RA. *Dansk Biologisk Station, Johannes Schmidts embedsarkiv*: Korrespondance om åleforskning (1908–1933) E28: 1908–1913. Letters from G. Hansen to J. Schmidt, 28 March and 15 April 1913.

72 RA. *Dansk Biologisk Station, Johannes Schmidts embedsarkiv*: Fiskeriundersøgelser og skildpaddemærkning i Dansk Vestindien (1914–1916) E30.

73 Schmidt J. Marking experiments with turtles in the Danish Eest Indies. *Meddelelser fra Kommissionen for Havundersøgelser: serie Fiskeri.* 1916; 5(1); RA. Komm. Danmarks Fiskeri- og Havundersøgelser: Forhandlingsprotokoller (1900–1960) N4: 113th meeting, 7 January 1916.

74 Frazer NB, Ladner RC. A growth curve for green sea turtles, Chelonia mydas, in the US Virgin Islands, 1913–14. *Copeia.* 1986(3):798–802; Bjorndal KA, Bolten AB, Chaloupka MY. Green turtle somatic growth model: Evidence for density dependence. *Ecol Appl.* 2000;10(1):269–282

75 Meylan AB. Status of the hawksbill turtle (eretmochelys imbricata) in the Caribbean region. *Chelonian Conservation and Biology.* 1999;3(2):177–184

76 RA. Dansk Biologisk Station, Johannes Schmidts embedsarkiv: Korrespondance om åleforskning (1908–1933) E28: 1908–1913. Letters from G. Hansen to J. Schmidt, 28 March and 15 April 1913.

77 RA. Dansk Biologisk Station, Johannes Schmidts embedsarkiv: Korrespondance om åleforskning (1908–1933) E28: 1908–1913. 'Beretning om Fiskeriet i Dansk Vestindien (Kladde)'.

7.5 Lead marks from Caribbean sea turtles *Schmidt was the first to publish studies on Green Turtles (Chelonia mydas) based on marking experiments. These lead marks were returned to Copenhagen after the turtles had been recaptured in The Danish West Indies (RA: Dansk Biologisk Station, Johannes Schmidts embedsarkiv. E30).*

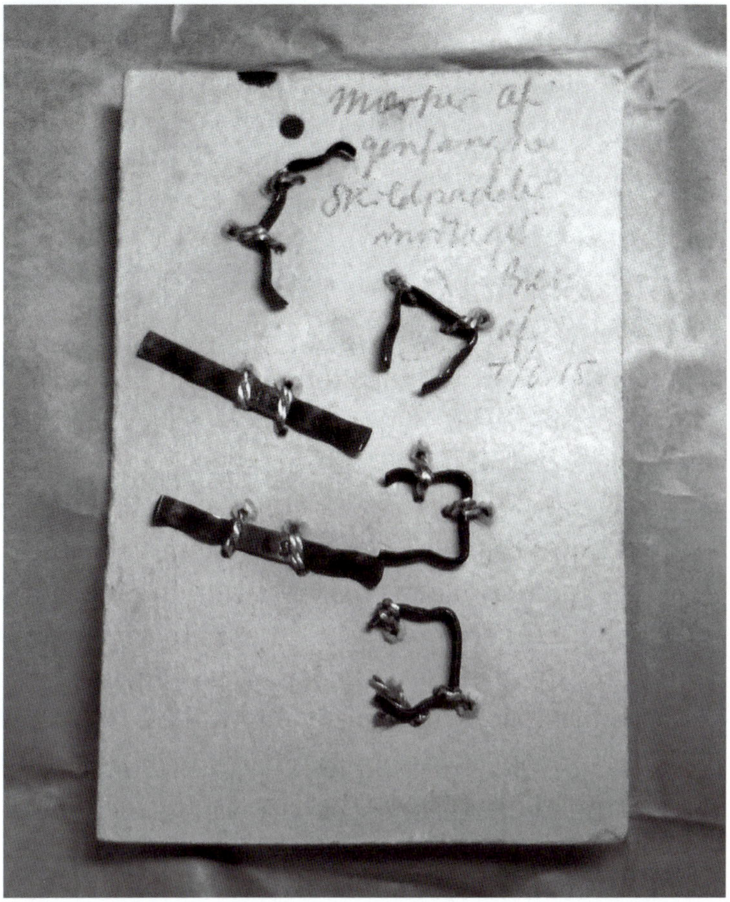

history of the fishes were needed as well as studies testing whether or not it would be feasible to keep the landed fish alive in dams, which was a practice used in Florida, or if as an alternative it was possible to set up a canning industry in the Danish West Indies.[78] The rearing of sponges was another hugely successful undertaking in Florida at the time.[79] This prompted Schmidt to also target such an investigation for the Danish Caribbean colony.

Based on the landings from 123 fishing vessels operating out of St. Thomas more than 10,000 individual specimen were investigated. More than half the fish had been caught with hand lines, while the rest were caught in pots.[80] Jespersen, Hansen and probably some local assistants must have worked very hard, determining the species and measuring the size of each of these more than 10,000 fish with common names such as red goatfish, grouper and old wife.

Nonetheless, the results of the fisheries investigation do not seem to have been published. One can only speculate why, but it is likely that the loss of the *Margrethe* was a significant setback, while the outbreak

78 RA. *Komm. Danmarks Fiskeri- og Havundersøgelser:* Forhandlingsprotokoller (1900–1960) N4: 96th meeting, 20 October 1913.

79 McClenachan L. Social conflict, overfishing and disease in the Florida sponge fishery, 1849–1939. In: Starkey DJ, Holm P, Barnard M, eds. *Oceans past: Management insights from the history of marine animal populations.* Earthscan; 2008:25–46.

80 RA. Dansk Biologisk Station, Johannes Schmidts embedsarkiv: Fiskeriundersøgelser og skildpaddemærkning i Dansk Vestindien (1914–1916) E30: 'Beretning om Fiskeriet i Dansk Vestindien (Kladde)'.

of World War I later in the same year, and the sale of the Danish West Indies to the USA in 1917 were disastrous with regards to the continuation of Danish marine research in the Virgin Islands. In the spring of 1914, however, hopes were still high.

Conclusion

It is well known in the literature that the voyage of the *Margrethe* in the fall of 1913 resulted in sampling of eel larvae, which enabled Schmidt to deduce that the breeding grounds of the freshwater eel were somewhere in the Sargasso Sea. However, many circumstances surrounding this important discovery have not been accounted for before now:

To properly assess the organizational background for the successful sampling of eel larvae, the expedition with the *Margrethe* needs to be looked upon in the context that the Sargasso Sea happened to be located within reach of the Danish West Indies. This means that Schmidt could draw on the infrastructure already in place at the Danish Caribbean colony. Moreover, what emerged is the possibility for Schmidt to successfully tie in his own eel research with a tradition for marine science investigations around the Danish West Indies. In previous scholarship the fisheries investigations around the islands and the tagging experiments of sea turtles have not received any attention in the literature on the history of marine science, in spite of the fact that Schmidt's tagging experiments with sea turtles remain the earliest reference point of its kind, and are still cited in current scholarship.

From an organizational point of view, the expedition proved how the tightly knit group of marine scientists in the Danish marine science commission could mutually support one another. Christian F. Drechsel as chairman of the commission is most likely the broker of the arrangement, whereby the marine scientists were able to borrow the *Margrethe* from the Vendsyssel Packing Co., whose owner, Mikkelsen Vendsyssel was a former business associate of Drechsel on their joint fishing adventure in Iceland twenty years earlier.

Finally, Schmidt's determination and willingness to take risks was an important factor not to be overlooked. The Atlantic cruise with the *Margrethe* was probably the most dangerous undertaking of all of Schmidt's 26 oceangoing expeditions, and the wreckage of the ship could easily have been fatal for all men on board.

8 Collecting funds for an Atlantic deep sea investigation

With Drechsel at the helm, efforts were directed at finding money from a number of private sponsors from the spring of 1914 onwards. Schmidt's former assistant, Poul Jespersen, wrote a brief account of the efforts in which he describes how Drechsel and Schmidt in spite of 'great difficulties, were able to generate funds and other kinds of support.'[1] This meant that in February 1917, more than 250,000 Danish Kroner had been collected for the purpose, and a formal expedition committee was inaugurated. Jespersen however, did not indicate what the difficulties consisted of, or *why* wealthy businessmen wished to fund marine science far away in the Atlantic Ocean, nor *who* they were.

Finding 'a rich man' post the *Margrethe*

Following the news of the loss of the *Margrethe,* the commission convened just after Christmas in 1913. The members agreed that until further notice from Schmidt, they would notify the Ministry of Agriculture only by telephone. K. Mikkelsen Vendsyssel though, had already been notified of the loss of his ship, and he had tentatively suggested that the commission could buy the motor ship, *Hvalen* as a replacement. Vendsyssel proposed that he and the Naval Ministry could donate 5,600 Kroner each, and other private donators were willing to provide 4,500 Kroner for the acquisition of *Hvalen,* while the state should contribute with 10–15,000 Kroner for equipping the vessel for oceanographic research. Although the commission was in need of another ship, they decided to decline the offer. At this stage they may have had enough of sailing vessels with motor engines, such as the *Margrethe*. Instead, the commission agreed to try and find sponsorship for covering the West Indies investigations with rented vessels, if Johannes Schmidt would approve this plan.[2]

1 Jespersen P. Togter med M/S 'Dana' (I) 1920–21. In: Winge Ø, Vedel Taaning Å, eds. *Naturforskeren Johannes Schmidt. Hans liv og ekspeditioner*. Copenhagen: Gyldendalske Boghandel Nordisk Forlag; 1947:80.

2 RA. *Komm. Danmarks Fiskeri- og Havundersøgelser*: Forhandlingsprotokoller (1900–1960) N4: 99th meeting, 27 December 1913.

Schmidt was back in Copenhagen in March and on the 17 March the commission was summoned to learn about the expedition's Caribbean accomplishments. Schmidt reported that the eel investigations were now so far advanced that only a small area of the vast Atlantic Ocean still awaited further investigation. The commission agreed that Schmidt should be allowed to pursue the quest to find the breeding grounds of the eel, and Drechsel proposed that his former business partner Mikkelsen Vendsyssel should be invited to a conference under the auspice of the commission, where he might indicate if he would support a continuation of the investigations. No doubt, the commission was conscious that the story of the pursuit of the eel had drawing power for potential sponsors, which could act as leverage for further expeditions. Mikkelsen Vendsyssel as well as the Danish commission played a part in setting up a fisheries and marine science congress as part of the *Baltiska utställningen,* an upcoming arts and crafts fair in Malmö Sweden. Schmidt, who was a member of the executive committee of the congress, was invited to talk about his eel investigations.[3] Mikkelsen Vendsyssel also heard Schmidt talk about his findings at the next meeting of the commission, for which he had an invitation. It was estimated that a new expedition to the West Indies would cost 10–20,000 Kroner, and Mikkelsen Vendsyssel offered to cover 10,000 jointly with the Naval Ministry.[4]

One outstanding issue was the insurance money to be collected for the loss of the *Margrethe*. The commission expected to receive the share of the insurance payment belonging to the Naval Ministry, which was in the order of 10,000 Kroner.[5] This may be one reason why the commission did not include further expeditions into the Atlantic, when they planned the marine science activities for 1915. In spite of the supportive attitude towards Schmidt's eel investigations, they were viewed as something extra on top of the normal activities geared at servicing the programme of the International Council.

Also, since Drechsel was chairing the Danish commission as well as holding the position of general secretary of ICES, his participation would have more or less tacitly assured the compatibility of the activities in both organizations.

A further issue of the research infrastructure should also not be overlooked. The commission was planning to send the *Thor* back into service in the Faeroese waters as well as in the seas around Denmark to conduct a series of plankton and fisheries surveys, the latter ones

3 RA. *Komm. Danmarks Fiskeri- og Havundersøgelser*: Forhandlingsprotokoller (1900–1960) N4: 102nd meeting, 17 March 1914.
4 RA. *Komm. Danmarks Fiskeri- og Havundersøgelser*: Forhandlingsprotokoller (1900–1960) N4: 103rd meeting, 26 March 1914.
5 RA. *Komm. Danmarks Fiskeri- og Havundersøgelser*: Forhandlingsprotokoller (1900–1960) N4: 105th meeting, 2 June 1914.

centered on measuring the length, weight, and age of commercially important species.[6] The deployment of the *Thor* would entail a range of activity limited by the *Thor's* need for frequent coaling.

Finally, it should not be overlooked that Johannes Schmidt was absent for more than a month in April 1914. Schmidt's wife, Ingeborg was holidaying in Egypt, when she fell ill. We know of the illness of Ingeborg Schmidt through Schmidt's correspondence with Drechsel, as both men were busy fundraising also outside of the framework of the Danish commission for marine science. In a letter to Schmidt from 6 April 1914, Drechsel wrote that he had a conversation with the East Asiatic Company, and they had been very forthcoming. This may be related to Schmidt's specific wishes for more samples to be taken by the company's commercial vessels crossing the Atlantic. However, Drechsel also told Schmidt that he had been searching for what he called 'a rich man'. So far he had not succeeded in raising money for a follow-up expedition.[7]

The Danish fisheries inspector, F. V. Mortensen was working on a draft budget for such an expedition. Also, Drechsel noted that there was no immediate rush, as the expedition would not be ready for the summer of 1914 anyway. The plan was to have an expedition in the Sargasso Sea area by early summer the following year. Drechsel expressed his hopes that no other nation would go there in 1914, and beat the Danes at the finishing line.[8] From 1914, Drechsel had become more important for the success of Danish and international marine science. At a meeting in 1913, the International Council had decided to raise Drechsel's salary as secretary general of the organization. Drechsel's new salary would be 8,000 Kroner a year, a very fine wage, and somewhat more than what Schmidt earned as director of the Carlsberg Labroatory. Drechsel was now in a position to relinquish his position as Captain (Port Director) of the Port of Copenhagen, and work fulltime organizing international and Danish marine science.[9] Little did Drechsel or any of his colleagues know that the onset of World War One a few months later, would curtail any attempt to launch another deep-sea expedition in the foreseeable future.

World War One – marine science in dry-dock

On 4 August of 1914, World War One began with The Allies – Great Britain, France and Russia on one side, and Germany and the Austro-Hungarian Empire on the other side engaged in war. In rapid succession, all

6 RA. *Komm. Danmarks Fiskeri- og Havundersøgelser*: Forhandlingsprotokoller (1900–1960) N4: 105th meeting, 2 June 1914.

7 KB. Johannes Schmidt 'Papirer'. Ut.374.VI.4. Blandet Korrespondance 1912–1928. Letter from C. F. Drechsel to J. Schmidt 6 April 1914.

8 KB. Johannes Schmidt 'Papirer'. Ut.374.VI.4. Blandet Korrespondance 1912–1928. Letter from C. F. Drechsel to J. Schmidt 6 April 1914.

9 Went AEJ. *Seventy years agrowing. A history of the international council for the exploration of the sea, 1902–1972*. Copenhagen: International Council for the Exploration of the Sea; 1972. p. 45.

major world powers joined the war on one side or the other, except the USA, which eventually joined the side of The Allies in the spring of 1917. The small state of Denmark managed to maintain a policy of neutrality throughout the war. The political freedom of movement however, was very limited for Denmark since the country bordered the German Empire and lies across the North Sea from Great Britain. The aim of the Danish foreign policy was to not provoke the dominant military and economic powers in Europe, Germany, and the UK, which were also major trading partners for Danish companies.

With World War One lurking at the doorstep of mainland Denmark, deep sea oceanographic research became almost impossible. Submarines and sea mines now posed a danger to any ship operating in open waters, while frequent fighting also took place on the open seas, in particular between the German and British high seas fleets. The Battle of Jutland in 1916 is still the largest naval battle in world history, and the firing of German and British cannons could be heard from the Danish west coast of Jutland.

Nevertheless, the partially successful expedition with the *Margrethe* had furthered the insatiable Atlantic appetite of the Danish marine scientists. Schmidt felt there was still a way to go before he would be fully able to chart the spawning grounds of the Atlantic eel, and describe the life history of the eel from their larval state to their subsequent migrations. To fulfill these ambitions Schmidt needed a specially equipped vessel, which could be used full-time for deep-sea investigations. Technologically, the ship had to be able to navigate freely in the sense that a level of improvisation was needed in light of what the individual hauls would reap from the deep sea.[10] Nonetheless, for Danish deep-sea aspirations within marine science, the trouble had only just begun.

ICES and the Danish commission during World War One

Since its inauguration in 1902, ICES had been of vital importance for the flourishing scientific environment in Denmark in the field of marine science. The International Council provided leverage for international collaboration as well as all the authority of an international expert community. Yet the outbreak of World War One seriously disrupted this work.

10 Jespersen P. Togter med M/S 'Dana' (I) 1920–21. In: Winge Ø, Vedel Taaning Å, eds. *Naturforskeren Johannes Schmidt. Hans liv og ekspeditioner*. Copenhagen: Gyldendalske Boghandel Nordisk Forlag; 1947:80.

In February of 1915, the German president of ICES, Fritz Rose informed the bureau in Copenhagen that on behalf of the German government he had to withdraw from further participation in ICES. The Vice President, Otto Pettersson from Sweden took over the presidency by correspondence, facing a series of practical and fiscal restraints on the international work. Thus for 1915 no contributions were called for from the member countries, and a reduced budget was drawn up. For the following years only the neutral states of Denmark, Norway, Sweden and The Netherlands contributed fully to ICES, while Great Britain paid their fees for the fiscal year 1916–1917. In light of these restrictions on the scale of ICES work, many activities were cancelled. In the Baltic Sea though, where there was little active naval warfare, some investigations continued.[11]

The offices of the Danish commission for marine science were situated on the same floor of the same building vis-á-vis the International Council. Therefore, information sharing will have been substantial, and not everything will have been recorded in writing. However it is known that, Otto Pettersson from the International Council was invited to join the Danish commission's first meeting following the outbreak of World War One. In the minutes only one sentence was devoted to the international work, and this stipulated that the difficult conditions surrounding the war in Europe would not hinder Denmark, Norway and Sweden in conducting investigations into herring in the fall of 1914. However, a mere 400 Kroner was set aside for this activity, which appears to have been more of a diplomatic showcase from the respective government funders.[12]

As the year came to end, the commission faced substantial budget cuts due to falling funds from the state. During the financial year 1912–13 for instance, the commission had disposed of 73,300 Kroner to cover the various activities plus an additional 4,700 Kroner intended as Denmark's contribution to the international council. The cost of running the *Thor* was budgeted at 30,000 Kroner or more than a third of the total.[13] In late November 1914, however, the Naval Ministry told the commission that they wished to make use of the *Thor* for patrolling The Sound during the winter. The commission agreed to this, and probably this was more of a government order than a polite request.[14] In any case, as the war dragged on in the following years, it became impossible for the commission to argue that they still needed the c. 30,000 Kroner annually for the *Thor*, when they could not conduct high seas fisheries investigations

11 Went AEJ. *Seventy years agrowing. A history of the international council for the exploration of the sea, 1902–1972.* Copenhagen: International Council for the Exploration of the Sea; 1972. p. 51–52.

12 RA. *Komm. Danmarks Fiskeri- og Havundersøgelser:* Forhandlingsprotokoller (1900–1960) N4: 106th meeting, 21 September 1914.

13 RA. *Komm. Danmarks Fiskeri- og Havundersøgelser:* Forhandlingsprotokoller (1900–1960) N3: 69th meeting, 6 February 1911.

14 RA. *Komm. Danmarks Fiskeri- og Havundersøgelser:* Forhandlingsprotokoller (1900–1960) N4: 108th meeting, 23 November 1914.

anyway. Thus for 1916 and 1917 the state funds for the commission were cut to 47,675 Kroner per year. 80% of this went on paying rent for the commission's office space, while some money was spent on publications. This left very little for investigations in the field.[15]

In a private note to Johannes Schmidt from June 1916, Drechsel wrote that he felt sure that the Danish government would continue to fund the Danish commission. Drechsel expressed the no doubt realistic view that for as long as ICES continued to exist, Denmark would take part in it, and as secretary general he would make sure, the international organization kept afloat during the war.[16]

One hundred years later, we now know that ICES is still very active indeed. In the middle of World War One though, this was hardly a foregone conclusion.

Schmidt's dismal search for money 1914–16

Schmidt's plan for renewing eel investigations after the outbreak of war was discussed by the commission in November 1914, where it was decided to postpone the plans until the war had ended and until the insurance issue of the *Margrethe* had been finally settled. The minutes clearly read as if everyone at the table anticipated the war to be over relatively soon. Meanwhile, Drechsel promised Schmidt that he would seek the support of the East Asiatic Company with regards a continuation of the project of taking oceanographic stations from ships of opportunity when crossing the Atlantic.[17] Just two months later Schmidt was able to report to the commission that one of the ships of opportunity of the East Asiatic Company had collected eel larvae with an average length of a mere 18 millimetres, indicating close proximity to the spawning grounds. The next year, in the winter of 1916 the timing would be right for undertaking another Atlantic expedition to conduct a thorough examination of the most important area in the Atlantic Ocean.[18]

Over all, future prospects looked dismal with regards to upholding a Danish oceanographic presence in the North Atlantic. While Schmidt was convinced that he had now located the breeding places of the fresh water eels to the Sargasso Sea, he needed more samples and more evidence before he was able to convince the rest of the World, what a marvellous discovery he was sitting on. He needed a new ship, but the

15 RA. *Komm. Danmarks Fiskeri- og Havundersøgelser:* Forhandlingsprotokoller (1900–1960) N4: 115th meeting, 15 April 1915.

16 KB. Johannes Schmidt 'Papirer'. Ut.374.V.1. Privat 1908–1916. Breve fra Skrivebordet. Letter from C. F. Drechsel to J. Schmidt 28 June 1916.

17 RA. *Komm. Danmarks Fiskeri- og Havundersøgelser:* Forhandlingsprotokoller (1900–1960) N4: 108th meeting, 23 November 1914.

18 RA. *Komm. Danmarks Fiskeri- og Havundersøgelser:* Forhandlingsprotokoller (1900–1960) N4: 109th meeting, 18 January 1915.

war was dragging on making ocean voyages dangerous. Moreover, ICES had become largely inoperative as the member countries were either directly at war with each other, Russia, Germany, Great Britain, Belgium and France, or trying to stay neutral, as was the case for The Netherlands and the Scandinavian countries.

In the summer of 1916, Schmidt seems to have been somewhat desolate. His close friend and colleague, Ove Paulsen wrote him a letter on 15 June asking Schmidt to join him on a cruise to Northern Norway in connection with the upcoming 16th Meeting of Scandinavian Natural Scientists in Kristiania. 'You need to be cheered up a bit,' Paulsen wrote, '…and if you find it too boring to be on a cruise, we could go ashore in a fishing village and make some measurements or another useful piece of work.'[19] Schmidt however, was suffering from bronchitis, so he never went to the meeting in Norway.[20] At this point he had not yet realized that he really suffered from asthma, which was triggered by the powder of dried daphnia he used to feed the fish in the aquarium of the Carlsberg Laboratory.[21]

To catch a goldfish

Two months later, Schmidt seems to have regained his strength. From the upmarket Ellekilde Beach Hotel in North Zealand, Schmidt was busily coordinating the fundraising effort with Drechsel, himself operating out of Copenhagen. From their correspondence, urgency springs to mind. 'Already yesterday, I was in action to promote the cause, we discussed in the evening of the day before yesterday,' Drechsel began his letter.[22] From the next four pages it becomes clear that they had just renewed their active interest in raising funds for an Atlantic deep sea expedition. In the words of Drechsel, he was trying to catch 'goldfish', wealthy men willing to spend a fortune on marine science. Drechsel is, however, wary that the kind of men they have in mind, would not be impressed by loose conversation. Rather they would need a solid written up proposal to back their wishes, and who was better in drafting such a plan than Schmidt, he asked rhetorically.

Once more, Drechsel had been consulting his vast network of Copenhagen business-men. This time he had begun his search in the large Danish bank, Privatbanken, where he had met with the CEO, C. C. Clausen. In addition to running a large bank, Clausen was also president of the

19 KB. Johannes Schmidt 'Papirer'. Ut.374.V.1. Privat 1908–1916. Breve fra Skrivebordet. Letter from O. Paulsen to J. Schmidt 15 June 1916.

20 KB. Johannes Schmidt 'Papirer'. Ut.374.V.1. Privat 1908–1916. Breve fra Skrivebordet. Letter from C. F. Drechsel to J. Schmidt 20 June 1916.

21 Winge Ø. Andre strenge anslaas – paa Carlsberg Laboratoriet. In: Winge Ø, Vedel Taaning Å, eds. Naturforskeren Johannes Schmidt. Hans liv og ekspeditioner. Copenhagen: Gyldendalske Boghandel Nordisk Forlag; 1947:78–79.

22 RA. Carlsbergfondets DANA-ekspeditioner. 1917–1973. 2–4. C. F. Drechsel to J. Schmidt, 11 August 1916.

Danish Chamber of Commerce. While he was not able to donate money himself, he recommended to Drechsel that a specific expedition committee should be formed as a visible platform that could be used to catch the attention of, and receive money from, the 'goldfish', and Clausen said that he was willing to help with setting up such a committee. Drechsel was now convinced that this was the right way to proceed.

There was however, on problem that worried Drechsel, and that was the timing of the proposed initiative. At the time of writing the news had just broken in Denmark that the government had secretly negotiated a sales agreement with the USA, whereby the USA would buy the Danish West Indies. From a Danish point of view, the government wanted to make sure the islands would not fall into the hands of Germany during the war, thereby forcing Denmark to enter too. From a US point of view, the acquisition was more economically motivated.[23]

In the days preceding Drechsel's letter to Schmidt, the Danish newspapers were filled with criticism that the government had tried to keep a lid on the sales negotiations with the USA, and on 10 August a heated debate on this issue commenced in the Danish parliament.[24] In the words of Drechsel from 11 August: 'Each and all in the country were preoccupied with the issue of the Danish West Indies, including the change of government ministers etc.'[25] He further pondered that if the idea of selling the islands was indeed overturned, this would influence the expedition plan as such, since this would facilitate a return to the research program, they had been working on at the time of the outbreak of the war. As long as the islands were part of Denmark, there was a better chance the government would support the cause financially. At any rate, Drechsel suggested that they keep their plan secret from the public for as long as there was no decision on the sale of the West Indies.

Finally Drechsel returned to the idea of a committee, and he argued they should approach Prince Valdemar as the first point of contact: 'If only he was in on the plan, the "goldfish" would be easier to find.'[26] Responding to Drechsel's proposal, Schmidt focused on how to best approach Prince Valdemar. In 1916, Valdemar was the paternal uncle of the Danish king, Christian X, and obviously a celebrity, who could draw public attention to the cause of the Atlantic expedition. More importantly perhaps, Valdemar was a close business ally of H. N. Andersen, the wealthiest business man in Denmark and owner of the international corporation, the East Asiatic Company, and also himself an officer in the Danish navy. Schmidt emphasized to Drechsel that if

23 Lidegaard, Bo, Overleveren. 1914–1945, Carsten Due-Nielsen, Ole Feldbæk & Nikolaj Petersen (eds.): Dansk Udenrigspolitiks Historie, vol. 4, Copenhagen: Gyldendal Leksikon, 2003. p. 82.

24 Lidegaard, Bo, Overleveren. 1914–1945, Carsten Due-Nielsen, Ole Feldbæk & Nikolaj Petersen (eds.): Dansk Udenrigspolitiks Historie, vol. 4, Copenhagen: Gyldendal Leksikon, 2003. p. 83.

25 RA. Carlsbergfondets DANA-ekspeditioner. 1917–1973. 2–4. C. F. Drechsel to J. Schmidt, 11 August 1916.

26 RA. Carlsbergfondets DANA-ekspeditioner. 1917–1973. 2–4. C. F. Drechsel to J. Schmidt, 11 August 1916.

he happened to speak to Valdemar before he did, then Drechsel should make sure to portray the expedition in very common terms and not become too technical in his way of explaining matters. Schmidt suggested a few arguments for Drechsel to use, such as the realization that the plan he had drawn up was very costly, but that was only because he wished to operate at such great depth in the ocean. Obviously a much smaller cruise would be a lot less costly, and indeed the riddle of the eel could be solved without this very large enterprise. However, if the expedition was marketed as something truly spectacular it would be easier to convince a larger audience. As an example of how not to continue fundraising, Schmidt mentioned that he had presented his plan to Emil Glückstadt already, but with no success. Glückstadt was the owner and CEO of the largest bank in Denmark at the time, Landmandsbanken, and widely considered one of the wealthiest and most influential capitalists in the country. The support from Glückstadt would have been highly important. However, as it turned out, Glückstadt had stated that he did not really understand much of what Schmidt had told him about his deep sea plans, and as he did not have sufficient interest in this problem, he would have to decline supporting the plan. Schmidt finally aired his hopes that Drechsel would be successful with his first moves, which no doubt would be the most difficult.[27]

An Atlantic deep sea expedition

Schmidt went about fleshing out his vision for an expedition so that it would capture the public imagination as well as the interest of potential private sponsors. He presented a sketch of how international marine science had been greatly advanced in the first decade of the 20th century right until the outbreak of the Great War. Thus, an international programme to conduct an Atlantic wide survey of the physical oceanography of the vast ocean had been adopted by ICES in 1913, just one year before the war. As nothing had come out of this, any country able to fill the role as a beacon of progress in ocean exploration would fulfill a 'cultural task of excellent and easily recognizable significance.'[28]

While this was a daunting task since it required the cooperation of many different countries, Denmark happened to be in a unique position to solve the issue, the draft prospectus continued. The numerous Atlantic and Mediterranean expeditions already undertaken, had turned

27 RA. Komm. Danmarks Fiskeri- og Havundersøgelser: A 22 Dana-ekspeditionerne 1917–1924, J. Schmidt to C. F. Drechsel, 21 August 1916.

28 RA. Komm. Danmarks Fiskeri- og Havundersøgelser: A 22 Dana-ekspeditionerne 1917–1924, J. Schmidt to C. F. Drechsel, 31 August 1916.

Danish marine scientists into both pioneers and veterans in ocean exploration. Moreover, it was emphasized that Denmark was in the fortunate position of being neutral during the war. Therefore, Denmark had a particular responsibility to carry out this expedition for the world, not only for the potential research possibilities but also as a means to re-establish understanding and cooperation between the nations once the war had ended. 'Denmark, spared of the toils of war, has profited from its neutral position, not least in the area of shipping, and taking up this cause would bring honour to our country, and it would bring an international organization based in Denmark even closer and ensure that Denmark would find a permanent place in the rank of nations conducting practical scientific investigations.'[29]

Schmidt then turned to the issue of finding a suitable vessel. He envisioned that something in the size range of 300–500 tons would do, and ideally the ship would be engine powered and could provide a far ranging working radius, so that those on board did not have to worry constantly about the distance to the next coaling station. The projected budget was around 250,000 Kroner, and if they were successful with the fundraising and finding a ship, the expedition could commence in August 1917 and last one year until August 1918. Considering the fact that Schmidt's own initial interest sprung from his urge to find out more about the eel, he was surprisingly hesitant in promoting this topic, perhaps out of fear of being a nerdy type only nurturing his own hobby horse. Only on the last page did Schmidt bring up the 'special assignment' of finally solving the riddle of the breeding place and early life history of the freshwater eel.[30]

As a final piece of advice to Drechsel, Schmidt urged him to make sure that the rich people he would enlighten on the proposal would treat this information as confidential. Schmidt wanted to make sure that the money and a ship were secured before going public. Schmidt's fears of a leak mainly related to what the Norwegians would do once they got wind of a Danish expedition. 'I am fully convinced the Norwegians – with all the money now flowing into that country – would do everything in their powers to get ahead of us.'[31]

As late as October the same year, there was no fixed agreement in place, at least with regards to the vessel. Vice Admiral Wandel, the retired former collaborator from the first years of the Commission for marine science, was in correspondence with Drechsel at this point, and it seems that Drechsel was seeking advice on the right ship to use for an

29 RA. Komm. Danmarks Fiskeri- og Havundersøgelser: A 22 Dana-ekspeditionerne 1917–1924, J. Schmidt to C. F. Drechsel, 31 August 1916.
30 RA. Komm. Danmarks Fiskeri- og Havundersøgelser: A 22 Dana-ekspeditionerne 1917–1924, J. Schmidt to C. F. Drechsel, 31 August 1916.
31 RA. Komm. Danmarks Fiskeri- og Havundersøgelser: A 22 Dana-ekspeditionerne 1917–1924, J. Schmidt to C. F. Drechsel, 3 September 1916.

upcoming expedition. Wandel remained skeptical telling Drechsel that under the current economic circumstances, it was unlikely that any ship owner would lend out a newly built vessel for the purpose of marine science. Wandel also advised that for safety reasons Drechsel should be looking for vessels of 450–500 tonnage or above.[32]

Confidential news from Drechsel

By 3 November 1916 everything had changed. The support from Andersen's East Asiatic Company was clear from a letter Drechsel wrote to the CEO of EAC, Schmiegelow following a conversation the day before with H. N. Andersen himself. Schmiegelow was responsible for a series of motor vessels to be built at the company's own shipyard in the town of Nakskov. Drechsel thanked Schmiegelow for sending along the blue print for the motorship under construction, which could serve the interests of marine science. Drechsel requested that a number of alterations were necessary to fit their needs. This included an additional number of cabins for the scientific crew, a number of winches for deep sea investigations and a number of smaller apparatuses necessary to bring onboard. Following a conversation with Andersen, Drechsel sent Captain Hansen, the trusted captain on most of Schmidt's expeditions, to Nakskov to have a closer look at the vessel.[33]

It is clear from the ensuing correspondence that this was news to Schmiegelow at the time, but in February the plans became much more concrete. Schmiegelow reported to Drechsel that now a four-masted schooner currently being built, had been secured for purposes including scientific investigations.[34] Drechsel on his side underlined how much he was hoping that the ship would be ready for a cruise no later than 1918.[35]

On 7 November 1916, Drechsel summoned the members of the Danish commission for marine science to a meeting in the offices of ICES. On top of the agenda, and the reason why Drechsel had ordered the meeting, was the item of 'Atlantic Ocean Expedition'. Drechsel informed confidentially that from private means a total sum of 250,000 Danish Kroner had now been raised for a deep-sea expedition as described in Johannes Schmidt's plan for a one year cruise. Now negotiations were ongoing with regards to finding a suitable vessel. The reason why Drechsel wanted to meet with the commission was twofold. First of all, he

32 RA. Komm. Danmarks Fiskeri- og Havundersøgelser: A 23 Dana-ekspeditionerne 1917–1924, C. F. Wandel to C. F. Drechsel, 4 October 1916.

33 RA. Komm. Danmarks Fiskeri- og Havundersøgelser: A 23 Dana-ekspeditionerne 1917–1924, C. F. Drechsel to C. Schmiegelow, 3 November 1916.

34 RA. Komm. Danmarks Fiskeri- og Havundersøgelser: A 23 Dana-ekspeditionerne 1917–1924, C. Schmiegelow to C. F. Drechsel, 9 February 1917.

35 RA. Komm. Danmarks Fiskeri- og Havundersøgelser: A 23 Dana-ekspeditionerne 1917–1924, C. F. Drechsel to C. Schmiegelow, 1 March 1917.

8.1 Christian F. Drechsel in his navy commander uniform *From 1908–1927, Chr. F. Drechsel was a very active Secrectary General of the International Council for the Exploration of the Sea. As a team Drechsel and Schmidt were hugely successful fundraisers and science organizers (Carlsberg Archives).*

wanted assurance from the government fisheries director, F. V. Mortensen that it would be possible to create a special expedition committee operating on its own, but in a form where the members of the marine science commission could join. Drechsel had in mind Prince Valdemar as Patron of this expedition. Secondly, Drechsel was seeking confirmation from the other scientists on the board, Martin Knudsen and A. C. Johansen, that it would be possible to find suitable apparatuses to be taken onboard which ever vessel was procured.[36]

While the availability of a ship was of paramount importance, it was equally important to find support for equipping the vessel with the right gear and for all the running costs of a large cruise from provision to the crew members to maintenance and replacements of fishing gear.

36 RA. Komm. Danmarks Fiskeri- og Havundersøgelser: Forhandlingsprotokoller (1900–1960) N4: 117th meeting, 7 November 1916.

Collecting cash funds

From January 1917, Drechsel opened a new ledger for the future expedition, where the first entry is on the 'income' column of the accounts. Emil Glückstadt, who was the head of the Landmandsbanken, paid a sum of 150,000 Danish Kroner to the cause and thus became the largest cash donor for the project.[37] The money was transferred on 30 December 1916 to a separate account in his own bank.[38] Thus at least for a while, the money was still inside the vault of the largest Danish bank.

Another 'gold fish' was landed as Andreas Erlandsen, shipowner and also owner of a large bunker firm selling coal out of the Port of Copenhagen, agreed to finance the expedition with a sum of 100,000 Kroner. Erlandsen did not commit all this money right away though, instead a Kjøbenhavns Handelsbank bank guarantee was issued in May 1919.[39] With Erlandsen onboard, the expedition was now in possession of sufficient funds to keep not one but as it turned out, two separate expeditions with two different vessels afloat until 1922.

As a third private donor, the lawyer, L. Zeuthen, promised to donate 10,000 Kroner to the cause. Writing to Zeuthen Drechsel was optimistic, and mentioned that he was still hoping to send off the expedition during August 1917, although this relied upon the right ship being ready by then. Meanwhile Zeuthen paid the sum he had promised to the commission's special bank account at the Landmandsbanken.[40]

In the spring of 1917, Drechsel was still searching for more money. On 9 March he received a letter from the chairman of the board of directors of the large industrial enterprise, Det Store Nordiske Telegraf-Selskab. Unfortunately, this company was not able to find the requested sum of 50,000 Kroner in spite of the honourable goal of the proposed project as the letter stipulated.[41]

Conclusion

It is well known that the outbreak of World War One was detrimental to Atlantic marine science, not least because of the immanent dangers from submarine warfare, and the breakdown of much pre-war international collaboration also became a real peril. In this most difficult political climate, Schmidt and Drechsel became comrades in arms pursuing the same goal of launching a spectacular Danish-led deep-sea expedi-

37 RA. Komm. Danmarks Fiskeri- og Havundersøgelser: A 22 Dana-ekspeditionerne 1917–1924, Revideret Koncept.

38 RA. Komm. Danmarks Fiskeri- og Havundersøgelser: A 22 Dana-ekspeditionerne 1917–1924, Den Danske Landmandsbank to C. F. Drechsel, 30 December 1916.

39 RA. Komm. Danmarks Fiskeri- og Havundersøgelser: A 22 Dana-ekspeditionerne 1917–1924, Kjøbenhavns Handelsbank to C. F. Drechsel, 30 May 1919.

40 RA. Komm. Danmarks Fiskeri- og Havundersøgelser: A 23 Dana-ekspeditionerne 1917–1924, C. F. Drechsel to L. Zeuthen, 22 January 1917; RA. Komm. Danmarks Fiskeri- og Havundersøgelser: A 23 Dana-ekspeditionerne 1917–1924, L. Zeuthen to C. F. Drechsel, 7 February 1917.

41 RA. Komm. Danmarks Fiskeri- og Havundersøgelser: A 23 Dana-ekspeditionerne 1917–1924, Board of Directors for Store Nordiske to C. F. Drechsel, 9 March 1917.

tion to the Atlantic Ocean. Both men served the interest of the commission for marine science, where they both held a seat. For Schmidt there was the added value that such an expedition could once again allow him to explore the Sargasso Sea and settle the issue of the breeding place and early life history of the Atlantic eel populations, which had by then puzzled him for more than a decade. For Drechsel there was the added value that such a Danish led enterprise might spur a renewed interest in international cooperation among other European countries and help re-establish ICES, which suffered during the war.

Previous commentary on the way in which these expeditions eventually came about has documented in passing that the East Asiatic Company supplied a vessel, the M/S *Dana*, and it is also known that a quarter of a million Danish Kroner were raised from private donations.[42] But for the first time, this chapter has told the story of how Schmidt and Drechsel went about catching their so-called 'gold fish', and it has also given names to these donors.

In some ways it can be said that Drechsel and Schmidt went searching for a new vessel, as soon as they heard the news that the M/S *Margrethe* expedition had been stranded – literally on a beach in the British parts of the Lesser Antilles. However, for the following almost three years they had no success with this endeavour. The war meant that substantial government subsidies for ocean exploration were out of the question, and there was little interest from potential private sponsors. In the early summer of 1916, for example, the inordinately wealthy banker, Emil Glückstadt had declared to Schmidt that he had little interest in marine expeditions, which he did not understand much about.

Then, in the summer of 1916 news broke that Denmark was about to sell their Caribbean colony, the Danish West Indies. This spurred Drechsel and Schmidt to draw up an ambitious prospectus of ten pages to be discussed with wealthy men from the corporate sector. The Danish shipping sector especially had profited enormously from shipping goods on ships from a neutral state, and their wartime profits broke all previous records.

Then suddenly around 1 November funding was secured for a Danish Schmidt-led expedition for when the war was over. The East Asiatic Company promised to make amendments to a ship they were building, to accommodate the needs of the marine scientists who were going to borrow the vessel for free for a period of three months. Emil Glückstadt of Landmandsbanken had now turned around completely to become

42 Jespersen P. Togter med M/S 'Dana' (I) 1920–21. In: Winge Ø, Vedel Taaning Å, eds. *Naturforskeren Johannes Schmidt. Hans liv og ekspeditioner*. Copenhagen: Gyldendalske Boghandel Nordisk Forlag; 1947:80.

the quiet, but all-important largest sponsor of an Atlantic expedition, giving 150,000 Kroner to the project. Additional sponsors included ship owner Andreas Erlandsen with 100,000 Kroner and lawyer Zeuthen with 10,000 Kroner.

Zeuthen and Erlandsen both earned themselves a seat on the expedition committee, which was set up when it seemed that the project would finally come to fruition following the end of World War One in November 1918. Thus, the Privatbanken CEO, C. C. Clausen's recommendation to award private sponsors with a seat on the expedition committee appears to have paid off. But apart from personal vanity, why would private enterprises support the prospects of international marine science at this point in time? The following chapter is devoted to teasing out the distinct national and corporate interests that large private companies used to justify their lavish spending.

9 Resistance from the sea – Danish marine scientists and the sale of the Danish West Indies

One year after war broke out; the USA was still a neutral power. The Americans though, feared that Germany would occupy Denmark and claim sovereignty also over the Danish West Indies. Therefore, the US government approached the Danish government, and the matter fell into the hands of foreign minister, Erik Scavenius. Scavenius is still a well-known figure in Danish history, and part of his reputation rests on the way in which he handled the affair of selling the Danish West Indies.[1] Belonging to the political party of Det radikale Venstre, Scavenius, he followed the implicit doctrine that Denmark's political actions were to be as non-provocative as possible. The idea was that if the great powers viewed Denmark as a harmless state, then the country would not be invaded by Germany. Claims to possess an offshore empire, even as modest as the Danish North Atlantic possessions could be seen by foreign powers as an attempt by the Danes to play out of their own small nation's league. Therefore, and because of the ensuing financial problems associated with running a colony in the Caribbean, Scavenius viewed the American approach as a welcome opportunity to get rid of the islands, and even be paid for letting them go.[2]

The Americans also proposed as part of the deal that the USA would acknowledge Denmark's supremacy of the entire island of Greenland. Upholding Danish supremacy over Greenland was later to become an integral part of Danish oceanographic expeditions to the North West Atlantic, but in 1916 it was a minor issue among Danish parliamentarians. The skeptic conservative MP Alexander Foss asked polemically, if possessing a huge chunk of ice on the North Pole was to be a mere spiritual asset, and foreign minister Scavenius replied that frankly he did not know if in the future, important geological assets would emerge.[3]

1 Sjøqvist, Viggo. *Erik Scavenius: Danmarks udenrigsminister under to verdenskrige. Statsminister 1942–1945*. Copenhagen: Gyldendal, 1973; Lidegaard, Bo, *Overleveren. 1914–1945*, Carsten Due-Nielsen, Ole Feldbæk & Nikolaj Petersen (eds.): *Dansk Udenrigspolitiks Historie*, vol. 4, Copenhagen: Gyldendal Leksikon, 2003.

2 Sjøqvist, Viggo. *Erik Scavenius: Danmarks udenrigsminister under to verdenskrige. Statsminister 1942–1945*. Copenhagen: Gyldendal, 1973. p. 203–218.

3 Anonymous. *Forhandlingerne om overdragelsen af de dansk-vestindiske øer til de amerikanske forenede stater 1915–1916*. Copenhagen: Udenrigsministeriet; 1916. p. 19–24.

Fearing what the Germans might think of the US initiative, Scavenius managed to keep negotiations secret to the public for more than a year. The heads of the various parties in the Danish parliament though, were informed, but wowed to keep it a secret to their fellow parliamentarians. Also, since they had substantial financial interests in the Danish West Indies, H. N. Andersen of the East Asiatic Company, Emil Glückstadt representing Landmandsbanken and plantation owner, Holger Petersen were all kept informed about the on-going diplomatic process. Andersen and Glückstadt both had a say on the terms of the final treaty.[4] The Dansk Vestindisk Bank for instance continued to thrive under US governance after 1917. By and large this bank was a profitable investment throughout its existence until the lease expired in 1935. In terms of accessing the huge and prosperous market on the American mainland, the Danish business entrepreneurs probably did not gain as much as they had hoped for after World War One, but this was hard to foresee in 1916.[5]

In addition the opinion of the Danish electorate also came to play an important role in the sale. The Kingdom of Denmark was gradually maturing as a democracy in these years, and in Mid-summer of 1916 news leaked to the press that the government was about to sell the Kingdom's West Indian assets to the USA. The leak was handled somewhat clumsily by foreign minister Scavenius and minister of finances, Edward Brandes, who both claimed ignorance as to such negotiations, while in effect the negotiations had been scheduled to become publicly known only a couple of weeks later.

According to Danish historians Sjøqvist and Lidegaard, this forms part of the reason why MPs from the dominant conservative and liberal parties were able to create a public stir of the whole affair, claiming that the government was not trustworthy. Much newspaper debate followed, so in order to regain control over the political process, three politicians from the opposition parties were asked to join the cabinet. A compromise was then reached, whereby the sale should be decided upon, only following the recommendation from the majority of a public referendum. In addition, a parliamentary commission was set up in October in order to provide a more solid ground for making the right decision with respect to the sale of the Danish West Indies.[6] Over the span of a few weeks the commission interviewed some thirty Danish businessmen, civil servants, politicians and scientists, and their oath sworn testimonies were published in transcript. One of the first names on the list of Danish experts was none other than that of 'Johannes Schmidt, Laboratory Director, Copenhagen.'[7]

4 Sjøqvist, Viggo. *Erik Scavenius: Danmarks udenrigsminister under to verdenskrige. Statsminister 1942–1945*. Copenhagen: Gyldendal, 1973. p. 203–218. Anonymous. *Forhandlingerne om overdragelsen af de dansk-vestindiske øer til de amerikanske forenede stater 1915–1916*. Copenhagen: Udenrigsministeriet; 1916. p. 27–29.

5 Worm Begtrup, Jens, *Dansk-vestindiske Nationalbank 1904–1935 – En undersøgelse af bankens oprettelse, drift og lukning, herunder dens ret til at trykke penge med dansk værdi på et amerikansk ejet territorium*. Unpublished MA-thesis, University of Copenhagen, 2008.

6 Sjøqvist, Viggo. *Erik Scavenius: Danmarks udenrigsminister under to verdenskrige. Statsminister 1942–1945*. Copenhagen: Gyldendal, 1973. p. 214–215.

7 Anonymous, ed. *Betænkning angiven af den i henhold til lov nr. 294 af 30. september 1916 nedsatte rigsdagskommission angaaende de dansk vestindiske øer*. Copenhagen: J. H. Schultz; 1916.

Johannes Schmidt's expert opinion

On the second day of questioning, October, 28 1916, Johannes Schmidt appeared in front of the commission to give his expert testimony on the state of fisheries in the Danish West Indies.[8] Asked about his general knowledge of the islands, Schmidt remarked that the expedition with the *Margrethe* in 1913, had provided him with first-hand knowledge mainly on the state of affairs in St. Thomas. In addition, Schmidt pointed to the field studies provided by Theodor Mortensen ten years earlier as well as studies Schmidt himself had conducted at the United States' Fisheries Department. Answering as an ambitious fundraising scientist, Schmidt emphasized that much more knowledge would be needed in order to come up with a precise assessment of the natural resources of the Virgin Islands.

With regards to the current state of fisheries, Schmidt remarked that the ocean surrounding the islands seemed to provide excellent conditions for a modern and economically sound fishery. Nonetheless, the current state of affairs regarding for instance the sea turtle fisheries was marked by managerial neglect, where a ban on collecting turtle eggs on the beaches was recommended. One key problem was the lack of international coordination on this issue. Schmidt suggested this was something, which could readily be put forward to the forum of the International Council (ICES).[9] During World War One however, ICES was largely inoperative with respect to international meetings.[10] The general situation for the other fisheries in the Danish West Indies was equally lacklustre. Most fish landed were undersized, and since no cooling facilities or other ways of preserving the catches were in operation, the fish would often be in a poor condition when finally marketed locally. Regardless of immediate obstacles for the West Indies fisheries, Schmidt was busy making plans for better times to come.

Having explained the main problems of the West Indian fisheries, Schmidt proceeded to describe ways to improve the fishing sector. The main problems were lack of technology, lack of know-how, widespread poverty and what he labeled 'the common sloppiness of the tropics'. The fishermen ought to introduce dams in order to keep the fish alive for days and weeks so that catches might be sold to more populous islands nearby, such as Puerto Rico. Schmidt thought that refrigeration with ice would be too costly in the tropical climate. Salting and drying the fish should be possible though.

8 Anonymous, ed. *Betænkning angiven af den i henhold til lov nr. 294 af 30. september 1916 nedsatte rigsdagskommission angaaende de dansk vestindiske øer.* Copenhagen: J. H. Schultz; 1916. p. 60–65.

9 Anonymous, ed. *Betænkning angiven af den i henhold til lov nr. 294 af 30. september 1916 nedsatte rigsdagskommission angaaende de dansk vestindiske øer.* Copenhagen: J. H. Schultz; 1916. p. 62.

10 Rozwadowski H. *The sea knows no boundaries. A century of marine science under* ICES. Seattle and London: ICES/University of Washington Press; 2002.

He added that while such measures could help satisfy local markets, the export of processed products such as canned fish was another way forward. Since an increasing number of ships went through the port of St. Thomas, it should be feasible to set up a distribution network catering for the Americas as well as for the European market. Schmidt's recommendations suggest that he was reaching out to large Danish ship owning companies such as the East Asiatic Company and The United Danish Steam Ship Companies (DFDS).

In order to finance and implement his plans, Schmidt shrewdly directed attention towards the political and financial arena. Schmidt argued that: '…it is necessary that the government takes these matters in its own hand. Without it, nothing will happen. The way, in which I would like to see this matter approached, is to have the Danish state fund a larger fishing vessel operational with sails as well as a motor engine and send it to the West Indies equipped with all kinds of fishing gear. In addition, a number of boats of various sizes are needed. We also need the company of skilled Danish fishermen, but the services of native fishermen should be sought extensively, in order for them to learn the new methods.'[11] Schmidt once again underlined that to change the current state of affairs properly, much more light had to be cast upon the possibilities for improving marine resource exploitation.

Additional comments addressed the position of Danish marine science in light of a possible sale of the Danish West Indies to the USA. According to Schmidt, it would be a huge and irreplaceable loss if the islands were to be sold; 'For many years the Danish West Indies has been a fruitful workplace for Danish natural historians,' and 'this position is in many ways the envy of other Scandinavian scientists.'[12]

'It is despicable to sell parts of one's country'

On the overall issue of selling the islands though, Johannes Schmidt's attitude was stalwart nationalist. 'I find it of great importance that Denmark does not sell the Danish West Indies,' Schmidt said. Paraphrasing this point several times in his testimony, Schmidt presented a whole number of reasons why the West Indies should remain Danish territory. First of all, he found it despicable to get rid of parts of the country. Secondly, he feared that a sale would set a dangerous precedence for a small and weak state like Denmark. Lastly, he emphasized that it might

11 Anonymous, ed. *Betænkning angiven af den i henhold til lov nr. 294 af 30. september 1916 nedsatte rigsdagskommission angaaende de dansk vestindiske øer.* Copenhagen: J. H. Schultz; 1916. p. 62.

12 Anonymous, ed. *Betænkning angiven af den i henhold til lov nr. 294 af 30. september 1916 nedsatte rigsdagskommission angaaende de dansk vestindiske øer.* Copenhagen: J. H. Schultz; 1916. p. 65.

very well be short sighted to get rid of the islands during a time of weak economic performance. No one in their right mind could rule out that in the more distant future, the islands might become an even better asset than what the islands had been during the age of the slave economy and plantation riches. In the future, a particularly profitable crop might be transplanted to the islands and turn around the entire situation.[13]

Schmidt concluded his answer by summing up that many weighty arguments speak out for not selling the islands. Instead Denmark should invest in the Danish West Indies, so as to bring the islands to a situation, where Danes and West Indians alike would be proud of the colony. Referring to his own previous visits to the USA, Schmidt commented that he was convinced that Denmark could do the job of modernizing the Danish West Indies better than the Americans. The coloured population of the islands would come to miss belonging to Denmark, and Denmark should feel obliged to protect the white population considering themselves to be Danes.[14] While Schmidt was frank, personal and direct in all his answers to the committee, others came to the witness stand mindful that they had substantial financial investments at stake in Denmark's last tropical colony.

Andersen and Glückstadt before the commission

The head of Landmarksbanken, Emil Glückstadt was also subpoenaed to testify before the Parliament's Commission on the West Indies. Unlike Schmidt, Glückstadt answered in a guarded and defensive fashion. Answering the question as to whether or not he would like to see the colonies sold to the USA, the mighty banker carefully avoided any remarks on the issue of patriotism. Sticking to the pros and cons in terms of new business opportunities, as well as the interest of preserving existing investments, Glückstadt told the committee that negative consequences arising from the greater political distance to Denmark might be outweighed by a greater access to the huge markets within the American economy.[15]

Probably, the careful attitude of Glückstadt was clever in light of the scale of investments that he on behalf of Landmandsbanken had placed in the Atlantic colony. In addition, Glückstadt along with H. N. Andersen had been accused in the media of interfering with the secret government negotiations with the USA, and both the accuser, Th. Adler-Lund and the two business men were quizzed as to their engagement prior to

13 Anonymous, ed. *Betænkning angiven af den i henhold til lov nr. 294 af 30. september 1916 nedsatte rigsdagskommission angaaende de dansk vestindiske øer.* Copenhagen: J. H. Schultz; 1916. p. 64.

14 Anonymous, ed. *Betænkning angiven af den i henhold til lov nr. 294 af 30. september 1916 nedsatte rigsdagskommission angaaende de dansk vestindiske øer.* Copenhagen: J. H. Schultz; 1916. p. 64.

15 Etatsraad, Bankdirektør E. Glückstadt, 16 November 1916, in: Anonymous, ed. *Betænkning angiven af den i henhold til lov nr. 294 af 30. september 1916 nedsatte rigsdagskommission angaaende de dansk vestindiske øer.* Copenhagen: J. H. Schultz; 1916. p. 355–357.

the time in January 1916, when the bilateral consultations had surfaced at cabinet level.[16] Glückstadt had every reason to sit tight in order to protect his investments at a time, when the outcome of the sale was still in the hands of the people and government in accordance with the upcoming referendum.

In any case, H. N. Andersen was the strong arm of the two, and on the same day as Glückstadt, Andersen testified to the commission that he had also not placed any illicit influence on the bilateral trade negotiations. He did admit though, that in January 1916, Andersen had asked the Minister of Foreign Affairs to secure the interests of the Danish West Indian Company in the event of a sale to the USA. Business interests in the port of St. Thomas clearly had an influence on Andersen's public statement. St. Thomas he said would likely be better off in the short term as a Danish trading port, than it would be as an American naval port at a time of war. Andersen went on to describe how the Virgin Islands should be expected to be profitable, even financially self-sustainable in the long run if it was to make sense to keep them in Danish hands. Otherwise they would merely have '...a right to exist under Danish rule from the point of view of national patriotism.' National sentiments we are led to believe, are thus secondary, even if he went on to argue that it is a societal duty to create the most favorable conditions for the trades and industries, which he labeled as the pillars of society.[17]

Andersen's influence on the decision to sell

It appears that while Andersen was strongly opposed to selling in 1902, he had changed his mind by 1915. In recent Danish historiography however, H. N. Andersen's attitude toward selling the West Indies has been interpreted quite differently. Viggo Sjøqvist was the first to note Andersen's change of mind in his 1973 biography of the minister of foreign affairs, Erik Scavenius.[18] Sjøqvist thoroughly analyzed events, and his conclusions were cited by later commentators such as historians Ove Hornby (1980) and Bo Lidegaard(2003).[19] In his 2007 biography of the Danish king Christian X, Knud J. V. Jespersen came to a similar conclusion. Jespersen notes that not only did H. N. Andersen back the idea of selling the islands in light of future business opportunities. Andersen even persuaded the Danish king, Christian X that selling the islands was a geopolitical necessity.[20]

16 Anonymous, ed. *Betænkning angiven af den i henhold til lov nr. 294 af 30. september 1916 nedsatte rigsdagskommission angaaende de dansk vestindiske øer.* Copenhagen: J. H. Schultz; 1916. p. 351–355; 355–357; 358–362.

17 Anonymous, ed. *Betænkning angiven af den i henhold til lov nr. 294 af 30. september 1916 nedsatte rigsdagskommission angaaende de dansk vestindiske øer.* Copenhagen: J. H. Schultz; 1916. p. 354.

18 Sjøqvist, Viggo. *Erik Scavenius: Danmarks udenrigsminister under to verdenskrige. Statsminister 1942–1945.* Copenhagen: Gyldendal, 1973. p. 203–218.

19 Hornby O. *Kolonierne i Vestindien.* Copenhagen: Politikens Forlag; 1980, p. 370; Lidegaard, Bo, *Overleveren. 1914–1945*, Carsten Due-Nielsen, Ole Feldbæk & Nikolaj Petersen (eds.): Dansk Udenrigspolitiks Historie, vol. 4, Copenhagen: Gyldendal Leksikon, 2003. p. 86–87.

20 Jespersen KJV. *Rytterkongen – et portræt af Christian 10.* Copenhagen: Gyldendal; 2007.P. 225.

In the 1980s, historian Ole Lange published two critically acclaimed volumes on H. N. Andersen, focusing on Andersen's role as a free agent in Danish diplomacy and foreign policy. Yet, somewhat surprisingly Lange makes no mention of Andersen's role in connection to the sale of the Danish West Indies.[21] Given the number of prominent publications on this issue, it is even more surprising that in his 1989 synthesis of Danish political history in the early 20th century, historian Lorenz Rerup claimed that Andersen's East Asiatic Company initiated the political movement *against* the sale in 1916.[22] Rerup refers to Ove Hornby for his section on the West Indies, but apparently Rerup overlooked Hornby's conclusion.

Then, as recent as 2007, historian Søren Ellemose stated in his monograph on the history of the East Asiatic Comapny that 'Andersen was strongly against selling.'[23] Ellemose, however, did not reach this conclusion from citing Rerup, nor did he cite Hornby or any of the above-mentioned commentators. Ellemose quotes historian Per Boje's book on Danish multinational companies as well as a number of older commemorative volumes on the history of the East Asiatic Company.[24] None of these volumes however, lends any support to the idea that Andersen was against selling the Danish West Indies in 1916. They are all silent on where he actually stood in 1916.[25]

The number of mistakes evident in recent Danish historiography on Andersen's attitude to selling the Danish West Indies seems unusual, but it serves to show just how counter intuitive Andersen's role seems to have been for some later commentators. Back in 1916, however, the lack of support from the wealthiest business-man in Denmark must have been a sizeable disappointment for the conservative and nationally minded groups that opposed the sale. Even without Andersen in the anti-sale fold, there were plenty of other dissenting voices in the fall of 1916.

Public debates and De Danske Atlanterhavsøer

The public arena was flooded with agitprop against selling the West Indies. Booklets were published carrying the opinion of influential people from various walks of life.[26] A new *ad hoc* society was even set up, *Dansk Vestindisk Samfund*, independently from *Foreningen De Danske Atlanterhavsøer*. In Dansk Vestindisk Samfund mainly conservative students aided by the landed elite were fundraising, agitating and publishing pamphlets.[27]

21 Lange O. *Jorden er ikke større… – H. N. Andersen, ØK og storpolitikken 1914–1937*. Viborg: Gyldendal; 1988.

22 Rerup, Lorenz, Tiden 1864–1914, p. 317.

23 Ellemose, Søren, *Kompagniet. – H. N. Andersens ØK 1884–2007*, Gylling: Jyllands-Postes Forlag, 2007. p. 86–87.

24 Ellemose, Søren, *Kompagniet. – H. N. Andersens ØK 1884–2007*, Gylling: Jyllands-Postes Forlag, 2007. p. 323.

25 Boje, Per, *Danmark og multinationale virksomheder før 1950*, Odense: Odense University Press; 2000. p. 39; Ernst PG. *Det store eventyr om H. N. Andersen og Ø. K.* Copenhagen: Frederik E. Pedersens forlag; 1937; Heinberg, Aage, *H. N. Andersen. Stifteren af Det Østasiatiske Kompagni*, Copenhagen: De Unges Forlag, 1952.

26 Hagemann G. A. *Danmark og dets vestindiske øer*. Copenhagen: J. H. Schultz; 1916; Hansen G. *Red vestindien! et forslag i den 11. time*. Copenhagen: H. Meyers Bogtrykkeri; 1916; Lehmann, Johannes (ed.). *For eller Imod. Udtalelser med Henblik paa det eventuelle Salg af de Danske Vestindiske Øer*. Copenhagen: Lehmann & Stage; 1916.

27 Anonymous, ed. *Sandheden om dansk vestindien. dansk vestindisk samfunds skrifter 1*. Copenhagen: H. Hagerups Forlag; 1916.

Also the director of Carlsberg Breweries and son of Carl Jacobsen, Vagn Jacobsen was asked to air his opinion. Vagn Jacobsen was not summoned to the commission, but his viewpoint appeared in a small book on pros and cons of the West Indies, which was published just before the referendum.[28] Vagn Jacobsen was nationally minded by default, and he would later become engaged in several projects concerning Danish national heritage.[29] In 1920 for instance, when new borders were drawn after World War One, Jacobsen agitated for Denmark to have a larger share, mainly the town of Flensburg, of Northern Schleswig-Holstein, than what the Danish government had agreed to. Unlike Johannes Schmidt however, Jacobsen viewed the Caribbean colonial possession mainly from an economic point of view. Jacobsen was reserved in his address, in which he focused on the success of on-going private initiatives in the Danish colony as the pivotal concern.[30] In this guarded remark, Jacobsen seemed to follow the attitude of the banker Emil Glückstadt, and probably Jacobsen was just as keen on maintaining a good business relationship with his retailers and beer consumers.

Sentiments in favor of selling the Danish West Indies however, also resonated in the more literary circles of high society. Emma Gad for instance, an influential playwright, socialite and the wife of Rear Admiral Gad, announced that: 'the black inhabitants of these islands *are* not our countrymen, but rather a glued-on unnatural appendix.'[31] This may come as a surprise, since Emma Gad was also active in De Danske Atlanterhavsøer, where she was a rather active board member of the Icelandic section of the society.[32]

As we will see however, the society of De Danske Atlanterhavsøer would split down the middle in its attitude towards the Danish West Indies. In light of the upsurge in agitprop, and considering the significance of the sale, it may be surprising that this rather prominent society was not directly involved in the on-going debate.

Marine scientists' agitation

As mentioned above, the news broke in August 1916 on the plan to sell the Virgin Islands, and within the West Indies section of the society, a board meeting was also held on the 6 September. The section chairman was government consultant on freshwater fisheries Chr. Løfting, who had been collaborating for years with the Danish commission for

28 Lehmann, Johannes (ed.). *For eller Imod. Udtalelser med Henblik paa det eventuelle Salg af de Danske Vestindiske Øer*. Copenhagen: Lehmann & Stage; 1916. p. 25–26.
29 Sander, Fr. 'Jacobsen, Vagn' in: Povl Engelstoft & Svend Dahl (eds.), *Dansk Biografisk Leksikon*, vol. 11, Copenhagen: J. H. Schultz Forlag; 1937. p. 336–338.
30 Lehmann, Johannes (ed.). *For eller Imod. Udtalelser med Henblik paa det eventuelle Salg af de Danske Vestindiske Øer*. Copenhagen: Lehmann & Stage; 1916. p. 25–26. p. 25–26.
31 Lehmann, Johannes (ed.). *For eller Imod. Udtalelser med Henblik paa det eventuelle Salg af de Danske Vestindiske Øer*. Copenhagen: Lehmann & Stage; 1916. p. 18.
32 Anonymous, 'Foreningens Generalforsamling 1916', in: *Atlanten*, vol. 4, (1915–1917), p. 449.

Marine science.[33] Løfting took the standpoint that it would be odd to be on the board of a society such as Foreningen De Danske Atlanterhavsøer, if one supported the sale of the West Indies.[34] This opening came to split the section board. Dr. Phil. Carl Rothe was one of the founding members of the society back in 1902, when he was against selling the Danish islands.[35] Now on the contrary, he was in favor of selling. In light of this, he decided to retire from his office on the board, and then left the meeting. The remaining four board members had invited four experts on the West Indies to hear their opinion. Johannes Schmidt and botanist Frederik Kølpin Ravn were two of the four experts, but none of them had been able to be present. Unable to draw a clear conclusion on what to do in relation to the pressing political developments, the board of the West Indies' section agreed to pass on the issue to the overall board of the society.[36]

The president of the overall society, Niels Neergaard was a member of parliament and one of the leaders of the liberal party, Venstre. As the former minister of finance Niels Neergaard was a high-ranking politician, and in 1920 he would form his own government and become prime minister. In 1916 Neergaard belonged to the parliament majority across party lines, which had recommended that The Virgin Islands were to be sold, and this was also the official policy of his own liberal party. On the 13 September the issue was discussed as the only item on the agenda at the society's overall board meeting. Neergaard argued that the society should not take a stand on the issue of the sale. This was opposed by a minority on the board, with Drechsel and professor of botany, Kølpin Ravn eager advocates of the society siding with those against selling the colony. Drechsel spoke warmly of the money now being made from the port of St. Thomas, but in the end, the majority decision became that the society as such should sit on the fence in relation to the matter of the sale of the Danish West Indies.[37] After this decision was taken, a few more board members left their office.[38] Others began plotting against Neergaard.

On 2 November, the director of the Danish Meteorological Institute, Carl Ryder asked the secretary of the society to forward a copy of the previous as well as the present statutes to De Danske Atlanterhavsøer.[39] Less than two weeks passed before the then museum curator and plankton zoologist Carl Ostenfeld, on behalf of a group of five natural scientists, returned the favor by submitting a rather polemic question about the agenda of the upcoming general assembly: 'What had the board

33 Poulsen EM. *Kommissionen for danmarks fiskeri- og havundersøgelser 1902–1952*. Copenhagen: C. A. Reitzel; 1952.

34 RA.10154. *De Danske Atlanterhavsøer*. 1904–1916 Forhandlingsprotokol for sektionen i Dansk Vestindien, bind 5, p. 59–64.

35 Læssøe Müller, Paul, 'Rothe, Carl', in: Povl Engelstoft & Svend Dahl (eds.), *Dansk Biografisk Leksikon*, vol. 20, Copenhagen: J. H. Schultz Forlag; 1941, p. 219–220.

36 RA.10154. *De Danske Atlanterhavsøer*. 1904–1916 Forhandlingsprotokol for sektionen i Dansk Vestindien, bind 5, p. 59–64.

37 RA.10154. *De Danske Atlanterhavsøer*. 1902–1917. Forhandlingsprotokol for foreningen, bind 4, p. 238–250.

38 RA.10154. *De Danske Atlanterhavsøer*. 1902–1917. Forhandlingsprotokol for foreningen, bind 4, p. 253.

39 RA.10154. *De Danske Atlanterhavsøer*. 1903–1919. Kopibøger med registre. bind 18. p. 169.

been doing in relation to the issue of the sale?', they asked[40] Carl Ostenfeld was accompanied by his two friends and colleagues from ocean explorations, museum curator and plant zoologist Ove Paulsen and Johannes Schmidt. Furthermore, the group consisted of Carl Ryder, who was also a veteran Arctic explorer and former navy officer, and finally Professor Kølpin Ravn was also in the group.[41]

9.1 Johannes Schmidt, Chr. Schmiegelow and Carl Ostenfeld in the Danish West Indies

From left to right Johannes Schmidt, CEO of the East Asiatic Company Chr. Schmiegelow and botanist Carl Ostenfeld are posing in a plantation on the island of St. Croix in 1914. Two years later, the two scientists were campaigning against the sale of the Danish West Indies to the USA. Meanwhile The East Asiatic Company's business interest had been secured in the secret negotiations between Denmark and the USA (Carlsberg Archives).

40 RA.10154. *De Danske Atlanterhavsøer*. 1903–1919. Kopibøger med registre. bind 18. p. 173–176.

41 Ostermann, H. 'Ryder, C. H.', in: Povl Engelstoft & Svend Dahl (eds.), *Dansk Biografisk Leksikon*, vol. 12, Copenhagen: J. H. Schultz Forlag; 1937. p. 485–486.

Support from Andreas de Richelieu

While Ryder and Ostenfeld may have taken center stage with regards the De Danske Atlanterhavsøer, Schmidt was active behind the scenes as well. At least this is what comes across from reading the letters Schmidt received from business-man and prominent conservative member of the upper house, the Landstinget, Andreas de Richelieu.[42] In the autumn of 1916 Richelieu had been one of the staunchest opponents of the sale of the Danish West Indies.[43] Richelieu and Schmidt had known each other well, at least from before Schmidt's trip to Siam in 1900.[44]

In any case, Johannes Schmidt maintained his relation to Richelieu upon his return to Denmark. According to the memoires of Schmidt's wife, Ingeborg, Richelieu gave 'an unusually funny speech' on the occasion of Schmidt's doctoral defense in December 1903.[45] Now, Schmidt surely felt gratitude for the help he had received from Richelieu over the years. In a draft letter dated 20 March 1919, Schmidt acknowledges the good advice, which he received from Richelieu 'several times over the years, (...) not least when I 10–11 years ago was about to organize the investigations in the Mediterranean and was in doubt, as to whether or not, I should rent a ship down there or use our own.'[46]

The two men also corresponded in the summer of 1916, when Schmidt received a letter, in which Richelieu acknowledges the receipt of two of Schmidt's scientific reports, one on the Flora of Koh Chang and one from the Kommissionen for Havundersøgelser. Richelieu found Schmidt's investigations to be 'marvelous both in the East and the West.'[47] This letter is dated 30 June, little more than one month before news broke of the government's plan to sell the Virgin Islands, so it could not be directly linked to this issue, more likely, Schmidt was probing Richelieu for ideas on how to fund his plans for the Atlantic deep sea investigations.

At this point in time, Richelieu had fallen out with H. N. Andersen, but he was still one of the most centrally placed business-men in Denmark and still a member of the board of directors of the East Asiatic Company. Richelieu was chairman of the board of directors of both the United Danish Steamship Companies (DFDS) as well as the largest Danish bank, Landmandsbanken. Both of these companies had invested heavily in the Danish West Indies. As a member of the board of directors of the East Asiatic Company, Richelieu had also been one of the

42 KB. Johannes Schmidt 'Papirer'. Ut.374.V.2. Privat og Blandet Korr. 1912–1927. Letter from A d. Richelieu to J. Schmidt 24 November 1916.

43 de Richelieu A. Landstingsmand Admiral A. de Richelieus Tale i Landstinget. In: Anonymous, ed. *Sandheden om Dansk Vestindien.* dansk vestindisk samfunds skrifter 1. Copenhagen: H. Hagerups Forlag; 1916:23–29.

44 Mortensen, T. Siam-Ekspeditionen, in: Winge Ø, Vedel Taaning Å, eds. Naturforskeren Johannes Schmidt. Hans Liv og Ekspeditioner. Copenhagen: Gyldendalske Boghandel Nordisk Forlag; 1947: 12–16.

45 Schmidt I. *Blade af Carlsbergs historie. Kaptajn S. A. van der Aa Kühle og hans liv på Carlsberg*. Copenhagen: Bianco Lunos Bogtrykkeri; 1956. p. 52.

46 KB. Johannes Schmidt 'Papirer'. Ut.374.VI.4. Blandet Korrespondance 1912–1928. Draft letter from J. Schmidt to A d. Richelieu 20 March 1919.

47 KB. Johannes Schmidt 'Papirer'. Ut.374.V.1. Privat 1908–1916. Breve fra Skrivebordet. Letter from A d. Richelieu to J. Schmidt 30 June 1916.

proponents for this company's substantial engagement in the Danish West Indies.[48]

Unlike Andersen, Richelieu joined the movement opposing the sale of the Virgin Islands, and he was active in the Dansk Vestindisk Samfund, where one of his speeches on the subject was also published. Richelieu had good reason to preserve the business interests of those companies where he presided over the board, but in full concordance with Johannes Schmidt, Richelieu argued primarily from a devout conservative viewpoint, emphasizing that Denmark had a moral obligation to further the economic and social wellbeing of the Danish West Indies.[49]

However, in a letter to Schmidt in November 1916, Richelieu is somber in his attitude. He thanks Schmidt for having supported the cause in sending some unspecified material to Richelieu, which will 'serve our purpose just fine'.[50] He reports to Schmidt that currently there is a large majority in the Landstinget in favor of selling the islands, but all hopes should not be forsaken. He speculated that the upcoming election might reveal a public swing in favor of keeping the islands.[51]

The scientists' no-confidence motion

Nevertheless, on December, 14 the referendum took place, and almost 65 percent of the votes were in favor of selling the Danish West Indies. Some districts however, saw a majority against the sale. This was the case in the more nationally minded areas such as well-to-do areas of Copenhagen, in districts near the German border and in areas where maritime occupations were particularly important.[52] The intransigent attitude within the conservative party was maintained for another week, and a small group of Danish marine scientists now took center stage in a direct political move to assess blame for the defeat at the polling stations.

On the evening of the December, 19 1916, tensions were heated at the general assembly of the society of *Foreningen De Danske Atlanterhavsøer*. A majority of the Danish people had just cast their votes in favor of selling the islands to the USA, and of the four sections of the society, the one responsible for the Danish West Indies had just become irrelevant. On the night of the general assembly of *De Danske Atlanterhavsøer* however, Neergaard was in dire straits. Neergaard's stand on the West Indies

48 Kaarsted T. *Admiralen. Andreas de Richelieu – forretningsmand og politiker i Siam og Danmark*. Odense: Odense Universitetsforlag; 1990. p. 220–221.

49 de Richelieu A. Landstingsmand admiral A. de Richelieus tale i landstinget. In: Anonymous, ed. *Sandheden om Dansk Vestindien. Dansk Vestindisk Samfunds Skrifter 1*. Copenhagen: H. Hagerups Forlag; 1916:23–29.

50 KB. Johannes Schmidt 'Papirer'. Ut.374.V.2. Privat og Blandet Korr. 1912–1927. Letter from A d. Richelieu to J. Schmidt 24 November 1916.

51 KB. Johannes Schmidt 'Papirer'. Ut.374.V.2. Privat og Blandet Korr. 1912–1927. Letter from A d. Richelieu to J. Schmidt 24 November 1916.

52 Kraft O. B. *Ung mand undervejs – erindringer fra årene 1912–1926*. Copenhagen: Nyt Nordisk Forlag Arnold Busck; 1958. p. 113.

had alienated a great number of more nationalist conservative minded members of *De Danske Atlanterhavsøer*, and tensions had been growing in the autumn months of 1916. The general assembly became high noon for settling scores in the society, and the society's leading natural scientists moved to keep Neergaard accountable for what they saw as a serious neglect of the president's responsibilities.

Kølpin Ravn addressed the item on behalf of the rebellious group of scientists, when he posed the question: 'Has the board of the society taken any action in relation to matters concerning the sale of the Danish West Indies?'[53] Then a heated discussion ensued for what looks like more than one hour of debating. Kølpin Ravn and the other scientists claimed that Neergaard had not done anything to promote the cause of the society. Neergaard and a majority of board members had moved not to be involved at all, and in his defense Neergaard said that he felt it would have been an untimely politicization of the society if the board had acted in direct relation to the topic of selling the islands. The scientists and several other delegates at the general assembly found this to be a poor excuse. One commented polemically that it was an odd paradox to have an active and influential politician as president, if the society was to be thought of as an apolitical forum. Tensions rose as Neergaard threatened to step down if the assembly was not confident he had done a proper job, and following yet more debate, the general assembly moved to vote on exactly this issue. At this point 37 votes were cast in favor of the no-confidence motion, while 29 or 30 expressed their approval of the president. Following the majority decision, Neergaard as well as several other board members stepped down from office.[54]

The ousting of Niels Neergaard became an act of retaliation. In close collaboration with his colleagues, and also his personal friends, Ove Paulsen and Carl Hansen Ostenfeld, Johannes Schmidt had asserted his authority as a scientist in the front line of Danish North Atlantic geopolitics. While the fate of the islands had now been decided in the referendum, the nationally minded marine scientists were anguished over the loss of fertile grounds for field studies and the loss of what they saw as an important national asset. The heated discussion undoubtedly weighed in favor of a confrontation, but events on the night of December, 19 were hardly a co-incidence. On the contrary, the attack on Neergaard had been planned for at least eight weeks ahead of the general assembly, probably longer.

53 Anonymous, 'Foreningens Generalforsamling 1916', in: *Atlanten*, vol. 4, (1915–1917), p. 449.
54 Anonymous, 'Foreningens Generalforsamling 1916', in: *Atlanten*, vol. 4, (1915–1917), p. 449–472.

Tropics sold for gold and Greenland

On the next morning, December, 20, Conservative MP Alexander Foss proposed to the Parliament that perhaps only one or two islands should be sold, not all three, but only 15 percent of the votes in parliament backed this suggestion. The case was settled now. On January, 17 1917, the treaty was ratified by both countries, and on April, 1, the USA came into possession of the former Danish West Indies. Five days later the USA entered World War I on the side of the Allied forces.[55] Among government supporters of the sale, this was a sign that selling the islands had become inevitable. If the USA had not been able to buy the islands, they would have seized them anyway in order to keep the Germans out of the Caribbean.[56] The turmoil of the 1916 general assembly however, was to cast a shadow over Foreningen De Danske Atlanterhavsøer, and soon after the future of the society as such, would be at stake.

In spring of 1919 the society was fully reorganized and from the ashes of Foreningen De Danske Atlanterhavsøer rose a new society; *Dansk Samvirke*. Marine science remained prominent with Carl Ostenfeld as the society's first chairman.[57] Ostenfeld's prominence may be one reason why the new society came to serve the purposes of Danish marine science through the 1920s, as we will see later in this book.

Conclusion

Following the sale of the Danish West Indies, the group of marine scientists had every reason to doubt that the Danish government would sustain a Danish marine science presence in the Caribbean. Moreover, the war seemed to be dragging on with the USA still a few months away from declaring war on Germany. However, what looked somber from the outside may in fact have been a crucial turning point for the scientific dreams of Schmidt, Drechsel, Ostenfeld and the rest of the prominent Danish marine scientists at the time.

The turning point was that the events surrounding the sale of the Danish Caribbean possessions had brought to the table an informal maritime Danish Atlantic alliance consisting of the marine scientists, merchant capitalists, ship owners and leading bankers. They shared a belief in the importance of upholding a Danish presence in the Atlantic realm for financial, scientific and not least patriotic reasons. Next to

55 Lidegaard, Bo, *Overleveren. 1914–1945*, Carsten Due-Nielsen, Ole Feldbæk & Nikolaj Petersen (eds.): *Dansk Udenrigspolitiks Historie*, vol. 4, Copenhagen: Gyldendal Leksikon, 2003. p. 86–87.
56 Munch, P. *Erindringer 1914–1918 – Under Den første Verdenskrig*. Copenhagen: Nyt Nordisk Forlag Arnold Busck, 1961. p. 174.
57 Johansen, Arne P., *Foreningen De danske Atlanterhavsøer og Foreningen Dansk Samvirke: Privatarkiv nr. 10154 og 10279*, Copenhagen: 1974.

the business people, Danish marine scientists made up the one group that would stand to lose most, if Denmark's Atlantic Empire began to shrink. This is the main reason why the issues of selling the Danish West Indies brought into the open, the political will of Johannes Schmidt and his fellow Danish marine scientists. Thus, Schmidt's expert witness testimony to the government hearing reads like a consistent culturally conservative position, where Schmidt sees the West Indies colony as integral to Danish society, while he also remarks that the islands form an asset with future promises of economic progress given the right investments from the public and private sector.

Through agitation and diverse political moves, the marine scientists had flagged their will to defend the mutual interests of the informal alliance, and now the marine scientists turned to their friends in the private sector for assistance in sustaining Danish marine science. Drawing on archival material from the society itself, this is the first analysis of the political role of the society Foreningen De Danske Atlanterhavsøer in relation to marine science and the issue of the Danish West Indies. This arena brings to light, an alliance where leading businessmen and potential sponsors of Johannes Schmidt's projects positioned themselves during the period of intense agitation in the fall of 1916. In particular, the wealthiest business-man in the Denmark, H. N. Andersen played a peculiar role in 1916, as he was one of very few advisors to the government in matters concerning the sale. Andersen and EAC kept a low public profile while behind the scenes Andersen secured his business interests in the draft treaty between Denmark and the USA.

While evidence remains circumstantial, there is a striking similarity between the political and oceangraphical interest of Schmidt and colleagues, and the interest of those, who became the biggest sponsors of Schmidt's plan to launch a new Atlantic, and if possible global expedition during the months from August-November 1916. In the case of Richelieu (Landmandsbanken) they even campaigned together in the fall of 1916 advising that the Danish West Indies ought to remain part of Denmark.

10 The Atlantic expeditions with the M/S *Dana*

In the literature on Johannes Schmidt's Atlantic expeditions 1920–22 commentators note that there were three separate expeditions, one following after the other. As revealed in chapter eight, the main sponsors were attracted by Schmidt and Drechsel's original 1916 proposal for one large Atlantic deep-sea expedition. Nowhere however, have previous commentators provided a reason why the eventual outcome involved three different expeditions, and why it was that two different vessels were used on these Danish led expeditions.[1]

This chapter shows for the first time, how the dream of creating one large Atlantic expedition was altered through a sequence of events and changes of strategy from 1917 onwards. These eventually led to the three separate expeditions undertaken. In this chapter, it is revealed that two smaller expeditions, focused on the breeding sites of the eel, preceded the large expedition. This change of plan went hand in hand with the desire to keep a public and scientific interest in international marine science alive in the difficult political climate that prevailed towards and after the end of World War One.

'You will be dancing like a bear'

Perhaps Johannes Schmidt had dreamed about orchestrating a spectacular circumnavigation since childhood. This is hard to tell, but at the start of 1917 he had clearly set his mind on exactly that, possibly spurred on by the sudden inflow of a good quarter of a million Kroner in the fall of 1916. In January 1917, Schmidt forwarded a new plan to Drechsel, in which he aired his wish to divide the expedition plan in two. First he wanted to solve the riddle of the eel with a focused expedition in the Atlantic. Afterwards, he now wanted to go round the world for two years sampling the Pacific Ocean and the Indian Ocean.[2] His global ambitions

1 Previous publications on the Dana expeditions include; Wolff T. *Danske ekspeditioner på verdenshavene: Dansk havforskning gennem 200 år*. Copenhagen: Rhodos; 1967. p. 155–156; Nielsen, Kristian Hvidtfelt, Ries, Christopher Jacob and Østergaard, Kristian B., 'Danske Videnskabelige ekspeditioner i perioden', in: Hvidtfelt Nielsen, Kristian and Nielsen, Henry (eds.), *Viden uden Grænser 1920–1970, Dansk Naturvidenskabs Historie vol. 4*, Aarhus: Aarhus University Press; 2006. p. 285. Fort T. *The Book of Eels*. Great Britain: Harper Collins Publishers; 2002. p. 96–97. Winge Ø, Vedel Taaning Å, eds. *Naturforskeren Johannes Schmidt. Hans Liv og Ekspeditioner*. Copenhagen: Gyldendalske Boghandel Nordisk Forlag; 1947:39–40.

2 RA. Komm. Danmarks Fiskeri- og Havundersøgelser: A 22 Dana-ekspeditionerne 1917–1924, J. Schmidt to C. F. Drechsel, 30 January 1917.

did not, however, become reality for another decade. In the meantime the two men focused their attention on planning for those seaborne exploits previously agreed upon. Schmidt, however, once again succumbed to bronchitis in the spring of 1917, at which point Drechsel used the prospect of the expeditions to cheer him up: 'You will see, once we get this expedition up and running, you can go to sea again and get well immediately. You will be dancing home like a bear – one of the brown bears that is.'[3] However, as they had neither ship nor equipment ready during the spring of 1917, they had to put the water samplers, formalin, hemp nets etcetera they bought in 1917 and 1918 in storage.[4] Schmidt appears to have been coughing his way all through summer, when he was not trying to recuperate at various Danish beach hotels in search of fresh air.[5]

Drechsel was in charge of handling most of the practical acquisitions, which often proved a difficult task due to the ongoing war in Europe. Imports and exports were limited by government restrictions, and in England, from where they liked to buy the steel wire for deep-sea fishing, the metal industry was highly geared towards the needs of the armed forces. Nonetheless, most of the special equipment was either ordered or bought during 1917, and by the beginning of 1918, Schmidt was again bubbling with enthusiasm for the whole enterprise. In February 1918, even the idea of a circumnavigation surfaced again in a letter from Schmidt to the 'Dear Commander,' as he called Drechsel: 'I am more and more convinced that we should not limit ourselves to the North Atlantic. Rather we should make a big leap and conduct a circumnavigation just like the *Challenger* back in the 1870s, (but without any other attempts in the meantime), but with a much greater reward, which will result from our far superior technology. I am afraid the Commander finds this thought too bold, this was my first reaction too.'[6]

Schmidt continued, stressing that he did realize Drechsel would criticize him subsequently. Nonetheless, Schmidt tried to explain that the cost of equipment, such as the expensive steel wire would be more or less the same, except that some more stations should be taken along the way necessitating some additional spare parts. Schmidt continued by stating that the recent opening of the Panama Canal and the use of the Suez Canal would make the journey much shorter than previously possible. What was more such an expedition would appeal to a wider public than the quite specialized eel research in the Atlantic. A circumnavigation would bring honour to the nation and have a lasting impact,

3 RA. Komm. Danmarks Fiskeri- og Havundersøgelser: A 22 Dana-ekspeditionerne 1917–1924, C. F. Drechsel to J. Schmidt, 17 May 1917.
4 KB. Johannes Schmidt 'Papirer'. Ut.374.V.3. Kommandør Drechsels Papirer fra DANA-Ekspeditionen 1920. C. F. Drechsel to DANA-Committee 11 June 1919.
5 RA. Komm. Danmarks Fiskeri- og Havundersøgelser: A 22 Dana-ekspeditionerne 1917–1924, J. Schmidt to C. F. Drechsel, 27 July 1917.
6 RA. Komm. Danmarks Fiskeri- og Havundersøgelser: A 22 Dana-ekspeditionerne 1917–1924, J. Schmidt to C. F. Drechsel, 8 February 1918.

and the mighty collections would raise the standing of a Danish marine science institute, if one was created, to an absolute leading position. This is what happened to British marine science following the Challenger Expedition, Schmidt argued. Finally, Schmidt ended his attempt to persuade Drechsel with a look to the Pacific Ocean. According to John Murray who had been to the Pacific with the Challenger Expedition, the Pacific Ocean was the richest of the three great oceans, but so little explored.[7]

As nothing further came of the idea to go global at this point, we can assume Drechsel did not like it. In late May 1919, Schmidt tried his luck with Drechsel once more on this point, but again to no avail.[8] At any rate, the chance to get to dance like a bear only started to look realistic towards the end of 1918.

Peace and further planning

When the guns of World War One went silent on the 11 November 1918, Drechsel was the one incapacitated by illness. Two years had lapsed since the agreements were first made. Now as peace had broken, it was time to set the expedition plans in motion. As Schmidt mentioned in a note to Drechsel, the new motor schooner they had been promised by Andersen had now been built at the shipyard and was ready to go into service.

Other obstacles arose however, as Schmidt for once felt restrained by his day-time job at the Carlsberg Laboratory. He was afraid that he was going to lose his long time assistant, Øjvind Winge who looked likely to be offered a professorship. If this happened, Schmidt would have to spend more time overseeing the work at the laboratory.[9]

With regards to Drechsel he had undergone surgery in early December 1918 and was unable to make appointments himself with the owner of the East Asiatic Company, H. N. Andersen, and communicated in writing only. Drechsel apologized for not having approached Andersen right after the ceasefire, but thought Andersen would be pre-occupied with the revolution in Germany.[10] Drechsel hoped that Andersen would still be inclined to help international marine science with the loan of a motor schooner. But Drechsel was not sure that they would be able to acquire the right equipment at this point in time. To make sure that the equipment could be obtained at the right price, Drechsel inquired

7 RA. Komm. Danmarks Fiskeri- og Havundersøgelser: A 22 Dana-ekspeditionerne 1917–1924, J. Schmidt to C. F. Drechsel, 8 February 1918.

8 RA. Komm. Danmarks Fiskeri- og Havundersøgelser: A 22 Dana-ekspeditionerne 1917–1924, J. Schmidt to C. F. Drechsel, 31 May 1919.

9 RA. Komm. Danmarks Fiskeri- og Havundersøgelser: A 22 Dana-ekspeditionerne 1917–1924, J. Schmidt to C. F. Drechsel, 8 December 1918

10 RA. Komm. Danmarks Fiskeri- og Havundersøgelser: A 22 Dana-ekspeditionerne 1917–1924, C. F. Drechsel to H. N. Andersen, 10 December 1918

if they could halt the expedition plans until 1920.[11] To speed up the process though, it was decided that Schmidt would communicate with Schmiegelow. Schmiegelow replied that since they would not need the vessel until 1920, the East Asiatic Company would make use of their newly built schooner for freight in the meantime.[12]

Having secured access to the motor schooner and a tentative schedule, it was time to think about the expedition committee. Drechsel and Schmidt agreed that it would be desirable to have a sort of executive committee in addition to a larger expedition committee serving the purpose of representation.

Schmidt provided Drechsel with a comprehensive list of people to invite to join the committee along with the reasoning behind his choices. Prince Valdemar was to be a figure-head only, as Schmidt doubted he would take any interest in the daily business of the expedition. Then, the members of the Kommissionen for Havundersøgelser were all to be included; that is Drechsel, fisheries scientist A. C. Johansen, the hydrographer Martin Knudsen, the government Director of Fisheries, F. V. Mortensen, plankton zoologist Ostenfeld, and Schmidt himself.

The interests of the navy, the shipping industry as well as the fishing industry were accounted for with the inclusion of the Admirals Wandel and Evers, the Count Moltke, the ship owner Erlandsen, Mikkelsen Vendsyssel from the wholesales fisheries organization, and fisherman, Jensen representing the fishermen's organization.

Further, Schmidt wished to have someone on-board with legal skills such as the lawyer Zeuthen or the prominent judge Harald K. Koch. The Ministry of Agriculture ought to be represented as the expedition would fly their flag in addition to highlighting the importance of a future marine science institute. The Ministry of Foreign Affairs was to be represented on the committee in order to facilitate consular support from countries around the North Atlantic Ocean. The idea of creating a new marine science institute is also what prompted Schmidt to favour the inclusion of a representative from the Danish Parliament, preferably someone from the inter-parliamentary committee, MP Niels Neergaard for instance was someone whom they did not seem to bear grudges against.

The Carlsberg Foundation's Board of Directors should also be invited, and the name of the chairman, Kristian Erslev was brought up by Schmidt. This would be most advantageous in light of future applications for research funding, once the expected results were being readied for publication. 'They should also "loan out" me,' as Schmidt put it.[13] To

11 RA. Komm. Danmarks Fiskeri- og Havundersøgelser: A 22 Dana-ekspeditionerne 1917–1924, H. N. Andersen to C. F. Drechsel, 14 December 1918.

12 RA. Carlsbergfondets DANA-ekspeditioner. 1917–1973. 2–4. J. Schmidt to C. Schmiegelow, 10 December 1918; RA. Komm. Danmarks Fiskeri- og Havundersøgelser: A 22 Dana-ekspeditionerne 1917–1924, C. Schmiegelow to J. Schmidt, 19 December 1918.

13 RA. Komm. Danmarks Fiskeri- og Havundersøgelser: A 22 Dana-ekspeditionerne 1917–1924, J. Schmidt to C. F. Drechsel, 16 December 1918

avoid the committee becoming too large, Schmidt remarked that he was not too worried about the size as long as an executive body would take responsibility for actually managing the expedition.[14]

The final composition of the committee is interesting as it represented a set of highly influential people from what one might term the maritime sector in Danish society, and it was more or less a replica of the list of names Schmidt had proposed to Drechsel. The Danish Prince Valdemar became the protector of the committee, while Drechsel became chairman. Johannes Schmidt was there as the would-be leader of the expedition, and in the committee he was joined by long-time collaborators from Kommissionen for Havundersøgelser, fisheries scientist, A. C. Johansen, renowned hydrographer Martin Knudsen, Director of the Fisheries Office in the Department of Agriculture, F. V. Mortensen, and botanist Carl Hansen Ostenfeld. Schmidt's old champion Professor Eugen Warming was there representing the Department of Natural History at University of Copenhagen as well as the Carlsberg Foundation. The Danish navy was represented by Vice Admiral A. F. M. Evers. The judge, H. Koch was in the committee, and finally the ship-owner and chairman of the Board of Directors of the large Danish bank, *Handelsbanken*, A. Erlandsen was on the committee. From his seat, he could oversee the spending of 100,000 Kroner, which he had delivered on behalf of Handelsbanken.[15]

The terms of using the motor schooner M/S *Dana*

In mid-January of 1919, Drechsel and Schmidt met with H. N. Andersen and Schmiegelow in the imposing headquarters of the East Asiatic Company in the centre of Copenhagen to flesh out the terms of using the motor schooner, which was now nearing completion in the EAC shipyard. Following a discussion Drechsel tentatively labelled the ship *Dana* in the minutes he prepared after the meeting and circulated to Schmiegelow. Originally, the scientists had tried to have the schooner modified with a deck cabin to accommodate the need for laboratory analyses on-board. However, instead of modifying the accomodation substantially for one cruise, it was agreed that the EAC would cover the running costs of three months of expedition in the Atlantic Ocean. This included insurance premium, oil, and water. In view of the fact that the commission could then save their cash funds for a larger expedition on

14 RA. Komm. Danmarks Fiskeri- og Havundersøgelser: A 22 Dana-ekspeditionerne 1917–1924, J. Schmidt to C. F. Drechsel, 16 December 1918

15 RA. Komm. Danmarks Fiskeri- og Havundersøgelser: A 22 Dana-ekspeditionerne 1917–1924, Nørgaard to C. F. Drechsel, 15 December 1920.

a vessel of its own this was a good offer. Captain Hansen, crucial for his knowledge of deep sea surveys, would be in charge of navigating the schooner for the duration of the cruise, with his salary paid for by the EAC. Andersen agreed that the expedition could be timed so that the scientific part of the cruise could commence from Gibraltar in mid-March and continue three months thereafter. Once the three months were up, the vessel would most likely be near the American coastline, at which point those on board would send a telegram to the East Asiatic Company and await further instructions as to the where they wanted to receive the ship.[16]

16 RA. Komm. Danmarks Fiskeri- og Havundersøgelser: A 22 Dana-ekspeditionerne 1917–1924, C. F. Drechsel to C. Schmiegelow, 21 January 1919.

10.1 The schooner M/S *Dana*
The schooner M/S Dana was built by H. N. Andersen at his own shipyard during World War One. As part of his sponsorship of the Danish Atlantic deep sea expedition, Andersen fitted the vessel with a laboratory and more cabins than usual on such a freight ship (RA. Carlsbergfondets DANA-ekspeditioner. 1917–1973. 7a).

In February 1919, Drechsel and Schmidt began to draft a new expedition plan. As the names of the members of the expedition committee were now in place, Drechsel circulated a draft to Schmiegelow and H. N. Andersen at the East Asiatic Company, the most important sponsor for the expedition. Andersen commented in red in the margin of a few paragraphs dealing with the role of the EAC, and he suggested the following statement as a replacement: 'Considering the national interest associated with this expedition, the East Asiatic Company has placed the motor schooner (...), the most important component of the expedition, at the disposal of the expedition.' As for the name of the vessel Andersen for some reason did not wish to name the ship after the protector of the expedition, *Prince Valdemar*. The Danish business tycoon par excellence, wished to have something like *Dana* or *Dania*, which would underline the national character of the enterprise.[17] So it was, that this vessel, and a number of subsequent state owned research vessels came to bear the name, *Dana*.

'A New Danish Deep Sea Expedition' Press release

In the summer of 1919, preparations had reached the point, where the committee for the *Dana* expedition was ready to reveal the plans of the forthcoming expedition to the general public. The headline, 'A New Danish Deep Sea Expedition' was wired to the Danish public by the Ritzau Bureau press agency on the 17 June 1919.[18]

'Initiated by the Kommissionen for Havundersøgelser, private sources have supplied considerable amounts of money and a motor ship for a large Danish Deep Sea expedition, which will commence in the early summer of 1920. The main field of operation will be the central parts of the Atlantic Ocean north of the Equator, and the Director and Dr. Phil., Johannes Schmidt, who has led several oceanographic expeditions with the *Thor*, will be the leader of the new expedition.' Then the members of the *Dana* committee are listed in the press release followed by the note that the general public will be informed in greater detail in the fall of 1919.

It thus appears that the central role of the East Asiatic Company, Landmandsbanken, and the other sponsors was kept relatively quiet. The readers of the press release could guess, of course, that probably only a handful of Danish corporations would have the finances to

17 RA. Komm. Danmarks Fiskeri- og Havundersøgelser: A 22 Dana-ekspeditionerne 1917–1924, C. Schmiegelow to C. F. Drechsel, 7 February 1919.

18 KB. Johannes Schmidt 'Papirer'. Ut.374.V.3. Kommandør Drechsels Papirer fra DANA-Ekspeditionen 1920. Copy by C. F. Drechsel to DANA-Committee 17 June 1919.

support such an event, and the fact that His Royal Highness Prince Valdemar as the protector of the committee had '…personally made a considerable effort for the realization of the expedition.' would hint at the involvement of the East Asiatic Company, where he was a well-known figure head. The presence of wholesale merchant Erlandsen on the list of board members, of course would promote his name as well as that of the companies he was associated with.[19]

On the 1 November 1919 the committee convened in Schmidt's elegant surroundings, the lecture hall of the Carlsberg Laboratory. Here, Drechsel presented a short account of the expedition plans in English. Drechsel reasoned that it would be beneficial for the expedition in 'several ways' if foreign governments were notified of the Danish plans.[20] Schmidt on the other hand was wary that they should not give away too much information via the international press release. He urged Drechsel to only circulate it among the member states of ICES, with whom they had a good collaboration, as he was fearful of what would happen if the news of the planned Danish led expedition arrived in Monaco or in Italy. In particular Schmidt feared that sending out a map of the tentative cruise plan was a gift for potential competitors trying to solve the riddle of the eel.[21]

Schmidt's disagreement with Drechsel's tactics of alerting the international press partly resulted in Schmidt pressing for a speedy execution of the first of two expeditions, they had agreed on with the East Asiatic Company in spring of 1919. Schmidt had not yet published his definitive account of how the Sargasso Sea was really the breeding place of the Atlantic eel. Therefore, he was keen to avoid the situation of having to share his theory with another marine scientist, or a whole tema of marine scientists.

Towards the end of 1919 Schmidt became alarmed by news from French friends that fisheries scientists in France, Spain, and Italy had begun collaborating on fisheries investigations in the Mediterranean and in the Strait of Gibraltar. Allegedly Giovanni Batista Grassi, who had proposed that eel bred in the Mediterranean more than twenty years earlier, was now going to receive all the research material on eel collected by the new joint Mediterranean investigations. Schmidt was worried that once Grassi got hold of eel samples from these areas, he would give up his earlier idea of the Mediterranean breeding grounds and turn his attention to the Atlantic, just as Schmidt had done before him.

19 KB. Johannes Schmidt 'Papirer'. Ut.374.V.3. Kommandør Drechsels Papirer fra DANA-Ekspeditionen 1920. Copy by C. F. Drechsel to DANA-Committee 17 June 1919.

20 KB. Johannes Schmidt 'Papirer'. Ut.374.V.3. Kommandør Drechsels Papirer fra DANA-Ekspeditionen 1920. C. F. Drechsel to DANA-Committee 1 November 1919.

21 RA. Komm. Danmarks Fiskeri- og Havundersøgelser: A 22 Dana-ekspeditionerne 1917–1924, J. Schmidt to C. F. Drechsel, 3 November 1919.

Schmidt was wary of Grassi for two reasons. First of all there was personal animosity towards Schmidt, which surfaced in an article Grassi had published during the war, in which he commented on Schmidt's idea of the Atlantic Ocean as the true breeding ground for the eel. 'it is not completely ruled out that one day, I too will arrive at the conclusion that the eel does not breed in the Mediterranean, but the rewards of having demonstrated this to be true, should certainly not fall in the hands of Schmidt, since he has merely presumed this. Currently, a true demonstration is lacking, and the problem remains unsolved.'[22] To Schmidt this was a sign of Grassi's hatred. The second reason to worry was the fact that Grassi and the Italians indeed had the capability to exploit what the Danes had found out, head out into the Atlantic, and sweep up an easy victory preying on the hard work of the Northern Europeans. Based on the information collected from the *Margrethe* expedition and from various ships of opportunity, Schmidt had published the results in 1916 that the eel was breeding far out into the Atlantic Ocean, although he could not be as specific as he would have liked to have been.[23] Schmidt continued that naturally this would be a personal disappointment to himself, but more importantly, Danish marine science interests at large would suffer from losing this victory, otherwise in close reach. Even the international council would suffer from what would have been a 'feather in their hat' as well.[24]

While Schmidt presented the threats as real, he also had a solution. Instead of conducting one large expedition there should be two expeditions. First he wanted to conduct a focused expedition that located with more precision where the eel bred. The idea was to take over M/S *Dana* as she was, install a drum with a wire to take stations just targeting eel larvae, and then leave all the hydrographical and other deep sea fisheries investigations to a second more substantial expedition, akin to what they had hitherto planned. Finally, Schmidt pointed out that they had to act quickly, if this new plan was to materialize, and he proposed that probably this was also a better solution to the East Asiatic Company, since they did not need to do so many alterations to their freight ship, when it was only going to be used for the smaller trip collecting eel.[25]

Drechsel agreed to Schmidt's new proposal and immediately contacted Schmiegelow and Andersen at the East Asiatic Company. Fortunately, the East Asiatic Company went along with this proposal as well.[26] So it was that by early 1920, what was supposed to be one large deep sea exploration was now divided into two parts.

22 RA. Komm. Danmarks Fiskeri- og Havundersøgelser: A 22 Dana-ekspeditionerne 1917–1924, J. Schmidt to C. F. Drechsel, 1 January 1920.

23 Schmidt J. On the early larval stages of the fresh-water eels (anguilla) and some other north atlantic murænoids. *Meddelelser fra Kommissionen for Havundersøgelser: serie Fiskeri.* 1916;5(4):1–20.

24 RA. Komm. Danmarks Fiskeri- og Havundersøgelser: A 22 Dana-ekspeditionerne 1917–1924, J. Schmidt to C. F. Drechsel, 1 January 1920.

25 RA. Komm. Danmarks Fiskeri- og Havundersøgelser: A 22 Dana-ekspeditionerne 1917–1924, J. Schmidt to C. F. Drechsel, 1 January 1920.

26 RA. Komm. Danmarks Fiskeri- og Havundersøgelser: A 22 Dana-ekspeditionerne 1917–1924, C. F. Drechsel to J. Schmidt, 7 January 1920 & 9 January 1920.

Reconnecting with ICES

Many international organizations established before World War One had effectively ceased to exist by the time the guns went silent in November 1918. In the literature, the former ICES president, Arthur Went, stressed that for the survival of ICES during World War One, it was vital that the northern neutral countries in Scandinavia and The Netherlands placed the organization on the back burner until better times. Only by March 1920 did the International Council reconvene for a council meeting. The number of member countries had been reduced by then compared to the time before World War One. Germany was not welcomed straight away, and Russia was not in a position to participate, so only eight countries, Belgium, Denmark, Finland, France, The Netherlands, Norway, Sweden, and Great Britain were taking part in rebuilding of the international organization in 1920. The Swedish delegate and pre-war president, Otto Pettersson did not run for re-election, and was replaced by the British former Vice President, Henry Maurice.[27]

Went cited Dutch marine scientist, H. C. Redeke, for having emphasized how Drechsel's carefulness and zeal in his management was crucial for bringing the international council safely through the war.[28] As an example Drechsel skilfully used the prospect of being ready to launch the expedition right after the war in his attempts to keep ICES intact during the war.[29]

Nonetheless, a controversy between Schmidt and Drechsel from the autumn of 1920 reveals how the two men thought differently about marine science. Unlike Drechsel, Schmidt wished to keep the second large Atlantic expedition a Danish enterprise, so he was quite astonished when Drechsel told him that he was going to share the expedition with the other member countries inviting them to take part with their own vessels conducting complementary investigations. First, Schmidt rushed to send a telegram to Drechsel: 'Finding internationalization of the expedition incompatible with its national base, with promises for establishment of *Dana* committee and with my position as leader. Schmidt.'[30] Schmidt then wrote up a letter to expand on his viewpoint, hoping that he was not too late in trying to change Drechsel's opinion. To Schmidt it was of paramount importance to secure the expedition as a national Danish enterprise, and he backed this viewpoint with a reference to H. N. Andersen, as Schmidt had received the clear impression

27 Went AEJ. *Seventy years agrowing. A history of the international council for the exploration of the sea, 1902–1972*. Copenhagen: International Council for the Exploration of the Sea; 1972. p. 54.

28 Went AEJ. *Seventy years agrowing. A history of the international council for the exploration of the sea, 1902–1972*. Copenhagen: International Council for the Exploration of the Sea; 1972. p. 53.

29 Rozwadowski H. *The sea knows no boundaries. A century of marine science under* ICES. Seattle and London: ICES/University of Washington Press; 2002. p. 70.

30 RA. Komm. Danmarks Fiskeri- og Havundersøgelser: A 22 Dana-ekspeditionerne 1917–1924, J. Schmidt to C. F. Drechsel, 5 November 1920.

from the owner of the East Asiatic Company that it was the patriotic character of the expedition that had aroused his interest to support it.

Schmidt then turned to more practical reasons why the Danes should keep the expedition to themselves. Schmidt referred then to the cruise plan, where they stipulated that it was a sign of scientific quality that the investigations were continuously carried out by the same men in the same way on board the same vessel. In terms of science the French in particular were complete beginners with no training or experience. Schmidt ended his letter with a threat. While it was a matter of great significance, he himself would not stand in the way of changing the operation in the proposed manner. However, as the national character of the expedition was central to Schmidt, he would renounce his stake in the whole matter, if Drechsel went forward with such a proposal.[31]

In the end, nothing came of creating a concerted effort with multiple ships, but the *Dana* expedition did come to fly the flag of the International Council as a sign that they were fulfilling international obligations to marine science. From the perspective of ICES, there was good reason for Drechsel to turn the great *Dana* expedition into an international enterprise. Nonetheless, by late 1920, Drechsel and Schmidt had to turn their attention once again to the first two expeditions, the ones focused primarily on the riddles of the eel.

The 1920 and 1921 expeditions with M/S *Dana*

The scientific set up for the expedition with M/S *Dana* was quite similar to the *Margrethe* expedition almost eight years earlier. With Captain Hansen at the helm, the expedition benefitted not only from his navigational skills, but also his skills when it came to sorting material hauled on board from the deep sea. In the early spring of 1920 Hansen and Poul Jespersen, the only scientist on the *Margrethe* expedition was once more ready for adventure. First, they travelled to England to board the M/S *Dana*, and then on the 17 March they left England for Gibraltar. This is where the sampling for eel would commence. Their first stop was Madeira to pick up Schmidt, With Schmidt on board the expedition continued south to The Canary Islands in early April, and from the 11 April the course was set almost due west heading straight for the Sargasso Sea, where they arrived a good week later.[32]

31 RA. Komm. Danmarks Fiskeri- og Havundersøgelser: A 22 Dana-ekspeditionerne 1917–1924, J. Schmidt to C. F. Drechsel, 6 November 1920.
32 Jespersen P. Togter med M/S 'Dana' (I) 1920–21. In: Winge Ø, Vedel Taaning Å, eds. *Naturforskeren Johannes Schmidt. Hans liv og ekspeditioner*. Copenhagen: Gyldendalske Boghandel Nordisk Forlag; 1947:80–81.

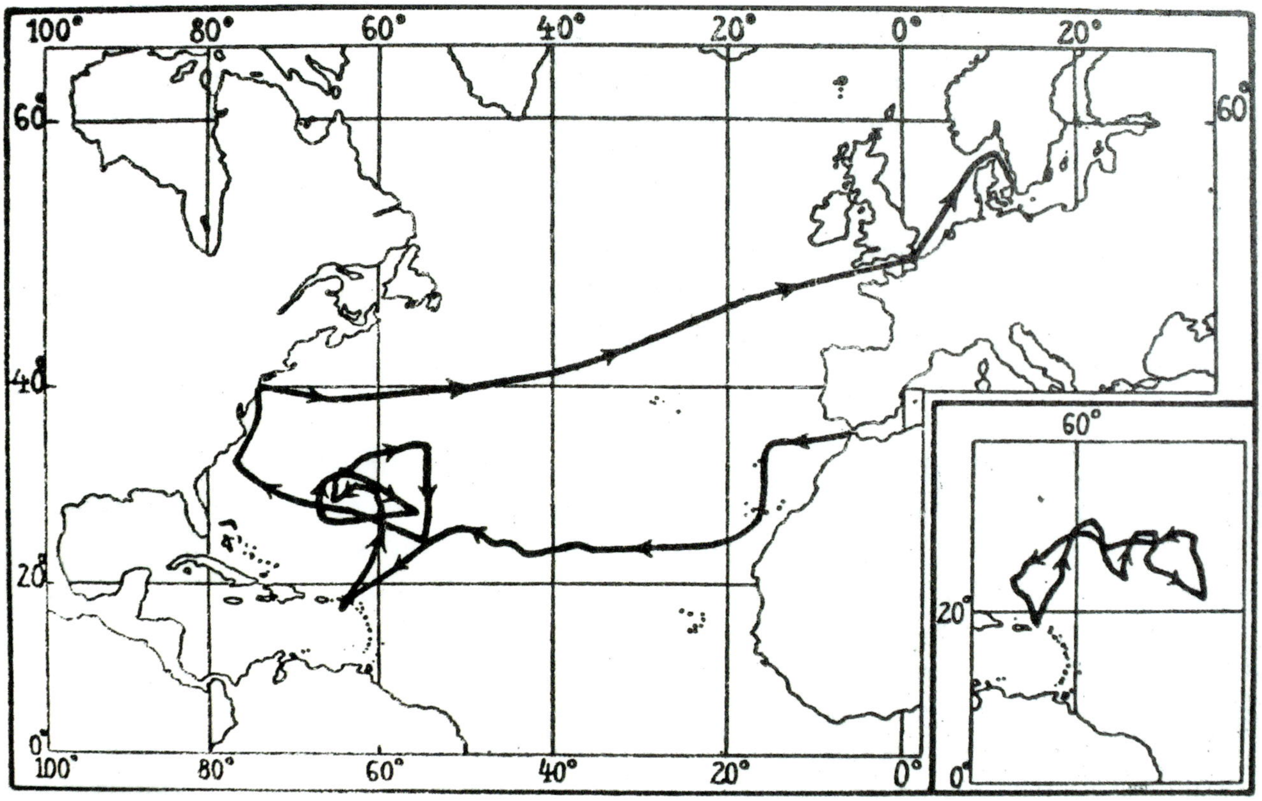

10.2 The two expeditions with M/S *Dana* in 1920 and 1921
The route of the first Atlantic expedition with the M/S Dana. The continuation in the spring of 1921 is inserted into the map (Winge Ø. & Vedel Taaning Å. 1947).

Jespersen had fond memories of this stretch of the trip. Soon after arriving in the Sargasso Sea they hauled on board the first larvae, small ones of between 11–20 mm. Every night the expedition members were kept busy with intensive fishing, when they would take several stations a night using a long wire with nets attached at different depths simultaneously. The tension and excitement rose, as they now felt really close to the actual breeding grounds of the eel.

But, just as excitement was rising, disaster struck. In the middle of an intensive chase for eel larvae a serious leak was discovered below deck in the ship's storage room. The crew started pumping out water, but the severity of the affair made Schmidt take a difficult decision. The expedition turned around and headed for a safe port, where repairs could be made. The choice was the town of Charlotte Amalie on the island of St. Thomas, now part of the US Virgin Islands. The scientific investigation continued on the way there, so another week passed before the *Dana* docked in St. Thomas. Every hour round the clock, the pumps were used to empty the hull, and to increase speed a portion of the ship's ballast was thrown out to sea. In St. Thomas, a further delay was encountered as the *Dana* had to wait twelve days before she could dock at the shipyard. By then however, local dockworkers took over the duty of pumping out water.[33]

Schmidt grew impatient and worried as the responsibility for the leaking vessel still had to be settled. In a letter to Drechsel, Schmidt was hopeful that the bulk of the cost of repairing the vessel would be covered by the insurance. He then stated that '…probably the East Asiatic Company is cursing the M/S *Dana* and the marine science investigations just as much as we are cursing the East Asiatic Company.'[34] The following week Schmidt suggested a way out of the calamity, namely that the time wasted due to the leaking vessel would be deducted from the amount of time, the East Asiatic Company had promised they would loan the *Dana* to the marine scientists. Schmidt thought he had wasted a total of three months in this way.[35] Apparently Schmidt's proposal went down well in Copenhagen. In early June Drechsel was able to inform Schmidt that the East Asiatic Company had now made the promise that the marine scientists could borrow the vessel once more at a later stage, for an additional three months of cruise time.[36]

Meanwhile, Schmidt and Jespersen dined out almost every night, as they were invited by some of the Danish dignitaries still residing in Charlotte Amalie.[37] The local bankers, for instance, were indirectly involved in the sponsorship of the expedition, as the Landmandsbanken and the East Asiatic Company still held on to their private investments in St. Thomas both in shipping and banking. But in spite of the hospitality they encountered, there is no doubt Jespersen and Schmidt were both dissatisfied with the fact that St. Thomas was now part of the USA instead of the Kingdom of Denmark. As Jespersen stated laconically; 'where we once found cocktails and rum, there is now icecream and sodawater.'[38] What Jespersen was hinting at in this case was not only the sale of the islands to the USA but a very concrete lack of rum due the prohibition of alcohol, which had just been introduced in the USA the same year. Commenting on the teetotal situation in St. Thomas, Schmidt informed Drechsel that the local inhabitants were strongly opposed to the embargo on alcoholic beverages. Then he added that in the interest of truth, he should note that all wells had not yet dried out. Imports were made from the nearby English West Indies.[39]

33 Jespersen P. Togter med M/S 'Dana' (I) 1920–21. In: Winge Ø, Vedel Taaning Å, eds. *Naturforskeren Johannes Schmidt. Hans liv og ekspeditioner*. Copenhagen: Gyldendalske Boghandel Nordisk Forlag; 1947:82–83.

34 RA. Komm. Danmarks Fiskeri- og Havundersøgelser: A 22 Dana-ekspeditionerne 1917–1924, J. Schmidt to C. F. Drechsel, 15 May 1920.

35 RA. Komm. Danmarks Fiskeri- og Havundersøgelser: A 22 Dana-ekspeditionerne 1917–1924, J. Schmidt to C. F. Drechsel, 22 May 1920.

36 RA. Komm. Danmarks Fiskeri- og Havundersøgelser: A 22 Dana-ekspeditionerne 1917–1924, C. F. Drechsel to J. Schmidt, 4 June 1920.

37 Jespersen P. Togter med M/S 'Dana' (I) 1920–21. In: Winge Ø, Vedel Taaning Å, eds. *Naturforskeren Johannes Schmidt. Hans liv og ekspeditioner*. Copenhagen: Gyldendalske Boghandel Nordisk Forlag; 1947:83.

38 Jespersen P. Togter med M/S 'Dana' (I) 1920–21. In: Winge Ø, Vedel Taaning Å, eds. *Naturforskeren Johannes Schmidt. Hans liv og ekspeditioner*. Copenhagen: Gyldendalske Boghandel Nordisk Forlag; 1947:83.

39 RA. Komm. Danmarks Fiskeri- og Havundersøgelser: A 22 Dana-ekspeditionerne 1917–1924, J. Schmidt to C. F. Drechsel, 22 May 1920.

As it turned out, the leaking motor schooner ended up costing the expedition a full month of idle waiting time on land. In June however, they were ready to embark on a new journey to the Sargasso Sea. After a few days of sailing, they were able to pull the nets in, once more filled with herring larvae. Considerable numbers of American (*Anguilla rostrata*) as well as European eel larvae were caught, but very few of them were now tiny. Most of the eel were large, suggesting that they had more or less missed the season when the eggs hatch into larvae. As Jespersen recalled, he and the ship's telegraph operator spent long days sorting baskets filled with samples from deep below, and to keep up with the pace of fishing, the captain and Schmidt himself also lent a hand with this part of the work. After a while, they resorted to only pulling out the eel from the baskets, conserving the rest of the catches. Going through the eel samples, Schmidt sat next to Jespersen and noted down the results from estimating the muscle segments to determine whether or not a larva belonged to the American or European eel. On numerous occasions several hundred larvae were caught in one haul, and a total of 6,069 larvae of European origin and 1,027 larvae of American eel ended in the hands of the team of scientists. A total of five weeks lapsed from the time they left St. Thomas following the repair, until the scientific part of the expedition came to an end in the Bermuda Islands on 4 July. From here on, Captain Hansen continued to Charleston in the USA to hand over the M/S *Dana* to the East Asiatic Company. Schmidt and Jespersen went back to Denmark via New York City.[40]

Comfortably installed at the Mount Alpin Hotel on Broadway, New York City, Schmidt had time to take stock of the findings of the 1920 expedition with the *Dana*. 'Considering the large amount of difficulties we have encountered, we have every reason to be satisfied with what we have accomplished, which as they say in English: Ends every discussion about the breeding places of the Eel.'[41] He still had to convince a learned audience though, through scientific publications in the right places.

Towards the end of the first expedition with the M/S *Dana*, Schmidt and Drechsel were able to take stock of their funds as well. In addition to the private donations, they also had access to the interest earned from the three years in which the money had been sitting in their bank account. Therefore, a total income of more than 276,000 Kroner was recorded by late October as Drechsel took stock of the first cruise with the M/S *Dana*. Meanwhile, the expenses amounted to a mere 65,000 Kroner,

40 Jespersen P. Togter med M/S 'Dana' (I) 1920–21. In: Winge Ø, Vedel Taaning Å, eds. *Naturforskeren Johannes Schmidt. Hans liv og ekspeditioner*. Copenhagen: Gyldendalske Boghandel Nordisk Forlag; 1947:84–86.

41 RA. Komm. Danmarks Fiskeri- og Havundersøgelser: A 22 Dana-ekspeditionerne 1917–1924, J. Schmidt to C. F. Drechsel, 9 July 1920.

so the expedition committee was still in control of 211,000 Kroner to be spent on deep-sea investigations.[42] Fortunately, as Drechsel reported to the expedition committee, the East Asiatic Company had promised that the scientists could make use of the vessel once more in the West Indies to make up for the technical challenges and delays which had somewhat hampered the first part of the expedition.[43]

The chance arose in the spring of 1921, as the motor schooner *Dana* was scheduled for shipping in the Caribbean in the winter months. This time however, only Captain Hansen went on the *Dana*. For two and a half months from the end of February until mid-May, another 52 stations were taken in the Sargasso Sea area. The pelagic nets were pulled out of the sea with samples on no less than 320 occasions during this second expedition with M/S *Dana*. Captain Hansen performed all the scientific duties of the surveys. He sorted the eel larvae from the nets, and then conducted a preliminary distinction between American and European eel larvae. This time, the collection of larvae included the missing months from the previous year, and a sizeable number of very tiny larvae could now be added to the growing collection of specimen at the Carlsberg Laboratory, which included more than 8,000 larvae from the two expeditions with the *Dana*.[44]

Conclusion

This chapter has presented the first explanation for why there was not just one large *Dana* expedition but a total of three different expeditions. While in hindsight they may appear like a natural progression, the reality was more complicated. Schmidt and Drechsel formed a match made in heaven, when it came to matters of fundraising. They were both very well connected in the international world of marine science, and Drechsel's connections in the corporate world centred in Copenhagen were probably unparalleled. However, when the money was secured in late 1916, the two headstrong men had to agree on how to spend the money.

Drechsel took seriously the fact that he was not only the chairman of the Danish commission for marine science, but also the secretary general of ICES. His actions provide an indication of his dedication to the revival of the international science programme following the end of World War One in November 1918. Drechsel's internationally minded idea to organize a large joint Atlantic expedition with France and The

42 RA. Carlsbergfondets DANA-ekspeditioner. 1917–1973. 1. Forhandlingsprotokoller. Regnskab 'Dana'-Ekspeditionen.

43 RA. Carlsbergfondets DANA-ekspeditioner. 1917–1973. 1. Forhandlingsprotokoller. Minutes 5th meeting of the Baord of the 'Dana'-Ekspeditionen, 20 November 1920.

44 Jespersen P. Togter med M/S 'Dana' (I) 1920–21. In: Winge Ø, Vedel Taaning Å, eds. *Naturforskeren Johannes Schmidt. Hans liv og ekspeditioner*. Copenhagen: Gyldendalske Boghandel Nordisk Forlag; 1947:85–86.

United Kingdom however, came to clash with Schmidt's more nationalistic approach to their common goal of sending a Danish research vessel far out into the Atlantic Ocean.

On the one hand Schmidt tried to convince Drechsel of the virtues of launching a circumnavigation instead of merely investigating the Atlantic Ocean. From the position of ICES, this would be a wasted opportunity, as the organization was active in the Atlantic Ocean only. On the other hand, Schmidt gave Drechsel the ultimatum that he would step down from running the planned expeditions, after he had learned of Drechsel's plans to involve other countries and invite them to send out their own research vessels to compliment the work of the Danish enterprises. Schmidt's emphasis was on national honour and pride, which was apparent from his insistence that there would be no glory sharing with other countries. Drechsel, meanwhile, sought to serve his international obligations just as much as his national duties.

Nonetheless, Drechsel and Schmidt acted swiftly and in concordance. So, in early 1920 when Schmidt felt threatened by his old adversary Grassi, who he thought might launch a competing expedition from Italy, it was possible to split the proposed expedition discussed at earlier meetings, and so make sure that Schmidt's vessel was at sea before Grassi could launch his own investigations. Striking a deal with the East Asiatic Company, Drechsel and Schmidt ensured that the partly purpose-built motor schooner, the *Dana* could be borrowed from the East Asiatic Company. This allowed them to conduct focused eel investigations in the spring of 1920, before any other country could make a move. This trip was only partly successful due to technical problems, which the East Asiatic Company ended up paying for. As compensation to the scientists, Schmidt and Drechsel were granted the right to make use of the M/S *Dana* once again in the spring of 1921.

At the end of the second of three Atlantic expeditions, the eel investigations had been highly fruitful for Schmidt. Meanwhile, in Copenhagen Drechsel was at the helm of their most ambitious project yet: the great Atlantic Expedition with a new research vessel, the R/V *Dana*. This is the topic of the next chapter.

11 The great Atlantic expedition with the R/V *Dana*

Many results from the great Atlantic expedition with the R/V *Dana* 1921–22 are still cited in the current research literature. The use of 10 km long steel wire to facilitate the catch of fish and larvae in pelagic nets lowered at different depths became a technological novelty, partly responsible for the catch of unprecedented numbers of deep sea creatures such as anglerfish and the mollusc, Spirula. This was also the first time a marine science expedition travelled from the Caribbean to the Pacific Ocean via the Panama Canal, where the discovery in the Gulf of Panama of large anoxic areas in the Pacific Ocean was one of the more spectacular results.

This chapter presents for the first time, the accomplishments of the great Atlantic expedition in its political and economic context, unravelling the extent to which Schmidt and Drechsel agreed on how the expedition should be executed. Schmidt's wilfulness was such, that the entire expedition was affected by it, albeit largely successfully so, when looking at the expedition with the benefit of hindsight.

This third and last of the Atlantic expeditions from 1920–1922 was the one that really propelled Schmidt into the international limelight, and earned him a reputation as a first rate marine scientist. This was mainly down to the successful publication of his grand theory of the life history of the eel, which centred on establishing that the breeding place is somewhere in the Sargasso Sea. However, before the third expedition could get under way, a suitable ship had to be purchased.

Buying the R/V *Dana*

While the second stage of the first *Dana* expeditions was still in the pipeline in the autumn of 1920, a whole new research vessel was being acquired by the Danish state for the purpose of furthering Danish marine science. Again Drechsel was the man in the middle, travelling to Great Britain with the fisheries inspector, Mortensen to find the right vessel. Two expedition makers were assisted by the director and a navy engineer from the navy docks, both approved the vessel before it was bought. The future research vessel (R/V) was a two-year-old steam trawler, hitherto known as the *S/S John Quilliam.* Following an overhaul in Denmark, which involved converting it into a proper research vessel, she was renamed R/V *Dana*, or as some called her, *Dana II*. The state had granted a sum of 600,000 Kroner for the purchase as well as the refurbishing of the vessel.[1] This was a very substantial government investment at the time, and there is no reason to doubt the influence wielded in particular by Drechsel, who was able to showcase his expedition funds to anyone interested. More than 200,000 Kroner was still sitting in the Landmandsbanken waiting to be spent on a deep-sea voyage.

11.1 R/V *Dana* in the Gulf of Panama

The Great Atlantic Expedition with R/V Dana lasted 11 months, and midway through the expedition the ship started to show signs of fatigue (*RA. Carlsbergfondets DANA-ekspeditioner. 1917–1973. 6*).

The great Atlantic Expedition with R/V *Dana* became the most significant of Schmidt's expeditions to date. The expedition would last for more than ten months, and more scientists were on-board than ever before. Schmidt

1 RA. Carlsbergfondets DANA-ekspeditioner. 1917–1973. 1. Forhandlingsprotokoller. 5th meeting of the Baord of the 'Dana'-Ekspeditionen, 20 November 1920.

2 Jespersen P, Vedel Taaning Å. Den store atlantiske ekspedition med S/S 'Dana' (II) 1921–22. In: Winge Ø, Vedel Taaning Å, eds. *Naturforskeren Johannes Schmidt. Hans liv og ekspeditioner*. Copenhagen: Gyldendalske Boghandel Nordisk Forlag; 1947: 89.

3 Jespersen P, Vedel Taaning Å. Den store atlantiske ekspedition med S/S 'Dana' (II) 1921–22. In: Winge Ø, Vedel Taaning Å, eds. *Naturforskeren Johannes Schmidt. Hans liv og ekspeditioner*. Copenhagen: Gyldendalske Boghandel Nordisk Forlag; 1947: 93–94.

took part in the entire expedition this time. So did the zoologists Aage Vedel Taaning, Knud Hensch Stephensen, and the experienced Atlantic explorer Poul Jespersen. Physicist Johs Olsen and the ship's medical doctor N. C. Andersen assisted with the hydrographical measurements, and the ship's captain, Hansen assisted with the fisheries investigations as on previous occasions. In addition, Physicist and veteran of the Icelandic cruises with the *Thor*, J. N. Nielsen was on board for six weeks of investigations in the Strait of Gibraltar, and finally, as we will hear more about below, Schmidt's close friend and zoobotanist, Carl Ostenfeld joined the expedition for a couple of months in the winter of 1921–22.[2] One notable feature of the third *Dana* expedition with the R/V *Dana* was the 10,000 metres of steel wire on board that enabled samples to be gathered from a greater depth in tropical waters than in any other research programme.[3]

11.2 10 kilometres of steel wire to fish deeper than anyone before

Perhaps the most crucial piece of equipment on the Dana expeditions was the steel wire. With 10,000 meters of steel wire available on board, the research team of the Great Atlantic expedition was able to sample for larvae, plankton as well as pelagic fishes at a greater depth than anyone had attempted in tropical waters before. Prior to the trawls being lowered into the water the ship was halted. Then as the vessel gained speed again the steel wire with the trawls was unleashed via a drum (RA. Carlsbergfondets DANA-ekspeditioner. 1917–1973. 6).

By late summer of 1921 the long quest for a large scientific expedition intensified as the R/V *Dana* was made ready for the cruise scheduled to begin in ten months. On 24 August the commission for marine science – rather than the designated expedition committee – agreed on the instructions for the leader of the expedition. This meant that Drechsel and all the scientists Knudsen, Ostenfeld, Mortensen, Johansen, and Schmidt decided the scientific plan to be carried out. The first scientific objective to be met was to complete as far as possible the picture of the biology of the eel or in contemporary language, the life history of the eel. This included finding the location and extent of the breeding places, and locating the spawning season and the migration patterns of the larvae.[4]

Secondly, it was emphasized that another main objective was to provide an overview of the oceanographical qualities of the temperate and tropical zones of the North Atlantic Ocean, modelled on the ways in which the *Ingolf* expedition and the *Thor* expeditions had sought to investigate the waters around Iceland, Greenland, and the Faroe Islands. In this way the expedition should strive to observe and characterize the different waters of the Atlantic in order to find out, where the dominant ocean currents were flowing, and how the various layers of water were transported round the Atlantic Ocean. This included measurements along the tropical borders between the North Atlantic and the South Atlantic Ocean between Northern Brazil and West Africa to document the exchange of water between north and south. From Brazil the expedition was expected to chart a northern course towards the Sargasso Sea and if time allowed it, also further north to Newfoundland in Canada searching for evidence for how water is distributed back and forth along the Gulf Stream between the Caribbean at one end and the Mediterranean at the other end of the system.[5]

To find out, the expedition would measure the temperature, salinity, and nutrient contents, in addition to the living organisms, animals, and plants. The chief occupation of the participants was to find connections that linked the physical and biological conditions in the greater North Atlantic Ocean with the more specific conditions prevailing in the North Sea. The so-called pelagic or free swimming and free flowing parts, of marine life were to be prioritized over marine life at the bottom of the sea, the so-called benthic fauna.[6]

4 RA. Komm. Danmarks Fiskeri- og Havundersøgelser: A 22 Dana-ekspeditionerne 1917–1924, Instructions for Havundersøgelses Ekspeditionen 'DANA', 24 August 1921.
5 RA. Komm. Danmarks Fiskeri- og Havundersøgelser: A 22 Dana-ekspeditionerne 1917–1924, Instructions for Havundersøgelses Ekspeditionen 'DANA', 24 August 1921.
6 RA. Komm. Danmarks Fiskeri- og Havundersøgelser: A 22 Dana-ekspeditionerne 1917–1924, Instructions for Havundersøgelses Ekspeditionen 'DANA', 24 August 1921.

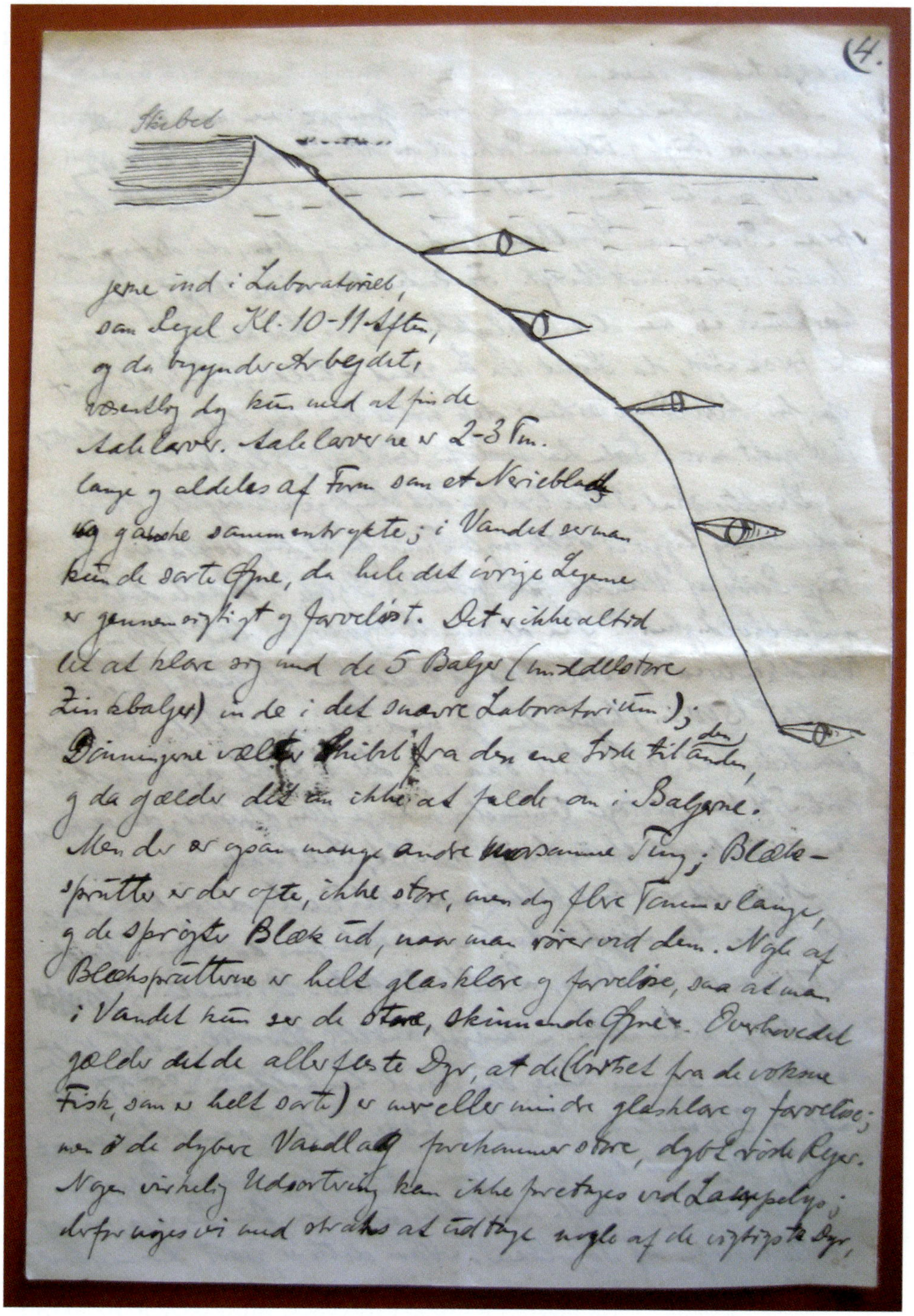

11.3 Sketch of deep sea pelagic trawling
Knud Hensch Stephensen kept a diary during the Great Atlantic Expedition with R/V Dana in 1921–22. On this page he has sketched how the pelagic trawls were fishing simultaneously at different depths (RA. *Stephensen, Knud Hensch*).

In other words this was an ambitious and far-ranging research programme. Estimating the total expenses ahead of the expedition was difficult considering the novelty of the whole enterprise, however, the expedition members had been able to estimate the duration, which they believed to be 10 months. This allowed for some adjustment along the way, depending on the degree to which unforeseen issues arose such as maintenance and replacement of the technical equipment, or the lack of cheap coal, which might necessitate a detour.

Kick off and Schmidt's nationalism

Right until late August provisions were being bought, but finally on the 30 August the R/V *Dana* left Copenhagen on a rainy and windy day. The expedition rounded Elsinore and continued through the Great Belt and the Western Baltic Sea, and into the Kiel Canal the next day.[7]

Even if Denmark had not been occupied during the war, there was considerable animosity towards Germany within the population. Schmidt displayed such anti-German sentiments in 1921 during a brief encounter with his longstanding German colleague and Professor of zoology, Ernst Ehrenbaum. On the 31 August 1921 the R/V *Dana* reached the Kiel Canal in Northern Germany connecting the Baltic Sea with the North Sea, and at noon they stopped in the town of Rendsburg. This was a rare chance for Schmidt to meet face-to-face with Ehrenbaum, with whom he had corresponded frequently over the years. Ehrenbaum came on board the *Dana* with some samples as gifts to the Danish expedition, and the zoologist and member of the expedition Knud Hensch Stephensen commented on this in his diary: 'He is an immensely kind man, not at all Prussian and Dr. Schmidt presented me to him.'[8] Hensch Stephensen and Ehrenbaum spoke in German, but as soon as Schmidt joined them the conversation switched to English. Hensch Stephensen noted that he learned from other members of the expedition that Schmidt under no circumstances would speak German, not even with a nice man like Ehrenbaum. Ehrenbaum travelled with the *Dana* expedition all the way through the Kiel Canal and he would sometimes address other members on board in German. When he was about to be ferried back ashore on a pilot vessel, he said farewell in English and wished the expedition best of luck. Once he was back on solid German soil, he repeated all of his greetings, but now in German. Hensch Stephensen had the clear impression that Ehrenbaum felt more at ease, when he returned ashore. Confiding to his diary that he was not very fond of things German, Hensch Stephensen still felt sorry for Ehrenbaum, who would have been quite aware of the fact that all the Danish

7 Jespersen P, Vedel Taaning Å. Den store atlantiske ekspedition med S/S 'Dana' (II) 1921–22. In: Winge Ø, Vedel Taaning Å, eds. *Naturforskeren Johannes Schmidt. Hans liv og ekspeditioner*. Copenhagen: Gyldendalske Boghandel Nordisk Forlag; 1947: 90.

8 RA. Stephensen, Knud Hensch: 'Dagbog fra Dana-ekspeditionen til Det karibiske Hav 1920–1922'. 31 August 1921.

scientists understood German quite well, yet chose to address him in English.[9]

Schmidt once again showed himself to be a Danish nationalist to the bone. He was by no means just playing to the prevailing political tunes, when trying to uphold the flag of Denmark in the face of international marine science.

Further diplomatic duties followed over the course of the next week, when they stopped in England and Wales for coaling, and en route they tried some of the new instruments and fishing gear to make sure that they worked and were likely to still be working when the expedition reached Portugal and Spain in September 1921, where the investigations would have their first focal point.[10]

The Strait of Gibraltar was a target area once more, as the team sought to verify the influx of eel larvae from the Atlantic Ocean into the Mediterranean through this narrow stretch of water between Spain and North Africa. In the strait itself the depths are relatively shallow compared to the waters outside and inside the Strait of Gibraltar. In this way the strait serves as a barrier between the Atlantic Ocean and the Mediterranean, prohibiting the entry of many deep-sea species. The eel however, was able to migrate freely between shallow and deep waters, and therefore it should not be obstructed from entering the Mediterranean.[11]

There was reason to be satisfied with the over all accomplishments during the first months at sea, apart from a torn trawl wire, of which an extra 4,000 metres were bought as a replacement at the end of 1921.[12] The rather expensive trawl wire was indispensable for pelagic trawls and the technological clout of this expedition, but as there was a misfit between the drum and the wire, the wire was quickly worn out, when it was reeled in. In the middle of December, the wire snapped, fortunately as Schmidt reported, it happened several hundred metres down the water column, so no one was injured, but five pelagic trawl nets, custom made for the trip were lost.[13]

Another dark cloud on the horizon was the developing disagreement between Schmidt and Drechsel over the charted course of the expedition.

Pacific dreams

As mentioned above, the instructions for the scientific programme had already been planned when the core group of scientists convened for a

9 RA. Stephensen, Knud Hensch: 'Dagbog fra Dana-ekspeditionen til Det karibiske Hav 1920–1922'. 31 August 1921.

10 RA. Stephensen, Knud Hensch: 'Dagbog fra Dana- ekspeditionen til Det karibiske Hav 1920–1922'. 6–16 September 1921.

11 Jespersen P, Vedel Taaning Å. Den store atlantiske ekspedition med S/S 'Dana' (II) 1921–22. In: Winge Ø, Vedel Taaning Å, eds. Naturforskeren Johannes Schmidt. Hans liv og ekspeditioner. Copenhagen: Gyldendalske Boghandel Nordisk Forlag; 1947: 90.

12 KB. Johannes Schmidt 'Papirer'. Ut.374.I.3. J. Schmidt to C. F. Drechsel, 19 December 1921.

13 KB. Johannes Schmidt 'Papirer'. Ut.374.I.3. J. Schmidt to C. F. Drechsel, 14 December 1921.

final debriefing three days before the cruise was set to commence from Copenhagen. Schmidt however, used the opportunity to air a long desired dream of his, to extend the planned programme to include a stop over in the Pacific Ocean. Perhaps, Drechsel was paying attention to other matters, since the issue was not settled at the final meeting before the expedition started, but already on the 1 September as the R/V *Dana* made its way into the North Sea, Drechsel was getting worried that he could no longer control Schmidt and the expedition to the extent he wanted to, so he wrote a letter to 'Dear Dr. Schmidt,' as he tended to address the leader of the expedition. In this letter he asked for a more in depth explanation for Schmidt's sudden desire to cross the Panama Canal and head into the Pacific. Drechsel was worried about the expenses, the duration of such a trip, and not least he worried that this would serve a purpose separate from the core area of interest of ICES – the Atlantic Ocean. How would Schmidt meet all of these expectations, if he also wanted to go to the Pacific? In theory Drechsel declared that he found it interesting, to pursue this opportunity, but he would have to discuss it with the board, and he also raised the concern that the Danish Ministry of Foreign Affairs would have to ensure free passage through the Panama Canal, so the ship would not have to pay the same toll as other ships did.[14]

Schmidt received Drechsel's request in Portugal, from where he was able to expand on his ideas in a letter dated 12 September. Schmidt reminded Drechsel that they had originally considered a big plan to go round the World, but the financial situation had required them to focus on the Atlantic Ocean only. Schmidt's idea would be to compare the fauna in these two great oceans, something which had otherwise been possible only in the case of the Challenger Expedition in the 1870s. While Schmidt indicated that it was too early to say whether the opportunity would arise, he wished to have the possibility to go to the Pacific if time allowed. Looking at the programme for the expedition, Schmidt found that a short trip through the Panama Canal would be feasible on the first half of the trip, before they were set to commence the search for eel larvae in the spring of 1922. Schmidt finished his plan with the request that he would like to stay out for as long as possible, preferably at least until the end of August 1922.[15]

14 KB. Johannes Schmidt 'Papirer'. Ut.374.I.3. C. F. Drechsel to J. Schmidt, 1 September 1921.

15 KB. Johannes Schmidt 'Papirer'. Ut.374.I.3. J. Schmidt to C. F. Drechsel, 12 September 1921.

11.4 Schmidt in R/V *Dana*'s laboratory. *Johannes Schmidt in the laboratory on board R/V Dana (Carlsberg Archives).*

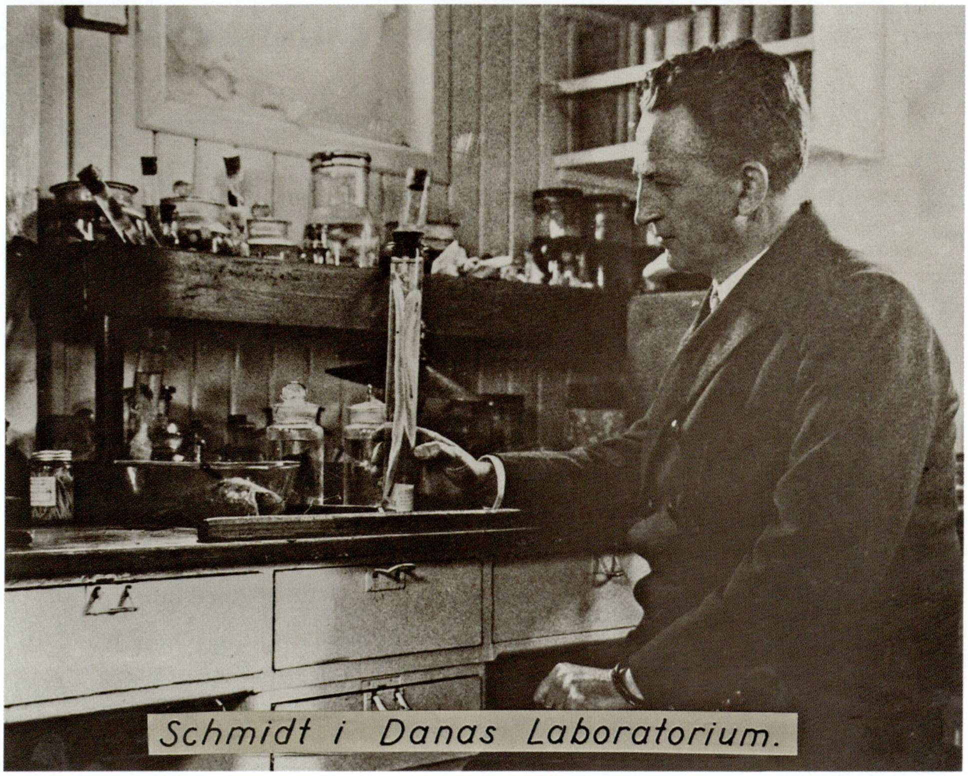

Perhaps Schmidt realized that he needed not only Drechsel's support, but also that of members of the commission in general. At any rate, Schmidt's letters to Drechsel were now addressed to the 'Kommissionen for Havundersøgelser.' Drechsel for one noticed this difference, and changed his letters accordingly: 'I don't mind upholding a more formal tone in our correspondence as you seem to prefer it, he wrote' From then on, Drechsel addressed his letters 'to the leader of the expedition.'[16]

Drechsel informed Schmidt via letter that he had proposed the idea of extending the expedition to the Pacific to the board of the expedition committee, and they were sympathetic towards it. However, the cost of using the recently completed Panama Canal was the outstanding issue. Drechsel and the board had reached an agreement where it was only feasible for Schmidt to cross into the Pacific Ocean under the condition that The US government would let the ship pass without paying the tolls expected of other vessels.[17] Drechsel learned in the middle of November that this was not possible following an official inquiry to the United States' Chargé d'Affaires in Denmark, John C. Wiley.[18] The Danish commission for marine science therefore agreed that Schmidt would have to give up his idea of going to the Pacific as part of the on-going expedition.[19]

16 KB. Johannes Schmidt 'Papirer'. Ut.374.1.3. C. F. Drechsel to J. Schmidt, 29 September 1921.
17 KB. Johannes Schmidt 'Papirer'. Ut.374.1.3. C. F. Drechsel to J. Schmidt, 28 October 1921.
18 KB. Johannes Schmidt 'Papirer'. Ut.374.1.3. J. C. Wiley to C. F. Drechsel, 10 November 1921.
19 KB. Johannes Schmidt 'Papirer'. Ut.374.1.3. Minutes 160th meeting of Kommissionen for Havundersøgelser, 18 November 1921.

Carlsberg Foundation as a strategic resource

The Board of the *Dana* Committee represented by Drechsel now turned to the Carlsberg Foundation. Since September 1921, Ostenfeld had replaced Eugen Warming on the Board of Directors of the Carlsberg Foundation.[20] They were both members of the *Dana* expedition Committee as well as the Commission for Marine Science, hence, the connection was close, and Drechsel sent a short application to the Carlsberg Foundation on the 3 December.[21] Already on the same day, 3 December 1921 Drechsel was able to announce to the *Dana* Committee that at the request of Professor Ostenfeld the Carlsberg Foundation's Board of Directors had agreed to cover expenses – up to 10,000 Danish Kroner – associated with *Dana's* passage through the Panama Canal.[22] For Ostenfeld this was a win-win situation. As one of the World's leading plankton scientists, the deep seas of the Eastern Pacific Ocean as well as the artificial Gatun Lake of the Panama Canal were virgin territories waiting to be sampled and sifted for new oceanographic evidence.

If Ostenfeld had sat on the Board of Directors of the Carlsberg Foundation today, he might have thought that there was a conflict of interest in donating funds for a project, so deeply associated with his own research. However, times were different in 1921, and for Ostenfeld and the expedition, there was no time to waste.

No doubt relieved, Schmidt aired his views on board the R/V *Dana*. According to Knud Hensch Stephensen's diary, Schmidt speculated that the company running the Panama Canal had declined the Danish request for free passage only after having consulted with the US Fisheries Department in Washington D. C. Schmidt took this as a sign that the Americans were uncomfortable having a Danish team of researchers doing a job, which they ought to have carried out themselves. As a sort of retaliation, they then sought to prevent the Danes from reaping the low hanging fruits too easily.[23]

In any case, only four days later, on the 7 December, Ostenfeld left Copenhagen to join the expedition waiting for him in St. Thomas. By way of Esbjerg, London, Liverpool, New York, and San Juan in Puerto Rico, Ostenfeld arrived at St. Thomas on the morning of the 31 December 1921. 'This is a great advantage for us', was the phrase Stephensen trusted to his diary. 'Ostenfeld is very lively, and as he is extremely straightforward and knows most people already, he immediately plays in tune with the rest of the ensemble.'[24] The members of the expedition

20 Petersen, Johannes, *The Carlsberg Foundation*, Copenhagen: Bianco Luno; 1956, p. 92.

21 CFD. 1921–22. No. 73. Application from C. F. Drechsel, 3 December 1921.

22 KB. Johannes Schmidt 'Papirer'. Ut.374.V.3. Kommandør Drechsels Papirer fra DANA-Ekspeditionen 1920. Circular from C. F. Drechsel to DANA Committee 3 December 1921.

23 RA. Stephensen, Knud Hensch: 'Dagbog fra Dana-ekspeditionen til Det karibiske Hav 1920–1922'. 3 December 1921.

24 RA. Stephensen, Knud Hensch: 'Dagbog fra Dana-ekspeditionen til Det karibiske Hav 1920–1922'. 7 January 1922.

11.5 The Panama Canal in January 1922
The R/V Dana went through the Panama Canal in January 1922 as perhaps the first research vessel since the opening of the canal in 1913. This was not part of the agreed cruise plan, when the expedition began, but the surveys conducted in The Pacific Ocean turned out to be very successful from a scientific point of view ((RA. Carlsbergfondets DANA-ekspeditioner. 1917–1973. 6).

waited for Ostenfeld, and two days later the expedition continued across the Caribbean. One week later, they first caught sight of the town of Colon in Panama, and the next day 10 January, the *Dana* went through the Panama Canal.[25] Moneywise this was a pleasant surprise. Contrary to the information they had gathered earlier, the cost of the passage through the Panama Canal was 150 US Dollars each way, not 500 as they had thought.[26]

25 KB. Johannes Schmidt 'Papirer'. Ut.374.V.3. Kommandør Drechsels Papirer fra DANA-Ekspeditionen 1920. Travel account from C. H. Ostenfeld to DANA Committee 5 April 1922.
26 RA. Stephensen, Knud Hensch: 'Dagbog fra Dana-ekspeditionen til Det karibiske Hav 1920–1922'. 14 January 1922.

Probing the Pacific

Over the next eight days the expedition was busy taking hydrographic as well as biological samples from the Pacific Ocean. The stations taken were focused on the deep waters to the south of the Gulf of Panama off the Punta Mala peninsula. In addition a few stations were taken near shore in more shallow waters. In his cruise report, Ostenfeld revealed that 'the Pacific Ocean gave us some particularly interesting results'.[27] With regards the hydrography of the area, samples revealed that the salinity was considerably lower in the Pacific Ocean than in the Atlantic Ocean, while the temperature was somewhat lower in the Pacific.[28]

From the collected water samples the expedition members made the ground breaking discovery that in the Pacific there is an intermediary layer of sea water, c. 500 metres below the sea surface, which is practically without any oxygen. A few years later the *Dana* would return to the Pacific and make the discovery that this phenomenon occurred in huge sections of the Pacific Ocean.[29] Johannes Schmidt and his collaborators on board the R/V *Dana* in 1922 became the first researchers to document this phenomenon,[30] which is still referred to in contemporary marine science.[31]

Ostenfeld further noted that there was a high density of plankton. This suggested to the expedition members that in that particular area of the sea there were shifts in temperature that promoted a vertical upward drift of water. Such upward drifts are often accompanied by growth in the primary production.[32] Today this is described as an upwelling system.

The 10,000-metre steel wire allowed the expedition members to collect biological samples from considerable depths. Thus in the Pacific Ocean, the pelagic trawls were sampling at a depth of down to 4,500 metres. Ostenfeld summarized the results of these trawls by stating that the species richness seemed to be much greater in the Pacific than in the Caribbean and in the open seas of the North Atlantic. Moreover, diversity and numbers of animals found in the Pacific were also much greater than in the North Atlantic.[33]

According to Ostenfeld, the Pacific sampling was completed without any accidents. He did note, however, that rather surprisingly the expedition encountered a number of poisonous sea snakes while fishing near shore, but only one snake was actually caught by the pelagic nets.[34]

On 20 January the R/V *Dana* went back into the Panama Canal, and the ship anchored in Lake Gatun for a couple of days. Most of the mem-

27 KB. Johannes Schmidt 'Papirer'. Ut.374.V.3. Kommandør Drechsels Papirer fra DANA-Ekspeditionen 1920. Travel account from C. H. Ostenfeld to DANA Committee 5 April 1922.

28 KB. Johannes Schmidt 'Papirer'. Ut.374.V.3. Kommandør Drechsels Papirer fra DANA-Ekspeditionen 1920. Travel account from C. H. Ostenfeld to DANA Committee 5 April 1922.29

30 Jespersen P, Vedel Taaning Å. Den store atlantiske ekspedition med S/S 'Dana' (II) 1921–22. In: Winge Ø, Vedel Taaning Å, eds. *Naturforskeren Johannes Schmidt. Hans liv og ekspeditioner*. Copenhagen: Gyldendalske Boghandel Nordisk Forlag; 1947: 93.

31 Schmidt, J. 'On the contents of Oxygen in the ocean on both sides of Panama', *Science*, 1925; 61, 1588, 592–593.

32 Thamdrup, B., Bristow, L. A., Canfield, D. E., Dalsgaard, T., Brabandere, L. d., Revsbech, N. P., 'Om iltsvind, bakterier, og det marine kvælstofkredsløb', paper presented at *17. Dansk Havforskermøde*, Roskilde Universitet, 2013.

33 KB. Johannes Schmidt 'Papirer'. Ut.374.V.3. Kommandør Drechsels Papirer fra DANA-Ekspeditionen 1920. Travel account from C. H. Ostenfeld to DANA Committee 5 April 1922.

34 KB. Johannes Schmidt 'Papirer'. Ut.374.V.3. Kommandør Drechsels Papirer fra DANA-Ekspeditionen 1920. Travel account from C. H. Ostenfeld to DANA Committee 5 April 1922.

35 Jespersen P, Vedel Taaning Å. Den store atlantiske ekspedition med S/S 'Dana' (II) 1921–22. In: Winge Ø, Vedel Taaning Å, eds. *Naturforskeren Johannes Schmidt. Hans liv og ekspeditioner*. Copenhagen: Gyldendalske Boghandel Nordisk Forlag; 1947: 93.

36 Ostenfeld CH, Nygaard G. On the phytoplankton of the Gatun Lake, Panama Canal. *Dansk Botanisk Arkiv*. 1925;4:1–16.

37 Coesel PFM, Alfinito S. *Staurastrum multinodulosum* (desmidiaceae), a little-known plankter from a crater-lake in Lazio (Italy). *Algological Studies*. 2006;122(1):1–6.

bers of the expedition saw this as a welcome break from the intensive work of taking daily stations, and using the ship's dinghies they explored the vast artificial lake, where bare trunks of trees and bushes were still sticking out of the waters as relics of the former hillside vegetation that covered the area before the valley was flooded from 1907–13.[35]

Even during this break Ostenfeld was busy taking samples of the phytoplankton in the lake. At the time this was the world's largest artificial lake, and it was fed entirely with water from the surrounding rain forest. The water was completely fresh with no chlorine salt in it. Hitherto some plankton studies had been published based on some of the major freshwater lakes of Africa and Europe, but this was only the study of a freshwater lake in Latin America. Ostenfeld found that the order of green algae known as desmids were particularly prominent in the lake, the *Staurastrum* type especially.[36] This is another result of this expedition, which is still cited in the academic literature on the phytoplankton in lake systems.[37]

11.6 Schmidt in front of a ringnet in 1922

The smile on Johannes Schmidt's face is no co-incidence. This picture was taken on the last day in The Pacific Ocean. The hydrographical measurements showed a markedly different marine ecosystem in the Gulf of Panama on the Pacific side of the canal than what they had encountered on the Caribbean side. Species richness was greater on the Pacific side, and c. 500 metres below the sea surface they discovered a huge anoxic layer of seawater (Carlsberg Archives).

Back on the Atlantic side of the Panama Canal, Johannes Schmidt must have been proud. His and Ostenfeld's stubborn instance on passing through the Panama Canal into the Pacific Ocean had been justified. It was high noon for deciding whether or not to include a Pacific leg into the *Dana II*-expedition, when Ostenfeld secured the 10,000 Kroner donation from the Carlsberg Foundation. At yet another critical moment in the history of Johannes Schmidt's career as an expedition leader, he had been able to pull the right strings in his vast network. In this case it was the close connection to the well-placed Carl Ostenfeld, which enabled him to maximize the reach of the *Dana II*-expedition.

What the two friends and colleagues were not able to realize of course, was that their results from probing the Pacific in 1922 would still be cited almost a century later. There is little doubt, however, that for Schmidt and Ostenfeld, their week-long deep water explorations in the Pacific Ocean prompted them to consider returning to the World's largest ocean on a future expedition.

Drechsel's disappointment

Meanwhile back in Denmark, Drechsel seems not to have been entirely at peace with the recent events, during which his ability to control the expedition from Copenhagen had been challenged by Schmidt and Ostenfeld.

Drechsel was getting worried that he did not receive as much information from Schmidt as he had hoped. In mid-December Drechsel sent a letter to Schmidt in which he stated that he hoped to soon receive an explanation as to why the charted course across the Atlantic had deviated so substantially from the plan in the instructions. Drechsel had received only a short message describing how health concerns over landing in the state of Para in northern Brazil, accompanied by a depressed mental state among the crew had prompted Schmidt to change the course and leave the South American shoreline earlier than planned. Drechsel let him know that he could not report this to the board just now, especially as Schmidt was indicating that the focal area of the investigations was now the waters around the West Indies, somewhat away from the plan.[38]

38 KB. Johannes Schmidt 'Papirer'. Ut.374.I.3. C. F. Drechsel to J. Schmidt, 17 December 1921.

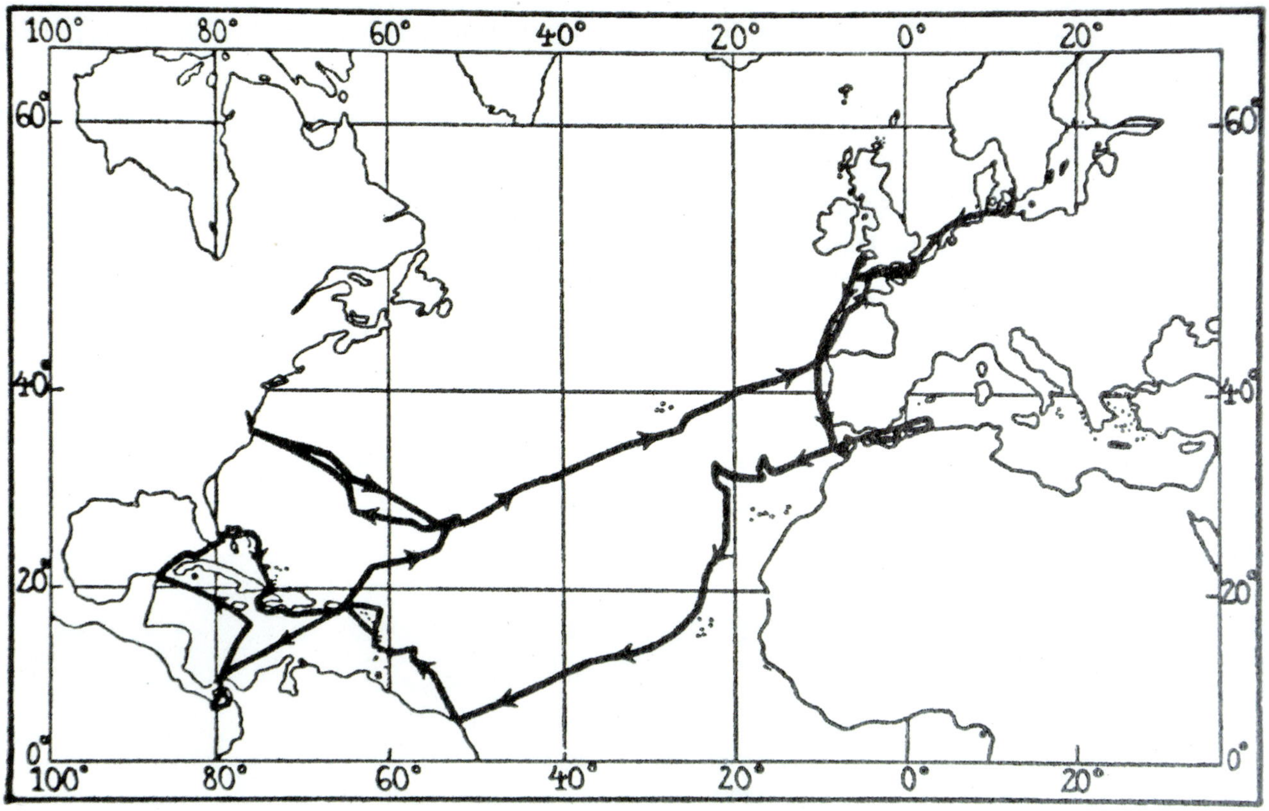

11.7 The Great Atlantic Expedition with R/V *Dana*, 1921–1922
The route of the great Atlantic expedition with the R/V Dana. An 11 months long cruise extended past the Panama Canal in to the Pacific Ocean (Winge Ø. & Vedel Taaning Å. 1947).

Still impatient in a letter from January 1922, Drechsel informed Schmidt that he had hoped to receive some more information on possible results from the first parts of the expedition, including the first Atlantic crossing. 'To me, it is a disappointment that we have gone so far away from the program, (...) but from a distance I am incapable of making any judgement about it.'[39] Drechsel was clearly frustrated that he had no news to tell: 'You must have something to report, which I can present the board – I almost don't care if it is a lie or the truth.'[40]

Becoming ever more impatient, Drechsel sent Schmidt another formal letter in February with a rare complaint about his conduct in relation to a particular publication. Clearly offended, Drechsel proclaimed his astonishment that although he had been sent no report what so ever about the results obtained so far during the expedition, it had now been brought to his attention that an article had recently appeared in the English science journal, *Nature*, where Schmidt as the author informed the

39 KB. Johannes Schmidt 'Papirer'. Ut.374.1.3. C. F. Drechsel to J. Schmidt, 5 January 1922.

40 KB. Johannes Schmidt 'Papirer'. Ut.374.1.3. C. F. Drechsel to J. Schmidt, 5 January 1922.

readers about research results obtained in the Strait of Gibraltar as part of the expedition.[41]

This prompted Drechsel to inform Schmidt that under no circumstances did he possess any right to publish results from the expedition in this way. Drechsel then referred to an enclosed copy of the protocol of the Danish commission for marine science, where it was stated in a meeting in 1912 that according to the opinion of the chairman, who was none other than Drechsel himself, the results obtained by the works of the commission should be published first in the publication series of the expedition, and further that results should not be published anywhere else without the consent of the members of the commission.

41 KB. Johannes Schmidt 'Papirer'. Ut.374.1.3. C. F. Drechsel to J. Schmidt, 20 February 1922.

11.8 R/V *Dana* in dry-dock

R/V Dana in dry dock for maintenance. During the winter months 1921–22, the port of Charlotte Amalie in the island of St. Thomas was the main centre for the expedition (Carlsberg Archives).

No doubt taunting Schmidt with an echo of his own words, Drechsel emphasized that it was particularly problematic that Schmidt had published his results in a foreign journal first, 'considering the national character of the expedition, which is not changed by the fact that the works are conducted following an internationally approved program, and partly for funds directed at the international marine science investigations.'[42] In a letter to Ostenfeld, Schmidt declared that under no circumstances would he accept that his correspondence with *Nature* should be communicated via Drechsel: 'I have dealt with this magazine many times over the years, and at my age, it would be really awkward to have someone else act as my butler in such a matter'.[43]

From the rather extensive correspondence preserved between Drechsel and Schmidt, it does not appear that Schmidt ever gave a direct answer to Drechsel's complaint, but in the following months it appears that Schmidt did take greater care when writing letters back to Copenhagen. In any case, Schmidt's letters informed about the successes from the brief visit to the Pacific. However, the time lag from when a letter was sent from for instance Colon in Panama until it reached Copenhagen was probably significant and unreliable.[44]

Homecoming

Towards the end of May 1922, all of the abovementioned groundbreaking results had been hauled on deck in the Caribbean, the Gulf of Panama, and not least in the Sargasso Sea. Then on the 30 May, the expedition left Bermuda to work its way back to Copenhagen, although with a few stops along the way for the sake of coaling and diplomacy. The longest leg of the expedition was the stretch from Bermuda to the Azores Islands in the Mid-Atlantic. Schmidt reported back to Drechsel that the eel larvae now being caught off Bermuda had the expected length according to their age and position of the station. The Azores seems to have been a welcome break from the daily routines onboard the steamer, and Schmidt and the other scientists were treated to a natural history excursion hosted by the local marine station, which incidentally had visitors from Norway and Belgium at the same time, as the marine scientists Professor Helland-Hansen and Professor Damas were themselves on an expedition on a motorship, M/S *Armauer-Hansen*. By this point, 10 months into the expedition, the crew of the *Dana* seems to

42 KB. Johannes Schmidt 'Papirer'. Ut.374.1.3. C. F. Drechsel to J. Schmidt, 20 February 1922.

43 KB. Johannes Schmidt 'Papirer'. Ut.374.1.2. J. Schmidt to C. H. Ostenfeld, 20 March 1922.

44 KB. Johannes Schmidt 'Papirer'. Ut.374.1.3. J. Schmidt to C. F. Drechsel, 2 January 1922; KB. Johannes Schmidt 'Papirer'. Ut.374.1.3. J. Schmidt to C. F. Drechsel, 20 January 1922; KB. Johannes Schmidt 'Papirer'. Ut.374.1.3. J. Schmidt to C. F. Drechsel, 5 February 1922.

45 KB. Johannes Schmidt 'Papirer'. Ut.374.1.3. J. Schmidt to C. F. Drechsel, 16 June 1922.

46 KB. Johannes Schmidt 'Papirer'. Ut.374.1.2. J. Schmidt to H. Boesgård, 26 May 1922; KB. Johannes Schmidt 'Papirer'. Ut.374.1.2. H. Boesgård to J. Schmidt, 12 June 1922.

47 Jespersen, P, Vedel Taaning, Å. Den store atlantiske ekspedition med S/S 'Dana' (II) 1921–22. In: Winge Ø, Vedel Taaning Å, eds. *Naturforskeren Johannes Schmidt. Hans liv og ekspeditioner*. Copenhagen: Gyldendalske Boghandel Nordisk Forlag; 1947: 104.

48 Pietsch TW. *Oceanic anglerfishes. extraordinary diversity in the deep sea*. Berkeley & Los Angeles: University of California Press; 2009, p. 11.

have become restless and Schmidt made references to the serious troubles he was having with the crew. From Schmidt's writings it seems that the crew was prone to drunkenness, and some had been arrested ashore on the Azores. 'They are becoming more and more difficult to control, as the end of the expedition comes closer,' Schmidt concluded.[45]

Nonetheless, the expedition was indeed coming towards an end, and this entailed a number of celebrations. Following a series of fishing stations in the Bay of Biscay the R/V *Dana* arrived in Plymouth to be received by representatives from the Plymouth Marine Laboratory. Then they crossed The Channel to an official welcome by the French city of Le Havre. Schmidt made sure that the Danish news reporter, Hans Boesgaard from the *Berlingske Tidende*, covered the reception in Le Havre. Boesgaard, followed the *Dana* on the last stretch back to Copenhagen, and placed the story about the French reception prominently in his newspaper.[46] Then on the 11 July 1922, the R/V *Dana* returned to Copenhagen following its ten and a half month long expedition.[47]

Deep sea anglerfish and spirula

As mentioned previously, one of the great technological novelties on board the R/V *Dana* was the 10,000-metre long galvanized steel wire. This enabled the *Dana* expedition to fish with nets further down the water column than anyone had ever done before in the off-shore waters of the temperate and sub-tropical parts of the Atlantic Ocean. The fisheries for eel for instance took place at night, when conditions were more promising than in daylight, but this was irrelevant when investigating depths of 6,000 metre or more. So far below there is no daylight and the animals there live in complete darkness. The temperatures are near freezing this far down, the biomass is low, and the pressure enormous from the kilometres of water weighing down on the deepest parts of the ocean.[48] Out of this darkness however, the *Dana* explorers were able to pull a number of fish, which had never been scientifically described before, perhaps most notably the strange ceratioids, commonly known as anglerfish.

Very little was known about these deep-sea creatures before the *Dana* expedition. The trip to the Gulf of Panama was particularly profitable when it came to finding specimen of anglerfish, an unexpected success of charting the depths of this otherwise completely uncharted

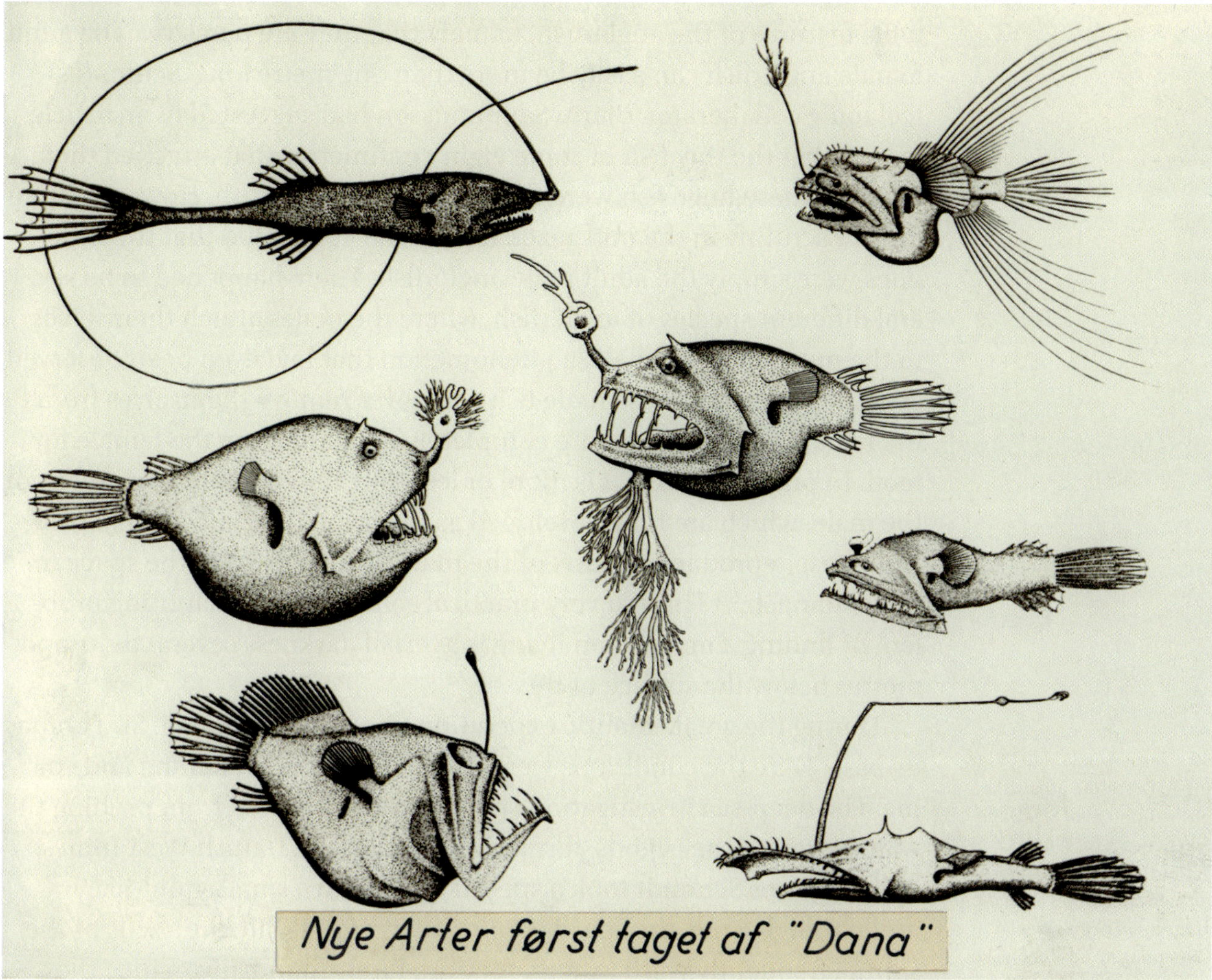

11.9 Anglerfish from the great Atlantic expedition
The great Atlantic expedition, 1921–22 with the R/V Dana brought home 26 new species of anglerfish. These creatures live their whole life in complete darkness using the phosphorescent extension of the forehead to attract food items (Carlsberg Archives).

section of the Pacific Ocean. After his return to Europe, Schmidt convinced Charles Tate Regan, director of the Zoological Department of the British Museum to work up the characteristics of the specimen taken by the *Dana*. The material then enabled Tate Regan to describe nine new groups of anglerfish, adding to the sixteen different groups already known from before the *Dana* expedition. The great Atlantic expedition with the R/V *Dana* brought home 26 new species of anglerfish. Prior to the return of this expedition in 1922 only 25 species were known.[49]

49 Jespersen P, Vedel Taaning Å. Den store atlantiske ekspedition med S/S 'Dana' (II) 1921–22. In: Winge Ø, Vedel Taaning Å, eds. *Naturforskeren Johannes Schmidt. Hans liv og ekspeditioner*. Copenhagen: Gyldendalske Boghandel Nordisk Forlag; 1947: 95.

Tate Regan was also able to draw conclusions about one of the most notable features of the anglerfish, namely that they are parasitic. The adult female anglerfish can easily be more than one metre long. Schmidt's Icelandic collaborator, Bjarni Sæmundsson had suggested in an article in 1922 that the tiny fish of some eight centimetres that attached themselves to these huge fish were the offspring of anglerfish. However, upon further scrutiny in the mid-1920s Tate Regan concluded that the little ones were simply the adult male anglerfish. There happened to be several different species of anglerfish, where the males attach themselves to the much bigger female, a phenomenon that had even been observed at the larval stage. These male fish can never remove themselves from the female, and are therefore completely dependent on the female for food. In practical terms it is more or less only the reproductive organs of the male, which are fully developed as they would be in a normal fish. And these reproductive parts of the male take up most of the space in their stomach.[50] This is a very practical solution to the daunting problem of finding a mate when living in eternal darkness several thousand metres below the surface of the sea.[51]

During the great Atlantic expedition, the R/V *Dana* made St. Thomas its base. Here the conditions were favourable for provisioning and coaling. The deep-sea investigations however, also profited from probing the waters around the islands, formerly known as the Danish West Indies, and Johannes Schmidt took a special interest in a small squid-like mollusc of 7–8 cm, the spirula (*Spirula spirula*). While the shells of the spirula frequently wash ashore, they are rarely found alive, since they dwell in the deep sea. Before the *Dana* investigations very few specimen were known, but between 1921–22 close to a hundred were caught in the pelagic trawls at depths between 2–300 metres down to 2,000 metres. The spirula has no close relatives among other living animals, but it bears some resemblance to extinct molluscs known from fossil deposits of the Jurassic and Cretaceous Periods, some 201–66 million years ago. Schmidt placed the live spirula in an aquarium on-board the R/V *Dana* and made the first ever observations of the movements of this creature, which turned out to be surprisingly mobile. The spirula can regulate its buoyancy through an air-filled chamber in its inner shell. Furthermore, Schmidt recorded that the spirula is capable of moving in any direction through the ejection of water.[52]

50 Jespersen P, Vedel Taaning Å. Den store atlantiske ekspedition med S/S 'Dana' (II) 1921–22. In: Winge Ø, Vedel Taaning Å, eds. *Naturforskeren Johannes Schmidt. Hans liv og ekspeditioner*. Copenhagen: Gyldendalske Boghandel Nordisk Forlag; 1947: 96–97.

51 Pietsch TW. *Oceanic anglerfishes. extraordinary diversity in the deep sea*. Berkeley & Los Angeles: University of California Press; 2009, p. 277.

52 Jespersen, P, Vedel Taaning, Å. Den store atlantiske ekspedition med S/S 'Dana' (II) 1921–22. In: Winge Ø, Vedel Taaning Å, eds. *Naturforskeren Johannes Schmidt. Hans liv og ekspeditioner.* Copenhagen: Gyldendalske Boghandel Nordisk Forlag; 1947: 91–92; Schmidt, J. Live Specimens of Spirula, *Nature* 110, 788–790 (9 December 1922).

11.10 Anglerfish with males attached to the female
Some of the angler-fish showed parasitic behaviour. The British zoologist Tate Regan was able to demonstrate that the smaller ones that attached to the large ones were actually life partners. The small males attach themselves to the female becoming entirely dependent on the female (Carlsberg Archives).

Total expenditure

When the expedition returned to Denmark in the summer of 1922 the expedition committee began to take stock of events at a meeting in August. The minutes reveal that Drechsel 'made some remarks' about the route chosen by the vessel, but over all, the committee seems to have been pleased with the outcome. Schmidt gave a presentation of the scientific outcome, while Drechsel provided an account of the total expenses from all three *Dana* expeditions. From the point of view of fundraising, more money had been raised for Danish marine science than ever before. For the first cruise with the M/S *Dana,* the expedition spent c. 65,000 Kroner, while the actual cost covered by the East Asiatic Company through the loan of the vessel, was not accounted for. For the second trip with M/S *Dana* in 1921 as well as the large Atlantic expedition with the R/V *Dana* a total of 325,000 Kroner had been spent. Thus, the running costs of sending out the three expeditions amounted to 390,000 Kroner. This was significantly more than the 265,000, which Drechsel and Schmidt had been trying to raise funds for during the busy autumn of 1916, when the fiscal foundation for the *Dana* expeditions had been laid. In addition to the private funds, the Danish commission for oceanography had spent around 86,000 on the deep-sea expeditions in 1921–22, and by August 1922, a further 14,500 Kroner had been spent from the commission's money. The acquisition of the R/V *Dana* was estimated at 750,000 Kroner, some 150,000 Kroner more than what had first been granted by the government. The grand total expenditure in August 1922 amounted to approximately 1,140,000 Kroner for the three expeditions. Further expenses were still expected however, as the sorting of samples would take another year and add 6–11,000 Kroner to the total.[53]

The costs of the R/V *Dana* would be written off in coming years as the vessel served the needs of marine science in various ways, but it is striking that Drechsel listed the acquisition of the new *Dana* in direct relation to the Atlantic expeditions. No doubt, Drechsel and his fellow explorers saw an intimate connection between the spectacular prospects of Schmidt's expedition plans and the purchase of a replacement vessel for the old *S/S Thor*, where one factor functioned as leverage for the other.

To the total expenses of the expedition however, should be added the amount of money granted by the Carlsberg Foundation for the *Dana* expedition. There were the obvious donations including the money provided to cross the Panama Canal and try out the Pacific, however the Carlsberg also provided other funding for follow-on projects that took place throughout the 1920s. As the expeditions had brought home a whole string of new species and large collections in general, there was

53 KB. Johannes Schmidt 'Papirer'. Ut.374.I.3. '1921–1924.' Minutes 11th meeting of the Dana Expedition Committee, 26 August 1922.

considerable interest in working on these collections. When Schmidt sent specimens abroad it is hard to track how much funding was associated with preparing this material. In the case of the anglerfish discovered in the deep sea, Tate Regan from the British Museum worked on it for many years thereafter.

When someone in Denmark needed research funding for something out of the ordinary, it was common that they turn to the Carlsberg Foundation. On at least four occasions for instance, the zoologist P. L. Kramp from the Zoological Museum was successful in applying for a total of 5,800 Kroner from the Carlsberg Foundation to work on dissecting the jellyfish collected by the Atlantic *Dana* expeditions.[54] Aage Vedel Taaning, who also took part in the large Atlantic deep-sea expedition, was granted a total of 17,500 Kroner to investigate the deep-sea fishes from the expeditions, whereas Schmidt himself was granted 1,200 Kroner in 1927 to investigate the findings on pelagic fish.[55] J. P. Jacobsen of the Pharmaceutical University was successful in 1925, when he applied for the sum of 2,000 Kroner to conduct investigations into the hydrographical material from the Atlantic.[56] In total, The Carlsberg Foundation spent 36,500 Kroner on the various expenses associated with the three Atlantic expeditions.

These are the expenses directly concerned with and targeting the expeditions, but there were also indirect subsidies associated with the fact that by the early 1920s, Johannes Schmidt had geared most of his laboratory at Carlsberg into serving the interest of marine science. The samples from the expeditions were piling up in the Carlsberg Laboratory, where Schmidt's assistant Wilhelm Ege was in charge of most of the actual dissection of the eel larvae and other marine species brought back from the deep-sea expeditions.

The ageing of the freshwater eel in open sea

While the expedition led to ground breaking results within hydrography as well as the discovery of new species from the deep sea, Schmidt's main interest was still the freshwater eel, and now was the time to finally set the record straight. In the spring of 1922, Schmidt had been writing up his most significant research paper on the eel, between taking stations on board the R/V *Dana*. Already before Schmidt embarked on the great Atlantic expedition of 1921–22, he had more or less worked

54 Spang-Hanssen E. *Carlsbergfondets Understøttelser 1876–1936*. Copenhagen: Bianco Lunos Bogtrykkeri A/S; 1937. p. 103.

55 Spang-Hanssen E. *Carlsbergfondets Understøttelser 1876–1936*. Copenhagen: Bianco Lunos Bogtrykkeri A/S; 1937. p. 141, 216 & 219.

56 Spang-Hanssen E. *Carlsbergfondets Understøttelser 1876–1936*. Copenhagen: Bianco Lunos Bogtrykkeri A/S; 1937. p. 94.

up his grand theory of how the eel breeds in the Sargasso Sea and later migrates to North America and Europe. The voyage to Europe was the most spectacular in that it took three years before the eel larvae had transmorphed into a glass eel ready to enter the freshwater systems in Europe as a yellow eel, then later again turn into the silvery glinsing eel, which eventually left European waters once and for all to migrate back to breed in the Sargasso Sea, where the whole process would start again. While Schmidt had been convinced of the validity of his own theory for at least a decade by the early 1920s, he still had to launch his theory, internationally, so that it could then be accepted beyond reasonable doubt. To this end he needed a high profile publication in a main language like English or German. Although, Schmidt also got the article published in Danish by the Commission for Marine Science.

There was one journal he was particularly interested in, which was the prestigious journal published under the auspice of the Royal Society in London, namely the *Philosophical Transactions of the Royal Society of London. Series B, Containing Papers of a Biological Character*. In 1911 Schmidt had been turned down by this journal, when he had first proposed his theory that the eel was breeding far away from the European Continent. Eventually, Schmidt published the same material in German in *Internationale Revue der gesamten Hydrobiologie & Hydrographie*, but he still bore a grudge against the more prestigious British journal.

Schmidt was anxious to have Tate Regan, whom he trusted to be on his side, help him publish a significant paper on the life history of the eel, and in a letter to Tate Regan at the British Museum, Schmidt revealed some of the frustration he had felt in 1911: 'Discussing where the paper ought to be published, we found it natural to ask the R. S. (Royal Society) of London to do it because the paper by Grassi on the same subject was published there.'[57] When Schmidt's submission failed to elicit a response after a couple of months, he approached one of the editors, Professor McIntosh, who informed Schmidt that while he did not know exactly what had transpired there was definitely some opposition to Schmidt's paper. Therefore, as Schmidt explained it to Tate Regan, he was advised by McIntosh to withdraw the contribution so as not to risk the embarrassment of an outright rejection from the journal. 'This was my first – and last – experience of that kind and the whole matter made a somewhat painful impression on me. I was much surprised, indeed, and found it strange that the Royal Society could accept Grassi's paper and award him the Darwin Medal, while my report which

57 KB. Johannes Schmidt 'Papirer'. Ut.374.VI.2. J. Schmidt to G. Tate Regan, 30 June 1922.

set forth a number of facts which – although not agreeing with Grassi's conclusions – were arrived at through several years' systematical work received short shrift. Now, nearly ten years have elapsed since that affair and continual researches have accumulated a much larger material of observations which show that I was in the right. I should therefore consider the publication of my present paper by the R. S. of London a kind of satisfaction.'

Within the next couple of years Schmidt indeed had reason to be satisfied. In a string of high profile publications from 1922 and 1923, he published his grand theory that the Sargasso Sea was the breeding place for all the freshwater eels in the Atlantic Ocean.[58] The full argument was laid out over 30 pages in the *Transactions of the Royal Society*, where Schmidt meticulously listed the research conducted over the previous decades from the publication of Grassi's paper in 1896 to his own 'accidental' catch of an eel larvae in 1904 with a Petersen young fish trawl, and the subsequent upsurge in interest, not least from the Danish parts of the international marine science community, to track the life history of the eel. Schmidt charted the average size of larvae found at different latitudes in the Atlantic Ocean at a given time of the year. In this way he could use Petersen's method, which used clusters of specific average sizes to indicate the different year classes. Hereby, Schmidt could draw a rough sketch of circles showing the distribution of the year-classes 0, 1, and 2. Two sets of circles emerged, one flowing towards North America, the other towards Europe. However, the crucial nexus of both circles was the Sargasso Sea.[59]

Film, fame, and medals

When the International Council convened in Copenhagen in July of 1921, the organization was still trying to get back on its feet following the unfortunate years during World War One. One of the decisions during the 1921 meeting however was to recommend that Johannes Schmidt assumed leadership over international investigations of the common eel, with the intention of completing the work begun before the war, when Schmidt first headed up the ICES eel committee.[60] The hard work involved in executing the first part of the Atlantic expedition and planning the second larger expedition was paying off in terms of influence and honours. The abovementioned publications in *Nature* and

58 Schmidt JS. Die laichplätze des Flußaals. *Internationale Revue der gesamten Hydrobiologie und Hydrographie*. 1923;11(1-2):1–40. Schmidt J. The breeding places of the eel. *Philosophical Transactions of the Royal Society of London.Series B, Containing Papers of a Biological Character.* 1923;211:179–208; Schmidt J. Breeding places and migrations of the eel. *Nature*. 1923;111:51–54.

59 Schmidt J. The breeding places of the eel. *Philosophical Transactions of the Royal Society of London.Series B, Containing Papers of a Biological Character.* 1923; 211: 179–208.

60 Went AEJ. *Seventy years agrowing. A history of the international council for the exploration of the sea, 1902–1972*. Copenhagen: International Council for the Exploration of the Sea; 1972. p. 58–59.

in the *Transactions of the Royal Society* gave Schmidt the best possible scientific credentials, but Schmidt also wanted his ideas to reach a more general public so he wrote newspaper articles, and from the 1921–22 expedition onwards, he also was involved in making motion pictures relating to his work.

In December 1922, Schmidt approached the film distribution company, Overseas Film Trading Company Inc., with a proposal to show a 30 minutes long film from his recent Atlantic expedition, which he wished to market for a larger audience. The fate of going commercial with this film is unknown, but Schmidt did show his film at a lecture he gave at the Royal Geographical Society in January 1923, and later it was his intention to show the film in France as part of a lecture series he was giving to learned societies in France.[61] For the latter occasion, Charles Rabot had helped him translate the accompanying texts, which would have been crucial for any commentary to accompany a silent movie.[62]

There are seemingly no film rolls preserved from Schmidt's great Atlantic expedition with the R/V *Dana*, and it looks as if neither the commission nor the expedition committee discussed film projects with Schmidt. There is no doubt, however, that Schmidt made an effort to film his voyage. A letter to the Danish Consular General Laub in St Thomas discusses some filming that took place in The Sound, a few nautical miles from Copenhagen on a chilly day in October. All the crew members from the expedition were assembled on the R/V *Dana*, where they were freezing in the light tropical clothing that they wore to suggest that they were hauling-in nets from the abyss in the West Indies, not the rather shallow strait between Denmark and Sweden.[63]

Schmidt's archives do not reveal where he got the idea to film his deep-sea investigations on board the R/V *Dana*, but there are at least two prime candidates for his inspiration. An obvious candidate was his own brother-in-law, Kay van der Aa Kühle (1877–1925), the owner of a cinema and the director of several thrillers such as 'The daughter of the Night' from 1915.[64] Another source of inspiration would have been the Danish Arctic explorer and collector of Inuit ethnographic materials, Knud Rasmussen (1879–1933). Rasmussen was a pioneer in the use of cinematic cameras to document his expeditions. His films coupled with a number of bestselling travel accounts from Arctic voyages made Knud Rasmussen an Arctic hero and much more famous in Denmark than Johannes Schmidt.[65]

61 KB. Johannes Schmidt 'Papirer'. Ut.374.1.4. J. Schmidt to S. Madsen, 6 December 1922; KB. Johannes Schmidt 'Papirer'. Ut.374.1.4. S. Madsen to J. Schmidt, 8 December 1922.

62 KB. Johannes Schmidt 'Papirer'. Ut.374.1.4. J. Schmidt to C. Rabot, 22 November 1922.

63 KB. Johannes Schmidt 'Papirer'. Ut.374.1.4. J. Schmidt to Captain Laub, 15 November 1922.

64 http://www.dfi.dk/faktaomfilm/person/en/119178.aspx?id=119178

65 Barfoed, Niels, *Manden bag helten*, Copenhagen: Gyldendal, 2011.

Schmidt's rising fame even made it to the barber shop reading lists, as 'the adventure about the eel and the deeds of *Dana*'s men on an 11 month long cruise' was portrayed over several pages in the Danish men's magazine, *Vore Herrer* in 1922.[66] Then in the spring of 1923, Schmidt approached the manufacturer of porcelain, Den Kongelige Porcelænsfabrik to have them produce a memorial plate in porcelain commemorating the Atlantic expedition.[67] The plate later became one of several popular motives in Danish households.[68]

In any case, Schmidt soon had the chance to present his film at one of the most prestigious venues, namely the 1923 conference of the British Association for the Advancement of Science held in Liverpool. According to the conference programme Schmidt presented 'the '*Dana*' Expeditions and their Work on the Life-history of the Eel, (lecture with films).' No one else at this conference seems to have conveyed their message through the use of motion pictures, thus Schmidt was quite ahead of his time. As a further testimony Schmidt was honoured at the conference, where the University of Liverpool made him an Honourable Doctor.[69] The same year, 1923, Schmidt also received the so-called Weldon Memorial Prize from the University of Oxford.[70] This bronze medal however, was presented to him in recognition of his efforts in bio statistics, that is his research into genetics, discussed in chapter 5.

In the autumn of 1923 Schmidt received notice that he had been awarded the official French decoration of *Officer in the Legion D'Honneur*.[71] In a letter to his friend, the French geographer and Arctic explorer, Charles Rabot, Schmidt gave an account of how he looked upon the institution of medals and decorations. Just a few days before Schmidt had heard a lecture by the Danish physicist Niels Bohr at the Danish Royal Academy of Science and Letters. Schmidt wrote that Bohr '...was a young man, but most likely the one physicist in the World who had the brightest future. Just a few days before, he (Bohr) had received the Nobel Prize, for which he was an obvious candidate.'[72] Schmidt then continued that his own wife Ingeborg was regretful that there was no Nobel Prize for biologists and marine scientists, as she thought Schmidt deserved to receive such an honour too. Pondering this idea Schmidt informed Rabot, somewhat politely, that he completely disagreed with Ingeborg in this matter, and furthermore he was not a fan of the idea that science should be the subject of prizes and awards. Science should be above such phenomena, but as he told Rabot, he nonetheless realized that such awards were very popular.[73]

66 Anonymous, 'Eventyret om Aalen og DANA's Mænds Gerning paa det 11 Maaneders Togt,' in: *Vore Herrer*, 7. årgang (16), 17 August 1922.

67 KB. Johannes Schmidt 'Papirer'. Ut.374.V.2. Privat og Blandet Korr. 1912–1927. J. Schmidt to F. Dalgas, 29 March 1923.

68 I am thankful that my colleague Professor Poul Duedahl, Aalborg University remembers that this as one of several porcelain plates from the living room of his grandparents. Today the plate is an expensive collectors item.

69 KB. Johannes Schmidt 'Papirer'. Ut.374.VI.3. Preliminary Programme, BAAS Liverpool meeting, 1923.

70 KB. Johannes Schmidt 'Papirer'. Ut.374.I.4. University of Oxford to J. Schmidt, November 1923.

71 KB. Johannes Schmidt 'Papirer'. Ut.374.I.4. J. Schmidt to C. Rabot, 23 November 1922.

72 KB. Johannes Schmidt 'Papirer'. Ut.374.I.4. J. Schmidt to C. Rabot, 22 November 1922.

73 KB. Johannes Schmidt 'Papirer'. Ut.374.I.4. J. Schmidt to C. Rabot, 22 November 1922.

While we have no access to the inner thoughts of Johannes Schmidt, he does appear to have exaggerated his lack of interest in recognition from learned societies and states around the world. At any rate, he appears to have been most pleased each time his hard earned successes as a scientist were noticed by his peers as well as heads of states. Moreover, he was very keen on recommending both his own staff and his international peers for various Danish decorations. Thus in July 1922, as the great Atlantic expedition was near completion, Schmidt approached the Danish Chargé d'Affaire in Paris, Kammerherre Bernhoft with a plea to have him recommend four French scientists for the appointment as Commander of the Order of the Dannebrog. Charles Rabot was one of them.[74] This was just one of many letters, Schmidt sent to the Danish ministries. So, not only did Schmidt become a frequent recipient of medals and other decorations, he was also busy securing such anointments for his international network.[75]

Conclusion

This chapter has shown for the first time, the true cost of the Danish Atlantic expeditions from 1920–1922, as well as the sources of funding. Johannes Schmidt and Drechsel were central to this fund raising effort, and it is apparent that they had become big spenders by the early 1920s. It may be difficult to estimate the monetary value of the help they received from the East Asiatic Company which loaned them the M/S *Dana*, but if they were to rent a similar vessel themselves and pay the running costs of insurance and maintenance, not least the repair in St. Thomas following the leak, they could have easily spent another 100,000 Kroner. The total expenses therefore, amount to approximately 1,300,000 Kroner, including funds from the Danish state, private companies, and the research endowments of the Carlsberg Foundation. All the vessels used for Schmidt's expeditions were fitted to the highest standards for the era in terms of fishing gear and water samplers for hydrography, but the real technological novelty of the third and largest Atlantic expedition was the use of a 10 km long steel wire for fishing with pelagic gear that went deeper than ever before. Although the first steel wire had to be partly replaced midway through the expedition, the team of explorers were nonetheless able to reap unprecedented results using this fishing technique.

74 KB. Johannes Schmidt 'Papirer'. Ut.374.I.4. J. Schmidt to Kammerherre Bernhoft, 19 July 1922.
75 KB. Johannes Schmidt 'Papirer'. Ut.374.I.4. Various letters from 1922–1923.

The main scientific output of Schmidt's Atlantic expeditions has been known since the results were first published, mostly in the 1920s. The discovery of new species including a number of anglerfish, and the catching of many live specimens of the spirula mollusc are among the most important findings, while the surprise discovery of large anoxic areas in the Eastern Pacific Ocean has had far-ranging significance for the current understanding of the amount of biomass and nutrients in the World's oceans. Almost needless to say, Johannes Schmidt's most famous discovery, the breeding place and early life history of the Atlantic eel also came about as a result of the three Atlantic expeditions.

However, what this chapter has touched upon for the first time, is how the scientific results were heavily influenced by the political and personal motivations of Johannes Schmidt and Christian F. Drechsel in particular. Both men agreed that the Atlantic Ocean was the most important target of the three expeditions. This fitted Schmidt's interest in solving the riddles of the eel, and it fitted Drechsel's responsibility to revive ICES following the end of World War One.

Thus far, their interests had complimented each other, but the above chapter has shown how Schmidt was able to pull other strings to serve his additional interest in exploring the Pacific Ocean, a detour which was not part of the cruise plan, agreed upon less than a week before the R/V *Dana* steamed out from the Port of Copenhagen. The international council were primarily serving the interests of European fisheries, and these fisheries operated only in the Atlantic Ocean, and certainly not in the Pacific Ocean. Drechsel therefore grew suspicious that Schmidt was dodging his duties, discarding parts of the investigations he was supposed to carry out in the Atlantic Ocean. In fact he did stray off the charted course quite a lot, and eventually managed to create a successful detour out into the Pacific Ocean, thanks to the help of his friend Carl Ostenfeld, who had just taken up a seat on the Board of Directors of the Carlsberg Foundation.

This was not the circumnavigation of the World, which Schmidt had dreamed about when he and Drechsel were making expedition plans during World War One, yet his successful diversion into Pacific waters is yet another testimony to Schmidt's perseverance and his will and ability to improvise from the conditions laid out ahead of a large ocean expedition. Pride is another sign of Schmidt's character which surfaced onboard the R/V *Dana*, and in particular his strong sense of national pride, if not outright nationalism, as noted by his colleague Knud Hensch

Stephensen, when they met with Ehrenbaum in the Kiel Canal in 1921. National pride went hand-in-hand with Schmidt's personal pride. Schmidt's personal pride was at stake in the way he dealt with the publication of his eel research in the *Transactions of the Royal Society*. In this instance he made sure that the journal, which had rejected his theory on the breeding grounds of the eel back in 1911, came to admit their earlier mistake, at least indirectly, when it published his full account in 1923.

At any rate, Schmidt now started to receive all the international scholarly recognition most scientists could dream of, even if his wife Ingeborg reminded him that receiving the Nobel Prize would have been appropriate, if such an honour was to be bestowed on scientists in the life sciences. The filming of the Atlantic expeditions – carried out as a re-enactment in October 1922 in the shallow waters of The Sound – testifies that Schmidt also had the ambition to publicize his achievements to a larger audience both inside and outside the world of science.

Following the publications, presentations, and publicity surrounding the landmark discovery of the breeding place of the freshwater eel, Schmidt quickly renewed his appetite for more marine science. As shown in chapter 6 Schmidt's interest in research even remotely related to the scientific world of improving the world's beer, had subsided completely by the early 1920s. Instead, Schmidt assumed a greater responsibility within the resurgent *International Council for the Exploration of the Sea*, ICES.

12 Marking cod and territory in the North Atlantic

Following the triumphant return from the three Atlantic expeditions, Schmidt turned his focus to the North-western parts of the waning Danish empire, the waters around Greenland, Iceland, and the Faroe Islands. While the temperature arguably was cooler in the Northern North Atlantic than in Caribbean waters, the issues at stake were just as hot as in the waters around the Danish Caribbean possessions: in particular the sovereignty of Greenland. Schmidt found himself once more at the intersection of a complex situation blending marine science and politics.

Denmark and Norway were contesting which nation should gain sovereignty over Greenland. Meanwhile ICES was hoping its member states could come together to conduct a uniform and systematic international collaboration to increase knowledge on both physical and biological characteristics of the great commercial fisheries in the North Atlantic Ocean surrounding Greenland – the largest island in the World,

In this chapter we will follow how Schmidt became caught up in the diplomatic wrangling. On the one hand Schmidt was,working as an internationally minded cooperative scientist, while on the other hand he held the deep-seated belief that the small state of Denmark had a significant role to play the North Atlantic Ocean. For Christian Drechsel the situation was no less confused, as he was wearing two hats simultaneously, as the Chairman of the Danish commission and also as the Secretary General of ICES, where Norway was as much a member state as Denmark.

To investigate this, it is necessary to review the central role Schmidt and Drechsel played in the resurrection of ICES after World War One, aided in the field by the next generation of marine scientists such as Anton Bruun and Aage Vedel Taaning. Also worth mentioning is the ten year long conflict between Denmark and Norway over the issue of Greenland sovereignty, which was fought by the means of scientific exploration on land – and as we shall see, also at sea.

However, in the existing literature, nothing has been written about the role played by marine scientists working not chiefly in Greenland, but definitely in charge of exploring what was, and still is, Greenland's most obvious natural resource:, marine life. This chapter is by no means a full investigation of this topic. Such an investigation would warrant a full research project of its own. Some interesting observations arise however, from looking at the issues at stake from the viewpoint of one of the key actors in the waters surrounding Greenland, namely Johannes Schmidt. Schmidt, who in turn came to reap some of his most wide ranging research results from taking part in the international investigations in the 1920s.

The second part of this chapter therefore reviews the main scientific results of the fisheries investigations in the waters around Iceland and the Faroe Islands. From the first pelagic trawls in 1903 until the early 1930s there was a tremendous increase in the knowledge of the life history of commercial fishes in this area, and Schmidt's way of handling the political and economic situations that intermingled with his scientific reasoning, played a central part in these achievements.

Denmark in ICES

Denmark retained a strong position in ICES after the war vis-à-vis Drechsel's dual position as Secretary General of ICES and his job as Chairman of the Danish *Kommissionen for Havundersøgelser*. Drechsel remained Secretary General until his death in March of 1927 at the age of 73. From 1927, Denmark could no longer hold on to the influential position of general secretary, but the headquarters remained in Copenhagen, and the Danish government appointed none other than Johannes Schmidt as the Danish delegate to the council meetings of ICES. As a crowning achievement within ICES, Schmidt was appointed vice-president of ICES in 1932.[1]

Within the Danish commission, Drechsel was in the cockpit until he retired in 1925. A. V. Mortensen, Director of Fisheries in the Department of Agriculture took over the position. The following year Mortensen oversaw a change of name of the commission to *Kommissionen for Danmarks Fiskeri- og Havundersøgelser*. At this time two representatives of the fishing industry, fishermen Jensen and Henriksen were both made members of the commission. Apart from this, the men in charge of deposing Petersen in 1908 all held on to their seats in the commission from

1 Knudsen M. De internationale havundersøgelser. In: Winge Ø, Vedel Taaning Å, eds. *Naturforskeren Johannes Schmidt. Hans liv og ekspeditioner*. Copenhagen: Gyldendalske Boghandel Nordisk Forlag; 1947: 112.

1909 and throughout the 1920s. Martin Knudsen, Carl Ostenfeld, A. C. Johansen, and off course Johannes Schmidt were the ones in charge.[2] In the period before World War One, the International Council had been instrumental in orchestrating a concerted effort from several countries. Valuable ties had been formed between marine scientists all over Northern Europe, but the war had breached some of these links.

In 1922, Schmidt had flown ICES' flag with the R/V *Dana*. The next year he was about to launch a bigger commitment with the international council. At the annual council meeting of ICES in 1923, Schmidt was appointed chairman of a newly constructed working committee, the North-Western Area Committee, to oversee and coordinate fisheries investigations in the areas more or less covering the waters around Iceland, Greenland, and the Faroes. While partly Danish territory, these waters had an international significance for commercial fisheries since so many of the most important European fisheries took place in the North Atlantic Ocean in or adjacent to these waters. As a true veteran of the ground-breaking cruises with the *Thor* in the 1900s, Schmidt was ideally suited for this new position, and he immediately took responsibility for picking up where the investigations had been left off with the start of World War One.[3]

Schmidt's previous Working Committee on the eel was a small enterprise, largely run by him self, whereas the new Working Committee resulted in a large multinational research effort. In effect the Danish part of the study of the commercial species of the North Atlantic had been kept on the back burner for an entire decade from 1913–1923. The Great War had obviously played a part in this decision, but also the R/V *Dana's* use in the intensive search for the eel in the preceding years cannot be ignored. Between 1924–1927 and again in 1931, however, Denmark was able to send the *Dana* to the North Atlantic in support of ICES' investigations. This was as much a test of the scope of international cooperation as a research trip.

2 Knudsen M. De internationale Havundersøgelser. In: Winge Ø, Vedel Taaning Å, eds. *Naturforskeren Johannes Schmidt. Hans liv og ekspeditioner*. Copenhagen: Gyldendalske Boghandel Nordisk Forlag; 1947:110–111.

3 Knudsen M. De internationale Havundersøgelser. In: Winge Ø, Vedel Taaning Å, eds. *Naturforskeren Johannes Schmidt. Hans liv og ekspeditioner*. Copenhagen: Gyldendalske Boghandel Nordisk Forlag; 1947:110–111.

Collaboration – and 'The Norse issue'

The Danish effort from 1924 onwards was echoed by a substantial effort by several other ICES member states. From Great Britain, the Scottish research vessel, the *S/S Explorer* was sent north to Iceland in 1925 and the Faroes every other year throughout the 1920s. The German members,

4 Vedel Taaning Å. Fiskeriundersøgelser i nordlige have efter krigen. In: Winge Ø, Vedel Taaning Å, eds. *Naturforskeren Johannes Schmidt. Hans Liv og Ekspeditioner*. Copenhagen: Gyldendalske Boghandel Nordisk Forlag; 1947:113–114.

5 Vedel Taaning Å. Fiskeriundersøgelser i nordlige have efter krigen. In: Winge Ø, Vedel Taaning Å, eds. *Naturforskeren Johannes Schmidt. Hans Liv og Ekspeditioner*. Copenhagen: Gyldendalske Boghandel Nordisk Forlag; 1947:113–114.

meanwhile, worked on hydrographical measurements with the R/V *Meteor*. In 1930 for instance, the German investigations explored the Denmark Strait, the waters in between Greenland and Iceland, but the Germans kept a low profile with regards the actual fisheries investigations.[4] The French investigations were more substantial, which was also a reflection that France previously had profited from cod fisheries around Iceland, and the French section of the ICES areas to be investigated were situated just south of the Danish sections of the North Atlantic Ocean. The French investigations led by Schmidt's friend Eduard le Danois, were largely carried out with the support of French navy vessels. One expedition ended disastrously for the French though, as the M/S *Pourquoi-pas* sunk during a storm in the Faxe Bay, just west of Iceland's main city of Reykjavik.[4]

12.1 The plankton net is hauled in
The plankton net is hauled in. This was the most tightly wowen net operated from the Dana (RA. Carlsbergfondets DANA-ekspeditioner. 1917–1973. 5).

The Norwegian interests in the area were substantial, as the Norwegian fisheries' interests included the so-called bank fishery around the Faroe Islands and eastern Greenland as well as the herring fisheries just north of Iceland. This meant that the Norwegian marine scientist, Johan Hjort, was keen to explore these waters, not least with the research vessel, R/V *Michael Sars*. Thus in 1924 for instance, the *Michael Sars* travelled as far as the Davis Strait between Greenland and Canada.[6]

6 Vedel Taaning Å. Fiskeriundersøgelser i nordlige have efter krigen. In: Winge Ø, Vedel Taaning Å, eds. *Naturforskeren Johannes Schmidt. Hans Liv og Ekspeditioner*. Copenhagen: Gyldendalske Boghandel Nordisk Forlag; 1947:115.

12.2 Fisheries surveys on board the R/V *Dana*

A snapshot from the daily routine of conducting fisheries surveys. A flatfish is being measured and notes are taken (*RA. Carlsbergfondets DANA-ekspeditioner. 1917–1973. 5*).

From a political point of view it is safe to say that the Danish research effort was soon ramped up to equal its Norwegian counterpart. Setting aside the cooperative international spirit, it should not be forgotten that at the time of these research expeditions both Denmark and Norway were trying to secure their interests in Greenland. Schmidt referred to this as the 'Norse issue'.[7] In the 1920s there was more at stake than science and fisheries investigations, namely the sovereignty over Greenland.

As we learned in an earlier chapter, the political and economic rights to exploit Greenland had been part of the treaty, when the USA and Denmark agreed on the terms of sale of the Danish West Indies, which came into effect in 1917. In return, Denmark's sovereignty over Greenland was ratified, in the sense that the American government stated they would not resist the Danish government's attempt to expand its political and economic interests in Greenland. In an international context however, there was one big problem with this agreement. The treaty was bilateral.

7 CA. *Carlsbergfondet, Carlsberg Laboratorium.* E08. 1924. Letter from J. Schmidt to V. Henriques, 10 April 1924.

Thus, the other countries with North Atlantic interests, principally the UK, Norway, and Germany were not consulted. From an American point of view this made sense in light of the then dominant American foreign policy strategy, the Monroe Doctrine. According to this policy, adopted in 1823, The USA reserved the right to view any attempt from European powers to expand their interest on either of the American continents as a threat to American national security. In other words, North and South America lay within the sphere of interest of the USA, and as such they did not need to consult with for instance the Brits about it.[8]

One other reason for concern from a Danish point of view was the fact that much of Greenland was still fairly unexplored in the 1910s. The first Danish trading posts and missions were set up in Greenland in the early 1700s, yet they were all based along the coastline of the Southern third of the Greenland peninsula, and mainly on the West coast. In addition there were a few scattered and isolated Inuit settlements along the shoreline of this, the largest island on the planet Earth. So, was all this really for the Danes only, if they never used it?

The Danish government realized that jurisdiction over Greenland was precarious, but the peace negotiations following the end of World War One posed a welcome chance to frame the Danish sovereignty in an international context. The British delegation however, did not like the idea, as they were negotiating also on behalf of Canada, who might have its own future interest in exploiting Greenland territory. Nonetheless, as the Danes pushed on and secured bilateral approval from a number of other countries such as Italy, Japan, and France, not least through the tacit approval of the USA, the British finally caved in. But there was indeed one country left who refrained from approving Danish supremacy over the entire of Greenland, and that was the neighbouring state of Norway.

Throughout the 1920s Norway actively pursued a policy of securing economic interests in the Northern parts of the North Atlantic Ocean. Svalbard, for example, became Norwegian territory after World War One, but in some parts of Norway people dreamt of creating a modern North Atlantic community centred in Norway. This dream found a historical foundation in the ancient claim that the Vikings from Norway, not from Denmark, settled the islands in the North Atlantic. It was then a widely spread idea in Norwegian society that Norway had wrongly been separated from its North Atlantic possessions, when the two kingdoms were split apart following the Napoleonic Wars.[9]

8 Lidegaard, Bo, Overleveren. 1914–1945, Carsten Due-Nielsen, Ole Feldbæk & Nikolaj Petersen (eds.): Dansk Udenrigspolitiks Historie, vol. 4, Copenhagen: Gyldendal Leksikon, 2003. Pp. 117–179.

9 Lidegaard, Bo, Overleveren. 1914–1945, Carsten Due-Nielsen, Ole Feldbæk & Nikolaj Petersen (eds.): *Dansk Udenrigspolitiks Historie,* vol. 4, Copenhagen: Gyldendal Leksikon, 2003. Pp. 179–182.

One way in which Norway claimed its interests in Greenland was through scientific enterprise. Thus in 1922, a Norwegian expedition wintered in East Greenland in collaboration with the Norwegian meteorological institute, where they established a telegraph station. The Danes protested, but the Norwegians stayed there.[10]

Then in 1924, another conflict broke out, as an agreement was reached between Denmark and Norway, whereby Norwegians were granted a special right to trade, set up scientific stations, and do business in the eastern part of Greenland, but neither country was really satisfied with the terms, and both countries sought to assert the right to the eastern territories in the following years.[11] Denmark launched no less than five scientific expeditions between 1924 and 1930. They typically dealt with cartography and geological and zoological surveys. This was however, also a way of asserting one's sovereignty.[12]

Later, in the years 1931–33, the situation with Greenland became politically acute, as Norway simply decided to occupy parts of Eastern Greenland. From 1931, an area roughly the size of Iceland was occupied between the 71st and 75th Northern parallel by groups of Norwegian hunters, and in 1932, another stretch of land was seized by the Norwegians on the Southeastern most part of Greenland between the 60th and 63rd parallel. As the two countries were unable to resolve this situation bilaterally, they agreed to place the issue on trial at the Permanent Court of International Justice in The Hague. Here, the disagreements were settled finally on the 5 April 1933, when the Court ruled in favour of the Danish position. While this was seen as a slap in the face in Norway, the Norwegians nonetheless accepted the terms and withdrew from the occupied territories in East Greenland.[13] Within a few years, Denmark, Norway, and most other countries in Europe were faced with much bigger threats from Nazi Germany, and the issue of sovereignty over Greenland gradually faded from the Scandinavian political arena. In 1924 however, Norway and Greenland was the talk of the town in marine science circles in Copenhagen.

Marking cod and territory in 1924

In spring of 1924, Schmidt asked the laboratory's Board of Trustees for a leave of absence to join the Danish investigations around the Faroe Islands and Iceland in his capacity as member of the Danish commission.

10 Blom I. *Kampen om Eirik Raudes Land. Pressgruppepolitikk i grønlandsspørgsmålet 1921–1931*. Oslo: Gyldendal Norsk Forlag; 1973.
11 Lidegaard, Bo, Overleveren. 1914–1945, Carsten Due-Nielsen, Ole Feldbæk & Nikolaj Petersen (eds.): *Dansk Udenrigspolitiks Historie,* vol. 4, Copenhagen: Gyldendal Leksikon, 2003. Pp. 267–268
12 Blom I. *Kampen om Eirik Raudes Land. Pressgruppepolitikk i grønlandsspørgsmålet 1921–1931*. Oslo: Gyldendal Norsk Forlag; 1973.
13 Lidegaard, Bo, Overleveren. 1914–1945, Carsten Due-Nielsen, Ole Feldbæk & Nikolaj Petersen (eds.): *Dansk Udenrigspolitiks Historie,* vol. 4, Copenhagen: Gyldendal Leksikon, 2003. pp. 267–270.

Schmidt emphasized the importance to his chairman, Henriques, as he explained that the investigations had been cut short by the war, and as a consequence, Schmidt's new assistants had very little experience when it came to oceangoing cruises and the work associated therewith. Therefore, Schmidt wished to join them to make them accustomed to the ways of the work. Moreover, as Henriques no doubt was aware, this was a particularly important task in light of the political tensions with Norway – 'The Norse Issue.'[14] The result was a three months long leave-of-absence in the summer months of 1924.

Schmidt's preparations were comprehensive. This time, they were by no means restricted to the acquisition of provisions, fishing gear, and fine-tuned instruments, as he seems to have taken his role in Danish foreign policy very seriously. To take one example, Schmidt alerted the Danish press to the expedition.[15] Here, a brief account of the expedition plan was presented. The primary destination was the waters around Iceland and the Faroes, and the work scheduled was to be seen as a continuation of the decadal long effort from the Danish commission for marine science, which added to the internationally approved research programme under the auspices of ICES. Therefore, the work would be executed in collaboration with ships from England, Scotland, and France. The focus was to be on five commercially important species, cod, haddock, plaice, herring, and halibut, and there was also a particular more localized aim, which was to promote knowledge of relevance to the fisheries in Iceland and the Faroe Islands. Schmidt finished his press release with a plea: 'Taking into consideration the fact that this expedition will take place in an area which is undergoing national dispute, it would probably be very desirable if the Danish press took an interest in this work, which is carried out by means supplied by the Danish state.'[16] The elephant in the room – Norway, was not mentioned once. There was a reason for leaving out the Norwegians. From the beginning of 1924, the Danes had no clear idea of what the Norwegians were up to with regards to ocean exploration in the North Atlantic.

In the autumn of 1923, the International Council had met in Paris to agree on the plans for the upcoming year, and hopes were high that the international collaboration would soon be fully operational. Norway however, was not present at the meeting, and they did not feature in the official programme. In January of 1924, Drechsel sent a note to Johan Hjort in Oslo to learn more about the Norwegian interests, in particular in relation to the upcoming coordinated hydrographic research

14 CA. *Carlsbergfondet, Carlsberg Laboratorium*. E08. 1924. Letter from J. Schmidt to V. Henriques, 10 April 1924.

15 KB. Johannes Schmidt 'Papirer'. Ut.374.V.2. Privat og Blandet Korr. 1912–1927. Press release by J. Schmidt 22 April 1924.

16 KB. Johannes Schmidt 'Papirer'. Ut.374.V.2. Privat og Blandet Korr. 1912–1927. Press release by J. Schmidt 22 April 1924.

concerning the inflow of water from the North Atlantic Ocean into the North Sea and Norwegian Sea through the Faroe Canal. A coordinated Danish and Scottish effort was scheduled for 1924, and Drechsel had also learned that the Norwegian hydrographer, Bjørn Helland-Hansen was going on a similar cruise in the summer of 1924. Drechsel therefore inquired if it would be possible, at least informally to collaborate on the different hydrographical plans.[17]

Johan Hjort replied right away, in a polite but brief letter, where he aired no indications as to what his own plans were for the upcoming season in relation to the International Council's programme. Hjort wrote that he had been 'working with considerations concerning Norway's position, and would return to this matter as soon as the circumstances allowed this.'[18] Then he stressed that in the meantime he took for granted that no appointments were made between institutions or private men in Norway, which in any way could hinder his efforts. Hjort then agreed that it suited him well to have the upcoming meeting of the council bureau in London in April.[19] In other words, Hjort was evading Drechsel's questions. Probably Hjort was uncertain as to who's horse Drechsel was backing, the International Council or national Danish interests.

In any case, both Drechsel and Schmidt continued to communicate with Norwegian hydrographer, Helland-Hansen. Helland-Hansen and Schmidt re-assured each other that they would help each other as much as possible in the common interest of increasing knowledge about the currents in the North Atlantic Ocean. Schmidt and Helland-Hansen (1877–1957) were the same age, and they in particular seem to have been close to each other, disposing of their titles when addressing each other.[20] Bjørn Helland-Hansen had previously worked closely together with Johan Hjort until c. 1910, and in the early years of ICES, Helland-Hansen had contributed to statistical studies of gadoid fishes in the North Atlantic alongside Schmidt.[21] Not long after though, Helland-Hansen had established his own research independently of Hjort, which focused mainly on physical oceanography.[22] In fact, Hjort and Helland-Hansen never collaborated again on a noteworthy scale.[23]

In March, Schmidt had formulated the expedition plan for the summer cruise with the *Dana*, but from Norway there was still no news as to their plans in relation to fisheries investigations. In so far as hydrography was concerned, Schmidt had appointed the ship's medical doctor, Andersen, to be the on board expert in charge of hydrographical

17 KB. Johannes Schmidt 'Papirer'. Ut.374.1.5. C. F. Drechsel to J. Hjort, 22 January 1924.

18 KB. Johannes Schmidt 'Papirer'. Ut.374.1.5. J. Hjort to C. F. Drechsel, 29 January 1924.

19 KB. Johannes Schmidt 'Papirer'. Ut.374.1.5. J. Hjort to C. F. Drechsel, 29 January 1924.

20 KB. Johannes Schmidt 'Papirer'. Ut.374.1.5. B. Helland-Hansen to J. Schmidt, 20 January 1924; KB. Johannes Schmidt 'Papirer'. Ut.374.1.5. J. Schmidt to B. Helland-Hansen, 23 January 1924; KB. Johannes Schmidt 'Papirer'. Ut.374.1.5. J. Schmidt to B. Helland-Hansen, 25 January 1924; KB. Johannes Schmidt 'Papirer'. Ut.374.1.5. B. Helland-Hansen to J. Schmidt, 30 January 1924; KB. Johannes Schmidt 'Papirer'. Ut.374.1.5. J. Schmidt to B. Helland-Hansen, 1 February 1924.

21 Schwach v. *Til havs med vitenskapen. fiskerirettet havforskning 1860–1970*. [Dr. Philos.]. Oslo: Universitetet i Oslo; 2011. p. 199–200.

22 Schwach v. *Til havs med vitenskapen. fiskerirettet havforskning 1860–1970*. [Dr. Philos.]. Oslo: Universitetet i Oslo; 2011. p. 204–217.

23 Schwach v. *Til havs med vitenskapen. fiskerirettet havforskning 1860–1970*. [Dr. Philos.]. Oslo: Universitetet i Oslo; 2011. p. 199–200.

measurements. At this stage Schmidt also told Drechsel that he would continue his conversations on the hydrography with Helland-Hansen, as he would do with any other hydrographer.[24] Throughout 1924, Helland-Hansen and Schmidt continued to exchange information on what each other did, unabated by the political turmoil between Denmark and Norway.[25]

24 KB. Johannes Schmidt 'Papirer'. Ut.374.1.5. J. Schmidt to C. F. Drechsel, 26 March 1924.

25 KB. Johannes Schmidt 'Papirer'. Ut.374.1.5. J. Schmidt to B. Helland-Hansen, 4 April 1924; KB. Johannes Schmidt 'Papirer'. Ut.374.1.5. J. Schmidt to B. Helland-Hansen, 10 April 1924; KB. Johannes Schmidt 'Papirer'. Ut.374.1.5. J. Schmidt to B. Helland-Hansen, 29 April 1924. KB. Johannes Schmidt 'Papirer'. Ut.374.1.5. J. Schmidt to B. Helland-Hansen, 1 May 1924; KB. Johannes Schmidt 'Papirer'. Ut.374.1.5. J. Schmidt to B. Helland-Hansen, 3 September 1924.

12.3 Captain Hansen and Doctor Andersen operating a water sampler
The captain of R/V Dana, Georg Hansen is holding a water sampler, while the ship's medical doctor and scientific assistant, Dr. Andersen is scrutinizing the bottle. During 1924, when Denmark and Norway were intense political rivals, Schmidt upheld a close collaboration with Norwegian hydrographer, Bjørn Helland-Hansen, with whom the Danes exchanged information obtained from the water samplers (*RA. Carlsbergfondets DANA-ekspeditioner. 1917–1973. 5*).

With Hjort it was different. As late as 31 May, Drechsel had not received further information as to the plans and whereabouts of Johan Hjort's research vessel, the *S/S Michael Sars*. All he knew was that the R/V *Michael Sars* was scheduled to do marine science investigations around Iceland and the Faroe Islands, partly paid for by the Norwegian government, partly by a number of Norwegian whaling companies. In addition, a fishing company from the Norwegian fishing port of Aalesund was to embark on a fishing expedition around the Greenland West coast.[26]

Once again, Schmidt was trying to make the most of a political situation involving the North Atlantic. He was already trying to persuade the Danish Ministry for Agriculture to have the Danish state decorate two of his close collaborators in the North Atlantic, Bjarni Sæmundsson in Iceland and A. Øster in the Faroe Islands. Bjarni Sæmundsson in particular had been of immense importance to the Danish commission for marine science, as he was the 'eyes on the ground' so to speak, who knew about the local fisheries and fishes in his own waters, and there were good reasons to have them both knighted. In a letter to the cabinet secretary, Anthonius Krieger, who was responsible for official decorations, Schmidt aired his view that the time was absolutely right. The outgoing Minister for Agriculture, Thomas Madsen-Mygdal had not been in a mood to help Schmidt, due to a change of government, so Schmidt was trying to have Krieger arrange a meeting with King Christian X, prior to the departure of the R/V *Dana*. This was an old custom, Schmidt said.[27] Later that year, Sæmundsson did receive his knighthood, so Schmidt was successful, but what is striking here, is Schmidt's motivation for the decoration. 'Without exaggerating the significance of our upcoming expedition I dare say that it will be characterized as a manifestation of Danish interests in the areas of investigation. I would very much like to contribute to this, in particular as I have learned in a few days that the Norwegians – as a demonstration – will launch their own official marine science expeditions to the area, where we – according to tradition as well as following the authorization of the Paris meeting of the International Council for the Exploration of the Seas – are in charge.'[28] As a way of stepping up to the Norwegians, Schmidt told Krieger that he intended to give a number of lectures in Iceland and in the Faroes to highlight the fact that the Danish marine science contingent had been present in the North Atlantic for many years, before the Norwegians decided to intrude on these territories, and to show that Danish marine scientists intended to continue their work in the North Atlantic Ocean. The decorations

26 KB. Johannes Schmidt 'Papirer'. Ut.374.I.5. C. F. Drechsel to J. Schmidt, 31 May 1924.

27 KB. Johannes Schmidt 'Papirer'. Ut.374.V.2. Privat og Blandet Korr. 1912–1927. J. Schmidt to Kammerherre A. Krieger, 18 April 1924.

28 KB. Johannes Schmidt 'Papirer'. Ut.374.V.2. Privat og Blandet Korr. 1912–1927. J. Schmidt to Kammerherre A. Krieger, 18 April 1924.

for Sæmundsson and Øster therefore were to ensure that the leader of the Norwegian North Atlantic expedition, Johan Hjort could not drive a wedge between the Danes and the Icelanders and Faroese people.[29]

29 KB. Johannes Schmidt 'Papirer'. Ut.374.V.2. Privat og Blandet Korr. 1912–1927. J. Schmidt to Kammerherre A. Krieger, 18 April 1924.

Thus expectations were mounting for the trip to the North Atlantic, not least with the junior members of the expedition. While Schmidt was a frequent and prolific writer, he was matched on the 1924 cruise by the youngest member of the research crew, who kept his own diary, a student of natural history by the name of Anton Frederik Bruun (1901–1961).

12.4 Anton Bruun on board R/V *Dana*
Anton Bruun seated on board R/V Dana. In the spring of 1924, the 22-year old Anton Bruun went on his first research cruise with Schmidt to Iceland and The Faroes. Bruun testified to his diary that he was completely star struck when he came on board, but he was a quick study and soon rose to become a trusted collaborator also on Schmidt's circumnavigation, 1928–30. After World War II, Bruun became head of the next Danish led expedition around the World, when he commanded the Galathea II-expedition ((RA. Carlsbergfondets DANA-ekspeditioner. 1917–1973. 5).

Bruun would later become an influential zoologist and a master expedition leader himself. Much later in life Bruun would launch his own circumnavigation, the Galathea II Expedition (1950–1952), and in 1961, shortly before his unexpected death he became the first president of The Intergovernmental Oceanographic Commission of UNESCO. In fact, in his obituary many years later, American fisheries scientist, Carl Hubbs noted that the ocean cruises with Schmidt in the 1920s made Bruun emulate the great leader of the *Dana* expeditions: 'In due time Anton Bruun became Johannes Schmidt's scientific successor in several fields, particularly in oceanic exploration, the life-history, distribution, and systematics of various oceanic fishes, including eels, and marine science.'[30] Having suffered from polio in his childhood, Bruun was physically unable to take up farming, as the family tradition would have had it, and instead he settled on natural history. Although, his physique will doubtless have been tested during lengthy research cruises.[31]

Bruun was just a 22 year old student of natural history at the university in Copenhagen, when he wrote in his diary that '…the bomb exploded, or the orange fell into my turban…' on the morning of the 12 April. Bruun had a hangover from a party the previous day, when his fiancé, Magda came to see him and told him that his fellow student Ms. Gertrud Ostenfeld would like to see him the next morning in class at university. Apparently, she was carrying a message from her father, Carl Ostenfeld. Bruun went to see Ostenfeld straight after class and discovered that he had been handpicked to join a three month summer cruise with the *Dana* commencing in early May the same year. Thrilled with excitement Bruun cancelled all other plans for the summer and began preparing for his first international research cruise.[32]

In late April the researchers started packing the *Dana*. The research vessel headed off to sea before Bruun and Schmidt would commence their part of the journey, and on the 25 April, a press conference was held on board the vessel. Bruun appeared more star struck than prone to seasickness, as he had his first encounters with the famous 'Dr. S', as he named Schmidt in the diary. 'My first encounter with my "illustrious boss" occurred when he joked with me for not having introduced myself to him!'[33] A couple of weeks later, Bruun, Poul Jespersen, and Schmidt both left Copenhagen on a passenger ship, the *S/S Island*, to join the rest of the expedition in Scotland before heading out for their North Atlantic adventure. During the trip to Scotland, Bruun appeared anxious to get into conversation with his new boss, and during dinner one night he

30 Hubbs CL. Death notice: Anton Frederik Bruun. *Copeia*. 1962;1962(2):481–482
31 Hubbs CL. Death notice: Anton Frederik Bruun. *Copeia*. 1962;1962(2):481–482
32 RA. Anton Fr. Bruun. Danaekspeditionerne 1924–1930. Dagbog fra 'Dana' Expeditionen 1924.p. I–IV.
33 RA. Anton Fr. Bruun. Danaekspeditionerne 1924–1930. Dagbog fra 'Dana' Expeditionen 1924. I.

succeeded: 'Dr. S. was very quiet the whole day, but after I had gallantly(!) poured coffee, lit his cigarettes etc., he became more communicative and he, Jespersen and I had a fairly pleasant talk in the evening.'[34]

In mid-May, the cruise began in honest, as the team was assembled in Leith close to Edinburgh. As someone unaccustomed to overseas travel, Bruun was impressed with the landscapes and foreign folklore they encountered in port, and he appeared equally enthusiastic about the catches they hauled on board, be it halibut, cod or as in one case, a four meter long Greenland shark (*Somniosus microcephalus*), which had swallowed a couple of saithes (*Pollachius virens*) whereupon it had become entangled in the fishing gear. Schmidt is mentioned in passing mainly, and Bruun noted with some surprise that '…apparently Dr. S. does not lift one finger himself in the daily routines on board.'[35]

A few days before though, Bruun was duly impressed, as the ship learned through a telegram to the Orkney Islands that Schmidt had been named Corresponding Fellow of the Académie de Sciences in France. A couple of days later Schmidt threw a festive dinner to celebrate his new status: 'The lobster came on board as we celebrated Dr. S.'s appointment with a fine dinner: Lobster soup, followed by tenderloin steak, rum pudding and coffee. To accompany this Dr. S. gave champagne and red wine and cognac and cigars. Then we lifted anchor and steamed off into the misty night with the wind behind us.'[36]

At some point Bruun added an extra entry in his diary entitled '*Dana*'s crew', where he gave his impression of Captain Hansen, who '…is a stout-hearted type seaman from the island of Langeland (thick Dialect!), but surprisingly knowledgeable about the methods and goals of the expedition, a very hard worker and very popular with the men.'[37] Clearly Bruun had captured the main qualities that made Hansen an invaluable asset first for the *Thor* expeditions and then the *Dana* expeditions.

Considering how many thousands of sea miles the various vessels covered it is impressive how few technical and navigational problems they encountered. Yet one time, in the summer of 1927, the *Dana* had a close call in terms of wrecking in the middle of the night stumbling on under water rocks not far from Reykjavik. At a first glance Captain Hansen was cursing the map they were using, but on closer inspection it turned out that he had not been clear in his head because of the morphine powder he was consuming to get rid of shingles (*Herpes zoster*).[38] Bruun surely learned a great deal from being on his first cruise in 1924, and frequently during the summer cruises around Iceland he noted in

34 RA. Anton Fr. Bruun. Danaekspeditionerne 1924–1930. Dagbog fra 'Dana' Expeditionen 1924. 2.
35 RA. Anton Fr. Bruun. Danaekspeditionerne 1924–1930. Dagbog fra 'Dana' Expeditionen 1924. 27 May.
36 RA. Anton Fr. Bruun. Danaekspeditionerne 1924–1930. Dagbog fra 'Dana' Expeditionen 1924. 21 May.
37 RA. Anton Fr. Bruun. Danaekspeditionerne 1924–1930. Dagbog fra 'Dana' Expeditionen 1924. IV.
38 RA. Anton Fr. Bruun. Danaekspeditionerne 1924–1930. Diary from 1927. 4 August.

the diary that he went ashore with Schmidt and Hansen for social calls and interviews with local dignitaries and connoisseurs of fish such as Bjarni Sæmundsson.[39] This seems to be a clear indication that Bruun's work was appreciated and valued by both captain and expedition leader, although Bruun did not write about it himself, and as a 22 year old rookie, he probably did not fully realize that he was now an apprentice of the sorcerer of Danish marine science. Schmidt revealed in a note to Drechsel that the 'student Bruun was a skilled and good help.'[40]

The social calls and media interviews, which Bruun mentions, most likely revolved around the political aspects of the expedition. On the 18 June, Schmidt gave a lecture in Reykjavik about the Danish plans for marine science investigations, which was attended by no less than 450, according to his own count, including all the Icelandic Ministers and other local authorities. Schmidt further had the impression that, what Hjort, and thereby Norway was really after in Iceland was the herring fishery off the North coast of Iceland.[41]

Meanwhile, to the West of Iceland, Aage Vedel Taaning made his way to Greenland on board the *S/S Islands Falk*, a government navy vessel in service of the fisheries inspection. This was not part of the original plan for the expedition, but Schmidt was improvising. Drechsel was informed first by telegram saying that the ship's Captain West and Schmidt as expedition leader both recommended that Taaning was transferred to the *Islands Falk*. The reason given by Schmidt was that all the investigations could be carried out more rigorously when someone with Taaning's skills was on board, as opposed to leaving the navy personnel to fish themselves. Not least in a situation, where the Norwegians might make their own investigations. Additionally, Schmidt clearly had been moved by the political circumstances. He had just learned from the Danish Consular General in Iceland that Johan Hjort was headed for Iceland to conduct his own conversations with Bjarni Sæmundsson about the state of affairs of Icelandic fisheries.[42]

Sitting in his Copenhagen office Drechsel more or less just had to accept what Schmidt was executing on his own in the North Atlantic. He was too far away to do anything about it. In a prompt reply to Schmidt's telegram, Drechsel revealed in a letter that he – and the government Director of Fisheries, Mortensen, were both quite astonished to hear about this change of plans, but since Schmidt had made an agreement with the captain of the *Islands Falk*, they both acknowledged Schmidt's actions, as they trusted that Schmidt had good reason to do as he did.[43]

39 RA. Anton Fr. Bruun. Danaekspeditionerne 1924–1930. Dagbog fra 'Dana' Expeditionen 1924.

40 KB. Johannes Schmidt 'Papirer'. Ut.374.I.5. J. Schmidt to C. F. Drechsel, 2 August 1924.

41 KB. Johannes Schmidt 'Papirer'. Ut.374.I.5. J. Schmidt to C. F. Drechsel, 18 June 1924.

42 KB. Johannes Schmidt 'Papirer'. Ut.374.I.5. J. Schmidt to C. F. Drechsel, 13 June 1924.

43 KB. Johannes Schmidt 'Papirer'. Ut.374.I.5. C. F. Drechsel to J. Schmidt, 13 June 1924.

The Danish commission for marine science and the Danish Naval Ministry did have a prior agreement that the *Islands Falk* should try to assist Schmidt's investigations, but only in so far as this did not interfere with the other tasks of the inspection vessel, and only through the use of fishing gear borrowed from the *Dana*. Although the instruction was implicit on this point, it was not the intention that scientific personnel from the *Dana* should be transferred to the *Islands Falk*.[44] Moreover, this was clearly not coordinated with Drechsel. Poul Jespersen, who had to leave the expedition early, was questioned by Drechsel upon his return to Copenhagen. According to Jespersen, Drechsel did not seem to understand, why Schmidt all of a sudden managed to execute the work on board the R/V *Dana* without Taaning's assistance.[45]

In the marine science history literature, sending Taaning to Greenland in 1924 has been noted as having to do with the fact that there were reports of large abundances of cod in the area.[46] While this may be true, it is only part of the story. By sending Taaning to Greenland, Schmidt had in fact enacted his own foreign policy initiative, without prior consultation with anyone, except for the Captain of the receiving vessel. Meanwhile the summer expedition steamed ahead, the abovementioned agreement on Greenland was reached between the governments of Denmark and Norway, which allowed Norway to trade and conduct business in the eastern parts of Greenland.

In a private letter to Schmidt, Drechsel revealed that he was nothing short of outraged: 'It is such a scandal, which we probably in various ways will feel the impact from. Unfortunately, there is nothing else to do but to lower our heads a bit more than we have done already.'[47] Drechsel was clearly offended from a patriotic viewpoint, and felt that the new agreement would decrease Denmark's prestige at a more general level as well. Drechsel feared that it would push the Faroese in the direction of further independence from Denmark.[48]

At one point Schmidt also enrolled the Danish Minister of Foreign Affairs, Carl Moltke in his diplomatic endeavours. Schmidt heard a rumour that Moltke might be travelling to the Faroe Islands, so he was sent an open invitation to join the R/V *Dana* for a while as this would be a fairly convenient way of getting around in between the many small islands making up the Faroe Islands. He could easily have his own cabin, Schmidt noted. Regardless of whether or not the Minister would actually come to the North Atlantic, Schmidt no doubt saw this as a welcome chance to promote the patriotic virtues of the cod marking

44 KB. Johannes Schmidt 'Papirer'. Ut.374.I.5. Axel Topsøe-Jensen to C. F. Drechsel, 22 May 1924.
45 KB. Johannes Schmidt 'Papirer'. Ut.374.I.5. P. Jespersen to J. Schmidt, 21 June 1924.
46 Jakobsson J. Ocean travellers. ICES *Marine Science Symposia*. 2002:215: 335
47 KB. Johannes Schmidt 'Papirer'. Ut.374.I.5. C. F. Drechsel to J. Schmidt, 7 July 1924.
48 KB. Johannes Schmidt 'Papirer'. Ut.374.I.5. C. F. Drechsel to J. Schmidt, 7 July 1924.

experiments and hydrographical measurements: 'On this cruise I have strived to – next to the investigations – to do as much propaganda as possible for Danish marine science and for Denmark as such through lectures and by other means. Wherever possible, I have sought to distort the plans of our Norwegian colleagues, for instance by sending one of my assistants to Greenland on board the *Islands Falk*.'[49]

On board the *Dana* the competition with the Norwegian research leader, Hjort was noticed by the young Anton Bruun, who no doubt had some busy watches, following the absence of both Jespersen and Taaning. In his diary, Bruun wrote how the *Dana* was fishing between Skålefjord and Trangivåg in The Faroes. Bruun remarked how they were thoroughly investigating the Skålefiord making six different stations, since this place was in Schmidt's words 'a threshold fjord' and such a marine delicacy, which should not be left over for Dr. Hjort!'[50]

The determination of Schmidt's actions is echoed in his report back to his boss at the laboratory, Henriques. Schmidt was pleased to inform that the first half of the North Atlantic cruise had been successful both politically and scientifically. 'I think, I have been successful in thwarting the Norwegian plans for Greenland within the area of marine science.'[51]

When communicating with Drechsel, Schmidt was more direct in revealing what had been done – at sea level, to further Danish interests. In accordance with the internationally approved working programme, the expedition had investigated the spatial distribution and the growth rates for the younger year classes of cod, haddock, plaice, and halibut. However, an additional effort had been made to mark cod around the Faroe Islands, as one of the members of the expedition, who was Faroese himself, had been marking cod from a locally based trawler. Perhaps in an effort to underline how serious Schmidt was taking the political threat from Norway, he underlined in his report to Drechsel that he had spent no time this summer on purely theoretical scientific investigations, meaning that he had not been searching for eel larvae so far. The focus was on the practical, fisheries investigations.[52]

Two weeks later Schmidt sent another report to Drechsel, in which he informed the chairman that he had held several meetings with representatives for the Faroese fishermen as well as with Members of Parliament elected in the Faroe Islands. He also added information from Taaning in Greenland, who had reported to Schmidt that he had been marking 100 large cod with tags, while expecting to mark more to come. In addition he remarked that there was a lot of cod there, which would

49 KB. Johannes Schmidt 'Papirer'. Ut.374.1.5. J. Schmidt to Count C. Moltke, 14 July 1924.

50 RA. Anton Fr. Bruun. Danaekspeditionerne 1924–1930. Dagbog fra 'Dana' Expeditionen 1924. 2 August 1924.

51 CA. *Carlsbergfondet, Carlsberg Laboratorium*. E08.1924. Letter from J. Schmidt to V. Henriques, 7 July 1924.

52 KB. Johannes Schmidt 'Papirer'. Ut.374.1.5. J. Schmidt to C. F. Drechsel, 2 August 1924.

provide substantial catches to the Norwegian fishermen operating in the area.[53]

With regards to the Norwegians, Taaning painted a rather negative image. He noticed that both whalers and fishing vessels were 'swarming.' The fishermen also went ashore whenever they pleased, robbing the nests of birds, even though the Greenlanders have put a preservation order on them. They erected cairns ashore, and fished quite well at sea. Some warnings had been issued by the *Islands Falk*, but they had not arrested anyone. Johan Hjort's expedition had not been observed.[54] At the end of July, news of the Norwegian fishing effort reached the Danish newspaper, *Berlingske Tidende* who reported 'Fabulous Amounts of Fish in Greenland.' This was the result of the abovementioned expedition from the Norwegian town of Aalesund. Following a seven week long expedition, the steamer *Stormugulen*, (*The Great Mogul*) had arrived in England with 30 tons of halibut. According to the leader of the expedition, the future prospects were rich.[55]

Danish and Norwegian marine scientists and fishermen alike continued to roam the Greenland waters in subsequent years, and as noted previously the issue of access to Greenland's resources remained unsettled until 1933. This section therefore, merely highlights the events in one particularly intense year of national and scientific rivalry between Norway and Denmark.

The life history of commercial species

From the *Islands Falk,* Taaning took stations with a pelagic trawl to look for young cod and cod eggs, but found none.[56] The marking was done through the assistance of Inuit who were paid one half Danish Kroner for every cod that they caught from their kayaks, which the local fishermen then delivered to Taaning, who marked them with a small piece of metal on their gill cover before releasing them again.[57] 'So little has been done up here,' Taaning remarked in a report to Schmidt. 'I could work here for years and not weeks!'[58] Taaning, however, did return after some weeks, but others continued to mark cod in West Greenland in subsequent years building on the results obtained via the expeditions with the *S/S Thor* from 1903 onwards.

From the beginning, the North-Western Area Committee decided to focus investigations on five different species, Cod, Haddock, Halibut,

53 KB. Johannes Schmidt 'Papirer'. Ut.374.1.5. J. Schmidt to C. F. Drechsel, 14 August 1924.

54 KB. Johannes Schmidt 'Papirer'. Ut.374.1.5. A. V. Taaning to J. Schmidt, 2 July 1924.

55 KB. Johannes Schmidt 'Papirer'. Ut.374.1.5. News clip from Berlingske Tidende, 26 July 1924.

56 KB. Johannes Schmidt 'Papirer'. Ut.374.1.5. A. V. Taaning to J. Schmidt, 2 July 1924.

57 Vedel Taaning Å. Fiskeriundersøgelser i nordlige have efter krigen. In: Winge Ø, Vedel Taaning Å, eds. *Naturforskeren Johannes Schmidt. Hans Liv og Ekspeditioner*. Copenhagen: Gyldendalske Boghandel Nordisk Forlag; 1947: 118.

58 KB. Johannes Schmidt 'Papirer'. Ut.374.1.5. A. V. Taaning to J. Schmidt, 2 July 1924.

Plaice, and Herring. In addition to investigating the life history of these species, the international research effort aimed to support investigations of in-direct importance such as hydrographical measurements, plankton investigations, and studies of life on the seabed, the so-called benthic studies.[59]

In hindsight, many scientific breakthroughs may appear as the consequence of logical extensions of previous work, where more gaps of knowledge are covered. In the case of the life history of commercial fisheries, the international effort was literally marking out otherwise blank spots on the map of marine life. Yet, even in 1924, at which point Schmidt had been working in the area for more than 20 years, some of the results still came as a surprise.

The life history of the Greenland cod poses one such example. In one of his reports to Drechsel from the summer of 1924, Schmidt was speculating about the cod, which Taaning was then in the midst of marking in Greenland. This cod Schmidt thought for sure belonged to the same cod family as those fish that thrived in the waters around Newfoundland or elsewhere in Canada or the USA. Schmidt thought that Greenland nonetheless ought to be included in the Danish part of the North Atlantic investigations, the North-Western Area. If that was the case then closer connections could be formed to the USA and Canada, but as Schmidt also indicated to Drechsel, this matter should be dealt with utmost caution considering the viewpoints of the domestic Danish authorities on Greenland as well as Norway. For the time being Schmidt merely hoped that Taaning would mark as many cod as possible, and that some of them would make their way to Newfoundland or further down the Atlantic coast of North America to be caught. In brackets he added that '(I have little doubt that the Greenland cod are only passing by, that they do not breed around Greenland ...)'.[60] This hunch of Schmidt's as it was, underlines the extent of surprise, when a few years later one of the cods tagged by Taaning suddenly emerged on the deck of a fishing vessel in Iceland.

Five years into the work of the North-Western Area Committee, Schmidt was able to draw important conclusions about the distribution of this commercially important species in the North Atlantic. In several publications Schmidt addressed the issue of where to go fishing, and the prospects of extending the international fishing effort beyond the traditional grounds around Iceland and the European continental shelf. Pointing to a map showing all the Danish research stations of the

59 Vedel Taaning Å. Fiskeriundersøgelser i nordlige have efter krigen. In: Winge Ø, Vedel Taaning Å, eds. *Naturforskeren Johannes Schmidt. Hans Liv og Ekspeditioner*. Copenhagen: Gyldendalske Boghandel Nordisk Forlag; 1947:113.

60 KB. Johannes Schmidt 'Papirer'. Ut.374.1.5. J. Schmidt to C. F. Drechsel, 2 August 1924.

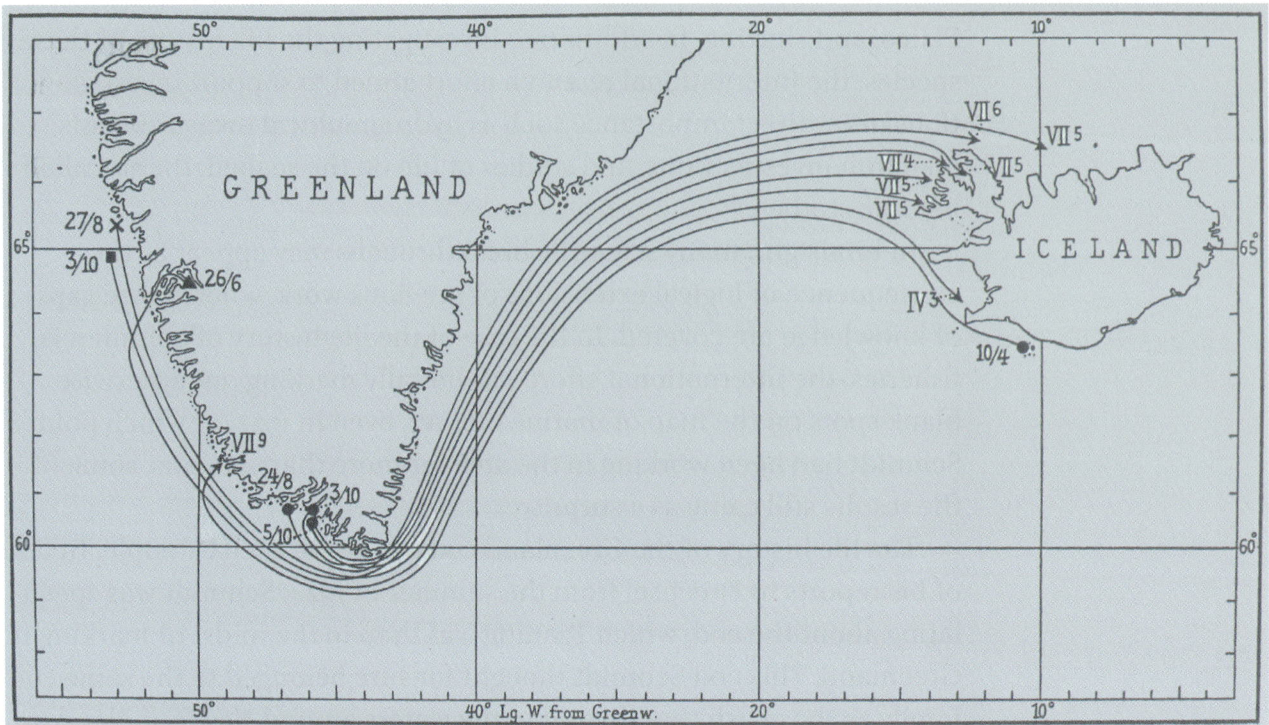

Fig. 11. Cod marked at Greenland recaught at Iceland, and vice-versa.
× Marked in 1924, ▲ in 1926, ■ in 1928, ● in 1929. IV Recaptured in 1927, VII in 1930; the index numbers alongside the years show the months of recapture, 3– March, 4– April, etc.

12.5 Cod marked at Greenland recaptured in Iceland
One of the most surprising results from the 1924 expedition to Iceland, The Faroe Islands and Greenland was the discovery that cod marked in Greenland would later turn up in Iceland. Schmidt had expected them to migrate towards Canada and the USA instead (Jensen, A. & Hansen P., 1931 p. 37.).

preceding 25 years of research, Schmidt stated that '…our northern food-fishes are restricted to the continental shelf and the Atlantic slope or its immediate neighbourhood. Within this – relative to the enormous extent of the ocean – narrow belt these fishes live the whole of their lives from egg to adult fish.'[61] This was of course of immense practical importance to the fishermen trying to catch the fish and it would within a few years be equally important for fisheries managers trying to optimize the amount of fish, which could be sustainably caught within a giving fishing season.

Schmidt then underlined the scientific importance of the recent findings in relation to more theoretical viewpoints on fish migrations, thereby tackling one of the core issues behind the organization of ICES from 1902 onwards. 'The main result that the great areas of the ocean

61 Quoted in: Jakobsson J. Ocean travellers. ICES *Marine Science Symposia.* 2002:215:340.

are empty of food-fishes agrees with the fact that most of the species so far closely investigated have proved to be divisible into several distinct races or populations, each living its life separate from the others.' Finally though, Schmidt remarked that for two species, the situation was different: The freshwater eel and the conger eel (*Conger vulgaris*).[62]

The more finely tuned the analyses, the more removed the results became from the theory, hitherto prevailing, which was the so-called polar migration theory. The first written evidence for the polar migration theory seems to be a Hamburg Mayor, Anderson, who proposed in 1746 that all the herring in the North Atlantic Ocean had their 'home' far away from humans, in the Arctic.[63] According to Danish theologian and natural historian, Erik Pontoppidan, this was a feature of the herring instilled by godly intervention. Each year, when the season was right, the herring migrated to their designated destinations along the Atlantic coasts of North America and Europe. In this way it was possible for people to gain access to the herring resources.[64]

While the religious background may have faded over the years, the theory of the polar home persisted for a good hundred years until the latter half of the 19th century, when German marine scientist, Friedrich Heincke concluded that the herring in the sea consisted of several distinct populations, with distinct geographic dispersion. Heincke used thousands of specimens in quantitative studies.[65] When the vast offshore areas of the North Atlantic were explored through ICES initiatives in the early 1900s, the life history of fishes was a significant aspect of why the international community decided to focus on the establishment of Committee A, the fish migration committee.[66]

In 1909, Schmidt summarized the findings from the round of ICES initiated expeditions in the North Atlantic, in his book chapter on, 'The distribution of the pelagic fry and the spawning regions of the gadoids in the North Atlantic from Iceland to Spain,' – as he stated in the subtitle 'based chiefly on Danish investigations.' Here Schmidt was able to rock the established theory substantially as he found that 'the spawning regions of a species are not identical with its distribution as a whole.' What he noticed was the tendency for most species to congregate in certain smaller areas to spawn, which he attributed to the sexual preference of spawning in a distinct environment, where for instance salinity and temperature was right.[67] This means that there was some kind of system in the way fish lived their lives. This gave further impetus to the downfall of the polar migration theory.

62 Quoted in: Jakobsson J. Ocean travellers. ICES *Marine Science Symposia.* 2002:215:340.
63 Jakobsson J. Ocean travellers. ICES *Marine Science Symposia.* 2002:215:340.
64 Pontoppidan E. *Det første forsøg paa norges naturlige historie vol. 1–2.* Copenhagen: 1752–53.
65 Sinclair M, Solemdal P. The development of 'population thinking' in fisheries biology between 1878 and 1930. *Aquatic Living Resources.* 1988:1(3):189–213.
66 Jakobsson J. Ocean travellers. ICES *Marine Science Symposia.* 2002:215:331.
67 Schmidt J. The distribution of the pelagic fry and the spawning regions of the gadoids in the North Atlantic from Iceland to Spain. *Rapports et procès-verbaux des Réunions du Conseil International pour l'Exploration de la Mer.* 1909:10(4):150.

12.6 Halibut caught by R/V *Dana* in 1924 *Halibut caught by the R/V Dana in 1924. Twenty years after Schmidt first found the larvae of halibut in the North Atlantic Ocean, the different life stages were subject to intense investigation (*RA*. Carlsbergfondets* DANA*-ekspeditioner. 1917–1973. 5).*

Following the intense work of the North-Western Area Committee in the 1920s Schmidt was able to draw the following conclusion, finetuning his observations from the 1909 paper: 'Our northern food-fishes are restricted to the continental shelf and the Atlantic slope or its immediate neighbourhood. Within this – relative to the enormous extent of the ocean – narrow belt these fishes live the whole of their lives from egg to adult fish.' Then Schmidt added that there were indeed two very interesting exceptions to this pattern, namely the eel (*Anguilla vulgaris*) and the conger eel (*Conger vulgaris*), who had radically divergent life histories, as we have seen in previous chapters.[68]

68 Qouted from Jakobsson J. Ocean travellers. ICES *Marine Science Symposia.* 2002:215: 340.

Continuing along this line Schmidt put the final nail in the coffin of the polar migration theory when he stated: 'The main result that the great areas of the ocean are empty of food-fishes agrees with the fact that most of the species so far closely investigated have proved to be divisible into several distinct races or populations, each living its life separate from the others. As the species are gradually taken up one after another for finer, racial analysis there becomes less and less room for the extended migrations earlier believed in. Here again, however, we must make exceptions of the eel and the conger, especially the former.'[69]

It was only in 1968, however, that this pattern was given a theoretical name when it was referred to as the so-called migration triangle in Harden Jones' book, *Fish Migration*, where the author states that it is a recurrent phenomenon that fish live their full life within a triangular geographical space with one distinct spawning area, another distinct nursery area where the young fish would feed and grow, while the third corner was characterized as the wintering area, where the fish would be relatively inactive.[70] The migration triangle was later modified by David Cushing in 1982 and David Secor in 1999, but the basic structure of what Schmidt summarized in 1928 still holds sway in contemporary marine science.[71]

Tagging on to the genetic studies

One of Schmidt's enduring interests with regards to the racial studies was the cod. During a five year period from 1924–1929, Schmidt and his collaborators collected more than 100 samples of cod, most of which consisted of more than 100 specimens of cod.[72] In total the racial investigation on cod was based on no less than c. 20,000 specimens of cod from 114 different samples taken at sea, the so-called *stations*.[73]

Such a large scale sampling was made possible through the work in two different institutions, were Schmidt was in the cockpit, ICES and the Carlsberg Laboratory. The practical implications of the international cooperation through ICES are evident from looking at where the stations of cod were taken, and by whom. The *Dana* was responsible for many of the stations, but in a 1930 article on the cod, Schmidt also thanked colleagues as far away as North America, where professor A. G. Huntsman in Ottawa, Canada and Henry O'Malley from the US Fisheries Commission in Washington D. C. had both submitted material to

69 Quoted from Jakobsson J. Ocean travellers. ICES *Marine Science Symposia*. 2002:215: 340.

70 Jakobsson J. Ocean travellers. ICES *Marine Science Symposia*. 2002:215:330–342.

71 Secor, D. H., Historical roots of the migration triangle, ICES *Marine Science Symposia*. 2002:215:323–329.

72 Vedel Taaning Å. Fiskeriundersøgelser i nordlige have efter krigen. In: Winge Ø, Vedel Taaning Å, eds. Naturforskeren Johannes Schmidt. Hans Liv og Ekspeditioner. Copenhagen: Gyldendalske Boghandel Nordisk Forlag; 1947:121.

73 Schmidt J. Racial Investigations: 10. The Atlantic Cod (*Gadus callarias* L.) and local Races of the same. *C R Trav Lab Carlsberg*. 1930:18(6): 3.

Schmidt's research.[74] A typical entry would read: 'St. 16. Southwest Iceland, off Landeyjasandur (63 degrees, 34 minutes North; 20 degrees, 25 minutes West). R/V *Explorer*. Station 40. 27–28 June 1925. Ottertrawl'.[75]

Working towards a common end, this Scottish vessel the *S/S Explorer* sent their cod to the Carlsberg Laboratorium for Schmidt's research team to work up the material at his own laboratory, the other important institution for the 1920s cod investigations. In an article from 1930, Schmidt mentions how N. R. Poulsen who was a regular employee at the laboratory treated the material in formalin, if it had not already been done so before arrival. Then another assistant, Esther Hansen dissected the fish, while the two fisheries scientists not fully employed by the Carlsberg Laboratory, Vedel Taaning and A. C. Strubberg oversaw some of the work. The statistical work associated with the thousands and thousands of vertebrates and dorsal counts was done by a Dr Kirstine Smith.[76] From the annual reports submitted by Schmidt to the Board of Trustees, it reads that more or less everyone employed at the laboratory was helping out with the fisheries investigations.[77] This means that virtually all of the 35,600 Danish Kroner which was spent annually on salary alone in the Physiological Department of the Carlsberg Laboratory in the latter half of the 1920s went into to supporting the international research agenda of ICES.[78]

Finally, it should not be overlooked that Schmidt had seemingly unlimited access to publishing in the Carlsberg Laboratory's own journal, the *Comptes-Rendus du Laboratoire Carlsberg*. In an era, when printing was cumbersome and in relative terms much more expensive than today, this was no insignificant aspect of Schmidt's marine research empire. Schmidt was once more orchestrating research on a very large scale. In reality Schmidt was uniquely positioned in the international world of marine science in the 1920s, in the sense that he had the necessary skills and experience, the finances, the laboratory infrastructure, and the skilled manpower to set him apart as the one doing genetic stu dies on wild fish populations long before anyone else.

The main result of Schmidt's effort in this was the discovery that the cod populations of the North Atlantic did not belong to a single population. Rather the cods in the North Atlantic formed a mosaic of mutually different populations. Some of these thrived in a greater geographical area, while other populations were much localized, especially in near shore areas. The techniques used were the same as those Schmidt had pioneered with the eelpout during World War One. The average number

74 Schmidt J. Racial Investigations: 10. The Atlantic Cod (*Gadus callarias* L.) and local Races of the same. *C R Trav Lab Carlsberg*. 1930:18(6): 3.

75 Schmidt J. Racial Investigations: 10. The Atlantic Cod (*Gadus callarias* L.) and local Races of the same. *C R Trav Lab Carlsberg*. 1930:18(6): 36.

76 Schmidt J. Racial Investigations: 10. The Atlantic Cod (*Gadus callarias* L.) and local Races of the same. *C R Trav Lab Carlsberg*. 1930:18(6): 3; CA. Carlsbergfondet, Carlsberg Laboratorium, E31. Endelige budgetter og noter 1914–1972.

77 CA. Carlsbergfondet, Carlsberg Laboratorium. E09. 1928. Letter from J. Schmidt to Board of Trustees, 17 October 1928; CA. Carlsbergfondet, Carlsberg Laboratorium. E09. 1929. Letter from v. Ege to v. Henriques, Annual Report 1929; CA. Carlsbergfondet, Carlsberg Laboratorium. E09. 1930. Letter from J. Schmidt to v. Henriques, 19 October 1930.

78 CA. Carlsbergfondet, Carlsberg Laboratorium, E31. Endelige budgetter og noter 1914–1972.

of vertebrae, and for cod also the average number of rays from the second dorsal fin, were compared across groups of cod caught in different areas, revealing noticeable hereditary differences.[79]

Generally speaking, results indicated clearly that cod near to Atlantic North America possessed more vertebrae on average than the cod thriving closer to Europe. By the same method the Greenland cod was then found to be much more closely related to the European cod than to the ones in Newfoundland. This was important as it supported the results of the marking experiments whereby it was shown that cod initially marked in southwest Greenland would frequently end their lives on a fishing trawl off the coast of Iceland. Similarly as for the eel-pout studies, it was now found that the cod living in coastal areas had a smaller average vertebrae count than the ones living their life in open sea. Finally, the study showed that there was a correlation between the number of vertebrae and the ocean temperature. When a relatively high temperature was found, the number of vertebrae was relatively low and vice versa.[80]

The importance of nurseries

The 1924 cruise to Iceland also gave Schmidt and colleagues important news on the state of fisheries inside and outside of the 3-mile zone. Inside the 3-mile zone the waters belonged to Iceland territory, whereas the areas outside were in international waters, where every nation could fish. According to international agreement it was illegal to trawl within 3 miles of the coastline all around Iceland, and especially the Faxe Bay in the west near Reykjavik as this was a particularly important nursery for young plaice, halibut, and haddock. The *Dana* conducted fisheries surveys, trawling the area both within and outside of the 3-mile line, and the difference in fisheries abundance was staggering. With regards the plaice there were 7 times more plaice found inside than outside.[81] While much of the actual research would have to wait for the expedition to return to the desktops in Copenhagen, Schmidt realized immediately from what was hauled on board, that there was a big difference between the areas inside and outside Icelandic territory. This is highlighted in a report to Drechsel from 4 July.[82]

This was a remarkable sign of how much impact fishing had on the stock of fish growing in the bay. Once more, Schmidt's old friend, Bjarni

79 Schmidt J. Racial Investigations: 10. The Atlantic Cod (*Gadus callarias* L.) and local Races of the same. *C R Trav Lab Carlsberg.* 1930;18(6): 26–27.

80 Schmidt J. Racial Investigations: 10. The Atlantic Cod (*Gadus callarias* L.) and local Races of the same. *C R Trav Lab Carlsberg.* 1930;18(6): 26–27.

81 Vedel Taaning Å. Fiskeriundersøgelser i nordlige have efter krigen. In: Winge Ø, Vedel Taaning Å, eds. *Naturforskeren Johannes Schmidt. Hans Liv og Ekspeditione*r. Copenhagen: Gyldendalske Boghandel Nordisk Forlag; 1947:125.

82 KB. Johannes Schmidt 'Papirer'. Ut.374.1.5. J. Schmidt to C. F. Drechsel, 2 August 1924.

Sæmundsson was helping out with the Icelandic surveys, and both from Denmark and Iceland there was a continued effort to increase knowledge on the importance of knowing more precisely how much fish could sustainably be caught in Faxe Bay. This included an extensive measurement of what the plaice and other fishes fed on in the bay area. Thus once more, Petersen's bottom sampler showed its worth, as it was used to grab the mud from the seabed for dissection and subsequent investigation of resident species.[83]

Following Schmidt's death the work continued, and in 1937, the then Icelandic envoy to Denmark, Sveinn Bjørnsson brought the issue to the table of ICES. This work was brought to a relative standstill by the outbreak of World War Two, but soon thereafter measurements were taken to limit the amount of fishing in the Faxe Bay following international recommendation from ICES.[84]

Conclusion

This chapter has centered on the fisheries investigations in the North Atlantic with regards to the main commercial fish species such as cod, plaice, and halibut. Schmidt played a central role in the exploration of the life history of these and other species from the beginning of the international collaborations in ICES and into the early 1930s, a period which saw a tremendous growth in knowledge of where these species were spawning, where they fed, and where lived their adult lives. From a theoretical point of view this research paved the way for a gradual shift in paradigm for how fish migrate, from the 18th century idea of the polar migration theory, towards the contemporary idea of a migration triangle. This was also tied in to what Schmidt called his racial investigations, where large scale investigations into the hereditary features of the cod proved the existence of many distinct populations of cod living in more restricted areas within the wider North Atlantic Ocean.

The migration pattern of the cod living off shore along the West coast of Greenland was one of the more spectacular discoveries emerging from an effort spanning several decades, and one discovery which came about as a direct consequence of the political agenda of marking not only cod in the North Atlantic, but also marking a Danish geopolitical presence and interest in the territory in and around Greenland, Iceland, and the Faroe Islands.

83 Vedel Taaning Å. Fiskeriundersøgelser i nordlige have efter krigen. In: Winge Ø, Vedel Taaning Å, eds. *Naturforskeren Johannes Schmidt. Hans Liv og Ekspeditione*r. Copenhagen: Gyldendalske Boghandel Nordisk Forlag; 1947:125.

84 Vedel Taaning Å. Fiskeriundersøgelser i nordlige have efter krigen. In: Winge Ø, Vedel Taaning Å, eds. *Naturforskeren Johannes Schmidt. Hans Liv og Ekspeditione*r. Copenhagen: Gyldendalske Boghandel Nordisk Forlag; 1947:126–127.

This chapter has served as a first inquiry into the state of Danish marine research in the face of the political crisis between Denmark and Norway over Greenland's sovereignty, which spanned more than a decade from c. 1922–1933. Looking into the strategic and tactical considerations of the Danish commission for oceanography in 1924 there is good reason to conclude that the substantial marine science output emerging form the intense study of the commercially important species in the North Atlantic Ocean, may in fact have been reinforced by Danish and Norwegian political interests in the area. Thus it is plausible that the tagging experiments conducted as part of the North-Western Area Committee would not have been carried out as far away as the West coast of Greenland, if it had not been for the immediate concerns to keep the Norwegians at bay.

The presence of marine science was recognized as a symbol of soft diplomacy as the research efforts were directed towards Greenland. Schmidt in particular, but actively supported by Drechsel and government officials in Copenhagen, sought to exploit the activities of marine research as a means to stake out Danish interests. In this particular case it is fair to state that Danish fisheries science was a spearhead of Danish foreign policy towards Norway. In addition to the actual research Schmidt unfolded a considerable public relations effort lecturing and giving interviews to local stakeholders of the fishing communities and official authorities in Iceland and the Faroe Islands.

Schmidt and Drechsel both portrayed Johan Hjort as the great antagonist in this. The reports from Taaning and Schmidt read like reports from a ghost chase, with Hjort's *Michael Sars* as a sort *Flying Dutchman*; something that they only ever heard about, but never quite located. Schmidt clearly saw an advantage in promoting his own research agenda as a means to thwart Norwegian geopolitical interests, but it is safe to say that his feelings towards Hjort were not just a showcase. The two marine scientists Taaning and Bruun both worked alongside Schmidt in the summer of 1924, and their writings testify to the fact that Schmidt's animosity towards Norway and Johan Hjort was part of everyday affairs on board the *Dana*.

In any case, it is noteworthy that the collaboration between Schmidt, Drechsel, and another Norwegian marine scientist, Bjørn Helland-Hansen continued throughout this crisis, and seemingly without serious interruption. In the case of hydrography therefore, it looks as if the common interest in science as well as personal friendship overruled the calamities of international political animosity.

While the government paid for the North Atlantic expeditions as part of its international obligations towards ICES, it is striking that once more Schmidt's own laboratory was a vital asset for his own main interests in marine science. The combination of the Carlsberg Laboratorium and the Danish marine science commission under the auspice of ICES made Schmidt the man in the middle of international fisheries science in the 1920s. Schmidt rose to prominence in ICES and his home and laboratory in Copenhagen came to function as a sort of fisheries science embassy whenever the ICES delegates were in Copenhagen. In reality Schmidt was uniquely positioned in the international world of marine science in the 1920s, in the sense that he had the necessary skills and experience, the finances, the laboratory infrastructure, and the skilled manpower to set him apart as the one doing large scale genetic studies on wild fish populations.

Under Schmidt's chairmanship of the North-Western Area Committee, remarkable and important fisheries science was carried out in the North Atlantic in the 1920s, yet his mind was still pondering about the eel, and he still had the desire to go round the World.

13 Organizing the Carlsberg Expedition around the world

According to Schmidt's own published writings his quest to circumnavigate the World began in the spring of 1927 following a number of conversations with Ostenfeld. The initial motivation was twofold. Having explored the Atlantic Ocean since the beginning of the 20th century, Schmidt and his peers had greatly expanded knowledge, not only of the elusive eel, but also of a whole array of scientific problems. Therefore, Schmidt proposed that such a circumnavigation would on the one hand cement the Danish position as a world leader in eel related marine science, and on the other, it would be a welcome chance to compare the flora and fauna of the Atlantic Ocean with the two other large oceans on the planet, The Pacific and the Indian Ocean. According to Schmidt a seed had been sown during the Great Atlantic Expedition 1921–22, which led to a brief excursion into the Gulf of Panama in the Pacific Ocean. In particular, Schmidt wanted to investigate the pelagic life, and add to the knowledge already obtained from research in the Atlantic Ocean.[1]

In his writings, Schmidt emphasized the colossal effort it would take to plan, execute, and follow up on such an undertaking. Schmidt also let the readers know that for a long time this kind of undertaking was far beyond the boundaries of what was accomplishable. He went on to note that fortunately, Denmark had in the Carlsberg Foundation, an institution with both the means and the foresight amongst its leaders, to address and solve scientific problems and patriotic duties. Without such an organization, a small county like Denmark would have been left powerless in such matters as exploration, in Schmidt's opinion.[2]

Schmidt wrote about this in 1931 in the aftermath of having already accomplished such a circumnavigation. As one would expect, Schmidt's departure date was set at 1927, when the plans that eventually led to the circumnavigation were first presented. This may be one reason why previous commentators on Schmidt's expeditions, including Schmidt

1 Schmidt J. Indledning. In: Schmidt J, ed. Dana's Togt omkring Jorden 1928–1930. Copenhagen: Gyldendal; 1932:9–10.

2 Schmidt J. Indledning. In: Schmidt J, ed. Dana's Togt omkring Jorden 1928–1930. Copenhagen: Gyldendal; 1932:10–11.

himself, have presented them as one long succession of ever more successful attempts to carry out ever more ambitious endeavours.[3]

This chapter goes behind the scenes to reconstruct how Schmidt's proposed circumnavigation, which was indeed of great international scientific importance, was planned, executed, and completed. As it turned out, the reality of the preparatory phases were complex and marked by blind alleys, although the eventual success of the expedition came to overshadow both earlier failed attempts to raise funds for a circumnavigation, as well as years, in fact more than a decade of careful preparation.

Eels east of Suez

When Johannes Schmidt had successfully ended the third Atlantic expedition with *Dana* in 1922, he had already accomplished much more than most marine scientists would ever do in a lifetime. It seems logical in a way that his global aspirations came about as a next step following the Atlantic successes. According to the commemorative volume of the Carlsberg Oceanographic Expedition around the World, *Dana's Togt Rundt om Jorden*, the plan was conceived in collaboration with Carl Ostenfeld as late as the 'beginning of 1927', while the dream of venturing into the Pacific had been nurtured by Schmidt's and Ostenfeld's brief visit to the Panama Canal on the R/V *Dana* in 1922.[4] Anton Bruun, recalling the global voyage in a 1947 book chapter, wrote that he first heard talk of sailing around the World in 1924 when he was on his own first cruise with Schmidt.[5]

In Chapter 8 however, it was revealed that Schmidt had global aspirations as early as the summer of 1916, when he and Drechsel were fishing for 'goldfish' among shipping tycoons and bankers. While it is not possible to time the exact moment, when Schmidt decided that he needed to go global to pursue his dreams of charting the World's eel populations, the letters he wrote and received testify that even in 1913 he mentioned the eels East of Suez. In April 1913 Schmidt received what reads as a reply to a letter, Schmidt had sent to the German State Bureau of Fisheries asking about the state of affairs with eel fisheries in German East Africa, present day Tanzania. In the bureau's reply to Schmidt the author speaks highly of the prolific eel fisheries, – more than 6.5 million eels had been caught in the previous year – and also about the presence

3 Schmidt, Johannes, 'Indledning' in: Schmidt J, ed. *Dana's Togt Omkring Jorden 1928–1930*. Copenhagen: Gyldendalske Boghandel; 1932, p. 9; Fort T. *The book of eels*. Great Britain: Harper Collins Publishers; 2002; Winge Ø, Vedel Taaning Å, eds. *Naturforskeren Johannes Schmidt. Hans Liv og Ekspeditioner.* Copenhagen: Gyldendalske Boghandel Nordisk forlag; 1947; Wolff T. *Danske ekspeditioner på verdenshavene: Dansk havforskning gennem 200 år.* Copenhagen: Rhodos; 1967; Nielsen KH, Ries CJ, Østergaard KB. 'Danske videnskabelige ekspeditioner i perioden.' In: Hvidtfelt Nielsen K, Nielsen H, eds. *Viden uden grænser 1920–1970, dansk naturvidenskabs historie vol. 4*. Aarhus: Aarhus University Press; 2006:283–302.

4 Schmidt, Johannes, 'Indledning' in: Schmidt J, ed. *Dana's Togt Omkring Jorden 1928–1930*. Copenhagen: Gyldendalske Boghandel; 1932, p. 9.

5 Bruun AF. Til New Zealand og Tahiti for at fange Aal. In: Winge Ø, Vedel Taaning Å, eds. *Naturforskeren Johannes Schmidt. Hans Liv og Ekspeditioner.* Copenhagen: Gyldendalske Boghandel Nordisk forlag; 1947: 132.

of elvers, or glass eel in Tanzanian waters.[6] This was a continuation of Schmidt's effort to present a biogeographical chart of all the freshwater eels in the World. In 1909 Schmidt had published an overview of where in the Atlantic Ocean the eel was living. Schmidt made a brief excursion from the Atlantic in this paper, as he included references he had obtained from two Danish consular in South Africa. They were able to tell him that the freshwater eel was found on the East coast of South Africa in the Natal province facing the Indian Ocean. For the rest of the World, information was quite scarce. Schmidt was bold enough to number this report as 'volume I.', a solid indication that he was planning to create a similar overview for the rest of the World.[7]

In the early 1920s Schmidt was anything but satisfied with having proposed a solid theory for the Atlantic eel. The life history of the freshwater eels in the Indian Ocean and the Pacific Ocean were far from fully understood. The rest of world's eel were still there to be investigated, thus Schmidt was going global, but such a desire added substantial challenges compared to those he had already overcome, and which had safely secured him a solid entry in the future textbooks on marine science and fisheries science. One new challenge was logistical. How was Schmidt supposed to conduct research on the other side of the planet? Secondly, and linked to this was the question of finances; who on Earth would be willing to finance his research? The fecundity of the eel in the Pacific South Seas would hardly reap insights of direct relevance to the fishing industry in any of ICES member states, all of which were European.

In 1923, Schmidt approached the readers of the *Danmarksposten*, the monthly news magazine published by the society for Danish expatriates, *Dansk Samvirke*, where Ostenfeld presided as chairman.[8] This was an ingenious way of reaching out globally in a time, where this magazine and not much else from Denmark was distributed around the globe. Perhaps it was Ostenfeld's idea in the first place. The year before Ostenfeld had entertained his readers with a popular article on the great Atlantic expedition with the R/V *Dana*, so Danes abroad already knew about the Danish deep sea accomplishments.[9] Schmidt explained on the cover page of the magazine that he needed the help of the magazine's readers, as he badly needed samples of eel from many parts of the Pacific and Indian Oceans.

As the eels would eventually seek freshwater streams and brackish lagunes etcetera, Schmidt envisioned that volunteers could catch and preserve specimen for him. He informed his readers that they should

6 KB. Johannes Schmidt 'Papirer'. Ut.374.V.1. Privat Korrespondance 1902–1916. Letter from *Staatliche Fischereidirektion*, Hamburg to J. Schmidt, 26 April 1913.

7 Schmidt J. On the distribution of the fresh-water eels (*Anguilla*) throughout the World: I. Atlantic Ocean and adjacent regions, a bio-geographical investigation. *Meddelelser fra Kommissionen for Havundersøgelser: serie Fiskeri*. 1909;3(7). 121–177.

8 Schmidt, J. Aale-Spørgsmålet i det Indo-Pacifiske Omraade. Vil Dansk Samvirke's Medlemmer Bidrage til dets Løsning?, in: *Danmarksposten*, (1923), 12: 215–216.

9 Ostenfeld CH. 'Dana'-ekspeditionen. *Danmarksposten*. 1922;3(8):125–127.

not hesitate to write to him at his Danish address if they had caught something of interest to him. He did mention though, that the best way of preserving the eel was to place them in the solvent formaldehyde. Smaller specimen were preferred over large fish, as the smaller ones were easier and cheaper to handle and to ship to Europe. Schmidt of course promised to pay for the costs associated with obtaining the specimen.[10]

Schmidt had included a map of the world with indications as to the state of knowledge about eels in the major oceans of the world. Generally speaking Schmidt informed his audience that the case of the Atlantic eel was well covered, and for large parts of the World there was no indication that freshwater eels of any sort were present. This included the West coast of the American continents and the countries bordering the South Atlantic Ocean, which were therefore irrelevant to the study. The most interesting areas were the shoreline along the entire Indian Ocean and the Indo-Pacific area to the easternmost South Sea Islands such as Tahiti. In the northern Pacific Ocean the Japanese freshwater eel was relatively well studied, but Schmidt put question marks as to where else the eel was found in the northern Pacific. Were they present in Pacific Siberia, China, Western and Southern Australia? And how come it is only in Japan that a large economically important fishery for eel takes place, in the same way as it does in Europe and North America?[11] The list of unknown territories was daunting. This was about to change, when in 1925 Schmidt authored the sequel to his 1909 paper on the bio-geographical distribution of the eel in the Atlantic Ocean, another lengthy report covering the Indo-Pacific region.[12]

Recreation and work around the World

In any case, Schmidt seems to have embarked on two new strategies from 1925. First of all, he wanted to see for himself what was in the Pacific Ocean, so he started to plan a simpler voyage for himself without a research vessel. Secondly, he looked abroad to scout for potential sponsors in the USA in particular. The solution to fulfilling both strategies was to go round the World, which he did in 1925–1926 with his wife Ingeborg and Ostenfeld's daughter Gertrud. As we saw in Chapter 6 this journey was aided substantially by his superiors, the Board of Trustees at the Carlsberg Laboratorium. They acknowledged the importance

10 Schmidt, J. Aale-Spørgsmålet i det Indo-Pacifiske Omraade. Vil Dansk Samvirke's Medlemmer Bidrage til dets Løsning?, in: *Danmarksposten*, (1923), 12: 215–216.

11 Schmidt, J. Aale-Spørgsmålet i det Indo-Pacifiske Omraade. Vil Dansk Samvirke's Medlemmer Bidrage til dets Løsning?, in: *Danmarksposten*, (1923), 12: 215–216.

12 Schmidt J. On the distribution of the freshwater eels throughout the world. part II. indo-pacific region. *Det Kongelige Danske Videnskabernes Selskabs Skrifter*. 1925;8(10(4)):329–382.

of this trip for the sake of Schmidt's health, a fact supported by the 10,000 Danish Kroner the trustees agreed to provide for the trip, a sum of money Schmidt which had not been spent on the previous year's budget. The costs of bringing Ingeborg along were paid for from her family fortune.[13]

Thus on Christmas Day 1925, Schmidt's party went on board a steamship taking them first to Marseille in southern France, then through the Suez Canal bound for Australia. Via Sri Lanka, Fremantle and Melbourne they reached Sydney towards the end of January 1926. Here, Schmidt organized the collection of eel samples from Australia, before they continued their journey to New Zealand. The European settlers in New Zealand apparently had little appreciation for the eel, but for the Maori population it was considered an important source of protein. From a base in Wellington Schmidt travelled inland up the river Wanganui to catch his own specimen aided by the local population of mainly Maori from the village of Pipiriki.[14] From his collections Schmidt eventually was able to establish that two different types of eel co-existed in New Zealand, which he referred to as the *Anguilla aucklandi* and the *Anguilla australis*. The first one is today known as *Anguilla dieffenbachia*.[15] Publishing his results in 1928 he drew his conclusions on the basis of nearly 1,500 specimens from more than twenty different samples from both the North Island and the South Island.[16]

From New Zealand the Schmidt couple and Gertrud Ostenfeld travelled on to what was anticipated as the exotic highpoint of the journey round the World, Tahiti. Ingeborg Schmidt kept a diary, 'intended for those back home' as she noted on the cover page of her manuscript, which was never published.[17] Ingeborg was full of excitement, as they approached Tahiti: 'We came on Monday. In the glittering sunlight, and accompanied by music on deck, the ship slid slowly through the natural gate in the coral reef and anchored in the picturesque landing dock. I stood on deck completely taken aback by all of this beauty. We passed a small island with palms, mirrored in the clear water. I have never seen the ocean so transparent blue as here, and the shoreline with palm trees and the view across to the rock island, Morea was so adorable that we could not get tired from staring at this little warm Paradise, we had now reached following the rough ride across the sea from New Zealand.'[18] During their stay, the image of having reached Paradise on Earth was spoiled somewhat according to Ingeborg Schmidt, but only to the extent that when something was not as positive as she had expected

13 CA. *Carlsbergfondet, Carlsberg Laboratorium*. E08. 1926. Letter from V. Henriques to Board of Trustees, 26 November 1926.
14 Bruun AF. Til New Zealand og tahiti for at fange aal. In: Winge Ø, Vedel Taaning Å, eds. *Naturforskeren Johannes Schmidt. Hans liv og ekspeditioner*. Copenhagen: Gyldendalske Boghandel Nordisk forlag; 1947:132–136.
15 Aoyama J. Life history and evolution of migration in catadromous eels (genus Anguilla). *Aqua-BioScience Monographs*. 2009;2(1):1–42.
16 Schmidt J. The freshwater eels of New Zealand. *Transactions and Proceedings of the Royal Society of New Zealand*. 1928:379–388.
17 RA. *Schmidt, Ernst Johannes, biolog*: Materiale vedr. ekspeditioner og rejser (1876–1933) 1: Blade fra min Dagbog fra en Rejse til TAHITI. Marts – April 1926.
18 RA. *Schmidt, Ernst Johannes, biolog*: Materiale vedr. ekspeditioner og rejser (1876–1933) 1: Blade fra min Dagbog fra en Rejse til TAHITI. Marts – April 1926.

it to be, she wrote it off as the fault of the European colonisers. Her observation of the mixing of races, the moral decay, and the depraved behaviour brought about by the white settlers however, did not distort the overall impression.

In the fall of 1926, when they were all back in Denmark Ingeborg gave an interview to the leading Danish ladies magazine, *Tidens Kvinder*, in which she did her best to convey her perception of a true paradisiac experience in the South Seas. She described how, 'Everything you need for your household, you just go and pluck from the trees or the land or the sea.'[19] Under the headline of 'Tahitian Hunt for Eel,' The article ran over two full pages with several illustrations showing the Danish contingent in white tropical clothing, while the indigenous Tahitians were portrayed via two postcards each showing a young woman clad in traditional clothing and adorned by colourful flowers and pearl necklaces. The pictures were accompanied by texts saying 'typical Tahitian beauty' or 'the little Tahitian girl is naturally modest, always showing herself in civilised and modest attire.'[20]

For the Schmidt couple, eagerly trying to drum up interest in a circumnavigation among potential sponsors, this article in *Tidens Kvinder* will have counted as a flawless public relations activity. The article explicitly referred to some of the most popular images of the South Seas in Danish and European culture at the time, the colourful painting by Paul Gauguin, who had married a well-to-do Danish lady before re-settling in Tahiti, and secondly the writer of travel accounts, Pierre Loti. In turn they were all playing on a much older vision of Tahiti as Paradise Lost, which first arose from the 18th century travel accounts of James Cook and Louis-Antoine de Bougainville.

There is no reason to believe that Johannes Schmidt's viewpoints differed substantially from his wife's on the people and nature of Tahiti, however Schmidt himself appears to have been kept busy with his work on the resident eel populations during their stay. Within the relatively short time of six weeks, which they spent there, Schmidt collected hundreds of specimen, as well as engaging local hired helpers to find eel for him at different sites in the rivers and lagoons around Tahiti. Once more, he was successful in unravelling a substantial part of the life history of eel populations. As it turned out, there were no less than three different types of freshwater eel around Tahiti, each thriving in their own niches. One of them for instance, was forcefully making its way upstream past small waterfalls to a significant altitude.[21]

19 Debora, 'Paa Aalejagt til Tahiti', *Tidens Kvinder*, 30 September 1926, 6–7.
20 Debora, 'Paa Aalejagt til Tahiti', *Tidens Kvinder*, 30 September 1926, 6–7.
21 Bruun AF. Til New Zealand og tahiti for at fange aal. In: Winge Ø, Vedel Taaning Å, eds. *Naturforskeren Johannes Schmidt. Hans liv og ekspeditioner*. Copenhagen: Gyldendalske Boghandel Nordisk forlag; 1947:132–136.

13.1 Lunch in Papeete in 1926
For Ingeborg Schmidt the visit to Tahiti was the highlight of the circumnavigation in 1925–1926. According to her diary as well as a later interview she was completely taken aback by the exoticism in the South Seas. Here she is seated in the middle flanked by Gertrud Ostenfeld and Johannes Schmidt (RA. Schmidt, Ernst Johannes, biolog. 1)

Fundraising in the USA

Leaving Tahiti behind, Schmidt was gearing up for the second part of his plan – fundraising for a full scale research cruise around the World. In previous commentary on Schmidt's work there is little information to be gained on his visit to the USA in 1926. Perhaps this is due to the fact that this part of the trip was not as successful as he had hoped.

At first Schmidt had sought advice from Adam Bøving, a Danish born zoologist, who had immigrated to the USA in 1911 and now served in the Bureau of Entomology. Schmidt no doubt knew Bøving from his time in Denmark, and approached him in 1925 with a request about how to approach potential American funding agencies. Bøving recommended that Schmidt should find a way to meet representatives from the Rockefeller Foundation in person. During the 1910s and 20s the US charitable foundation, The Rockefeller Foundation became internationally famous for its support of dedicated and problem oriented research centres around the World, and from at least the spring of 1925, Schmidt entered into conversation with its International Education Board, hoping to land a grant large enough to support his interests in the eels of the Indo-Pacific area.

The most important man was Dr. Wickliffe Rose, President of The International Education Board of the Rockefeller Foundation in New York City. Bøving had several recommendations including the importance of trying to find a way to decorate the American officials he was going to work with 'To make things glide better,' as Bøving put it. Bøving was very optimistic and reminded Schmidt that Niels Bohr and August Krogh had both recently been successful with their engagement with the Rockefeller Foundation.[22] Schmidt may very well have been inspired by the other prominent Danish scientists who had made an effort to attract the attention of the Rockefeller Foundation. The Nobel Laureate in medicine, August Krogh was working simultaneously along similar lines and had managed to secure funds for the creation of the Rockefeller Instituttet in Copenhagen, which opened in 1928 focusing on research in medical physiology and biochemistry. When Schmidt first approached them in April 1925, the response was negative, although the foundation acknowledged having an interest in both Denmark and in marine science.[23]

Therefore, as Schmidt was planning a trip to the USA in 1926, there was good reason to renew the contact with 'The Rockefeller Men,' as they were called by the Danish Minister (Chargé d'affaires) Constantin Brun at the Royal Danish Legation, when writing to Schmidt. In January 1926, Brun approached Wickliffe Rose. Brun simply asked if the board would be interested in supporting another expedition led by Dr. Schmidt, whom Rose would surely know of already. From the letter it appears Brun did

22 KB. Johannes Schmidt 'Papirer'. Ut.374.II.4. A. Bøving to J. Schmidt, 7 March 1925

23 KB. Johannes Schmidt 'Papirer'. Ut.374.II.4. Wickliffe Rose to S. Sørensen, 14 April 1925.

not present any solid arguments as for why they should do so other than the stating the fact that it was important.[24] In any case, Wickliffe Rose of the Rockefeller Foundation immediately replied back that at present they were not about to enter into support of this field of research.[25]

When Schmidt landed in San Francisco he was able to pick up several letters concerning his fundraising tour, including a letter from the Danish Legation in Washington DC. Once Schmidt made it to the east coast, it was arranged for him to give a lecture at the Smithsonian Institution in Washington. In addition Brun was planning to throw a dinner in Schmidt's honour, where an A-list of US dignitaries would be invited. In addition to Rose from the Rockefeller Foundation, the list included prominent American natural scientists such as a marine scientist Henry Bigelow from Harvard University, Thomas Hunt Morgan, from Columbia University who would later be awarded the Nobel Prize in medicine, Captain James P. Ault, representing another major research endowment, the Carnegie Institution of Washington, the head of the Smithsonian Institution, Charles D. Walcott, and several other high ranked officials from the US Navy and the US Department of Agriculture – and not least Adam Bøving.[26]

Bøving clearly worked hard to promote his former countryman in his attempt to impress the Americans. Bøving sent several updates to Schmidt's poste restante address in San Francisco in April and May 1926. On the 8 May, he wrote to Schmidt that no less than 1,000 invitations had been sent out advertising his lecture in Washington, where the Smithsonian Institution was now co-hosting the event with the Carnegie Institution also in Washington. One mishap though, was the fact that the Rockefeller Foundation did not wish to have Schmidt lecture in New York City, where they were based, since it was too late in the academic season to plan such an event. Bøving tried to comfort Schmidt saying that it was for the better, since everything was on schedule in Washington, which was preferable than a badly organized lecture in New York. Nonetheless, Bøving was probably disappointed on behalf of Schmidt, as he blamed the local Danish Consular General in New York for having done absolutely nothing to promote the cause. Then Bøving moved on to the next item, the promotion of Schmidt as an honourable fellow of the Washington Academy of Sciences. Bøving informed Schmidt that he was trying to have him promoted in time to receive this honour in immediate connection with his lecture in early June the same year.[27] By the time that Johannes and Ingeborg Schmidt and Gertrud

24 KB. Johannes Schmidt 'Papirer'. Ut.374.II.4. C. Brun to Wickliffe Rose, 13 January 1926.

25 KB. Johannes Schmidt 'Papirer'. Ut.374.II.4. Wickliffe Rose to C. Brun, 22 January 1926.

26 KB. Johannes Schmidt 'Papirer'. Ut.374.II.4. C. Brun to J. Schmidt, 8 May 1926.

27 KB. Johannes Schmidt 'Papirer'. Ut.374.II.4. C. Brun to J. Schmidt, 8 May 1926.

Ostenfeld boarded a steamer in New York to take them back to Europe, the honourable fellowship was the only substantial outcome of the American leg of their circumnavigation. Clearly the arranged business dinner and the scheduled meetings did not turn out the way Schmidt had hoped.

Following his return to Denmark, Schmidt was clearly frustrated that nothing had come of his trip. Considering his previous health problems he may have feared that time was running out for him. At the age of 49 he was far from being old, but perhaps did not have the physical strength he had possessed a few years previously. Schmidt blamed Brun. Something did not go as planned. It appears that someone at the Danish Legation in Washington DC forgot to send an invitation to Wickliffe Rose at the Rockefeller Foundation.[28] Instead Schmidt travelled to New York City to meet Wickliffe Rose, but Rose was not available at that particular moment.[29] This was a huge disappointment for Schmidt. Nine months later, Schmidt still felt the need to remind Brun that this was an act of inexcusable negligence.[30] Brun though, defended himself by stating that Schmidt indeed had got the chance to meet and talk to Rose, as he had been present at the dinner in Washington hosted by Brun on behalf of the Royal Danish Legation, and indeed Rose had received an invitation for the lecture as well, although admittedly at the last minute.[31]

Brun continued to contact Rose at the Rockefeller Foundation and in January 1927, Rose replied that marine science was indeed a growing field of great importance, but he also underlined that they were not likely to arrive at any conclusions in the near future. Towards the end however, he indicated that 'we can only hope to continue discussions with the men who must take a lead in this country in developing the field.[32] Schmidt replied that Brun should indeed proceed in this direction. Nothing more was to be gained through direct access with Rose, the approach was to come via American colleagues.[33]

While Schmidt was the uncrowned champion when it came to fundraising for research in Denmark, it looks as if both he and not least the Danish Legation in Washington DC still needed to learn the ropes when dealing with foundations in the USA. Nonetheless, through a concerted effort in 1926 and the spring of 1927 Schmidt and Brun managed to have their American connections knighted by the Danish state in recognition of their significance for a small country in Northern Europe. Members of the Rockefeller Foundation's International educational Board however, were not able to receive such honours, so Rose had to decline.[34]

28 KB. Johannes Schmidt 'Papirer'. Ut.374.II.4. J. Schmidt to C. Brun, 28 February 1927.

29 KB. Johannes Schmidt 'Papirer'. Ut.374.II.4. C. Brun to J. Schmidt, 22 July 1926.

30 KB. Johannes Schmidt 'Papirer'. Ut.374.II.4. J. Schmidt to C. Brun, 28 February 1927.

31 KB. Johannes Schmidt 'Papirer'. Ut.374.II.4. C. Brun to J. Schmidt, 21 July 1927.

32 KB. Johannes Schmidt 'Papirer'. Ut.374.II.4. Wickliffe Rose to C. Brun, 8 January 1927.

33 KB. Johannes Schmidt 'Papirer'. Ut.374.II.4. J. Schmidt to C. Brun, 28 February 1927.

34 KB. Johannes Schmidt 'Papirer'. Ut.374.II.4. C. Brun to J. Schmidt, 21 July 1927.

In any case a new possibility now emerged. Following a half year long journey round the World, the solution lay just across the street from his own house, on the other side of Carlsbergvej. Here, the kettles at Carlsberg were brewing at unprecedented rates and the surplus was mounting in the vaults of the owner, The Carlsberg Foundation.

Approaching the Carlsberg Foundation

There are two key players to understand, how Schmidt in 1927 was able to land a huge grant of 500,000 Kroner from the Carlsberg Foundation, These are Carl Ostenfeld and the recent Chairman of the Board of Directors, Anders Bjørn Drachmann. Ostenfeld was an important stakeholder in Danish marine science, and having a seat on the foundation's Board of Directors, was crucial for the Carlsberg Foundation's decision to fund Schmidt's expedition. Nonetheless, Ostenfeld still had to depend on the good will of the other members of the Board of Directors, chiefly Anders Bjørn Drachmann (1860–1935), who had replaced Kristian Erslev as chairman of the foundation in 1926. Drachmann was a professor of classical philology at Copenhagen University, famous for pioneering the use of modern methods of text criticism on ancient records. Judging from a brief encyclopaedic entry, Drachmann was shy of public exposure, although carrying himself in a dignified way. He was also described as having a sense of justice, and a conscientious and careful work ethic, someone who wielded his considerable influence in a discreet fashion.[35] Although, Drachmann came from a similar bourgeois and intellectual background as that of Johannes Schmidt, their personalities were worlds apart, and these differences were to surface when Schmidt was at the opposite side of the world.

In the summer of 1927 though, all hopes were high, as Schmidt continued to look for the core funding for his expedition, also in the United States. The Danish Minister in Washington, Brun was still trying to unlock the vaults of the Rockefeller fortune. Austin Clark of the Smithsonian Institution in Washington was advicing on how to best approach the Rockefeller Foundation by way of introduction from important people in American science. Clark suggested to go through Vernon Kellogg at the National Research Council, John Merriam who was President of the Carnegie Institution and finally Waylan Vaughan, Director at the Scripps Institution of Oceanography in California.[36]

35 Adler A. Drachmann, Anders Bjørn. In: Engelstoft P, Dahl S, eds. *Dansk Biografisk Leksikon*. 3rd ed. Copenhagen: Gyldendal; 1980:28–30.

36 KB. Johannes Schmidt 'Papirer'. Ut.374.11.4. Austin H. Clark to C. Brun, 16 July 1927.

Meanwhile, from his summer holiday at Tisvildeleje, then a couple of hours north from Copenhagen, Ostenfeld updated Schmidt on recent events. Ostenfeld had just paid a visit to Drachmann's summer quarters, where they had discussed the plans for the great expedition. Drachmann was very enthusiastic, and was particularly taken with the idea of the Carlsberg Foundation taking on the entire expedition so that it became the foundation's own expedition. Money was not a cause of great concern, and Ostenfeld reckoned that he and Schmidt could count on Drachmann to give his wholehearted support. Ostenfeld indicated that typically Drachmann would be anything but petty-minded once he had set his heart on a project.[37]

Schmidt and Ostenfeld then proceeded to secure the interest of the Danish state to back the circumnavigation. The first step was to convince the Fisheries Director, F. V. Mortensen on the importance of such a trip. Schmidt succeeded in this to the extent that Mortensen drew up a budget in August 1927, based on the expedition making the best use of the government research vessel the R/V *Dana*. This was not uncontroversial, as the *Dana* would have to be taken out of its regular service for fisheries research in Danish waters. A sum of 500,000 was the result, but before approaching the Carlsberg Foundation, Schmidt wanted to be certain that they could count on the *Dana* for two years' of research. To this end, Schmidt and Mortensen entered into conversations with the Prime Minister of Denmark, Madsen-Mygdal. From this meeting Schmidt was able to send Ostenfeld the encouraging message that the Prime Minister was unconditionally in favour of this proposal.[38]

Then, Ostenfeld spoke again with Drachmann about the expedition plan, presenting a budget of 500,000 Kroner. The chairman was 'not significantly surprised about the size of the proposed grant application.'[39] Thus, Drachmann decided to place the proposed expedition on the agenda at the upcoming board meeting scheduled for the following week on the 7 September.[40] Schmidt was summoned to present the plan to the board and he reported that the Danish state had already promised to make available the steamship and research vessel, the *Dana*. Favouring what they heard, the board then decided to have a site visit on board the *Dana*, scheduled for the following Saturday afternoon.[41]

After this meeting things fell into place in quick succession. On 12 October, the Board of Directors decided to grant the sum of 500,000 Danish Kroner for '...an oceanographical expedition to the Indo-Pacific area, with a special emphasis on investigating the whereabouts of

37 KB. Johannes Schmidt 'Papirer'. Ut.374.I.7. 'Korrespondance 1927–1929'. C. H. Ostenfeld to J. Schmidt, 5 August 1927.
38 KB. Johannes Schmidt 'Papirer'. Ut.374.VI.1. 'Dana-Expeditionen, 1928–30', J. Schmidt to C. H. Ostenfeld, 25 August 1927.
39 KB. Johannes Schmidt 'Papirer'. Ut.374.I.7. 'Korrespondance 1927–1929'. C. H. Ostenfeld to J. Schmidt, 1 September 1927.
40 KB. Johannes Schmidt 'Papirer'. Ut.374.I.7. 'Korrespondance 1927–1929'. C. H. Ostenfeld to J. Schmidt, 1 September 1927.
41 CFD. Carlsbergfondets Direktions Protokol, vol. 6, 7 September 1927.

the larvae of species of freshwater eel in the area.'[42] In addition it was decided to establish an expedition committee with relevant parties involved, including representation from the Carlsberg Foundation, a role which fell to Ostenfeld. Schmidt, who had kept his search for money from the United States on the back burner for a while sent a note to the Danish Charge d'Affaire, Brun already on the 16 October that he could now call off the search, as a good solution was found through the Carlsberg Foundation and the Danish Government.[43]

One suggestion by Ostenfeld that was approved was the possibility of the brewery handing over a larger share of the surplus to Section B of the Carlsberg Foundation, i.e. the statutory body issuing research grants, if extra funds for the expedition were required.[44] Here, it is important to note that the Board of Directors of the Carlsberg Foundation also represented the controlling part of the shares of the Carlsberg Breweries, including the chairmanship of the brewery's corporate board. Thus to a large extent, Drachmann and the other four members of the board were partly in control of turning up and down the flow of money from the beer taps according to their expected needs of expenditure in the Carlsberg Foundation. The liberal attitude of the Board of Directors to Schmidt's proposals reflects that the earnings from the brewery rose formidably during the latter half of the 1920s, more than doubling the foundation's fortune in the course of a few years.[45]

During the following months, Schmidt and Ostenfeld were both active in formalizing their connections to the various state departments as well as in the corporate and scientific worlds, which would ultimately lead to the formation of the expedition committee. It was also discussed whether a larger ship than the *Dana* would be available, but this proved not to be the case. These developments prompted several members of the Board of Directors to raise the issue to Ostenfeld that the expedition committee was a superfluous arrangement, which might endanger the idea that this was the Carlsberg Foundation's own expedition. To this, Ostenfeld stated that he would make sure, that the foundation was visible.[46]

42 CFD. Carlsbergfondets Direktions Protokol, vol. 6, 12 October 1927.
43 KB. Johannes Schmidt 'Papirer'. Ut.374.II.4. J. Schmidt to C. Brun, 16 October 1927.
44 CFD. Carlsbergfondets Direktions Protokol, vol. 6, 12 October 1927.
45 Glamann K. Carlsbergfondet. Copenhagen: Rhodos; 1976. p. 206–7.
46 CFD. Carlsbergfondets Direktions Protokol, vol. 6, 23 November 1928 & 13 February 1928.

A census of marine life – the cruise plan presented

By 21st century standards, the core interest group acted very quickly. The first formal meeting of the Committee for the circumnavigation was held 9 May 1928 at the Carlsberg Foundation, less than two months before the expedition commenced from the Port of Copenhagen. Ostenfeld was elected chairman of the expedition committee, Vice-Admiral Amdrup from the Danish Navy as committee Vice President. Jointly with Schmidt, Fisheries Director Mortensen and Admiral Rechnitzer, formed an executive board for the expedition. As a man of many trusted duties in the Danish world of research, Ostenfeld underlined that he saw himself appointed by the Carlsberg Foundation to this new position. The same foundation which had put 500,000 Danish Kroner at the disposal of Professor Johannes Schmidt to conduct a two year expedition studying the biology of the eel in The Pacific Ocean as well as in The Indian Ocean. The Danish state loaned the expedition the marine science vessel, the R/V *Dana*, fully equipped. Finally, the navy loaned some of its seamen to act as crew-members on board the research vessel, and the Naval Ministry paid their salaries.[47]

In light of the diverse set of results the expedition produced, it is striking to find a summary of Johannes Schmidt's vision as it was presented to the expedition committee. Schmidt presented the task of the forthcoming expedition, as having '…the main objective to study the *Anguilla* family in the Pacific and Indian Oceans.'[48] This was a special task demanding a considerable amount of effort on the behalf of the members of the expedition who were charged with the task of logging all the migration routes and biology of all species of eel the expedition committee wished to be charted. Schmidt emphasized that the Carlsberg Laboratory had already carried out a lot of work on eel samples derived from the Pacific Ocean and the Indian Ocean, not least as a consequence of Schmidt's own circumnavigation two years earlier. Therefore, the assembled team was unusually well equipped to undertake such a vast survey on the other side of the Earth. No doubt the life history of the world's eel populations was the selling point of the expedition. Although, other scientific aspirations are also duly mentioned in the accompanying expedition plan presented to the committee.[49]

Well aware of history, the grandeur of the expedition was framed against the background of the first Danish marine science circumnavigation of world, the westbound voyage of the navy corvette, the *Galathea* in 1845–47, as well as the British Challenger Expedition of the 1870s. Now, in the 20th century, Schmidt argued, the revolutionary advances in marine science technology lent themselves to renewed global enterprises.

47 CFD. Vedr. DANA II, 1828–1930. 103/1928. Minutes, DANA Committee, 1st meeting, 9 May 1928.
48 CFD. Vedr. DANA II, 1828–1930. 103/1928. Minutes, DANA Committee, 1st meeting, 9 May 1928.
49 CFD. Vedr. DANA II, 1828–1930. 103/1928. Expedition Plan, DANA Committee, 1st meeting, 9 May 1928.

Schmidt and his expedition committee then drafted the eastbound course, where the first major port of call would be Spain, then Portugal and further through the Strait of Gibraltar into the Western parts of the Mediterranean. Thereafter, Madeira, the Azores, and the West Indies were stopovers across the North Atlantic Ocean before reaching the Panama Canal and the Pacific Ocean. The expedition would then chart a course taking them to Tahiti, the Fiji Islands, and New Caledonia, then a French dependency, before reaching the waters around New Zealand and Eastern Australia. Then, the plan would be to conduct a more detailed survey off Eastern Australia for a couple of months. In March of 1929, they were scheduled to continue northwards, heading for Japan and China. Later on, the Malayan and Indonesian Archipelago would be paid a visit. In the fall of 1929, the idea was to conduct a proper investigation of the Indian Ocean, charting a line from Java to Madagascar. From Madagascar, the route would proceed northwards along the East coast of Africa, through the Red Sea and back into the Mediterranean. In the plan presented, four areas around the world are given special attention. These were the Strait of Gibraltar, the Strait of Bab-El-Mandeb, the sea around New-Caledonia, and finally, the Eastern Mediterranean. The date of homecoming for the expedition was set for 'the spring of 1930'.[50] As we will see below, neither the date of homecoming, nor the choice of these four focus areas, or the route around Africa would remain the same throughout the expedition.

An enduring feature of the expedition was physical oceanographic sampling. First of all, measurements of the ocean depths would be made thanks to the fairly novel instrument of an echo-sounding machine. The use of echo sounding had been pioneered in the beginning of the 20th century, and especially during World War I, great advances were made, so in the 1920s the first industrially produced models became available.[51]

Schmidt explained to the committee, how the echo-sounder was a device for measuring the speed of sound through the water column from the surface to the bottom and back again to the surface. The biggest known depth in the Pacific Ocean was estimated at around 10,000 metres, but for many parts of the World's oceans, the actual depths of the sea remained unknown.

Water samplers, pioneered and perfected on many previous cruises, would be used for measuring the salinity, the content of nitrate and phosphate, as well as the temperature of the sea as an approximation of

50 CFD. Vedr. DANA II, 1828–1930. 103/1928. DANA Committee, 1st meeting, 9 May 1928

51 Smed, Jens, 'Ekkolod og Asdic', in: *Naturens Verden*, 32, (1948), p. 177–183.

how the large currents flowed across the oceans. A further purpose of these measurements was to estimate the quantity of various nutrients, substances of vital importance for understanding the oceans' biomass – or the 'Urnæring' as Schmidt called it.[52]

How was this 'Urnæring', or biomass composed? A census of marine life was next up on the agenda. The plants and animals flowing freely in the water column, the so-called pelagic species, were to be investigated with large seines made of silk or hemp. The seines would have a round entrance, a ringed net up to 3 metres in diameter, which would then be lowered down the water column attached to a wire. This technology was very similar to that used by the preceding Atlantic deep sea expeditions with the *Dana*, but Schmidt emphasized rightly that this had never been done on such a scale across the Pacific Ocean and the Indian Ocean.

Contrary to this pelagic focus, the bottom of the sea was not to be investigated in such detail, not least because this would have been very time consuming. Although, according to the original plan put forward, Schmidt's long-time collaborator, Theodor Mortensen was to conduct such investigations in the area around New Caledonia, when the time schedule allowed. As Schmidt continued, he highlighted his many previous successes, such as his brief foray into the Pacific Ocean in 1922 on board the R/V *Dana*. The few samples collected indicated a richness of zoological treasures, which prompted Schmidt to suggest that new 'wonder creatures' could once again be hauled from the abyss. Then, Schmidt listed a number of more standardized but equally important tasks, such as examining the ocean depths, ocean currents, species abundance of plants and animals, as well as analyses of why the species abundance was so unevenly distributed across the oceans.

As a final set of remarks, Schmidt mentioned that the expedition would be host to a number of popular science lectures around the world. The expedition would carry along a copy of the so-called *Danmark-Filmen* to be shown in cinematic theatres, wherever this was appropriate.[53]

Planning equipment, engine, and crew

Once the funding was secured, less than half a year in total elapsed before the ship was ready to steam out from Danish waters. Some of the practicalities, for instance the crossing of the Pacific Ocean, remained

52 CFD. Vedr. DANA II, 1828–1930. 103/1928. DANA Committee, 1st meeting, 9 May 1928

53 CFD. Vedr. DANA II, 1828–1930. 103/1928. DANA Committee, 1st meeting, 9 May 1928.

a blind spot for both Schmidt and Ostenfeld. Ostenfeld had been to Australia in 1913, and Schmidt had just been round the World in 1925–26. Yet, sailing the steam trawler, the *Dana* across the largest ocean in the World had never been tried before. Schmidt therefore approached G. Schott from the Deutsche Seewarte, the German institute for maritime meteorology. Schott was very accommodating. He advised for instance that Schmidt should embark from Denmark in June of 1928 and try to be in Japan by June of 1929 to avoid as much of the typhoon season as possible. Being in Japan in the summer of 1929 would also enable the expedition to avoid the strong south-easterly winds in the Indian Ocean that occurred around this time.[54]

Schott was also able to provide advice on crossing such huge expanses of water with limited amounts of fuel. The *Dana* could only carry so much coal on-board for the steam boiler. Schott did not see any problems with the 3,012 nautical miles between the Sunda Strait in Southeast Asia and Mauritius in the western part of the Indian Ocean. The stretch from Panama to Tahiti on the other hand needed careful calibration if the 4,570 nautical miles were to be covered. Schott was well aware that the Galapagos Islands provided a natural history delicatessen for any marine scientist, but he also advised against coaling there, as this was too expensive and too time consuming compared to just passing by. In general though, the trade winds would be favourable to the expedition, and the cheapest insurance against running out of coal he could give, was to carry along sails to hoist if necessary.[55]

At the second meeting of the expedition committee on 8 June 1928, many practical details were still being settled. The echo sounder from the Atlas-Werke company in Bremen, Germany was now installed. A technician from the German company had travelled to Denmark and secured the installation. Since this was such a novel technology, he was contracted to travel with the expedition for the first couple of months to make sure that the echo sounder lived up to expectations. The machine needed to work reliably for two years, and from the other side of the planet, there was little chance to conduct serious repairs if it did not work as planned.[56] Just to be sure, and as a supplement, the expedition would carry with them 40 kilometres of piano wire, a strong steel wire. This was to be used for the old fashioned way of measuring depths, where a cannonball would be attached to one end of for instance 6,000 metres of piano wire, wound around a metre wheel. The cannonball would then be lowered down the water column until it reached the

54 KB. Johannes Schmidt 'Papirer'. Ut.374.I.7. 'Korrespondance 1927–1929'. G. Schott to J. Schmidt, 6 December 1927.

55 KB. Johannes Schmidt 'Papirer'. Ut.374.I.7. 'Korrespondance 1927–1929'. G. Schott to J. Schmidt, 6 December 1927.

56 KB. Johannes Schmidt 'Papirer'. Ut.374.VI.1. 'Dana-Expeditionen, 1928–30', Minutes, 2nd meeting of the Komiteen for Dana Expeditionen, 8 June 1928.

bottom. This whole process would take hours each time a measurement was taken, and additionally the ship had to lie still during the operation. The echo sounder however, was operational at full speed, and as the emitted sound travelled 1,500 metres per second, the whole process was over in a few seconds.[57]

Another leap forward technologically was use of wireless radio transmission. In all of Schmidt's previous expeditions, coordination between the ship and the expedition committees relied on finding a shore-based telegraph station. Now, using a radio station, the expedition on board the *Dana* was able to communicate directly from waters Southeast of New Zealand to Denmark, almost antipodal points on the globe, via the increasingly global grid of radio signal transmitters.[58]

All this hardware needed insurance. The ship itself needed insurance and for a premium of about 18,000 Kroner, Danish Lloyd had offered to insure the ship for a sum of 300,000 Kroner and the equipment on board for a further 100,000 Kroner.[59]

Then, there was the question of the 'software', or the crew of the expedition, which was drawn from three spheres of maritime life who did not usually work together. Schmidt was in charge of picking the right marine scientists. The Danish government was supplying the ship through the fisheries division of the Department of Agriculture, and finally the navy ministry supplied a number of sailors and ships' officers. As a veteran leader of scientific expeditions Schmidt had a clear understanding of which staff he needed for the circumnavigation.

This became evident, when one particular captain lieutenant of the navy, Fangsbøll, was enlisted. Schmidt was certain he would not fit into the assembly of civilians and military personnel, which Schmidt then explained to Fangsbøll's handling officer Admiral Rechnitzer at the Navy Ministry. In May 1928, Schmidt informed the admiral on some of his other thoughts on having civilians as well as military seamen on board the same vessel. Having a joint uniform was important to Schmidt. Since it was a civilian expedition with no apparent military aspect, Schmidt proposed the uniform to be more akin to what was normally found within the Department of Agriculture than the equivalent within the ranks of the navy. The main objective, Schmidt emphasized, was to have the different groups on board melt together into one entity, for the benefit of the expedition. Should an animosity arise between the civilians and the navy personnel, Schmidt feared that he would have to act as a middleman, and this was something that he wished to avoid.[60]

57 Bruun AF. Jordomsejlingen med S/S 'Dana' (II) 1928–30. In: Winge Ø, Vedel Taaning Å, eds. Naturforskeren Johannes Schmidt. Hans Liv og Ekspeditioner. Copenhagen: Gyldendalske Boghandel Nordisk Forlag; 1947:146.

58 Mou G. Lidt om Dana's Radiostation. In: Schmidt J, ed. Dana's Togt omkring Jorden 1928–1930. ; 1932:365–368.

59 KB. Johannes Schmidt 'Papirer'. Ut.374.VI.1. 'Dana-Expeditionen, 1928–30', Minutes, 2nd meeting of the Komiteen for Dana Expeditionen, 8 June 1928.

60 KB. Johannes Schmidt 'Papirer'. Ut.374.I.7. 'Korrespondance 1927–1929'. J. Schmidt to Rechnitzer, 12 May 1928.

With regards the particular officer in question, Schmidt had heard him state on several occasions that he would not be eating in the same ship's mess as this or that other person. To Schmidt this was an unacceptable behaviour from an upstart: '…in my opinion one should be careful about attempting to reform long accustomed traditions, as long as one has not acquired the skills expected to fulfil one's duties on board the ship.'[61] While separate officer's messes were the norm on larger navy vessels, this was not an option on the R/V *Dana*, a smallish steam trawler. Rechnitzer replied the same day, that he had held a meeting with Fangsbøll, who himself had expressed a reservation towards the prospect of going round the world with the *Dana*. This, therefore, proved to be an easy decision.[62]

Finally, the *Dana* was staffed with 26 seamen and scientists at a time.[63] For Captain, Georg Hansen was the natural first choice. However, this time he would have to stay at home, as he was recovering from a breakdown, and it was doubtful as to whether or not he would be able to command the ship at a later stage, as Ostenfeld put it in the minutes. Instead, a new captain Christensen, drawn from the ranks of the navy, would commence the voyage at the helm, and he had committed himself to the entire expedition, if needed. Second choice was the navy lieutenant Ries, while the second in command Captain Lieutenant Greve also was able to navigate the vessel, when required.[64] The Danish navy also supplied the vessel with a chief engineer of the steam engine and two assistant engineers, a navy boatswain and three seamen, all of whom would be mainly occupied with fuelling and maintaining the steamship. Apart from Captain Hansen who joined later, the civilians on board the *Dana* also included a steward, a cook, a cabin boy, one boatswain, two ship-mates, and two ordinary seamen.[65]

Then, there was the scientific staff of the *Dana*. Schmidt himself was to be on-board for the first month, but he was then scheduled to leave the ship and would not return to command the expedition until mid-March 1929. In the intervening period Schmidt's long-trusted marine scientist, Dr Poul Jespersen, who earned his stripes as the only scientist on board the unfortunate M/S *Margrethe*, 15 years earlier, would stay on board until the end of November 1929, when he would be replaced by another of Schmidt's former students, Aage Vedel Taaning. Taaning was also asked to take command of the expedition during the first three months of 1930, when Schmidt was again otherwise engaged. Anton Bruun was another member of the crew. As mentioned above he first

61 KB. Johannes Schmidt 'Papirer'. Ut.374.I.7. 'Korrespondance 1927–1929'. J. Schmidt to Rechnitzer, 12 May 1928.

62 KB. Johannes Schmidt 'Papirer'. Ut.374.I.7. 'Korrespondance 1927–1929'. Rechnitzer to J. Schmidt, 12 May 1928.

63 CFD. DANA. BS3 1. Dana-Expeditionen: Gratialer, 8 July 1930.

64 KB. Johannes Schmidt 'Papirer'. Ut.374.VI.1. 'Dana-Expeditionen, 1928–30', Minutes, 2nd meeting of the Komiteen for Dana Expeditionen, 8 June 1928.

65 CFD. DANA. BS3 1. Dana-Expeditionen: Gratialer, 8 July 1930.

13.2 Members of the expedition in Durban, 1930
The members of the expedition came from different backgrounds including the navy, civilian fishermen and scientists. To enhance the team spirit Schmidt had specially designed uniforms made. Here the expedition members are in Durban, South Africa in January 1930 (Carlsberg Archives).

joined Schmidt for the 1924 cruise to Iceland as a student.[66] This time, he would be there for the entire expedition, along with two natural history students, Helge Thomsen and Einar Nielsen. Finally, the ship's doctor, N. C. Andersen was also counted among the ranks of scientists. Having assisted Schmidt on several previous occasions, Andersen could also be entrusted with science related tasks on-board. When, midway through the expedition Andersen fell ill himself, he was replaced by a medical doctor from the navy, Strøyberg.[67]

66 RA. *Anton Fr. Bruun. Danaekspeditionerne 1924–1930*. Dagbog fra 'Dana' Expeditionen 1924.
67 CFD. DANA. BS3 1. Dana-Expeditionen: Gratialer, 8 July 1930.

Institutional arrangements

Although the circumnavigation was a Danish organized, Danish led, and Danish sponsored enterprise, Schmidt was keen to include ICES. As a token of appreciation of the international collaboration in ICES, the *Dana* was going to fly the flag of the international council.[68] This would also ensure that the expedition would receive a good share of collegiate attention and appreciation.

At the third meeting of the expedition committee in the summer of 1928, the CEO of the East Asiatic Company (EAC), Schmiegelow was welcomed as a member of the committee. Schmiegelow had brought with him a much-appreciated gift. The East Asiatic Company would handle all shipment of additional supplies to the ship, including equipment for the scientific surveys, free of charge. Moreover, Schmiegelow promised to take care of all homebound shipments of the natural history collections expected from the circumnavigation.[69]

The expedition would also need local currency while navigating far away seas and territories. To this end Schmidt was able to draw money from a credit account opened by the Carlsberg Foundation. From the foundation's Copenhagen headquarters, the chief accountant, the so-called *Kvæstur*, would then transfer additional money during the course of the expedition. In the more distant corners of the world, seen from Copenhagen, it happened that there were no feasible banking facilities, here the East Asiatic Company offered to help out with cash through their local representations.[70] In September 1929 for instance, the Copenhagen HQ of the East Asiatic Company informed the Carlsberg Foundation that 50,000 Danish Kroner had now been placed in the Singapore branch of East Asiatic Company at the planned disposal of the expedition.[71] In January 1930 the same procedure was used, when the South African branch of the East Asiatic Company handled the transfer of another 50,000 Kroner from the Carlsberg Foundation to the *Dana* expedition.[72]

On many occasions throughout his career Schmidt had sported a very explicit patriotism. He frequently used the opportunity to promote his scientific enterprises with adherence to the fact that this or that research result should be heralded as Danish. In this respect, the circumnavigation with the *Dana* was a showcase for Danish enterprise and ingenuity. In early 1928, Schmidt was in contact with the Danish Ministry of Foreign Affairs about ways in which the expedition could

68 Went AEJ. *Seventy years agrowing. A history of the international council for the exploration of the sea, 1902–1972.* Copenhagen: International Council for the Exploration of the Sea; 1972. P. 74–75.
69 KB. Johannes Schmidt 'Papirer'. Ut.374.VI.1. 'Dana-Expeditionen, 1928–30', Minutes, 3rd meeting of the Komiteen for Dana Expeditionen, 14 July 1928.
70 RA. Schmidt, Ernst Johannes, biolog: Materiale vedr. ekspeditioner og rejser (1876–1933) 2: m.m. 27 may 1929.
71 CFD. Vedr. DANA II, 1828–1930. 103/1928. Letter from Gotfred Johansen EAC to Carlsbergfondets Direktion, 17 September 1929.
72 KB. Johannes Schmidt 'Papirer'. Ut.374.VI.1. 'Dana-Expeditionen, 1928–30', J. Schmidt to Komiteen for Dana Expeditionen, 6 February 1930.

highlight and showcase aspects of Denmark. While the scientific investigation in itself was spectacular, it was also planned that the expedition should carry with it two promotional films. One was the so-called *Færø Filmen* showing images of rural life in the North Atlantic to be shown in port in Spain, while the second was the so-called *Danmark-Filmen*, which was to be shown in cinemas around the world.[73] The *Danmark-Filmen* was a two hour and forty minutes long documentary showing Denmark at its best. Touristic sites such as quaint streets and royal castles were presented alongside more modern sections dealing with science and industry, which together revealed a picture of a very technologically advanced modern society. The beer brewing processes at Carlsberg were part of the film for instance. Moving images from the lush countryside portrayed the advancements in Danish pork production, while Danish fisheries filled a number of sequences with stout fishermen at high seas as well as on market stalls.[74]

In the discussion with the Foreign Service Schmidt supplied a list of port towns, where the *Dana* was scheduled to arrive during the trip to meet the scientific goals of the expedition. Then the ministry came back with a revised suggestion for itinerary in cases where other towns and cities were more interesting from the point of view of Danish commercial interest. The ministry preferred St. Thomas to Martinique for instance, because of a potential interest in exporting Danish cement to the former colony. This was not altered though. Schmidt however, did call at Mombasa in Tanganyika, which the ministry preferred over Zanzibar, Schmidt's first choice. It was also mentioned at their meetings that Schmidt should be bestowed with additional duties such as that of scouting for Danish business opportunities abroad, but this was abandoned. Although Schmidt did receive something very important in return for his active interest in promoting Denmark. The expedition came to be regarded as a national expedition. Thereby, the *Dana* was a royal vessel for the duration of the expedition, and as such exempt from paying customs duties wherever she went.[75]

Schmidt also sought to promote the upcoming expedition to the leading Danish and international media hosting an international press conference at the Carlsberg Laboratory in May 1928. According to the laboratory guestbook sixteen different journalists attended and these included representatives of newspapers and press agencies in The Netherlands, Great Britain, Germany, as well as the leading Danish publicists.[76] As a testimony to how far the news of the expedition spread, the

73 KB. Johannes Schmidt 'Papirer'. Ut.374.I.7. 'Korrespondance 1927–1929'. Press Agency of the Department of Foreign Affairs to J. Schmidt, 16 April 1928.
74 Andersen, V. *Danmark-Filmen*, Copenhagen, 1925.
75 KB. Johannes Schmidt 'Papirer'. Ut.374.I.7. 'Korrespondance 1927–1929'., J. Schmidt to Presse Office of the Department of Foreign Affairs, 30 May 1928.
76 RA. *Carlsbergfondets DANA-ekspeditioner. 1917–1973*. 1. Gæstebog, 1922–1940.

De Indische Courant, covering the Dutch East Indies, ran a story over a full column soon after the *Dana's* departure. Thus it became known to the many expatriated Dutchmen in the Far East that '…the expedition was funded by the Carlsberg Foundation, an entity whose fortune is derived from the revenue of the well-known Carlsberg Brewery.'[77] Similarly, the *Kalgoorlie Miner* in Western Australia reported on the upcoming ocean adventure soon to commence from the opposite side of the planet.[78]

During the last week before the expedition commenced it was time to fill the *Dana* with stores – food with a long shelf life such as canned fruits, canned meat, canned fish and shellfish, marmalade, sausages, potatoes, flour, and sugar. More luxurious items such as chocolate, tea, coffee, sherry, and rum were also brought on to the research vessel, while tobacco appears to have been a personal item to bring on board.[79] Bottles of Carlsberg beer were presented to the expedition as a gift before departure.[80] In the late 1920s, all the canned items were global commodities really, whereas in practice fresh fruits were only regionally accessible. In Menado on the Sulawesi Island the crew were able to feast on pineapple, bananas, lemon, and papaya, but this was not the case at the start of the voyage in the temperate zones of Northern Europe.[81] Even more exotically, when coaling in the Seychelles in December 1929, the expedition crew's Christmas shopping included the acquisition of one live sea turtle, two live pigs, and one bottle of port wine.[82]

Conclusion

When looking at the fundraising, organizing, and planning of the Carlsberg Foundation's Oceanographic Expedition Round the World, 1928–1930, several aspects were different from all of Schmidt's previous expeditions, and the preceding negotiations have been explained here for the first time.

There is good reason to believe that Schmidt had been dreaming of exploring the life history of the World's eel already by the time of his 1909 publication of the biogeographical distribution of the eel. Nonetheless it was not until after the fulfilment of publishing the Sargasso Sea hypotheses concerning the Atlantic eel that he set his mind fully into furthering investigations in the Indian and Pacific Oceans. Great advances in New Zealand and Tahiti were made through his 1925–26

77 Anonymous, Oceanographische expeditie, in: *De Indische Courant*, 26 June 1928: 3.
78 Anonymous, Solving Ocean Mysteries – Scientific Expedition, in: Kalgoorlie Miner, Western Australia, 23 May 1928: 3.
79 CFD. DANA. BS3 2. Jens With Skibsproviantering, Copenhagen, 12 June 1928.
80 Thomsen H. Togtets længste stræk. In: Schmidt J, ed. *Dana's togt omkring jorden 1928–1930*. Copenhagen: Gyldendal; 1932: 68.
81 CFD. DANA. BS3 2. Het Manado Veem Scheepsleveranties, 5 July 1929.
82 CFD. DANA. BS3 2. Paul Lanier Ship Chandler, Mahé Seychelles, 13 December 1929.

circumnavigation on what was a combined research trip and recreational trip with his wife Ingeborg and Ostenfeld's daughter Gertrud as his research assistant. Once more, Schmidt was uniquely well positioned at the Carlsberg Laboratory and financially well-off through his marriage. What was less successful was the American leg of his journey around the World, as it did not fulfil his wishes to find US charitable research funding for his cause, although he was aided in his approach from the official Danish Legation in Washington DC and not least zoologist Adam Bøving.

In the end though, Schmidt had the good fortune to come to dispose of what was until then, the single largest donation from the Carlsberg Foundation, while the Foundation's involvement in the project was proportionately large. Obviously Schmidt's research initiative and his expedition plan had lots of merit. Once more, it was Schmidt's enduring interest in the eel, which had already made him world famous, which was the driving force behind the science programme of the expedition.

Nonetheless, the importance of having one of your closest personal friends as a board member should not be overlooked. Ostenfeld knew which buttons to push among the other professors on the Board of Directors. This was a time in the long history of the Carlsberg Foundation when interest in scientific expeditions and the accompanying fieldwork was relatively high. Research into the protocols of the foundation's board meetings has revealed that there was a certain element of vanity within the board with regards to their desire to have the circumnavigation be their expedition as opposed to it being the State's or some other foundation's project. Likewise, the actions of the board indicate a willingness to fund the expedition as a particularly expensive pet project, to the extent that they were willing to pull out additional money from the brewery if necessary.

Having said this, Schmidt was once more in the right place at the right time, and he was well connected in all relevant corners of society. Next to the Carlsberg Foundation, the East Asiatic Company was to play an important role, also financially, as a conveyer of goods and people to and from the ship in addition to providing vital connections around the world.

Important aid also came from the Danish government including the endorsement by the Prime Minister himself, both through the fisheries section of the Department of Agriculture as well as the Navy Ministry, and the diplomatic connections via the Foreign Service. For Schmidt, it was a novel addition to his portfolio to present promotional films about Denmark along the route, when visiting interesting ports. As he was an experienced science communicator, this was probably a welcome addition to his other public relations activities.

The organizational aspects of the circumnavigation also differed substantially from all previous oceanographic expeditions undertaken by Johannes Schmidt. For the previous large expeditions with the *Thor* and the *Dana I* and *Dana II*, Schmidt had referred to a committee chaired by someone senior. First it was Petersen, then Drechsel. Now, Schmidt referred to his own peer and friend, Ostenfeld. Unsurprisingly, the rest of the expedition committee was made up of interested parties.

Finally, the staffing of the vessel for the expedition shows that Schmidt was a keen observer of inter-relations between various sections of his crew, and also that he had a clear idea of how to compose and manage a crew to obtain the best results. This may not be a huge surprise in light of the fact that several civilians were veterans on Schmidt's cruises and thus had chosen to pursue yet another adventurous expedition.

14 Around the World in Two Years

Leaving Copenhagen

On the morning of 14 June 1928, it was time to say goodbye. According to one newspaper, the rain was pouring mercilessly on the Nordre Toldbod quay in Copenhagen, where the research vessel the R/V *Dana* was the scene of a ceremonious farewell to the 26 members of the *Carlsberg Foundation's Oceanographic Expedition Round the World*, the official name of the expedition. At 10 o'clock the couple of a hundred bystanders followed a religious Lutheran service ministered by a Pastor Hornbeck. Hornbeck gave his blessings to the expedition and expressed his wishes that each and every participant would return safely to Copenhagen.[1] The King of Denmark, Christian X, gave a short speech, wishing farewell.[2] Then it was Ostenfeld's turn to speak, followed by the committee vice president and Vice Admiral, Amdrup. Exactly one hour later, all friends and relatives were instructed to leave the ship.[3] Then the expedition members were left to the mercy of the pouring summer rain, a tried and tested vessel with brand new technology, and not least the will and desire, and planning of Johannes Schmidt.

After a couple of weeks, the expedition reached their first hot spot, the entrance to the Mediterranean. All the sophisticated equipment, developed and perfected over the past generation of Danish marine scientists was rolled out once more. Schmidt's old companion from the *Thor* expeditions, hydrographer, J. N. Nielsen took part in this first section of the voyage, where once more the salinity and temperatures were analysed at various depths around the Strait of Gibraltar. The results found during the *Thor* expedition were confirmed, namely that the so-called *Mediterranean Outflow* took high saline Mediterranean water out into the Atlantic at greater depth than the lighter and lower saline water flowing in from the Atlantic Ocean.

1 RA. Anton Fr. Bruun. *Danaekspeditionerne 1924–1930*. Avisklip, *Nationaltidende*, 14 June 1928.
2 Schmidt J. Indledning. In: Schmidt J, ed. *Dana's togt omkring jorden 1928–1930*. Copenhagen: Gyldendal; 1932:9–19.
3 RA. Anton Fr. Bruun. *Danaekspeditionerne 1924–1930*. Avisklip, *Nationaltidende*, 14 June 1928.

14.1 The start of the circumnavigation in 1928
On a rainy day, the 14 June 1928, the quay at Nordre Toldbod in Copenhagen was the scene of a ceremonious farewell to the 26 members of the Carlsberg Foundation's Oceanographic Expedition Round the World (*Carlsberg Archives*).

Another old companion Ove Paulsen was also there taking plankton samples. The large custom made ringtrawls were prepared for sampling. The seines were sewn from tightly woven hemp with a mesh size of barely 1 millimetre. To keep the nets open, an iron ring, 2 metres in diameter was attached to the front end of the net. Several of these ringtrawls were then attached to a heavy steel wire at various points, reflecting the desired depths at which they were to be deployed. Before the ringtrawls were lowered into the water the ship was halted. Then as the vessel gained speed again the steel wire with the trawls was unleashed via a drum. With the fine meshes, very small plankton could be found, though not the very smallest, which were extracted from the seawater through the use of a centrifuge.[4]

4 Paulsen O. I europæiske Farvande. In: Schmidt J, ed. *Dana's Togt omkring Jorden*. Copenhagen: Gyldendal; 1932: 19–38.

14.2 The scientists saying farewell in Copenhagen 1928
Time to say good bye. From left to right Johannes Schmidt, Poul Jespersen and Anton Bruun all look straight at the photographer. 26 years old, Bruun had just married before embarking on a two year long voyage around the World (Carlsberg Archives).

The expedition stopped at Madeira, and before crossing the Atlantic Ocean, the veteran contingent of Paulsen, Schmidt, and Nielsen all left the expedition. They had acted partly as mentors for the younger generation.[5] From now on, and until the early spring of the following year, Poul Jespersen was in command. More stations were taken in the Sargasso Sea in particular, sampling ever more testimony to solidify Schmidt's thesis on the breeding grounds of the eel.[6] The island of Martinique in the West Indies was chosen as a bunker station for coaling, and then the Panama Canal was reached. On the 2 September, the *Dana* went through the sluices and into the Pacific. According to Anton Bruun, the expedition began for real now, as the vessel entered fairly uncharted waters with respect to scientific surveys.[7]

5 Paulsen O. I europæiske Farvande. In: Schmidt J, ed. *Dana's Togt omkring Jorden*. Copenhagen: Gyldendal; 1932: 38.

6 Jespersen P. Vest paa til Panama. In: Schmidt J, ed. Dana's Togt omkring Jorden 1928–1930. Copenhagen: Gyldendal; 1932:39–56.

7 Bruun AF. Jordomsejlingen med S/S 'Dana' (II) 1928–30. In: Winge Ø, Vedel Taaning Å, eds. Naturforskeren Johannes Schmidt. Hans Liv og Ekspeditioner. Copenhagen: Gyldendalske Boghandel Nordisk Forlag; 1947:145.

14.3 Lake Gatun in 1928
In 1928 the R/V Dana steamed through the Panama Canal once more. The central part of the canal system is Lake Gatun, a flooded inland valley, where ghostlike dead trees were still visible above the surface of the lake (RA. Carlsbergfondets DANA-ekspeditioner. 1917–1973. 7a).

Across the Pacific

The previous visit to the Pacific with the *Dana II* Expedition of 1921–22 had reaped interesting results with regards to nutrient contents and hitherto unknown deep-sea fishes. This time was no exception, as the team caught three female deep-sea anglerfish (*Lophiiformes*), where the male is attached to the female as a small parasite, in addition to countless anglerfish where the male and female are not attached to one another. On the continental shelf of the Americas marine life was found to be abundant.[8]

Then, the *Dana* was coaling at Panama City for the longest stretch of the expedition, 22 days from Panama to Tahiti. With Jespersen following Schmidt's plan from home, they would not risk running out of coal, while stopping for surveys around the Galapagos Islands or the Marque-

8 Bruun AF. Jordomsejlingen med S/S 'Dana' (II) 1928–30. In: Winge Ø, Vedel Taaning Å, eds. Naturforskeren Johannes Schmidt. Hans Liv og Ekspeditioner. Copenhagen: Gyldendalske Boghandel Nordisk Forlag; 1947:144–145.

sas Islands. The coal storage was filled to the top with 175 metric tonnes, and in order to make room for more coal, an extra 15 tons were placed in bags on deck. Therefore, relatively few stations were taken in the Eastern and Central parts of the Pacific Ocean, except for the echo sounder, which was not dependent on the speed of the vessel or lack thereof. Bruun recalled that with respect to the main objective of finding eel larvae, they did not miss out on much during this 5,000 kilometres stage of the circumnavigation.[9]

Once the expedition came within sight of first the Marquesas Islands and later Tahiti, the fisheries for eel began for real, a nightly activity, where the ringtrawls would be hauled on-board and then emptied into buckets for sorting. On average though, coal rationing meant that stations were taken once every three days only.[10] The price of coal was a real issue in the Pacific, where for instance in Fiji, the price was much higher than anticipated, and the quality was often poor, which meant that it did not last as long as it should have when fuelling the steam engine.[11]

Jespersen appears to have acted very much in concordance with Schmidt and Ostenfeld's wishes, but the reports that Jespersen sent to Schmidt suggest that no set protocol was arranged for communicating back and forth between the ship and Copenhagen. By the time the ship had arrived in Tahiti it seems that overall, Jespersen was satisfied with how the expedition was progressing, but he did complain about a certain lack of discipline on the *Dana*, which was not as tidy as he felt it should be. In addition, he had felt obliged to act as middleman, when the ship's captain, Christensen and the officers of the marine, as well as the ordinary seamen did not get along. All in all though, Jespersen was pleased and he looked forward to meeting up with both Ingeborg and Johannes Schmidt in New Zealand just before Christmas 1928.[12]

9 Bruun AF. Jordomsejlingen med S/S 'Dana' (II) 1928–30. In: Winge Ø, Vedel Taaning Å, eds. Naturforskeren Johannes Schmidt. Hans Liv og Ekspeditioner. Copenhagen: Gyldendalske Boghandel Nordisk Forlag; 1947:144–146.
10 Ries C. Tropesejlads. In: Schmidt J, ed. Dana's Togt omkring Jorden 1928–1930. Copenhagen: Gyldendal; 1932:141–142.
11 KB. Johannes Schmidt 'Papirer'. Ut.374.I.7. 'Korrespondance 1927–1929'. P. Jespersen to J. Schmidt, 18 November 1928.
12 KB. Johannes Schmidt 'Papirer'. Ut.374.I.7. 'Korrespondance 1927–1929'. P. Jespersen to J. Schmidt, 18 November 1928.

Official representation in New Zealand and Australia

Johannes Schmidt accompanied by Ingeborg joined the expedition towards the end of 1928. In late October the couple boarded a cruise liner, the *S/S Rotorua* from Europe, which passed through the Panama Canal more or less in the wake of the *Dana*. According to the diary kept by Ingeborg during the cruise, they both enjoyed the somewhat more luxurious surroundings on this stylish first class cruise liner.[13] Upon arrival in New Zealand, Johannes took over keeping a diary. The tone became terse and factual with fewer personal impressions. Nonetheless, Schmidt did confide some information to his diary, a business calendar really, that presumably he never intended to have published.[14]

The diary entries confirm the image of Schmidt as a highly efficient scientist and traveller, who accomplished a wide range of organizational and scientific tasks simultaneously, even when operating on the opposite side of the planet. Mid-January 1929 appears to have been highly geared at public relations activities in New Zealand. The *Danmark-Filmen* was shown to a crowd consisting of many expatriated Danes and descendants of migrants from Denmark. In addition Schmidt was entertained at a whole string of civic receptions and other festive parties with local dignitaries. Speaking in front of the local government of New Zealand Schmidt was the official representative of the Kingdom of Denmark. In his speech he made eloquent comparisons between the Danish expedition and the state of Denmark in general: 'We know it is the custom for the official representative of a foreign power to come with the impressive dignity that is given by a ship of war. In Denmark we have no ships of war; we have no navy; – instead we have come in a science ship; and it adds to our appreciation of your welcome when we find that it is not considered inferior.'[15]

In between diplomatic duties, Schmidt also found time to direct the collection of eel from New Zealand rivers. It is very possible that he was now restless with regards the scientific programme of his trip. As soon as Schmidt and his wife had embarked on the next leg around the world, from New Zealand to Australia on the *S/S Wanganui,* Schmidt started working. Assisted by some of the ship's officers, Schmidt collected water samples of the sea temperature between Wellington and Sydney.[16]

In Sydney however, another string of diplomatic duties awaited the famous professor, and as Schmidt arrived in Sydney before the *Dana*, he could plan for her arrival. The research ship and the rest of the expedition meanwhile traversed the infamous so-called 'roaring forties,' the strong westerly winds between 40 and 50 degrees south, which gave the ship a fairly rough ride, a hindrance to much of the planned work. Bruun though, noted how the temperate waters around New Zealand reminded him a lot of what they knew from the North Atlantic Ocean, in the sense that there were fewer species than in the tropics, but much greater abundance of fish in the trawls.[17]

13 RA. *Schmidt, Ernst Johannes, biolog*: Materiale vedr. ekspeditioner og rejser (1876–1933) 2: m.m. Diary of I. Schmidt, 1928.

14 RA. *Schmidt, Ernst Johannes, biolog*: Materiale vedr. ekspeditioner og rejser (1876–1933) 2: m.m.

15 RA. *Schmidt, Ernst Johannes, biolog*: Materiale vedr. ekspeditioner og rejser (1876–1933) 2: m.m. 15 January 1929; KB. Johannes Schmidt 'Papirer'. Ut.374. 'Taler og Foredrag'. Speech, Town Hall, Wellington J. Schmidt, 15 January 1929.

16 RA. *Schmidt, Ernst Johannes, biolog*: Materiale vedr. ekspeditioner og rejser (1876–1933) 2: m.m. 1–18 January, 1929.

17 Bruun AF. Jordomsejlingen med S/S 'Dana' (II) 1928–30. In: Winge Ø, Vedel Taaning Å, eds. Naturforskeren Johannes Schmidt. Hans Liv og Ekspeditioner. Copenhagen: Gyldendalske Boghandel Nordisk Forlag; 1947:149

14.4 Reception with mayor of Wellington, New Zealand
The Danish diplomatic representations abroad were notified from the R/V Dana in advance to prepare for the arrival of the Danish globetrotters. Typically a formal event was thrown in honour of the expedition such as this one in Wellington, New Zealand hosted by city's mayor. The expedition would give interviews to the local press, and on more than 25 times Schmidt showed the Danmark-Filmen in cinematic theatres as a way to promote Danish culture and business (RA. *Carlsbergfondets* DANA*-ekspeditioner. 1917–1973. 7a*).

As the *Dana* arrived in Sydney Harbour a ten-day programme commenced. Obviously the crew would have time off for some days, but Schmidt was kept busy with official duties such as meeting with the governor of New South Wales and representatives from the port and city of Sydney. One week into the programme the *Danmark-Filmen* would be shown at the so-called King's Hall, while Schmidt was representing the expedition and Denmark at a lecture for the Royal Society of New South Wales.[18] Addressing fellow scientists, Schmidt once more drew a parallel between science and society at large. 'When we visit and meet with the scientific workers of other countries, we are not regarded as foreigners, but as Fellow Scientists. Science recognizes no Boundaries, no Class, and no Creed, all working to a common end, to increase knowledge and to uplift humanity.'[19] Some of Schmidt's remarks even made it into *The Sydney Morning Herald*: 'We come to your waters primarily in the cause of science, but if our sojourn amongst you will help towards uniting our two countries in greater bonds of friendship, then our mission will have fulfilled our very highest aspirations.'[20]

At any rate, the visit to Australia seems to have been a great public relations success. The Australian newspapers wrote plenty of stories covering the visit of the Danish ocean explorers. The arrival of Mr and Mrs Schmidt to Australia appeared in the *Sydney Morning Herald*, along with a full description of Schmidt's previous work with the eel. This built up expectations for the arrival of the actual research vessel in February.[21] Upon arrival of the *Dana*, the ship was described as 'a floating museum' with 'queer species of fish.' A fairly long article then presented the route of the expedition, the wonders of the echo-sounder as well as reports on one of the peculiar anglerfishes, which must have been displayed in sample bottles. Schmidt's affiliation to the Carlsberg Laboratory was also duly mentioned.[22] On the 7 February, the same newspaper ran another full column covering Schmidt's lecture at the Australian Museum.[23] As February drew to an end, Schmidt even made it into the gossip pages of the Perth based newspaper, the *Western Mail*. According to the newspaper 'officers of the Danish research ship, *Dana*, had only been in Sydney a few hours when they went into raptures over 'zee fascinating figures and legs of Sydney girls.'[24] Schmidt was quoted to have said that; '…The splendid physique of your maidens and youths as they frolic in the surf has impelled our deepest admiration. Australia is breeding a virile race, which we believe is not surpassed by any race in the world.'[25] This quote is likely just a hoax from the *Western Mail*, but it serves to exemplify just how newsworthy the Australians found this global expedition.

18 KB. Johannes Schmidt 'Papirer'. Ut.374.I.7. 'Korrespondance 1927–1929'. Danish Research Exploration Party. Tentative Programme, February 1929.

19 KB. Johannes Schmidt 'Papirer'. Ut.374. 'Taler og Foredrag'. Speech, Royal Society of New South Wales, Sydney J. Schmidt, 5 February 1929.

20 Anonymous, Marine Life. A Floating Museum. The Dana in Port. Queer Species of Fish. *The Sydney Morning Herald,* 6 February 1929 p 18.

21 Anonymous, Ocean Study. The Dana Expedition. Professor Schmidt Arrives, in: *Sydney Morning Herald*, 23 January 1929.

22 Anonymous, Marine Life. A Floating Museum. The Dana in Port. Queer Species of Fish. *The Sydney Morning Herald,* 6 February 1929 p 18.

23 Anonymous, Studying the eel. Professor Schmidt's Lecture, in: *Sydney Morning Herald*, 7 February 1929.

24 Anonymous, Mostly personal. People of the Week, in: *Western Mail*, Thursday 28 February 1929 p 4

25 Anonymous, Mostly personal. People of the Week, in: *Western Mail*, Thursday 28 February 1929 p 4

South East Asian travels spring and summer 1929

While the *Dana* was testing the waters east of Australia in the latter half of February, Schmidt was orchestrating the expedition from ashore along with his wife. In practical terms, Schmidt was unable to bring Ingeborg along on the *Dana*. Definitely, it would have been rather uncomfortable for the middle-aged lady to share a modest cabin with Schmidt in the sweltering heat of the South Sea summer. In addition, it would probably have been malapropos to have one's wife on-board for an extended period of time. This comes to mind as the most obvious reason why Schmidt did not board the *Dana* while she accompanied him on his travels. Ingeborg eventually travelled on to China on her own, boarding the *S/S Taiping* on 16 February 1929.[26] A visit to Bangkok followed, before Ingeborg Schmidt returned to Copenhagen in the summer of 1929. Ingeborg's name rarely entered Schmidt's expedition diary, but on the 3 July Schmidt noted that she had sent a telegram to let him know that she was now safely back in Copenhagen.[27] From then on she acted as the expedition's unofficial ambassador in the marine science circles in Europe. In early September for instance, the Scandinavian meeting of Natural Scientists was held in Copenhagen, and Ingeborg was invited to be present at the Banquet in the Forum Congress Centre and facilitate the exchange of telegrams. This was one of several occasions in which Ingeborg became involved in the passing back and forth of telegrams from the Seven Seas.[28]

Back at sea, the next phase of the *Dana's* expedition resulted in the first major deviation from the expedition plan as it had been presented at the Carlsberg Foundation headquarters in the spring of 1928. The idea was to go North East from Australia, across the Coral Sea and head for the Solomon Island Archipelago in search of eel larvae at great depths. Two problems had arisen that affected this plan. First there were the poor weather conditions, where the cyclone season made it difficult and dangerous to operate far from the shores of Australia. Second, the members of the expedition came to realize that there was a real risk that they might not be able to find suitable coal for the boiler in the Solomon Islands. Eventually the expedition members decided not to risk their safety, and so the plan was altered. Cautious of cyclones, the *Dana* steamed along the coast of Australia on the inside of the Great Barrier Reef and then through the Torres Strait between Australia and New Guinea.[29]

26 RA. *Schmidt, Ernst Johannes, biolog*: Materiale vedr. ekspeditioner og rejser (1876–1933) 2: m.m. 16 February, 1929.

27 RA. *Schmidt, Ernst Johannes, biolog*: Materiale vedr. ekspeditioner og rejser (1876–1933) 2: m.m. 3 July 1929.

28 RA. *Schmidt, Ernst Johannes, biolog*: Materiale vedr. ekspeditioner og rejser (1876–1933) 2: m.m. 1 September 1929.a

29 Bruun AF. Jordomsejlingen med S/S 'Dana' (II) 1928–30. In: Winge Ø, Vedel Taaning Å, eds. *Naturforskeren Johannes Schmidt. Hans Liv og Ekspeditioner*. Copenhagen: Gyldendalske Boghandel Nordisk Forlag; 1947:149.

14.5 Men from Manokwari in present day Indonesia
*To stir interest in the 2-year long circumnavigation Poul Jespersen wrote several articles for Danish newspapers along the way. His reports on the scientific work of the expedition, were accompanied by stories of cannibal encounters in New Guinea. A group of men from Manokwari in what is now Indonesia were used to illustrate his point (*RA*. Carlsbergfondets* DANA*-ekspeditioner. 1917–1973. 7b).*

Once the expedition was in the Banda Sea, south of the island of Amboina, the deep-sea investigations were in operation again, and depths of almost 7,000 metres were encountered. From here on, and for the next six months until November 1929, the expedition mainly operated in the central tropical waters of the Indonesian Archipelago from New Guinea and the Philippines, into the South China Sea and Gulf of Thailand, and onto the Indian Ocean, investigating the ocean trenches off Sumatra and Java.[30] The scientific results of these investigations are dealt with in chapter 16.

The promotional side of the expedition came into operation on several occasions when the expedition members took part in similar talks and presentations to those in Wellington and Sydney. In Bangkok, Thai-

30 Bruun AF. Jordomsejlingen med S/S 'Dana' (II) 1928–30. In: Winge Ø, Vedel Taaning Å, eds. *Naturforskeren Johannes Schmidt. Hans Liv og Ekspeditioner*. Copenhagen: Gyldendalske Boghandel Nordisk Forlag; 1947:149–151.

31 Bruun AF. Jordomsejlingen med S/S 'Dana' (II) 1928–30. In: Winge Ø, Vedel Taaning Å, eds. *Naturforskeren Johannes Schmidt. Hans Liv og Ekspeditioner*. Copenhagen: Gyldendalske Boghandel Nordisk Forlag; 1947:152–53.

32 RA. Schmidt, Ernst Johannes, biolog: Materiale vedr. ekspeditioner og rejser (1876–1933) 2: m.m. 15 May 1929.

land, for example, there was a substantial Danish contingent centred around the wide-ranging activities of the East Asiatic Company, which made a willing audience for the Danish talks and films. This Danish community also formed a natural entrepôt, and offered the possibility for the *Dana* to be serviced and have its boiler cleaned, and also for the ship to be restocked with provisions and coal. Schmidt found time to revisit the Koh Chang area. This is where he had first been thirty years earlier in the company of Theodor Mortensen, and which had formed the research area of most of Schmidt's doctoral work on mangroves.[32] On 15 May, Schmidt even met an elderly local man named Nai Sug in Koh Chang. He had been one of his hired helpers 30 years earlier.[33]

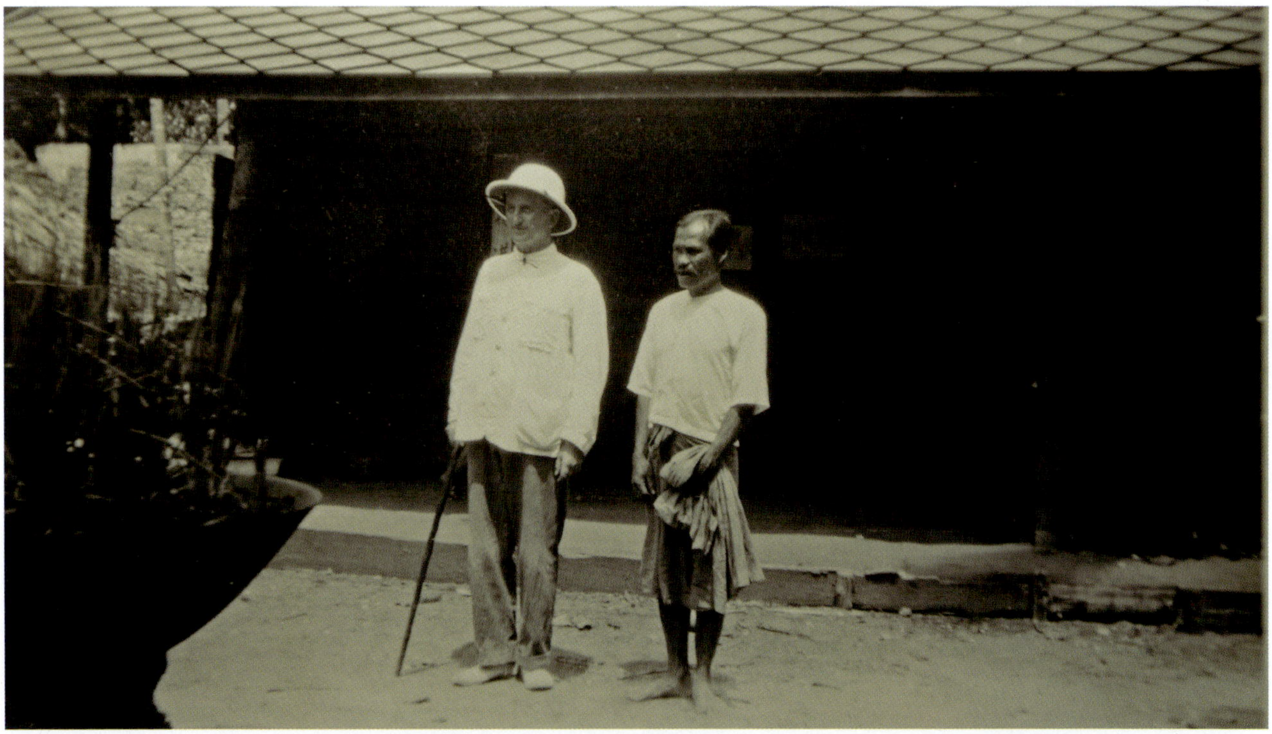

14.6 Schmidt in Koh Chang 1929
In April 1929, the R/V Dana expedition visited the island of Koh Chang in the Gulf of Thailand, where Schmidt had investigated mangroves three decades earlier. On the photo he has teamed up with his former assistant, Nai Sug who still resided on the island (RA. Carlsbergfondets DANA-ekspeditioner. 1917–1973. 7a).

Another part of the original plan had been to go to Japan to try and find the larvae of the Japanese eel, (*Anguilla japonica*). As the expedition headed north up the South China Sea and into the Yellow Sea in the

33 Bruun AF. Jordomsejlingen med S/S 'Dana' (II) 1928–30. In: Winge Ø, Vedel Taaning Å, eds. *Naturforskeren Johannes Schmidt. Hans Liv og Ekspeditioner*. Copenhagen: Gyldendalske Boghandel Nordisk Forlag; 1947:153–154.

early summer, this plan was scrapped as well. The expedition made it as far north as Shanghai, but with the danger of Typhoons and rough weather in general, Schmidt decided it was better to spend more time revisiting the Philippines and the Celebes Sea areas north of Sulawesi Island and New Guinea.[33] This would eventually lead to the most fruitful parts of the entire expedition, with regards to the eels, yet as we will see below it also pushed those on the ship close to their limits.

14.7 R/V *Dana* in Bangkok

*R/V Dana anchored in Bangkok. The building with the flag on top is the Asian headquarters of the Danish East Asiatic Company (*EAC*). The* EAC *shipped the ever-growing scientific collection from the Dana back to Copenhagen free of charge (Carlsberg Archives).*

Life on board

An experienced crew was in charge of the *Dana's* circumnavigation. Among the scientists on board, Schmidt had worked with Jespersen, Bruun, and Vedel Taaning before, on cruises in the North Atlantic. The ship's Captain Hansen, in command for the majority of the trip, was a true veteran from the days of the *Thor* working the North Atlantic before World War One.

The ordinary seamen on board were generally referred to as 'folkene', meaning 'the people,' and they are rarely mentioned in the surviving material from the expedition, which includes two diaries one kept by Johannes Schmidt, the other by Anton Bruun. This is a reflection of hierarchy more than anything else. There is no reason to believe Schmidt did not respect the common sailor working on his expeditions, but in line with tradition and hierarchy they simply belonged to separate worlds. Schmidt communicated with Captain Hansen with regards to matters of the whereabouts of the ship. Hansen in turn dealt with the crew on board.

Therefore, 'the people' only feature in Schmidt's diary when they fall out of line. This apparently occurred on the 11 June 1929. On this day the *Dana* had found a safe haven from stormy weather in Keelung on Formosa Island, present day Taiwan. Following a check-up from the resident quarantine doctors, the crew of the ship were able to help with the loading of 80 tons of coal. Meanwhile, Schmidt and Captain Hansen went ashore on a shopping spree, Schmidt recorded that he had his hair cut and washed in a Japanese style hair salon. Later that night, Captain Hansen had to go to the local police station. There had been a brawl between the crew of the *Dana* and 'the natives' as they were referred to in Schmidt's diary. Having freed the seamen from jail, Hansen came back with a limping fireman Lassen, who had fallen and injured his leg. Two others, 'Holger and the Boatswain', emerged with bruises to the head. In an entry for the 13 June, Schmidt laconically states that 'Two of Yesterday's giants, Boatswain and Holger turned up with head bandages, looking pathetic.'[34]

The next day it was the first anniversary of the expedition. Schmidt noted this in his diary, writing that it was celebrated with fine weather and the catch of an eel larva between Formosa and the Philippines. Schmidt also revealed to his diary that the expedition members were in need of a vacation in the fall of 1929. While in Manila Schmidt discussed

34 RA. *Schmidt, Ernst Johannes, biolog*: Materiale vedr. ekspeditioner og rejser (1876–1933) 2: m.m. 11–13 June, 1929.

35 Lidegaard B. *I kongens navn. henrik kauffmann i dansk diplomati 1919–1958*. Copenhagen: Samlerens Forlag; 1996.
36 RA. *Schmidt, Ernst Johannes, biolog*: Materiale vedr. ekspeditioner og rejser (1876–1933) 2: m.m. 20 June 1929.

holiday plans with the Danish diplomatic envoy to East and Southeast Asia, Minister Kaufmann. Kaufmann was a cosmopolitan minded diplomat and a connoisseur of both the perils and pleasures of Southeast Asia.[35] Kaufmann recommended Medan in Sumatra in the Dutch East Indies as an ideal place for relaxation.[36]

14.8 King Neptune appears when crossing the equator
When the expedition crossed the line of the equator King Neptune appeared to baptize those who had their first crossing of the line.

37 RA. *Schmidt, Ernst Johannes, biolog*: Materiale vedr. ekspeditioner og rejser (1876–1933) 2: m.m. 11 June 1929.
38 RA. *Schmidt, Ernst Johannes, biolog*: Materiale vedr. ekspeditioner og rejser (1876–1933) 2: m.m. 13 June 1929.

When it comes to the expedition members in Schmidt's own realm, in particular Captain Hansen, Schmidt appeared rather sensitive to their condition. 'Hansen looks tired these days,' is a typically concise remark from Schmidt[37] Schmidt and Hansen got along very well, for instance, *en route* to the Philippines they made bets as to how fast the *Dana* would steam in the next four hours. A beer is at stake.[38] Schmidt and Hansen were comrades in arms for more than twenty years, and the importance of having Hansen on board was not just a matter of running a ship. Hansen had developed the skills to distinguish fish larvae from one another. Therefore, as Anton Bruun noted, Schmidt and Hansen took turns identifying the content of the ringnets hauled on-board from the deep sea. The eel investigations were a night-time activity, and Schmidt, who was a night owl, was rarely in bed before early in the morning. However, if Schmidt had turned in, Hansen would be on guard, when a particularly interesting station was expected.[39]

During the months of July and August, the expedition entered a veritable hotspot for eel larvae southeast of the Philippines and into the Celebes Sea between Borneo and Sulawesi, where they worked incessantly, frequently in rough weather conditions. Then on the 15 August, Captain Hansen's health deteriorated with symptoms similar to what he had suffered in Saigon. Hansen was provided with opium and taken to bed. To keep him safe in bed, they skipped the port of Samarinda in the Makassar Strait and steamed on into the Java Sea between Borneo and the Island of Java, where the town of Surabaya was their next port of call.[40] A couple of days later, Schmidt recorded that Hansen was feeling better, but also that the Captain was embarrassed about his own situation.

On 20 August, the landed in Surabaya in East Java, and the participants all relocated to a mountain resort in the highlands, close to the active volcano, Bromo. While Hansen recovered and went walking in the cooler mountain air with Schmidt, the leader of the expedition himself appeared stricken by the strenuous nature of their most recent legs of the journey. Tired with a pounding heart-beat at night, Schmidt also succumbed to a cough and a toothache, and declined several social events on Java.[41] Nonetheless, the visitors from faraway Denmark appear to have made a lasting impression locally, since in February the following year, the Javanese newspaper, *Soerabaiasch Handelsblad* wrote an update on the status of the expedition, by then heading North through the South Atlantic Ocean.[42]

Having recovered, the members of the expedition continued, going east around Bali, they steamed into the Indian Ocean proper, probing the waters West of Java before re-entering the Java Sea through the Sunda Strait between Java and Sumatra. Here, they went ashore for a few days in present day Jakarta, which was then Batavia in the Dutch East Indies.

During the next month as the expedition relentlessly pushed on, exploring the waters west of Sumatra, which proved to be a veritable

39 Bruun AF. Jordomsejlingen med S/S 'Dana' (II) 1928–30. In: Winge Ø, Vedel Taaning Å, eds. Naturforskeren Johannes Schmidt. Hans Liv og Ekspeditioner. Copenhagen: Gyldendalske Boghandel Nordisk Forlag; 1947:152.

40 RA. *Schmidt, Ernst Johannes, biolog*: Materiale vedr. ekspeditioner og rejser (1876–1933) 2: m.m. 15 August 1929.

41 RA. *Schmidt, Ernst Johannes, biolog*: Materiale vedr. ekspeditioner og rejser (1876–1933) 2: m.m. 20–29 August 1929.

42 Anonymous. Gunstige resultaten der 'Dana'-expeditie. *Soerabaiasch Handelsblad*. 22 February 1930:2.

hot spot for eel larvae. With only very few days of rest ashore, the hard labour and long period at sea started to take its toll on the rest of the men on board. On the 23 September the young doctor Strøyberg, who had replaced Dr. Andersen in early August reported to Schmidt that three men, Lieutenant Greve, the young scientist Bruun, and ship mate, Gotfred are all 'tired', as Schmidt has noted – in brackets. The cure was to be a holiday the doctor had advised. 'Easy enough said', as Schmidt noted in his diary.[43] Three days later the ship landed in Emmahaven near Padang on the west coast of Sumatra.

Since Lieutenant Greve did not get well right away, Schmidt requested the Navy Ministry to have a replacement sent out. Then Schmidt, attentive to the needs of his men, decided to issue a holiday and pay the members of the expedition with a lump sum for their upcoming land leave. While the sailors remained on board, no less than three ship-mates were sent to the local hospital.[44] Then the various groups on board the ship were taken by car to the nearby highlands to recover from the voyage. On 5 October, the vacation was over and the ship left again, Schmidt no doubt anxious to return to his newly found hotspots for eel larvae. Two of the three sailors had returned to the ship from their visit to the hospital, while the third ship-mate, Gotfred, was discharged. For the expedition to continue travelling and working, two Malayan sailors were hired, but to Schmidt's frustration they suffered from seasickness as soon as the *Dana* sailed into open waters.[45]

As Lieutenant Greve was on loan from the navy, Schmidt also reported to Admiral Rechmnitzer that while his dutiful young officer was hospitalized, he did not wish to be without him for the remainder of the expedition. Schmidt did not think he suffered from a serious illness, but rather that he had been worn out from exhaustion. Schmidt was reluctant to dismiss Greve, even though it was costly to keep Greve in this Sumatran hospital and then in a hotel for recreation. First of all, Greve himself was very anxious to complete the circumnavigation. Secondly, Schmidt thought he was a vital person to keep a lid on a looming conflict among the navy crew. The next officer in line, Ries did not have the respect of the crew as such, and the naval personnel in particular held him in a very low regard. Schmidt himself found Ries lazy and lacking interest in the work on-board the ship. To Schmidt, the only positive side of him was the fact that he was a good navigator. Finally though, Schmidt emphasized that he was not in a position to interfere with animosities within the ranks of the navy seamen on-board. Captain

43 RA. *Schmidt, Ernst Johannes, biolog*: Materiale vedr. ekspeditioner og rejser (1876–1933) 2: m.m. 23 September 1929.

44 RA. *Schmidt, Ernst Johannes, biolog*: Materiale vedr. ekspeditioner og rejser (1876–1933) 2: m.m. 23 September 1929.

45 RA. *Schmidt, Ernst Johannes, biolog*: Materiale vedr. ekspeditioner og rejser (1876–1933) 2: m.m. 5 October 1929.

Hansen had sought to comfort Schmidt by saying that it had been like that on all ships he had commanded.[46]

When Schmidt wrote to Rechnitzer about the state of affairs with the navy personnel, he did not spare the young replacement doctor either. Schmidt had the impression that he was naïve and inexperienced, when the sailors became sick. According to Schmidt this often happened on long voyages, when the sailors wanted a break from the ship, resting in a soft hospital bed or going home instead. Moreover, Dr Strøyberg had not kept an eye out for the spread of venereal diseases. '…To put it mildly, this was now widely spread across the ship, since these blockheads have treatment too late.'[47] Schmidt appeared not to take any moral offense of these calamities; rather he was annoyed to find that it affected the productivity of the crew and ultimately the scientific outcome of the expedition.

Bruun's diary from this period is an echo of Schmidt's perspective. Bruun wrote frequent notes on how hard the conditions were, continuously rushing from place to place to take stations leaving very little time to rest and recover. Bruun welcomed that following the hospitalization of the three men Schmidt had finally realized that we needed a short break, which was granted in Emmahaven, Padang on the west coast of Sumatra.[48]

Emmahaven continued to function as a coaling station for the expedition throughout October, and on 1 November Lieutenant Greve came back from his hospital treatment in Batavia. 'Looks more healthy', as Schmidt noted.[49] In November the expedition moved north, paying a very brief visit to the Nicobar Islands North of Sumatra, where the Indian Ocean becomes the Andaman Sea. The Nicobar Islands had a mythical status in Danish maritime heritage, since these islands had been claimed a colonial possession by the Kingdom of Denmark from 1755–1848. Due to outbreaks of malaria the islands were never seriously exploited as a colony, but in 1846 the Danish Galathea Expedition had paid a visit to the islands while sailing round the World. Yet, due to malaria, this had been another tragedy with two thirds of the expedition party getting infected with malaria. In addition to being in a hurry, the perils of tropical diseases may be one reason why Schmidt only stopped for a couple of hours at the Nicobar Islands. Some of the scientists went ashore and collected what they could find between 5 PM and sunset. Then the *Dana* lifted anchor and steamed out, headed southeast for the long awaited holiday at Medan on the north-eastern shore of Sumatra above the entrance to the Strait of Malacca.[50]

46 KB. Johannes Schmidt 'Papirer'. Ut.374.I.7. 'Korrespondance 1927–1929'. J. Schmidt to Rechnitzer, 4 October 1929.

47 KB. Johannes Schmidt 'Papirer'. Ut.374.I.7. 'Korrespondance 1927–1929'. J. Schmidt to Rechnitzer, 4 October 1929.

48 RA. *Anton Fr. Bruun. Danaekspeditionerne 1924–1930.* Diary, 7–14 October 1929.

49 RA. *Schmidt, Ernst Johannes, biolog*: Materiale vedr. ekspeditioner og rejser (1876–1933) 2: m.m. 1 November 1929.

50 RA. *Schmidt, Ernst Johannes, biolog*: Materiale vedr. ekspeditioner og rejser (1876–1933) 2: m.m. 7 November 1929.

14.9 Brief visit to the Nicobar Islands

In November 1929 the expedition paid a brief visit to the Nicobar Islands North of Sumatra. Bruun (to the left) and Ries are looking out over Galathea Bay, named after the first Danish led scientific expedition round the World. The previous six weeks had been marked by relentless work to catch eel larva to the West of Sumatra, and the men were starting to look skinny and tired. For fear of contracting malaria they only stayed for a few hours (RA. *Carlsbergfondets* DANA*-ekspeditioner. 1917–1973. 7b*).

Landing in the port of Belawan near Medan, the expedition was greeted by a boat with the Danish flag. The expedition was welcomed by the local consular, Poul Ellehøj, a couple of representatives from the local Scandinavia Society, the new incoming consular as well as a Danish plantation manager, Nyhuus. Later the famous Danish expedition leader was taken by car to a suite of rooms in the Hotel de Boer, Medan.[51] As Schmidt was accustomed to high living, this was a luxurious classic hotel. The rest of the members of the expedition took turns going ashore for relaxation, while others kept watch at the ship. In the cooler mountain surroundings of Brastagi the rest of the expedition stayed at the Hotel Belvedere, a somewhat more Spartan looking hotel than Schmidt's. According to Anton Bruun however, this was a long awaited chance to get a good night's sleep in a proper bed.

51 RA. *Schmidt, Ernst Johannes, biolog*: Materiale vedr. ekspeditioner og rejser (1876–1933) 2: m.m. 10 November 1929.

The stay in Medan also left a good imprint on the locals. On the 11 November, The Dutch East Indies newspapers, *De Indische Courant, Bataviaasch Nieuwsblad* and *Het nieuws van den dag voor Nederlandsch-Indië* recorded the arrival of the Danish deep sea expedition.[52] *De Sumatra Post* had followed the *Dana*'s travels since January 1929, and for the visit to Medan, a reporter had been sent to the cinema to meet the Danes, as the *Danmark-Filmen* was shown in a cinema in Medan. The film was preceded by a welcome by Schmidt, which allegedly was very successful and had references to deep-sea investigations to the West of Sumatra. The newspaper report revealed that the contents of the film had nothing to do with the theme of the expedition. Nonetheless, the audience was treated to an interesting and enlightening display of Danish culture, society, and architecture. Finishing off with the Danish national anthem, followed by the Dutch counterpart, 'De Wilhelmina', the newspaper report concluded that 'it had been a perfect evening.'[53]

On the 15 November however, the stay in North Sumatra was over, and the local Danish contingent organized a similarly festive farewell in the port of Belawan.[54] One of the ex-patriated Danes, John Nielsen, who worked as a hospital manager in Medan, was so infatuated with the visiting expedition that he wrote a lengthy report about it for the Danish daily newspaper, *Politiken*. Here Nielsen described the *Dana* as a display of the essence of Danishness: 'Gloriously, the *Dana* has displayed the Danish flag, all around the world. Their tiny rooms on-board are stuffed to the brim with the results of eager and beautiful works of science. And perhaps it is the smallness, which makes the ship seem so Danish. The *Dana* is the very symbol that neither the ship's nor the country's small size matters. What matters is that under minuscule and modest circumstances, great and honourable results can be achieved.'[55] If an advertisement campaign for the expedition had been drawn up in advance, the prospect of promoting Denmark could hardly have been formulated more concisely, than John Nielsen did in this newspaper article.

Taking stock of the visit the next day, Schmidt has recorded how the first team visiting the Hotel Belvedere, had not behaved themselves very well, especially 'the chef and Kiel.' He further noted that the mood amongst the men on board was sombre. Following heavy drinking for two nights in Belawan, the men suffered from hangovers.[56] Clearly, the many months in tropical seas in a small sweltering steamboat, built for cold weather, was now taking its toll on the men on-board.

52 Anonymous notices: Het expeditie-vaartuig 'Dana', in: *De Indische Courant*, 1929, 11 November: 1; De 'Dana', in: *Bataviaasch Nieuwsblad*, 11 November, 1929: 2; and De'Dana', in: *Het nieuws van den dag voor Nederlandsch-Indië*, 11 November, 1929: 2.

53 Anonymous. De danmark-film. *De Sumatra Post*. 12 November 1929: 3; Deense Diepzee-Expeditie, *De Sumatra Post*, 22 January 1929.

54 RA. *Schmidt, Ernst Johannes, biolog*: Materiale vedr. ekspeditioner og rejser (1876–1933) 2: m.m. 10–15 November 1929.

55 RA. *Anton Fr. Bruun. Danaekspeditionerne 1924–1930*. Avisklip, *Politiken*, 16 February 1930.

56 RA. *Schmidt, Ernst Johannes, biolog*: Materiale vedr. ekspeditioner og rejser (1876–1933) 2: m.m. 16 November 1929.

14.10 Schmidt greated by Danes in Sumatra
A happy and tired man in Sumatra. Following one a half month of hectic search for eel larvae in The Indian Ocean, the expedition had unravelled the breeding grounds for several species of eel. When the expedition landed in the port of Belawan, Medan on northern Sumatra Schmidt was greated by ex-patriated Danes (RA. Carlsbergfondets DANA-ekspeditioner. 1917–1973. 7b).

The next legs of the voyage turned out to be much calmer, at least in practical terms, as they sailed across the Indian Ocean to Colombo in Sri Lanka, one of the key nodal points between the *Dana* and Copenhagen. In Colombo the East Asiatic Company had a resident agent, Mr Spence, while a consul Kai Nielsen represented the interests of the Danish state. Christmas gifts from home were awaiting the expedition there in the last week of November. In return, the *Dana* expedition shipped out of Colombo no less than 46 large boxes with scientific collections from the past couple of months on board a motor ship the M/S *Lalandia* bound for Copenhagen, via Bangkok.[57]

In between practical arrangements and garden parties, Schmidt was a prolific keeper of his diary during his weeklong stay in Colombo, making trips to Kandy in the interior of the island together with Captain

57 RA. *Schmidt, Ernst Johannes, biolog*: Materiale vedr. ekspeditioner og rejser (1876–1933) 2: m.m. 24–30 November 1929.

Hansen. Schmidt bought jewellery and antiquities, and it is clear that he was well during this part of the voyage: 'It is lovely here, peace and quiet in this idyllic place, where one can wonder about all the little things and situations, one encounters. It is also wonderful to sleep well in a proper bed and to have a hot bath before dinner.'[58]

With regards to the scientific members of the expedition, there is no doubt as to who was in charge of upholding appearances. During an official visit to the acting Danish consular in Surabaya, Schmidt recorded that Dr Jespersen had behaved in a rude manner. Schmidt then had a meeting in private with Jespersen about this. What transpired at this meeting was not revealed in the diary, but it did have the effect that the consular did not appear to say farewell to the *Dana*, when she embarked the following morning.[59] At this time, Jespersen was about to leave the expedition anyway, returning to Denmark in a matter of days, so perhaps the discipline was lacking.

In general though, Schmidt held Jespersen in very high regard. In a letter to Rechnitzer from the spring of 1929, Schmidt was regretful that Jespersen could not stay on board for the duration of the circumnavigation. He indicated that he did not know how to get by without Jespersen, as he was in possession of the invaluable ability to be seaworthy, a rare quality for a scientist. Thereto, he was an unusually neat person, well respected by everyone.

Towards the end of the circumnavigation Schmidt was also asked by journalists how everything was on board the *Dana* travelling for so long. Upholding a strict loyalty toward the expedition members, Schmidt told the readers of the *Berlingske Aftenavis* that everything had worked out splendidly. All 26 men were well, and the expedition has had virtually no illnesses at all. A tiny attack from malaria, that is all.[60]

The lure of ethnographica

Gradually during the expedition the travelling Danes filled the *Dana* not only with specimens from deep sea, but also ethnographical objects and even live animals. In the summer of 1929, this modern day Noah's Arc became home to a live crocodile bought in Menado. Schmidt recorded that the animal was already on board when he and Captain Hansen came back from a tourist trip in the outskirts of Menado.[61] Nonetheless, as they visited Sri Lanka in late November, it was Schmidt himself,

58 RA. *Schmidt, Ernst Johannes, biolog*: Materiale vedr. ekspeditioner og rejser (1876–1933) 2: m.m. 28 November 1929.

59 RA. *Schmidt, Ernst Johannes, biolog*: Materiale vedr. ekspeditioner og rejser (1876–1933) 2: m.m. 28–29 August 1929.

60 RA. *Anton Fr. Bruun. Danaekspeditionerne 1924–1930*. Avisklip, *Berlingske Aftenavis*, 30 March 1930.

61 RA. *Schmidt, Ernst Johannes, biolog*: Materiale vedr. ekspeditioner og rejser (1876–1933) 2: m.m. 4 July 1929.

62 RA. *Schmidt, Ernst Johannes, biolog*: Materiale vedr. ekspeditioner og rejser (1876–1933) 2: m.m. 30 November, 1929.

63 RA. *Schmidt, Ernst Johannes, biolog*: Materiale vedr. ekspeditioner og rejser (1876–1933) 2: m.m. 12 October 1929.

who ordered the acquisition of two live tortoises to be kept on board.[62] These animals added to the sea turtle already present on board, a creature, which during the fall of 1929 had learned to appreciate being fed with white sugar.[63] In the Seychelles, Schmidt paid a visit to the local botanical garden of Mahé, where the director donated two Aldabra giant tortoises (*Aldabrachelys gigantea*). The next day the ship bought a large green turtle (*Chelonia mydas*), which they carried with them only a few days until it was converted into lunch. Probably this 200 pound male turtle was too much to handle on deck for a longer period of time.

14.11 Crew members posing with an Aldabra giant tortoise
Crew members posing with an Aldabra giant tortoise (Aldabrachelys gigantea) from the Seychelles. Two giant tortoises survived the long journey to Copenhagen, where Schmidt kept them in the laboratory garden for a while.

Cockroaches were present on board as well. On 1 July 1929, Schmidt remarked that 'the cockroaches are now so numerous and hungry, they had started to bite. In addition, I have had the first encounter with bedbugs as I surprised one resting on my ankle; it was filled with blood. Phew!'[64]

Two black pigs also ended their lives on the dinner table on board the *Dana*, the second one as part of the Christmas dinner, which also included a rice pudding. Christmas 1929 was celebrated off the coast of present day Tanzania. Lots of telegrams where sent home, including a joint telegram to the radio news of Danmarks Radio, the Danish broadcasting corporation with seasonal greetings from the expedition to the People of Denmark. At 17.00 on 24 December the ship's laboratory boasted a Christmas tree with flags, two Christmas Carols were sung, and gifts from home were distributed. Then in the evening, the scientific investigations continued with the taking of hydrographical and fishing stations.[65]

Thus traditions form home were honoured abroad, yet gradually it seems as if the adventurous expectations of exoticism were fulfilled. Early on, in October 1928, the expedition caught sight of Tahiti, from the account told by Anton Bruun hopes were high that '…not least were we going to encounter the islands of the South Seas with their warriors, wild and tattooed, and the beautiful brown girls!'[66] Bruun though, was one of several expedition members who wrote of how disappointed they were about coming to Tahiti. Expectation no doubt had soared during the long stretch across the Eastern Pacific, and the reality of Tahitians living a rather plane and impoverished European lifestyle was a far cry from James Cook's bountiful paradise.[67]

From their travel accounts it appears as if New Guinea made up for this, at least in terms of providing thrill of excitement. Poul Jespersen was permitted to send home stories to be bought in Danish newspapers to give the expedition some PR along the route, and the horrific highlight seems to have been an encounter with cannibals in New Guinea, where they were shown a prison with inmates sentenced for having killed and eaten a couple of English missionaries – 'with skin and hair', as Jespersen told a Danish newspaper.[68] In Menado on Sulawesi Island, some of the crew members received a one metre long crocodile as a gift, which they then kept on deck for a month before killing it softly using chloroform as a sedative.[69]

Bearing in mind that this expedition took place in a colonial period of world history, it is quite clear from the recorded encounters with non-western culture and nature that the scientists were contemplating,

64 RA. *Schmidt, Ernst Johannes, biolog*: Materiale vedr. ekspeditioner og rejser (1876–1933) 2: m.m. 1 July 1929.

65 RA. *Schmidt, Ernst Johannes, biolog*: Materiale vedr. ekspeditioner og rejser (1876–1933) 2: m.m. 23–24 December 1929.

66 Bruun AF. Jordomsejlingen med S/S 'Dana' (II) 1928–30. In: Winge Ø, Vedel Taaning Å, eds. Naturforskeren Johannes Schmidt. Hans Liv og Ekspeditioner. Copenhagen: Gyldendalske Boghandel Nordisk Forlag; 1947:146.

67 Greve S, Nielsen E. Marquesas og Tahiti. In: Schmidt J, ed. Dana's Togt omkring Jorden. Copenhagen: Gyldendal; 1932:72–92.

68 RA. Anton Fr. Bruun. Danaekspeditionerne 1924–1930. Avisklip, Fyns Stifttidende, 3 October 1929; RA. Anton Fr. Bruun. Danaekspeditionerne 1924–1930. Avisklip, Fyns Venstreblad, 2 November 1929.

69 RA. Anton Fr. Bruun. Danaekspeditionerne 1924–1930. Avisklip, Nakskov Tidende, 11 January 1930.

even ranking the people they met according to preconceived notions of race and civilization. Schmidt himself was no exception. When the *Dana* landed in Taipeh on the island of Formosa, Schmidt was struck by the presence of the Japanese, who were then ruling the island. In a letter to the Director of the Foreign Service, Count Reventlow, Schmidt recorded that the Japanese port officials were scrutinizing the entire ship and the ship's papers, when they came on-board. Still though, Schmidt had found it impressive and fascinating that 'everything was organized in such a modern way in Formosa, without the aid of the white man. Apparently, Formosa is a country of great value to the Japanese, who govern with skills, but also in a rough fashion.'[70]

Apart from showing respect to the Japanese at Formosa, Schmidt sported a clear sense of Western superiority in this and other letters home, as well as in several entries to his diary. This was a perfectly normal outlook on the world in the late 1920s, and it was definitely echoed at the Department of Foreign Affairs. In a letter to Schmidt from April 1929, Reventlow extended his gratitude toward the way in which Schmidt is conducting the whole expedition: '...Under your leadership I am certain that the way in which the *Dana* is boasting our flag in far away places, as a badge of honour for a significant, pioneering, and civilizing work in the service of science, has the greatest national merit both in relation to the foreign nations that you encounter, and in relation to the expatriated Danes in various countries.'[71] Then the jubilant director continued by promising that Schmidt would receive an official decoration from the Danish state, when he returned to Denmark.

Conclusion

This chapter has provided the first documentation of how the *Carlsberg Foundation's Oceanographic Expedition Round the World* became a public relations exercise. As the expedition worked its way across the South Seas and through the Indonesian Archipelago to China and back to the Indian Ocean, Schmidt showed himself at helm of a public relations effort spanning his own previous research, the on-going work on-board the *Dana* and also the diplomatic interests of the small Danish state from a far corner of Northern Europe.

The diaries kept by Schmidt and Anton Bruun have both provided a detailed picture of the daily routines on board the *Dana*. Almost inevi-

70 KB. Johannes Schmidt 'Papirer'. Ut.374.I.7. 'Korrespondance 1927–1929'., J. Schmidt to E. Reventlow, 24 June 1929.

71 KB. Johannes Schmidt 'Papirer'. Ut.374.I.7. 'Korrespondance 1927–1929'. E. Reventlow to J. Schmidt, 9 April 1929.

tably conflict between some of men on-board erupted during the many months at sea, sometimes in challenging conditions. These instances are particularly helpful for deciphering some of the values dictating the way in which Schmidt ran his ship. The *Dana* was marked by clear distinctions between the different groups on-board. Schmidt was very aware of the need to create a sense of togetherness on the cruise through his choice of custom made uniforms across the civilian and naval ranks of crew members. In the name of efficiency he needed the different skillsets to come together, and he depended on his trusted veterans, Captain Hansen and the medical doctor, Andersen to lend a hand with the scientific sampling. He was also acutely aware it seems that there were parallel chains of command, which ought not to be violated. Schmidt himself would keep the scientists in check, the naval officers were responsible for the navy sailors, and the captain was responsible for the civilian sailors, if they did not comply with expectations, for example when they were involved in bar brawls or the contraction of venereal diseases.

The way in which the Danish explorers went about their encounters with foreign peoples and nature, informs us about a clear colonial attitude in their views of the world. Japanese people are admired for their efficiency in spite of their race, while the expedition's encounter with megafauna such as crocodiles and turtles were marked by a similar fascination of the otherness or exoticism of the tropics.

In spite of careful planning the weather and the availability of coal came to play a major restraining role for the expedition, and led to it abandoning two of four designated hotspot areas (Japan and Solomon Island Archipelago) outlined in the original plan for the circumnavigation scheduled.

15 Stretching the expedition, stretching ties to Carlsberg

Overspending and changing plans?

Schmidt ran the first year of the *Carlsberg Foundation's Oceanographic Expedition Round the World* with hardly any consultation with either the Expedition Committee in Copenhagen, or the Board of Directors of the Carlsberg Foundation, the actual funders of the voyage. Several thousands of documents have been preserved in total, yet Ostenfeld, who was chairing the expedition committee, and thereby fronting the biggest donation ever from the Carlsberg Foundation, appears to have been fairly passive with regards to monitoring the first year of the expedition. Ostenfeld is excused in so far as he and Schmidt were both in Denmark until mid-October 1928, during which period they, at least in theory, were able to manage the expedition between them without written communication. Nonetheless, only a handful of letters between Ostenfeld to Schmidt from between October 1928 to June 1929 have been preserved. In these, Ostenfeld was focusing on the upcoming marine science related grants from the Carlsberg Foundation as well as the Rask-Ørsted Fondet. Theodor Mortensen was granted money for a journey to the South Seas, where he was planning to team up with the *Dana* expedition. Short messages could be exchanged via telegram of course, but surprisingly, there is no reflection concerning the challenges associated with running the actual expedition.[1]

In early June 1929 however, Ostenfeld wrote a lengthy letter to Schmidt, in which he gently indicated that all the key players of the committee, Rechnitzer, Amdrup, Mortensen, and himself had agreed that it would be great if Schmidt could send them some more information about the scientific results from the expedition so far. Ostenfeld pointed out that it would be to the benefit of the expedition as well,

1 KB. Johannes Schmidt 'Papirer'. Ut.374.1.7. 'Korrespondance 1927–1929'. C. H. Ostenfeld to J. Schmidt, 25 November 1928.

since it was difficult to keep the media interest alive, when they knew so little 'from the meagre reports' about what Schmidt and his crew were working on.[2]

It was not until the summer of 1929 that a more formal request for information about the expedition was made. In this instance the request came from the foundation's chairman, A. B. Drachmann, and the reason for the interest seems to be threefold.

First of all, Ostenfeld was about to embark on a two and half month long journey to Canada. In Ostenfeld's absence, Drachmann would fill his position as the Carlsberg Foundation representative on the expedition committee, so he needed to be brought up to speed with events in the South Seas: Drachmann, for instance only learned of the exact composition of the expedition committee on 11 August 1929. At this point Ostenfeld described how there were three active members of the committee, while the rest were just 'filling'.[3] The active members included the committee vice president, Admiral Amdrup, who was knowledgeable about the economic aspects. Admiral Rechmnitzer acted as the liaison between the Naval Ministry and the expedition, whereas the Director of Fisheries, F. V. Mortensen formed the point of contact with regards to the equipment of the *Dana* vessel itself.

Secondly, Drachmann may well have been offended by a letter from Schmidt to the chief accountant of the Carlsberg Foundation dated 5 August 1929. While informing about the payment to the ship's medical doctor, Andersen, who was about to return to Denmark, Schmidt uses the opportunity, as he labelled it, to inform that during Ostenfeld's planned absence, he would carry out the business aspects of the expedition with the committee Vice President, Vice Admiral Amdrup.[4] Schmidt did so right away, sending a short update of surveys to Amdrup in late August.[5] As we will see below, Drachmann felt as if he had been snubbed, since he was the designated replacement for Ostenfeld as the foundation's representative and chairman, not the admiral.

Yet most decisive for Drachmann to take action, and what caused a real crisis midway through the expedition was the third issue of whether or not Schmidt and Ostenfeld were keeping sufficient track of the expedition's expenses. The chief accountant at the Carlsberg Foundation had become alarmed by the high running costs of the expedition, when he had looked at the accounts in the summer of 1929. Ostenfeld appears to have been surprised as well and wrote a letter to Schmidt, alerting him to the fact that just one year into the expedition

2 KB. Johannes Schmidt 'Papirer'. Ut.374.VI.1. 'Dana-Expeditionen, 1928–30', C. Ostenfeld to J. Schmidt, 2 June 1929.
3 CFD. Vedr. DANA II, 1828–1930. 103/1928. Letter from C. H. Ostenfeld to A. B. Drachmann, 11 August 1929.
4 CFD. Vedr. DANA II, 1828–1930. 103/1928. Letter from J. Schmidt to Carlsbergfondets Kvæstur, 5 August 1929.
5 CFD. Vedr. DANA II, 1828–1930. 103/1928. Letter from J. Schmidt to G. C. Amdrup, 20 August 1929.

400,000 out of the budgeted 500,000 Kroner had been spent by the time they had visited Shanghai.

Schmidt was alarmed as well, as he believed only about half the total allowed sum had been spent at this point. In Ostenfeld's absence, Schmidt replied to the Vice Admiral and vice president Amdrup, and he blamed the budget overspend on expensive coaling in the Pacific.[6]

Despite the shock of the expenses, Schmidt proceeded to outline his vision for how he would like to continue the expedition in such a way that it could obtain as many new results as possible. The experiences they had had with crossing the Pacific Ocean and not taking samples from every station due to the fuel situation had not been optimal from a scientific point of view. Therefore, Schmidt would rather not make the same mistake rushing across the Indian Ocean. Rather, his idea would be to go back to the waters around Celebes and into the Pacific north of New Guinea in October, only to return to Singapore or Batavia to prepare for the crossing of the Indian Ocean in January. He would then like to proceed via Sri Lanka to The Red Sea and into the Mediterranean in April. This would enable the expedition to conduct an intense study of the Mediterranean and the Strait of Gibraltar in May, June, and July, only to return to Copenhagen at the end of July. The main motivation for changing the course was that this would give the most complete study of the eel populations of the Indonesian Archipelago, because the expedition then would have visited the same areas at different seasons. Schmidt then informed Amdrup that if it was decided that he should return to Denmark earlier, then he could shorten the trip, but his personal experience from having worked in the Mediterranean before was that the conditions for taking stations would not be fruitful before April due to rough winter storms.[7]

Schmidt now also addressed Drachmann personally in a letter written in early September, in which he explained how the previous correspondence between the expedition and the organizers in Copenhagen had taken place, and he then apologized that he had not been sending more extensive information back to Copenhagen so far. In the future he would strive to provide more full accounts of scientific results and the whereabouts of the expedition. With regards the previous part of the expedition he was not in possession of the material needed to reconstruct everything that had happened, but assured Drachmann that all the scientific collections had been secured for the future and sent home. He also apologized that Drachmann and others had found out most of their information about the travels of the *Dana* and her crew

6 CFD. Vedr. DANA II, 1828–1930. 103/1928. Letter from J. Schmidt to Rechnitzer, 26 August 1929.
7 CFD. Vedr. DANA II, 1828–1930. 103/1928. Letter from J. Schmidt to Rechnitzer, 26 August 1929.

from newspaper articles written by members of the expedition, which bore no relevance to the science undertaken at sea.[8]

Then Schmidt proceeded to the economic situation. He repeated his report to Amdrup that by the month of May he thought everything was in good standing as he was then informed by the chief accountant of the Carlsberg Foundation that c. half the budget had been spent. He had then continued probing the waters in the Indonesian Archipelago with the hope of continuing this investigation until around January. The original plan to cross the Indian Ocean on one long stretch from the Strait of Sunda to Madagascar was pointless due to the lack of coaling opportunities. Schmidt then urged Drachmann to provide him with a comment on his suggested plan, so he could continue with scheduling the rest of the trip. Schmidt alerted Drachmann to the fact that if they had to rush home to save money and fuel, the scientific success of the entire expedition would be jeopardized. He proposed to Drachmann that the expedition should be given additional financial support so that that it could carry on beyond the spring of 1930, and preferably throughout the summer season. Schmidt must have been fairly assured he would receive extra funds to fulfil his dreams, as he did not bother to provide Drachmann with a suggested budget that described how he was going to use the money to ensure the *Dana* returned to Copenhagen with the best possible results.[9] The following day, Schmidt wrote another note, this time expanding on his latest plan to extend and prolong the expedition, but now in a more substantiated outline, that included more details about his research plan.[10] Realizing the time lag for letters to travel from the Dutch East Indies to Northern Europe, Schmidt urged Drachmann to send him a reply by telegram in as many words as needed to give Schmidt details of how he should proceed.

Indeed the letters only reached Drachmann's desk in Copenhagen on 6 October, but Drachmann was ready with a prompt reply to the leader of the expedition. This reached Schmidt as the *Dana* was heading towards the tiny atoll of Cocos Island some 700 km southwest of Sumatra, for a much needed coaling stop, Drachmann's words were not what Schmidt wanted to read: 'For financial reasons expenses cannot exceed 600,000 Danish Kroner. According to the Director of Fisheries (Mortensen) this means homecoming in January or February. Make plans accordingly.'[11]

While the telegram in itself was sparse in text, Drachmann expanded on his own frustrations with Schmidt in a letter posted to the leader of

8 KB. Johannes Schmidt 'Papirer'. Ut.374.VI.1. 'Dana-Expeditionen, 1928–30', J. Schmidt to A. Drachmann, 3 September 1929.

9 KB. Johannes Schmidt 'Papirer'. Ut.374.VI.1. 'Dana-Expeditionen, 1928–30', J. Schmidt to A. Drachmann, 3 September 1929.

10 KB. Johannes Schmidt 'Papirer'. Ut.374.VI.1. 'Dana-Expeditionen, 1928–30', J. Schmidt Memorandum, 4 September 1929.

11 RA. *Schmidt, Ernst Johannes, biolog*: Materiale vedr. ekspeditioner og rejser (1876–1933) 2: m.m. 8 October 1929.

the expedition three days later. Here, Drachmann turned to matters of organization, finances, and the course of the expedition respectively. With regards organization he wanted to make clear to Schmidt that he, Drachmann was now the chairman of the expedition committee, as Ostenfeld was absent in Canada. Financially, he aired his surprise that Schmidt had been surprised by how money had been spent, and blamed Schmidt for not being on the ball in relation to keeping an eye on the expenses. Under no circumstances was he to exceed the amount of 600,000 Kroner.

Then Drachmann continued to the more personal matter of the lacklustre level of information from the expedition. 'Within the Board of Directors, there has been a grave dissatisfaction for quite a while, that during the length of the expedition so far, we had not received one single report, neither with regards the scientific results, nor with respect to other matters concerning the expedition.' Drachmann expanded this situation over another page in his letter, only to soften it a bit towards the end, where he found it necessary to provide Schmidt with an assurance: 'The trust in you as the leader of a large enterprise, significant for science as well as for the foundation, is unaffected, and it will remain so.'[12]

Schmidt however, had only received the short telegram, when he replied via telegram on 15 October. His message was direct and blunt with frustration percolating between the lines. Schmidt acknowledged the news from Drachmann that the Carlsberg Foundation was not able to extend their donation beyond a total of 600,000 Kroner. 'I quite understand this and I express my sincere appreciation of how much the Foundation has sacrificed. As a responsible leader however, it is my duty to inform the committee that the shortening of the expedition, as telegraphed, will seriously worsen the working conditions to the extent that a responsible completion of the expedition it not judged to be feasible. Please, I urge you, the president and members of the expedition committee to ask the government for a contribution from the state, to allow the expedition to run for another three months...'[13] In an attempt to back his point of view, Schmidt highlighted the fact that the expedition had a national as well as a scientific significance. Moreover, as he emphasized, '...the ship now holds valuable installations such as the echo sounder paid for by the expedition, which will strongly increase the value of the ship as a research vessel.'[14]

At this point Drachmann summoned the expedition committee, a list for which he had only received on 10 October.[15] The Board of Directors

12 CFD. Vedr. DANA II, 1828–1930. 103/1928. Letter from A. Drachmann to J. Schmidt, 9 October 1929.

13 CFD. Vedr. DANA II, 1828–1930. 103/1928. Telegram from J. Schmidt to G. C. Amdrup, 15 October 1929.

14 CFD. Vedr. DANA II, 1828–1930. 103/1928. Telegram from J. Schmidt to G. C. Amdrup, 15 October 1929.

15 CFD. Vedr. DANA II, 1828–1930. 103/1928. Letter from G. C. Amdrup to A. Drachmann, 10 October 1929.

also discussed Schmidt's proposal to seek financial assistance from the state at a board meeting on 18 October, where Ostenfeld was not present. Here, an agreement was made that this was undesirable, and if this was to happen, it could only be possible through the Board of Directors, not through the designated expedition committee.[16]

Four days later the issue of applying for state support, as well as Schmidt's most recent proposal for re-charting the course of the *Dana* were both discussed at a meeting of the expedition committee. This was the fourth meeting of the expedition committee, yet the first meeting to be convened following the departure of *Dana* more than 15 months earlier.[17] The next day, as the Board of Directors gathered again, it was agreed to give Schmidt a firm deadline for when the ship was expected in Copenhagen, as well as putting emphasis on the fact that the Board had already provided an extra 100,000 Danish Kroner to the project.[18]

In late October, the decisions on funding and the duration of the expedition were communicated to Schmidt via another telegram from Drachmann. This stipulated that Schmidt should feel free to go south of Africa as part of his return to Europe, but it was out of the question to ask the Danish state for extra funds to cover excess spending. Therefore, the *Dana* had to be back in Copenhagen on the 1 March at the latest.[19] The next day though, Schmidt could take comfort in a telegram from his wife. Ingeborg Schmidt had personally spoken with Drachmann, Vice-Admiral Amdrup, and Henriques from the Carlsberg Foundation's Board of Directors, and was able to relay to her husband that they all, 'stipulated that the expedition is a success!'[20]

Perhaps this is what spurred Schmidt to approach the Board of Directors with yet another inquiry for additional financial assistance. On 28 October, Ingeborg Schmidt went to see Drachmann on H. C. Andersen Boulevard about the possibility of asking the Rockefeller Foundation for financial support. Drachmann then consulted the rest of the board, and the same night he phoned her back, stipulating that this was out of the question both from the point of view of the Carlsberg Foundation as well as the Rockefeller Foundation.[21]

Schmidt now took action to modify the expedition to get home in time, so a scheduled trip to Singapore, probing the Indonesian archipelago once more, was cancelled. Drachmann sought to tighten control over the expedition, as he wired a telegram to Schmidt, demanding that from now on he wished to receive all information with regards to the running costs of the expedition as well as a monthly status.[22]

16 CFD. Carlsbergfondets Direktions Protokol, vol. 6, 18 October 1929.
17 KB. Johannes Schmidt 'Papirer'. Ut.374.1.8. 'Korrespondance 1927–1929'. Minutes, 4th meeting of Komiteen for Dana-Expeditionen, 22 October 1929.
18 CFD. Carlsbergfondets Direktions Protokol, vol. 6, 23 October 1929.
19 RA. *Schmidt, Ernst Johannes, biolog*: Materiale vedr. ekspeditioner og rejser (1876–1933) 2: m.m. 25 October 1929.
20 RA. *Schmidt, Ernst Johannes, biolog*: Materiale vedr. ekspeditioner og rejser (1876–1933) 2: m.m. 26 October 1929.
21 CFD. Vedr. DANA II, 1828–1930. 103/1928. Memorandum A. B. Drachmann, 20 November 1929.
22 RA. *Schmidt, Ernst Johannes, biolog*: Materiale vedr. ekspeditioner og rejser (1876–1933) 2: m.m. 4 November 1929.

Meanwhile, Ostenfeld had returned to Copenhagen from Canada in the last days of October, where he found out that the business of the circumnavigation had 'swollen' as he described it in a private letter to Schmidt. Ostenfeld's clear ambition was to calm down Schmidt. 'You know Drachmann is actually a very friendly man, and his personal views on this matter are very much in line with my own opinion, but he does possess this unfortunate democratic minded tendency that everyone should have a say when decisions are made. I, myself perhaps fall in the other direction too much, thinking that once a man is bestowed with a task to fulfil then one should trust him to do so.'[23] Ostenfeld further emphasized that the suspicion arose from the fact hardly anyone on the Board of Directors or in the expedition committee had much knowledge about what was going on with the expedition.

Then Ostenfeld continued with the concrete issue that the running costs on a monthly basis were too high. Ostenfeld could not comprehend it himself, but he was surprised to learn from Schmidt's lengthy letter to Drachmann that Schmidt as the expedition's leader was surprised about the hefty spending. Whatever the cause, Ostenfeld suggested to Schmidt to look ahead. He should forget about applying for extra funds. The government would simply reject such a proposal from the Carlsberg Foundation, and besides the Board of Directors had every right to be confused about Schmidt's sudden urge to go south of Africa as well as into the Mediterranean, as research in the Mediterranean had not been part of the original cruise plan agreed upon in 1928. Finally, Ostenfeld gave his thoughts on how the Board of Directors would look upon a proposal to increase their own payments to the expedition. Ostenfeld flatly stated that Schmidt should be able to make do with the original 500,000 Kroner, plus the extra 100,000 he had been granted just recently.[24]

Schmidt had hoped that Ostenfeld would resume his presidency over the expedition committee, as soon as he had returned from his trip to Canada, but this was not to be, as Drachmann continued to chair the organizing team in Copenhagen. Although, Ostenfeld was now back in Denmark, Drachmann was now so involved with the whole expedition that it was best to have him continue in this capacity.[25] The continuation of Drachmann's presidency came about at his own request following oral consultation with the rest of the members of the Board of Directors.[26] Two days later, Drachmann wrote to the members of the expedition committee he was going to continue as chairman of the expedition committee for the time being. He concluded that an intimate

23 KB. Johannes Schmidt 'Papirer'. Ut.374.VI.1. 'Dana-Expeditionen, 1928–30', C. Ostenfeld to J. Schmidt, 1 November 1929.

24 KB. Johannes Schmidt 'Papirer'. Ut.374.VI.1. 'Dana-Expeditionen, 1928–30', C. Ostenfeld to J. Schmidt, 1 November 1929.

25 CFD. Carlsbergfondets Direktions Protokol, vol. 6, 6 November 1929.

26 CFD. Vedr. DANA II, 1828–1930. 103/1928. Memorandum A. B. Drachmann, 20 November 1929.

collaboration between the foundation and the committee was necessary, as long as the financial situation was not clear.[27]

Ostenfeld was, however, successful in influencing his fellow board members on one point, the ultimatum for the *Dana* to arrive in Copenhagen by 1 March. Ostenfeld had warned Schmidt that while Drachmann thought it was too rigorous to set this deadline as an ultimatum, the two 'Hs' on the board, would be very hard to move.[28] The two 'Hs' were Professor Valdemar Henriques and Professor Hjelmslev. Here the decision did go Schmidt's way. A majority of three board members, including Drachmann, agreed to let the expedition run longer than 1 March on condition that the remaining funds would last until the end. To do so, Schmidt was asked to submit a status when he got close to Africa that is to Mombasa, then in the British protectorate of Tanganyka.[29] In addition, when Drachmann met with Ingeborg Schmidt again on the 8 November, he told her that Schmidt was expected to arrive as soon after 1 March as possible in order to cut costs.[30]

In any case, Schmidt had begun scheduling his most recent plan to go South of Africa long before reaching Mombasa. Drachmann found out about this, when Count Reventlow, Director of the Danish Foreign Service came to see him twice in early November. Reventlow wanted to know, when *Dana* could be expected to land in a number of African ports. Schmidt had already informed the Ministry of Foreign Affairs that the expedition was going to visit Cape Town, St. Helena, the Cape Verde Islands, and Tenerife. Probably taken by surprise entirely, Drachmann was forced to put Reventlow on hold until further notice.[31]

It was not until the 10 November that Schmidt told Drachmann directly, via telegram from the *Dana* that he intended to go South of Africa.[32] The Board then decided to wire a text indicating that they did not see the purpose of the latest telegram, as the expedition had not yet reached the shores of Africa.[33] Having received this telegram, Schmidt placed a most bitter remark in his notebook; 'one of the continuously unpleasant telegrams (Arrogance!).'[34]

While, we can remain sure that Schmidt was personally offended, his frustration was well founded with regards to the possibility of planning ahead. The entire expedition depended on having reliable supplies of equipment ready when arriving in a new port, while it was just as important to ship home the ever expanding collection of glass jars and records of samples taken. Also, it was not always so easy to send messages by telegram. The telegram, which Schmidt wired to Drachmann on the

27 CFD. Vedr. DANA II, 1828–1930. 103/1928. A. B. Drachmann to Komiteen for Carlsbergfondets Dana-Expedition, 8 November 1929.
28 KB. Johannes Schmidt 'Papirer'. Ut.374.VI.1. 'Dana-Expeditionen, 1928–30', C. Ostenfeld to J. Schmidt, 1 November 1929.
29 CFD. Carlsbergfondets Direktions Protokol, vol. 6, 6 November 1929.
30 CFD. Vedr. DANA II, 1828–1930. 103/1928. Memorandum A. B. Drachmann, 20 November 1929.
31 CFD. Vedr. DANA II, 1828–1930. 103/1928. Memorandum A. B. Drachmann, 20 November 1929.
32 CFD. Vedr. DANA II, 1828–1930. 103/1928. J. Schmidt to A. B. Drachmann, 17 November 1929.
33 CFD. Carlsbergfondets Direktions Protokol, vol. 6, 18 November 1929.
34 RA. *Schmidt, Ernst Johannes, biolog*: Materiale vedr. ekspeditioner og rejser (1876–1933) 2: m.m. 21 November 1929.

10 November, for example, did not arrive in Copenhagen before the 17 November. In addition short telegram messages had their limitations and clearly some of the misunderstanding between Schmidt and Drachmann arose from the relatively long time lap between sending and receiving mail. Letters sent from Copenhagen by 25 November 1929 for instance, were expected to arrive in Mombasa on 15 December, 20 days later.

Nonetheless, Schmidt now finally gave up his first idea to change the course of the expedition and head back to Celebes and New Guinea as indicated in his letter to Amdrup from late August. In a letter sent from Colombo on 30 November, Schmidt revealed that the eel samples taken off the Sumatran coastline in November had been so prolific that a trip back to the western Pacific waters was now less important. Now, he was aiming solely at getting the acknowledgement from Drachmann that he could stay out past the previously stated deadline of 1 March.[35]

Shipping off the packages with the collections from the Indian Ocean had been delayed in Sri Lanka, but now, 1 December, the *Dana* was headed for Mombasa, Schmidt noted in a telegram to Drachmann. Thus, he really needed to know, if he could go to South of Africa. He had 38,000 Kroner left in cash, which could keep the expedition going until c. 1 February, but he needed advice to plan ahead the next stages of work on-board.[36] Drachmann responded immediately, that is on 3 December, indicating that Schmidt should feel free to advise a later date for arrival in Copenhagen, if this was necessary.[37]

What Schmidt did not know was that a similar, generous message was under way by mail. In a letter from 25 November, Drachmann reported to Schmidt that the monthly costs of the expedition so far, had been somewhat lower than previously understood. Therefore, there was good reason to believe that the remaining amount could last throughout the expedition, and since the arrival at Mombasa was delayed, the time limit of 1 March had become less realistic. He went on to suggest that the final date of arrival back in Copenhagen should then be set following consultation between Schmidt and the expedition committee.[38] The latter aspect, that Schmidt was to be consulted, was an important change. This opinion was shared by Martin Knudsen. At first, Martin Knudsen was quoted otherwise in the minutes, but following objections to the first minutes of the meeting, this was altered in Knudsen's – and Schmidt's favour.[39] In this way, the Board of Directors and the expedition committee came to concur.

35 KB. Johannes Schmidt 'Papirer'. Ut.374.VI.1. 'Dana-Expeditionen, 1928–30', J. Schmidt to Komiteen for Dana-Expeditionen, 30 November 1929.

36 KB. Johannes Schmidt 'Papirer'. Ut.374.VI.1. 'Dana-Expeditionen, 1928–30', J. Schmidt to Komiteen for Dana-Expeditionen, 1 December 1929.

37 CFD. Vedr. DANA II, 1828–1930. 103/1928. A. B. Drachmann to J. Schmidt, 3 December 1929.

38 KB. Johannes Schmidt 'Papirer'. Ut.374.I.8. 'Korrespondance 1927–1929'. A. B. Drachmann to J. Schmidt, 25 November 1929.

39 CFD. Vedr. DANA II, 1828–1930. 103/1928. M. Knudsen to A. B. Drachmann, 9 November 1929. CFD. Vedr. DANA II, 1828–1930. 103/1928. A. B. Drachmann to M. Knudsen, 11 November 1929. CFD. Vedr. DANA II, 1828–1930. 103/1928. M. Knudsen to A. B. Drachmann, 13 November 1929.

40 KB. Johannes Schmidt 'Papirer'. Ut.374.VI.1. 'Dana-Expeditionen, 1928–30', Circulair Drachmann to Komiteen for Dana-Expeditionen, 10 January 1930.

The duration, as well as the final route and the total allowable costs of the circumnavigation were settled finally towards the very end of 1929. On the 27 December, Schmidt wired a telegram to Drachmann stating: 'Going South around Africa. Home in June. Guarantee that the total allowed sum of 600,000 will not be exceeded'.[40] 'Accepted', was

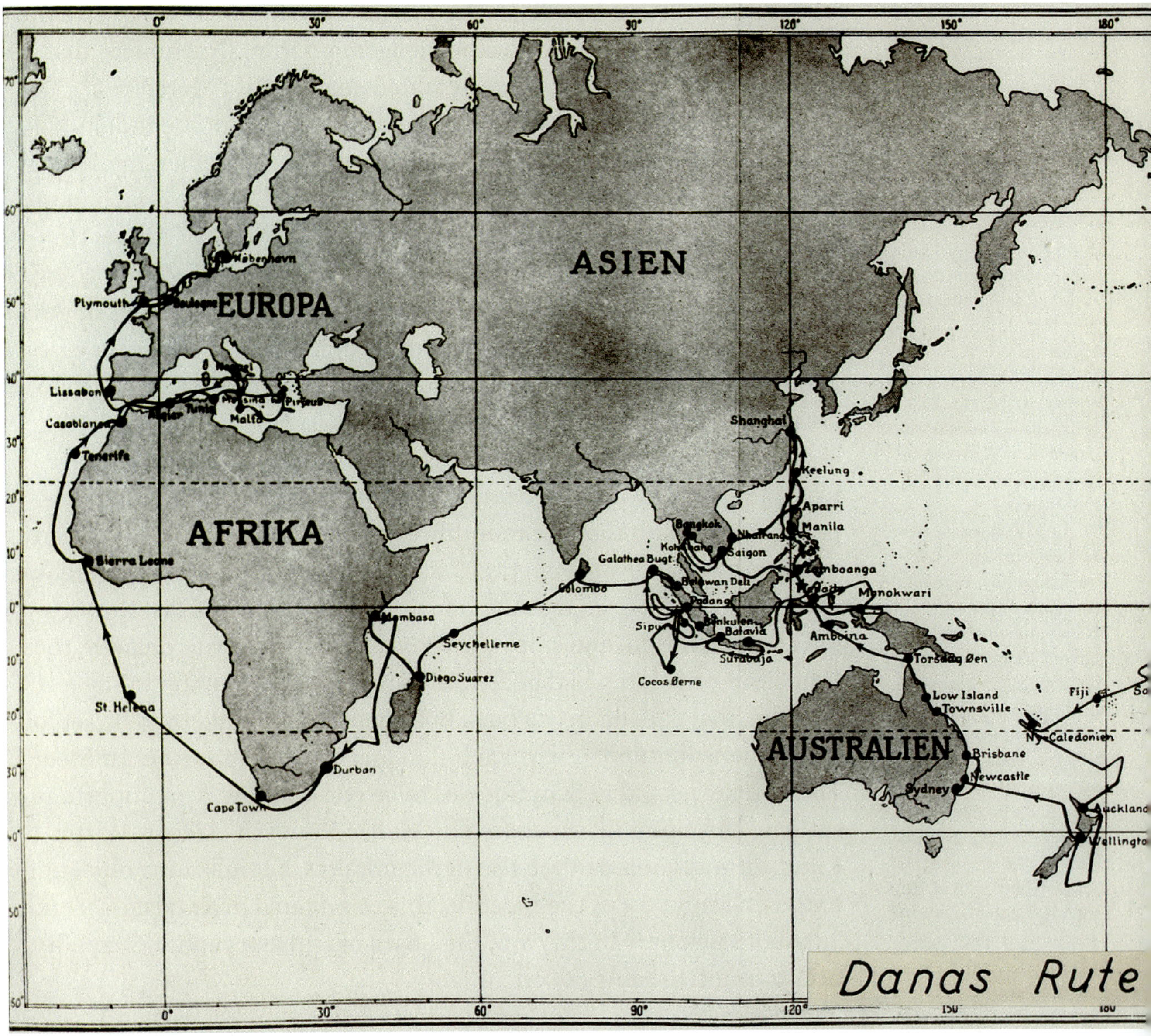

15.1 The route of *Dana*'s circumnavigation 1928–1930

The actual route of the circumnavigation is shown here. When the cruise plan was negotiated beforehand there was no mention of going South around Africa, nor was a visit to Sri Lanka part of the plan. Schmidt also had to give up a planned visit to the Coral Sea East of the Indonesian Archipelago as well as the scheduled visit to Japan (Carlsberg Archives).

Drachmann's short reply.[41] A few days later, Drachmann sent a wordier message to Schmidt by mail. Here, he expanded that on his own behalf he would be grateful if a significant sum was still left over, when the expedition came to an end. 'We could need the money, and there is no obligation to spend it all.[42]

For the benefit of the committee members, Drachmann then explained how developments since the month of October could justify the acceptance of these new terms: First of all, the planning of the homebound journey was to be settled, once *Dana* had left Colombo. This

41 KB. Johannes Schmidt 'Papirer'. Ut.374.VI.1. 'Dana-Expeditionen, 1928–30', Circulair Drachmann to Komiteen for Dana-Expeditionen, 10 January 1930.

42 CFD. Vedr. DANA II, 1828–1930. 103/1928. A. B. Drachmann to J. Schmidt, 4 January 1930.

happened one month later than previously scheduled. Secondly, Drachmann reported that by now the foundation was in possession of a more precise indication of whether or not the total costs of running the ship could be kept within the budget of 600,000 Kroner. Previous budgets based on expenses while traversing the Indonesian Archipelago, where good quality coal was in short supply, had set the monthly running costs too high. Therefore, Schmidt would still have c. 135,000 Kroner left for the remaining period of the expedition. However, Drachmann also indicated to Schmidt that he should still shorten the expedition, if the total costs looked like running above the remaining 135,000 Kroner.[43]

In this way, uncertainty about the direction of the expedition, the financial situation, and the timing of its arrival home in Copenhagen had been settled. Clearly, this was a huge relief to all parties involved. Schmidt's diary for instance, became much more lively and joyful during the first half of 1930.[44]

Going south round of Africa

New Year's Eve 1929 was spent in Mombasa, Tanganiyka. The long awaited arrival due to the delays in Colombo had caused the mail to pile up high, and the ship's crew rested for a few days, while the boiler was cleaned and new loads of coal brought on-board. The Expedition leader kept himself busy with a railroad trip from Mombasa to Nairobi in Kenya and back. Nairobi is situated close to 1,800 metres above sea level, where the river systems were home to a specific freshwater eel (*Anguilla labiata*). Now, Schmidt could secure the collection of specimen for future research. The Carlsberg Foundation was greeted with a postcard with giraffes. A real holiday for the expedition though, was not on the programme, as Schmidt thought Mombasa to be too dangerous due to the risk of malaria. Holidays then, were postponed to the next port of call, Durban in South Africa.[45] The visit to Durban also proved to be one of the most publicized and popular events of the entire cruise. The *Danmark-Filmen* for instance was shown in the Durban City Hall, where a crowd of 2,500 spectators witnessed the two and a half hours long display of Danish delights.[46]

As for the scientific parts of the cruise along the east coast of Africa, Schmidt was able to present equally important results. In the Mozambique Channel, the arm of the Indian Ocean located between Mada-

43 KB. Johannes Schmidt 'Papirer'. Ut.374.VI.1. 'Dana-Expeditionen, 1928–30', Circulair Drachmann to Komiteen for Dana-Expeditionen, 10 January 1930.

44 RA. Schmidt, Ernst Johannes, biolog: Materiale vedr. ekspeditioner og rejser (1876–1933) 2: m.m.

45 CFD. Vedr. DANA II, 1828–1930. 103/1928. J. Schmidt to Komiteen for Dana-Expeditionen, 31 December 1929.

46 CFD. Vedr. DANA II, 1828–1930. 103/1928. J. Schmidt to Komiteen for Dana-Expeditionen, 25 January 1930.

gascar and Mozambique, the expedition was able to take a number of successful stations. Schmidt deduced from this that the northern end of the Mozambique Channel was home to larvae from the tropical eel, (*Anguilla bicolor*), while in the southern parts of the channel the hauled larvae were from the more southern (*Anguilla mossambica*). Schmidt indicated that this was the first proof of the whereabouts of larvae from the latter sub-species.[47]

Meanwhile in Copenhagen, the Board of Directors convened again in mid-January, and once more the business of the circumnavigation was on the agenda. Having arrived at an agreement with Schmidt about the duration and the costs of the expedition, the Board decided that Drachmann could step down from his temporary position as head of the expedition committee, but he kept a grip on the expedition, as from then on the monthly financial reports from the *Dana* were to be sent to the Board of Directors for approval, before they were forwarded to the expeditions committee. The following month Ostenfeld reported to the Board that Schmidt had requested permission to take residence on the Canary Islands in the spring of 1930, while the *Dana* traversed the South Atlantic Ocean, on what was scheduled to be more a transport leg of the expedition. This was approved, seemingly without much praise.[48] Schmidt proceeded to book a ticket to Tenerife, where he would meet up with Ingeborg and spend some time relaxing away from the strenuous days on board the *Dana*.[49] He then wrote to Ostenfeld to thank him for all his help with the 'intermezzo' as Schmidt called his confrontation with Drachmann over when and how the *Dana* was to complete the expedition. Schmidt regretted that he had to deal with such issues at a point in time, when all of his energy was supposed to focus on one thing only – research. He underlined that in his mind there was no doubt that the expedition would have been able to solve more of the puzzles surrounding the life history of the eel in the Indonesian Archipelago if had not been for the quarrels and difficulties.[50]

Meanwhile Schmidt devised the plan for *Dana's* continued expedition into the Atlantic Ocean. Zoologist Theodor Mortensen had been promised that he could make his own surveys in the Indian Ocean, but due to the many changes to the original itinerary of the *Dana*, Mortensen never got the chance to work on board the ship in the ocean east of Africa. Now, the *Dana* was headed for the island of St. Helena in the middle of the South Atlantic Ocean with the sole purpose of fulfilling the wishes of, or rather ameliorating Mortensen. St. Helena though,

47 CFD. Vedr. DANA II, 1828–1930. 103/1928. J. Schmidt to Komiteen for Dana-Expeditionen, 25 January 1930.
48 CFD. Carlsbergfondets Direktions Protokol, vol. 6, 5 January 1929.
49 RA. *Schmidt, Ernst Johannes, biolog*: Materiale vedr. ekspeditioner og rejser (1876–1933) 2: m.m. 23 Janaury 1930.
50 KB. Johannes Schmidt 'Papirer'. Ut.374.VI.1. 'Dana-Expeditionen, 1928–30', J. Schmidt to C. Ostenfeld, 7 February 1930.

15.2 Ethnografica from around the World
The expedition also collected a sizeable amount of ethnographica during the two years of circumnavigating the World (Carlsberg Archives).

was also an important bunker station for coaling. In total Schmidt allowed only 21 scientific stations to be taken between Cape Town South Africa and Casablanca in Morocco.[51]

Return to the Mediterranean

In early April 1930 the *Dana* steamed into the Alboran Sea between Spain and North Africa, where Schmidt had suffered from torrential winter storms more than twenty years earlier during the *Thor's* first

51 KB. Johannes Schmidt 'Papirer'. Ut.374.VI.1. 'Dana-Expeditionen, 1928–30', J. Schmidt to Komiteen for Dana-Expeditionen, 8 February 1930.

Mediterranean expedition. On the 13 the April, the *Dana* arrived in Algiers, where a much needed cleaning of the ship's hull was to be carried out. While waiting for the drydock to be vacant – and approachable in the strong waves in port, Schmidt and the crew spent the next week socializing and sightseeing in Algiers, where Ingeborg had also awaited their company.[52]

Research wise, Schmidt revealed his plans for the final legs of the expedition in a couple of private letters to Ostenfeld. At this point in time, Schmidt's interests were fully devoted to the fate of the eel.

First of all, he felt a need to substantiate his findings from the *Thor* expeditions and the Atlantic expeditions with *Dana I* and *Dana II* in respect to the inflow of eel larvae to the Mediterranean. Still more larvae should be found, not least in the eastern half of the Mediterranean.

Secondly, Schmidt opened up a discussion that he himself had settled already in his publications of 1922 and 1923, namely the possibility of the Mediterranean being a separate breeding ground for the freshwater eel. While admitting to Ostenfeld that no evidence whatsoever had been found so far, Schmidt suggested that the Eastern Mediterranean basins might be home for a separate stock of eel. Referring to his findings in Sumatra, where the smallest larvae had been found at depths of more than 5,000 metres, and the Sargasso Sea, where larvae were also found at great depths, Schmidt proposed that the equipment on board the *Dana*, would lend the expedition a unique chance to investigate this. Presenting once more Italian marine science and politics as the antagonists, Schmidt continued his letter stressing how it was absolutely vital to carry out these investigations, not in the deep seas off Italy, but rather at great depths in the Ionian Sea. If the Italians or someone else someday found evidence to suggest a Mediterranean breeding place of the eel, then the Italians would surely present a politically motivated statement concerning the erroneous long-term labour of Danish marine scientists. As ever, Schmidt was highly aware of securing his own legacy.[53]

At this point Schmidt was still highly regretful about 'the embarrassing negotiations of last fall' as he called them. He pointed to the poor weather situation, as he emphasized how he was right in stating that arriving in the Mediterranean too early in spring would result in wasteful time weathering out storms.[54]

Schmidt was getting slightly unnerved by the prospects of a prolonged stay in Algiers and the frequent storms all over the Mediterranean, which would mean less time for conducting research. 'Il faut de la

52 RA. *Schmidt, Ernst Johannes, biolog*: Materiale vedr. ekspeditioner og rejser (1876–1933) 2: m.m. 13–20 April 1930

53 KB. Johannes Schmidt 'Papirer'. Ut.374.VI.1. 'Dana-Expeditionen, 1928–30', J. Schmidt to C. H. Ostenfeld, 16 April 1930.

54 KB. Johannes Schmidt 'Papirer'. Ut.374.VI.1. 'Dana-Expeditionen, 1928–30', J. Schmidt to C. H. Ostenfeld, 16 April 1930.

patience!' as Schmidt stated ironically. Probably he was not quite sure how Drachmann and the rest of the Board of Directors would react to the fact that he was not coming home just yet. In any case, he decided to test the water with Ostenfeld: 'Following the wording of the telegram, which I sent from Mombasa, and which was 'accepted' by Drachmann, I just had to be home in June, as long as the expenses were kept within the set limit. I hope that if needed, a solution could be found with regards to the crew from the navy, where some are due to enter regular service in the beginning of July.'[55]

Two weeks later, Ostenfeld responded to Schmidt's new proposal, clearly seeing that Schmidt was asking for yet another extension. On the idea of carrying out more tests to see if the eel bred in the Mediterranean, Ostenfeld answered that he was astonished, but curious to read about it.

> 'So there is really something new to this issue? I thought the Italian position was solely dictated by their wish to defend the great Grassi. Nonetheless, I will tell you one thing. If the *Dana* is not home before the 1 July, I cannot guarantee the position of the Carlsberg Foundation. Although, I do not have the wording of the Mombasa telegram in front of me, I know for sure that it has been an absolute expectation that the expedition arrives during the month of June. I was already amazed, when I learned that the board was prepared to let the expedition continue this long. I urge you not to time the arrival beyond this point.'[56]

Thus Ostenfeld showed a guarded interest in the scientific bait, but he did not chew on the hook at the end of the line. Ostenfeld for one was not going to give Schmidt any excuses not to come home in June.

As usual, Schmidt did not wait for a reply from Ostenfeld before proceeding with this latest addition to the cruise plan. In a letter to Admiral Rechnitzer from the Naval Ministry, Schmidt asked for permission to stay at sea until around 15–20 July, hereby borrowing the naval officers for a couple of additional weeks. Rechnitzer was also an ordinary member of the expedition committee, so Schmidt clearly broke the agreed chain of command by asking Rechnitzer directly instead of communicating via Drachmann. Schmidt also told Rechnitzer the story of the possibility of the eel breeding in the Levantine Sea. The boldest proposition though, was Schmidt's stating that he had not made a definite promise to Drachmann to be home by the month of June. Besides, he still had funds left from the 600,000 Kroner grant.[57] Finally, Schmidt

55 KB. Johannes Schmidt 'Papirer'. Ut.374.VI.1. 'Dana-Expeditionen, 1928–30', J. Schmidt to C. H. Ostenfeld, 17 April 1930.
56 KB. Johannes Schmidt 'Papirer'. Ut.374.I.8. 'Korrespondance 1927–1929'. C. H. Ostenfeld to J. Schmidt, 30 April 1930.
57 KB. Johannes Schmidt 'Papirer'. Ut.374.I.8. 'Korrespondance 1927–1929'. J. Schmidt to Rechnitzer, 6 May 1930.

58 KB. Johannes Schmidt 'Papirer'. Ut.374.1.8. 'Korrespondance 1927–1929'. J. Schmidt to F. V. Mortensen, 8 May 1930.

wrote to Mortensen, the Fisheries Director and also a member of the expedition committee on the 8 May. In this communication he presented a similar story and asked for permission to borrow a first mate through him, if he had to let go of the navy officers.[58] Neither of these requests prompted the response, Schmidt wanted.

Probably Ostenfeld was alerted to these letters, as he resorted to send Schmidt another message, an even longer letter, in which he listed a number of calamities arising from a late arrival. Most importantly, Ostenfeld urged Schmidt to keep his solemn oath and be back on time.

15.3 Bjarne Jørgensen brings the coffee
The ship's boy, Bjarne Jørgensen is bringing the coffee (RA. Carlsberg-fondets DANA-ekspeditioner. 1917–1973. 7c).

Ostenfeld neither could, nor would go to the Carlsberg Foundation to ask for another extension. Finally, Ostenfeld pointed out that the same foundation would be the only source of money to approach in relation to publishing the results arising from the expedition. Sometimes, Ostenfeld reminded Schmidt, it pays off to show some kindness and a little indulgency.[59]

In Schmidt's monthly report home from the 9 June, it is apparent that the Mediterranean part of the voyage centred on the life history of the freshwater eel. Schmidt included a map of the stations taken in the Mediterranean in support for this activity.[60] But, Schmidt was anything but satisfied. Ostenfeld's letter from 30 April arrived in the hands of the expedition leader only on 18 May, while the *Dana* was in the Athenian port of Piræus.[61] On 21 May Schmidt sent a brief, but bitter note to Ostenfeld written in the Patras Bay having just left Athens. '…I will ask you to remember that on this expedition, I have tolerated unjust and unfounded accusations from the point of view of a management that, putting it mildly, was never quite up to date, neither when informing me about my duties and rights, nor when judging my actions.'[62]

Thus in a brief statement concerning the surveys conducted in the Mediterranean, Schmidt more or less blamed the Carlsberg Foundation's Board of Directors for the meagre scientific returns. For scientific reasons, he said, he had placed great weight on examining the depths of the Levantine Sea, the easternmost Mediterranean basin between Turkey and Egypt. Then, he proceeded to lay out the final section of the circumnavigation. Steaming out into the North Atlantic, the team would take nutrient samples to compare with results from the Indian Ocean and the Pacific. One stop was arranged at the port of Boulogne sur Mer in northwest France, where an official French ceremony was to be held in honour of the Danish expedition.[63]

Meanwhile, Schmidt's theory of the Sargasso Sea as the breeding ground for the eel, was cemented in an interview he gave to the Napoli based newspaper, *Il Mattino*. Here it was stated clearly that the Danish expedition had established that the breeding grounds of the freshwater eels of Italy and the rest of Europe were far away at great depth near Bermuda.[64] Indeed, Schmidt would never return to probe the depths of the Mediterranean, and there is no indication that he ever again tried to question his world famous theory.

The rest of trip back home was dominated by diplomatic calls of duty and preparations for the homecoming festivities, which were in

59 KB. Johannes Schmidt 'Papirer'. Ut.374.I.8. 'Korrespondance 1927–1929'. C. H. Ostenfeld to J. Schmidt, 14 May 1930.
60 KB. Johannes Schmidt 'Papirer'. Ut.374.VI.1. 'Dana-Expeditionen, 1928–30', J. Schmidt to Komiteen for DANA-Expeditionen, 9 June 1930.
61 KB. Johannes Schmidt 'Papirer'. Ut.374.I.8. 'Korrespondance 1927–1929'. C. H. Ostenfeld to J. Schmidt, 30 April 1930.
62 KB. Johannes Schmidt 'Papirer'. Ut.374.VI.1. 'Dana-Expeditionen, 1928–30', J. Schmidt to C. H. Ostenfeld, 21 May 1930.
63 KB. Johannes Schmidt 'Papirer'. Ut.374.VI.1. 'Dana-Expeditionen, 1928–30', J. Schmidt to Komiteen for DANA-Expeditionen, 9 June 1930.
64 RA. Anton Fr. Bruun. Danaekspeditionerne 1924–1930. Avisklip, Il Mattino, 27 May 1930.

the hands of Jespersen. According to tradition, the *Dana* on homecoming should present herself in an English as well as a French port. First port of call, therefore, was Plymouth, home to the Plymouth Marine Laboratory, where the leader, E. J. Allen received the expedition accompanied by Henry G. Maurice, the President of ICES, and also the head of the Fisheries Department of Great Britain.[65]

Judging from the numerous newspaper articles surrounding the event, the French visit to Boulogne sur Mer was a great success. Ingeborg Schmidt had travelled to Boulogne alone, so the civic reception may have appeared as if the Danish royal family had suddenly emerged from the Strait of Dover. At any rate, the Danish journalist Boesgaard, who was on-board the *Dana* for the past week, gave a vivid account of the French visit supported by photographic documentation: School children were lined up in the streets waving small Danish flags, as Mr and Mrs. Professor Schmidt were paraded through town. At the town hall the French geographer, Louis Joubin presented Schmidt with the title of Commander of the National Order of the Legion of Honour, the highest distinction in the French Republic. Later, the Danish expedition sort of returned the favour when visiting a memorial site for World War One, where Ingeborg Schmidt placed a bucket of flowers 'in admiration and gratefulness for all that France had given to the world.'[66]

'To the honour of Denmark'

The timing of the expedition's return to Copenhagen was so close to the agreed deadline – end of June, that the end of Schmidt's circumnavigation came to rival the return of Phileas Fogg to the Reform Club salon in Jules Verne's *Around the World in Eighty Days*. A mere 12 hours before the Carlsberg Foundation's time limit was up, the *Dana* arrived in The Sound, first stopping briefly at Hamlet's castle, Kronborg at Elsinore. Here, the Danish royal dignitaries, Prince Valdemar and Prince Georg were both brought on-board along with Ostenfeld and Vice Admiral Amdrup to witness the last couple of hours down The Sound and into the Port of Copenhagen. In a great many Danish newspapers this was front-page news, and judging from the illustrations, several hundred spectators were lined along the Nordre Toldbod quays at 2 o'clock in the afternoon when the ocean explorers completed their circumnavigation.

65 Boesgaard H. De sidste Omdrejninger. In: Schmidt J, ed. *Dana's Togt omkring Jorden*. Copenhagen: Gyldendal; 1932:352–364.
66 RA. *Anton Fr. Bruun. Danaekspeditionerne 1924–1930*. Avisklip, *Berlingske Tidende*, 26 June 1930.

15.4 The *Dana* returns to Copenhagen in 1930
At 2 PM on the 30 June 1930 the R/V Dana arrived triumphantly to the Port of Copenhagen, making headlines on the cover of every daily newspaper in the country.

On a sunny afternoon, the expedition was welcomed by a cheering crowd led by the acting Prime Minister, Friis-Skotte and a number of heads of state departments and the Rector Magnificus of Copenhagen University.[67] Crowding the vessel itself, the welcoming party allegedly had difficulties keeping clear of the large sea turtles roaming the deck, but speeches went ahead as planned. The Board of Directors of the Carlsberg Foundation was there, and Ostenfeld thanked the expedition on behalf of the foundation as well as the expedition committee. Great results have been achieved, Ostenfeld said, and knowing the reputation of Johannes Schmidt, there was good reason to believe the results would continue to flow from follow-up work on the collected specimen.

According to the newspapers, Schmidt was moved, when it was his turn to address the crowd. Schmidt thanked everyone who had made the expedition possible. 'We have heard stronger words out in the world, but what we have heard today has gone straight to our hearts, because these are the words of our own country.'[68] Yet for even the finest equipped expedition, small things can tilt the entire operation. Therefore he was grateful towards 'that which dictates wind and weather, life and death,' for having brought the expedition back safely in their little 'nutshell of a boat.'[69] 'We are proud to have added to the glory and honour of Denmark.'[70]

67 KB. Johannes Schmidt 'Papirer'. Ut.374.IV.3. 'Avisklip', *Social Demokraten*, 1 July 1930.
68 KB. Johannes Schmidt 'Papirer'. Ut.374.IV.3. 'Avisklip', *Social Demokraten*, 1 July 1930.
69 KB. Johannes Schmidt 'Papirer'. Ut.374.IV.3. 'Avisklip', *Dansk Fiskeritidende*, 9 July 1930.
70 KB. Johannes Schmidt 'Papirer'. Ut.374.IV.3. 'Avisklip', *Scoial Demokraten*, 1 July 1930.

15.5 Ostenfeld addresses the expedition in 1930
Ostenfeld is thanking the expedition on behalf of the Carlsberg Foundation as well as the expedition committee. Schmidt arrived a mere 10 hours before the deadline set by the expedition committee.

The triumphant return reached a new highpoint on the 2 July when Crown Prince Frederik received the members of the *Dana* expedition at the Amalienborg Royal Palace. In his diary Schmidt noted that, 'The Crown Prince comes off neatly, and throws a nice speech, and bestows everyone with the Silver Medal of Merit in Silver with Clasps.'[71] Clearly satisfied and proud, Schmidt remarked that the delighted expedition members looked great in their white clothes. In the evening of the same day, Prince Valdemar threw a party at his own residence, Bernstorff Palace, where the officers and scientists of the expedition were invited along with a few members of the organizing committee.[72] Six days later, as Schmidt walked through the massive oak doors of the Carlsberg Foundation headquarters, the mood was shifting substantially.

71 RA. *Schmidt, Ernst Johannes, biolog*: Materiale vedr. ekspeditioner og rejser (1876–1933) 2: m.m. 2 July 1930.

72 RA. *Schmidt, Ernst Johannes, biolog*: Materiale vedr. ekspeditioner og rejser (1876–1933) 2: m.m. 2 July 1930.

15.6 The expedition in front of Amalienborg
The members of the expedition emerge from the Amalienborg Royal Palace each with a Silver Medal of Merit with Claps. The common seamen are still in their Dana uniforms, the scientists are in civilian clothes, while the navy officers are in their offical uniform (Carlsberg Archives).

Pelted with unpleasantness

On 8 July, the Chairman of the Board of Directors, A. B. Drachmann, received Schmidt in the stylish building on H. C. Andersen's Boulevard. According to Schmidt's own notes, he commenced the meeting by thanking Drachmann for all the trouble he had faced on accounts of the *Dana* expedition, and then a friendly conversation ensued. Finally though, Schmidt was then told that now Drachmann would issue a few statements, whereupon Schmidt was 'pelted with unpleasantness.'[73]

Drachmann also took note of this meeting, and wrote up a memorandum to the Board of Directors, and these 'few statements' clearly echoed in his account. First of all, Drachmann had found the organizing

73 RA. *Schmidt, Ernst Johannes, biolog*: Materiale vedr. ekspeditioner og rejser (1876–1933) 2: m.m. 8 July 1930.

Committee to be too large to manoeuver in any fruitful way. While, the board had formally approved the composition of the committee, it had come about as an arrangement directly between Schmidt and Ostenfeld, thus it was not to Drachmann's taste. Further, it was important for Drachmann to inform Schmidt that the financial problems themselves had direct bearing on the actions of the Board, not the committee. The Board had to step in at that point. Therefore, Drachmann was disappointed to learn how Schmidt had continued to correspond with the committee instead of referring to the directors, in a situation where the financial situation of the expedition was so dire. During the first part, the first year really, Schmidt had not taken action to produce a monthly economic status. The Board had granted Schmidt an additional 100,000 compared to the original budget, a sum which was to be spent in Australian waters. In the end, it was not spent on the intended purpose, as the expedition did not return to Australian waters as originally planned. Therefore, Drachmann was frustrated to observe, how Schmidt spent this money anyhow, instead of observing a higher measure of fiscal prudence.[74]

The least he could have done was to submit regular accounts of the scientific fruitfulness of the various steps of the circumnavigation. Again, according to Drachmann, Schmidt replied that it had never occurred to him that he should do so, and this was something that he had never done, so Drachmann decided to emphasize that it had indeed been the wish of the Board of Directors to follow the enterprise, which they had supported.

'Finally, I mentioned the festivities,' Drachmann continued. 'Apparently Schmidt was surprised to learn that we looked upon these critically. Had he known about that, he would have acted accordingly. To this, I added that the festivities would not take place at great length, since the arrival of *Dana* had not been reported in advance. Whatever the case may be, the Board of Directors found it more dignified if the members of the expedition would act more quietly.' Drachmann concluded his memo by noting how the whole conversation with Schmidt had been escorted '…by the full observation of friendly manners.'[75]

Nonetheless, Schmidt was so offended that two weeks later, he went back to the Carlsberg Foundation to see Drachmann. Schmidt then wrote in his diary how he had told Drachmann, that he had been insulted by his comments during the previous meeting.[76] Looking at the dispute today, it seems in many ways as if the issues arising between

74 CFD. Vedr. DANA II, 1828–1930. 64/1930. Memorandom of meeting with J. Schmidt 8 July, by A. B. Drachmann, 8 September 1930.

75 CFD. Vedr. DANA II, 1828–1930. 64/1930. Memorandom of meeting with J. Schmidt 8 July, by A. B. Drachmann, 8 September 1930.

76 RA. *Schmidt, Ernst Johannes, biolog*: Materiale vedr. ekspeditioner og rejser (1876–1933) 2: m.m. 22 July 1930.

Schmidt and Drachmann often related to sheer décor, formality, and procedure. Yet, clashes over procedure are usually cast in light of more profound elements of frustration, disagreement, and distrust, and this was also the case here.

However, no immediate conclusion seems to have been drawn from this meeting, and the following week, Ingeborg and Johannes Schmidt embarked on a long awaited four week long recreational trip to the spa waters of Clermont-Ferrand in the French Massif-Central.[77] Soon though, action would be taken to try and curb Schmidt's influence at the Carlsberg Laboratory. On the first day of September, the Board of Directors convened again, and as the meeting drew to an end, the circumavigation surfaced under the item of 'any other business.'

First, the innocent task of orchestrating a dinner in celebration of the return of the expedition was handled. Drachmann and Schmidt were to arrange this together with Ostenfeld, to the extent that Ostenfeld's ailing health allowed. Ostenfeld was absent at the meeting.

Perhaps Ostenfeld's absence was the reason why Professor Henriques, who also sat on the Board of Trustees of the Carlsberg Laboratory, dared to raise a rather bold question. In light of the fact that Johannes Schmidt had by now distanced himself so far from the core objectives of the Carlsberg Laboratory, Henriques proposed a change in Schmidt's job situation. Henriques therefore asked the Board if they would favour creating a new professorship in marine science at Copenhagen University, funded by the Carlsberg Foundation, although on the conditions that Schmidt himself was positive towards the idea. The other board members approved of this idea, with Drachmann raising one concern. He feared that if such a position was created for Schmidt, it would lead to the creation of a de facto Carlsberg supported 'Institute of Oceanography,' somewhat beyond the financial powers of the foundation. No one objected to the proposition since Schmidt was not conducting the expected beer related research at the Laboratory.[78]

Within the next few days Henriques approached Schmidt with the offer of a professorship. We know this from notes in Schmidt's diary, when he paid a visit to the hospitalized Ostenfeld on 8 September. While also a social call to his close friend, Schmidt noted that he received important news on the plans to separate him from his Directorship at the Carlsberg Laboratory, a position he had then held for two decades. According to Schmidt's diary, Ostenfeld had been affected negatively, 'in my disfavour,' by a previous visitor to his hospital bed, Profes-

77 RA. *Schmidt, Ernst Johannes, biolog*: Materiale vedr. ekspeditioner og rejser (1876–1933) 2: m.m. 30 July – 29 August 1930.

78 CFD. Carlsbergfondets Direktions Protokol, vol. 6, 3 September 1929.

sor Henriques from the Carlsberg Foundation. Ostenfeld told Schmidt that Henriques had been clumsy in his approach to Schmidt concerning the proposed transfer of position from the laboratory to a professorship in marine science. The intention had not been to lay him off, but to bestow on Schmidt this extraordinary professorship as an honour in light of the successful circumnavigation.[79] The above-mentioned memorandum of Drachmann's July meeting with Schmidt was in fact written up the same day, 8 September.

Clearly, it was high noon in the debacle between Schmidt and the entire Board of Directors. The Board was fed up with Schmidt's uncontrollable behaviour. The way in which the Board of Directors saw it, Schmidt did exactly as he pleased, no more, no less. He had now grown as big as a cuckoo in the nest of a small read warbler. The wealthy endowment enriched by the steady flow of beer from the pipes at the Carlsberg Breweries, had supported the growth of a world-class marine science empire. Now, this vagrant success on the seven seas was menacing, if not to the host itself, then seriously threatening the raison d'être of the Carlsberg Foundation. The statutes of the Carlsberg Laboratory were violated at the time of Schmidt's employment in 1910. Now, twenty years on, the rules had been bent to the point of breaking.

Nonetheless, nothing more came of this proposal, as Schmidt flat out rejected the proposition, so the Board of Directors seems to have caved-in for the remainder of Schmidt's directorship at the Carlsberg Laboratory. On his part, Schmidt had the memory of an elephant. The bitterness may have ceded gradually as the months went on, and he was able to recuperate from what must have been an arduous adventure. At the age of 53 and with a heart condition, Schmidt was hardly a young man anymore, and commanding a two-year circumnavigation would take its toll on anyone. During the fall of 1931, Schmidt was writing up and editing his and his colleagues' travel account of the great circumnavigation. The book came out as *Dana's Togt omkring Jorden*, (1932), a more than 360 pages long attempt to popularize the scientific findings of the expedition as well as a presentation of the expedition members' many impressions from visits on the Seven Seas.[80] In the introduction Schmidt showed some of the gratefulness and friendliness towards the Carlsberg Foundation, which Ostenfeld had encouraged him to put on display the year before. Yet, towards the end of the introduction the tone shifts, as Schmidt begins to justify some of his actions, in particular with regards to the course taken around the world. First of all, he refers

79 RA. *Schmidt, Ernst Johannes, biolog*: Materiale vedr. ekspeditioner og rejser (1876–1933) 2: m.m. 8 September 1930.

80 Schmidt J, ed. *Dana's Togt omkring Jorden 1928–1930*. Copenhagen: Gyldendalske Boghandel; 1932.

explicitly to Professor G. Schott from the Deutsche Seewarte who he says declared that the charted course was correct from a nautical point of view. Not knowing the above circumstances, this would be a superfluous statement in this sort of 'coffee table book'.[81]

Secondly, Schmidt recalls how the journey of an exploratory scientific expedition was different from most other ships sailing across the oceans. Normal ships will seek to go from point A to point B as fast as possible following the shortest route. A research vessel on the other hand, is supposed to operate in open waters far from the shore, seeking port only when necessary for coaling, cleaning the steam boiler, or for stocking up provisions. The programme can shift from day-to-day, depending on what results surface to the deck. In a way it is like whaling or fishing, as Schmidt put it. This creates excitement, the thrill of the chase one could call it. However, in combination with the dependency on coal supplies, it also leaves an uncertainty as to when the ship can arrive at set destinations, and it is difficult to plan ahead very far. At times we could not meet the agreed deadlines, and from time to time this seemed difficult to understand at home as well as abroad.[82] If the Carlsberg Foundation's Board of Directors read this, they would know what he was hinting at.

Meanwhile, outward appearances were held high throughout 1930, as Drachmann continued to orchestrate a dinner thrown in honour of the expedition. On the evening of 25 September, the restaurant Den Kongelige Skydebane was the scene of a spectacle involving Prince Valdemar, Prime Minister Stauning, and all the official representatives of foreign states in Denmark.[83] Schmidt gave a speech, where he once more expressed his gratitude towards the funding agency of the Carlsberg Foundation as well as the various government partners. He also included the corporate executive officers of the Carlsberg Breweries, Poul C. Poulsen, and Vagn Jacobsen, the grandson of the founder of the brewery, who were both present.[84]

Total costs

The costs of such a global enterprise fell into a number of different categories. The accounts for the expedition itself had been settled by the end of 1930, half a year after the *Dana* had returned to Denmark. The total expenditure on the accounts of the Carlsberg Foundation

81 Schmidt J. Indledning. In: Schmidt J, ed. Dana's Togt omkring Jorden 1928–1930. Copenhagen: Gyldendal; 1932:12.
82 Schmidt J. Indledning. In: Schmidt J, ed. Dana's Togt omkring Jorden 1928–1930. Copenhagen: Gyldendal; 1932:13–14.
83 RA. *Schmidt, Ernst Johannes, biolog*: Materiale vedr. ekspeditioner og rejser (1876–1933) 2: m.m. 25 september 1930.
84 KB. Johannes Schmidt 'Papirer'. Ut.374.VI.3. Speech, J. Schmidt, 25 September 1930.

amounted to 572,356.91 Danish Kroner.[85] A little more than half the expenditure was associated with the ship itself, such as maintenance and fuel. The salaries were paid to the three groups of men on board, the navy crew, the civilian sailors and officers, and finally the scientific staff. This totalled some 23 percent including the gratuity paid after the successful journey had ended (Figure 15.2). The remainder of the 600,000 Kroner budgeted for by the Carlsberg Foundation was spent during the following two years on writing-up and publishing some of the first results of the expedition.[86]

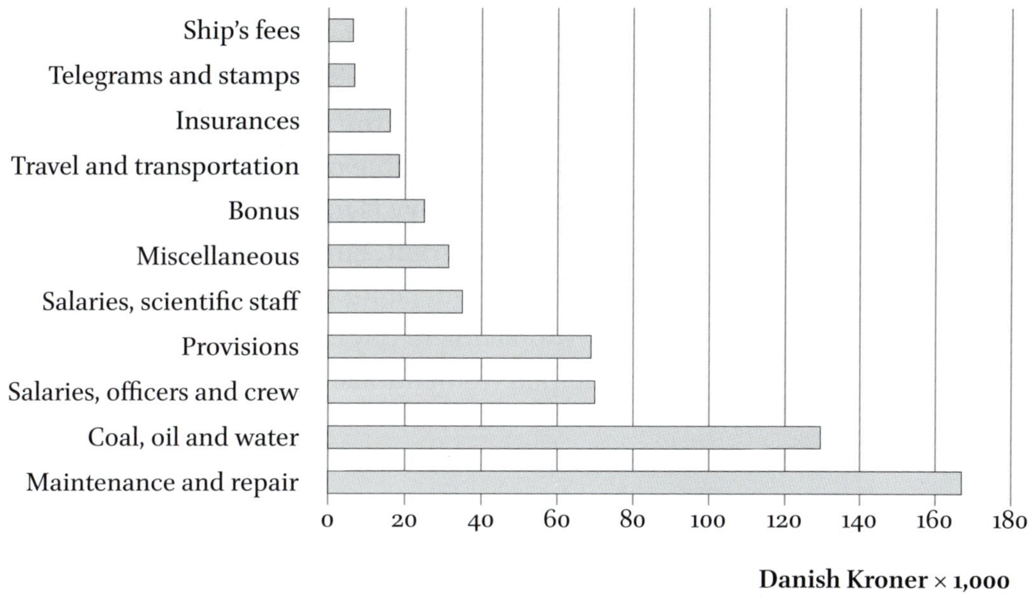

15.7 Carlsberg Foundation expenditure on the circumnavigation 1928–1930

This means that Schmidt did not go to the limit of his allowance from the Carlsberg Foundation, but obviously he did spend roughly 10–15,000 Kroner extra per month keeping the vessel at sea after 1 March, the first agreed date of return. Had he been home by then, he would have hardly exceeded the first grant of 500,000 Kroner.

The additional Carlsberg grant of 100,000 Kroner was mentioned in some of the Danish newspapers, such as the Vestsjællands Social Demokrat, which ran a story in November 1929 that the '*Dana* expedition is saved.'[87] The article noted that in addition to 100,000 Kroner extra from the Carlsberg Foundation, allegedly an unknown private person had followed the expedition with great interest, and had thus chosen to donate another 50,000 Kroner for its completion. This was a huge sum

85 CFD. DANA. BS3 1. Regnskab for S/S 'DANA' Expeditionen, 1/5–1928–31/3–1931.
86 CFD. DANA. BS3 1. Regnskab for Bearbejdelsen af 'Dana-Expeditionen', 24 October 1933.
87 RA. *Anton Fr. Bruun. Danaekspeditionerne 1924–1930*. Avisklip, *Vestsjællnds Social Demokrat*, 27 November 1929.

to receive although nothing is recorded about this in the central documents concerning the circumnavigation. There is nothing in Schmidt's or Bruun's diaries, in the memorandums from Drachmann, or in the substantial correspondence between Ostenfeld, Schmidt, Drachmann, Rechnitzer, and Reventlow. The claim in this newspaper article therefore appears to be an unjustified rumour.

What was true, however, was that the expedition could not have been carried out without the backing from three key government institutions and one private corporation. There are no accounts showing how much money the East Asiatic Company spent shipping the scientific collections from the *Dana*, but it will have defrayed costs of tens of thousands of Kroner. The value of borrowing the *Dana* from the Department of Agriculture can be estimated from the insurance sum paid for the expedition, where the ship itself with equipment was valued at 300,000 Kroner. The value of borrowing experienced navigators and machinists from the navy is more intangible, while their salary was paid out of the Carlsberg grant. Finally, the support from the Department of foreign affairs was crucial for establishing diplomatic liaisons around the world. Not having to pay custom fees in ports all around the world will have helped tremendously.

Conclusion

As the circumnavigation entered its second year, the public interest in the project was maintained, and the expedition was very successful in many ways right until the end 30 June 1930. As it turned out, the scientific results of the expedition were also highly significant. However, what the public was never informed about, and in fact has never been revealed before now, are the large numbers of problems that arose between the leading figures of the expedition project. Until now, it has not been revealed that one year into the expedition Schmidt had a substantial fight with the Carlsberg Foundation over the duration, expenses, and travel plan for the circumnavigation.

No doubt mistakes were made on both sides, and ultimately this soured relations between Schmidt and the chairman of the Carlsberg Foundation, Professor Drachmann. Schmidt and Ostenfeld were the main drivers behind the establishment of the expedition, and as they were both veterans of ocean exploration, they ought to have ensured a proper managerial control system between the expedition headquarters in Copenhagen and the actual expedition as it made its way round the World. In previous large expeditions Schmidt had profited from the now deceased Christian F. Drechsel to keep everything on a tight leash in Copenhagen, as Schmidt traversed the Atlantic Ocean. Ostenfeld and Schmidt may not have realized how much they depended on Drechsel. When Ostenfeld travelled to Canada and later fell ill, he failed to secure a knowledgeable replacement. The chief accountant and the rest of the Board of Directors of the Carlsberg Foundation had no prior experience with running such a complex operation as a two-year ocean exploration around the World.

Schmidt's idea in the middle of the journey, to extend the trip by three months, and add significantly to the length of cruise plan has not been revealed before, but it certainly added to the otherwise silent frustration at the way in Schmidt devoted his research efforts at the Carlsberg Foundation. It tells a lot about his status and determination, and also a lot about the intuitive approach to marine science at the time. This formed a personal dimension to his clash with Professor Drachmann, himself an expert in ancient text criticism. Drachmann must have been appalled by what he felt was a nonchalant way of spending the lavish funding from the foundation. When in September 1929, the Board of Directors decided to donate another 100,000 Kroner for the expedition they probably expected a little gratitude in return. Instead, they were met with more demands, bordering on blackmail from Schmidt.

The disagreement with the Board of Directors of the Carlsberg Foundation became the closest Schmidt would ever come to being fired from his job at the Carlsberg Laboratory. The proposal from the Board of Directors that Schmidt should be relieved of his duties at the laboratory and instead enter a Carlsberg funded position as professor at the University of Copenhagen was received by Schmidt as an attempt to depreciate and humiliate him. The fact that Schmidt got away with simply rejecting the proposal without the repercussion of actually being fired is a clear sign of how mighty Schmidt had become at Carlsberg and not least in the eyes of the chattering classes and the wider public in Denmark in 1930. Then again, how would it look if the Board of Trustees of the Carlsberg Laboratory simply fired Johannes Schmidt? The three professors at the helm of the laboratory board also made up a majority of three professors at the Board of Directors of the Carlsberg Foundation.

If they fired him at the laboratory, they would also scandalize the legacy of the largest single donation in the history of the foundation, the *Carlsberg Foundation's Oceanographic Expedition Round the World*. Schmidt indeed had become an uncontrollable force in the 'Carlsberg family', a sort of cuckoo in a warbler's nest.

16 Measures of impact and legacy

Today, Johannes Schmidt is far from a household name in Denmark, let alone the World. Many people around the World will have heard of Nobel Laureate and Danish physicist Niels Bohr, a few may have heard about H. C. Ørsted, the Danish scientist who discovered electro magnetism in the early 19th century, but virtually none will have heard of Schmidt, even though he was a bit of a science celebrity in his time. Schmidt's celebrity is one of the topics of the following chapter, in which the marine scientist is also portrayed as an active and dedicated researcher until only a few weeks before he died in February 1933. On top of his research, among other things, Schmidt managed to produce a motion picture and a classic travel account of the circumnavigation, which brought his work to a wider public. Although these endeavours, at least in part, were created as fundraising activities that would ensure the future of his vast collection of marine materials.

Following Schmidt's death, the Carlsberg Foundation stepped in to ensure the continuity of the collection. In the official historiography of the Carlsberg Foundation, however, Schmidt's name has not stood the test of time as well as his artefacts, and Schmidt faded gradually from the history books. That is until the beginning of the 21st century, when the rare discreet mentions were replaced by outright criticism of his directorship at the Carlsberg Laboratory.

The following sections will show how these writings contrast with how Schmidt is remembered in marine science circles. Marine scientists still herald Schmidt as the great founder of modern deep-sea marine research in Denmark, and his work is widely appreciated by researchers around the World. Furthermore, we shall see how Schmidt's grand theory of the Sargasso Sea made a lasting impact on what was to become the dominant geological theory in plate tectonics.

Popularising marine science

As 1930 came to an end, Johannes Schmidt had accomplished much more than most marine scientists would do in their entire lifetime. His laboratory was stacked with specimens and samples from the circumnavigation. Even the attic above the second floor was used as a zoological warehouse, where specimen jars were piled on hundreds of metres of shelving. But Schmidt was not finished, yet. He continued disseminating his findings at a seemingly steady pace. There was also a noticeable shift in his priorities following the return from the circumnavigation. Schmidt became increasingly interested in promoting marine science to the wider public. During the circumnavigation news clips were collected systematically from Danish newspapers. Literally hundreds of articles appeared in national and regional newspapers during the two years of the *Dana*'s voyage to remote areas of the planet, at least when looked at from Northern Europe.[1]

Schmidt had already dipped his toes into the world of film making when in 1922 he and his colleagues re-enacted work from the great 1921–22 Atlantic expedition, while wearing tropical clothing on a chilly October day on the Øresund – a few sea miles outside of Copenhagen. For the circumnavigation however, Schmidt was much more ambitious. This time the expedition had carried with them their own film camera and film rolls in order to record the atmosphere and action on board an oceangoing vessel, a world otherwise completely inaccessible to land-lubbers. Previously, those interested in the expeditions had relied on newspaper reports or published travel accounts, the latter being a quite popular genre in its own right in the first half of the 20th century.

Today, one celluloid copy of *The Carlsberg Circumnavigation round the World* is harboured in the archives of the Danish Film Institute in Copenhagen. The 20-minutes-long film consists of c. 20 different sequences. There is a very instructive introduction to the fishing gear, as it is being deployed in the sea, although this is supplemented by camera shots from what is clearly an aquarium, where the functioning of the water sampler is shown in action. As the *Dana* moves into the Pacific Ocean the focus of the film also changes. In the sections from the Pacific and Indian Oceans the film includes captions describing the local inhabitants of the Polynesian islands and Indonesia. The preserved edited film is a silent movie, and therefore the story accompanying the live pictures is told through still photos inserted between the

1 KB. Johannes Schmidt 'Papirer'. Ut.374.IV.3. 'Avisklip.'

different sequences. One version of this film was ready by December 1930.[2] But Schmidt clearly wanted to cater for an international audience so he eventually produced several different versions aimed at the French and English markets as well as those in his home country. Looking at the film today it can come across as colonial and orientalist, with its play on racial stereotypes. Thus, the sequence from Durban in South Africa shows black African men carrying coal on board the *Dana*. The accompanying Danish language text reads, 'A black day on board the *Dana* – negroes loading coal onto the deck.'[3] While the preserved film is far from the action packed films of today's Discovery Channel, Schmidt deserves credit for pioneering a genre, which found a large international audience only after World War II, when ocean explorers like Jacques Cousteau from France or the Austrian, Hans Haas brought the blue planet into the television sets of millions of viewers around the World.

In 1931 there was a large fishery fair, where all the latest technologies and products were put on show. The Danish Prime Minister, Thorvald Stauning who was also the minister for fisheries opened the fair, and to advertise the fair for the Danish people at large, Johannes Schmidt was asked to present his research and circumnavigation in a radio show, which was broadcast to the whole country.[4]

On top of these film and radio initiatives, Schmidt also ventured into the more traditional genre of writing a travel account published in Danish in 1932. Here, he and several colleagues who were part of the circumnavigation wrote a series of chapters making up an edited volume entitled, *Dana's Togt omkring Jorden*.[5] This travel account is fairly uneven in its structure, with the reader being treated to a very mixed collection of papers ranging from concise accounts of the functioning of the hydrographical research to some quite folkloristic sections in the chapters dealing with the islands in the South Seas, where several adjectives are used to describe the attractive brownness of the eyes of the local women.[6]

The last project

Perhaps it is tempting to see Schmidt's involvement in public relations activities as a sign of vanity or simply the joyful act of explaining what his research was all about to a larger audience than the relatively small readership of his journal articles. Alternatively they might be viewed as

2 KB. Johannes Schmidt 'Papirer'. Ut.374.VII.2. Ministry of Seafaring and Fisheries to J. Schmidt 13 December 1930.

3 Archives of the Danish Film Institute. *The Carlsberg Foundation's oceanographical expedition round the World 1928–30.*

4 KB. Johannes Schmidt 'Papirer'. Ut.374.V.6. Statsradiofonien to J. Schmidt, February 1931.

5 Schmidt J, ed. *Dana's Togt omkring Jorden*. Copenhagen: Gyldendal; 1932

6 Thomsen H. Togtets længste stræk. In: Schmidt J, ed. *Dana's Togt omkring Jorden 1928–1930*. Copenhagen: Gyldendal; 1932:57–71; Greve S, Nielsen E. Marquesas og Tahiti. In: Schmidt J, ed. *Dana's Togt omkring Jorden*. Copenhagen: Gyldendal; 1932:72–92.

16.1 Girl from Tahiti
Twenty two days after leaving Panama, the expedition arrived in Tahiti. Spurred on by eighteenth and nineteenth century travel accounts of Tahiti as being close to paradise on earth, Anton Bruun was slightly disappointed with real life Tahiti. Nonetheless, the fabled brown eyed Tahitians were documented on camera as well as in diaries and later accounts from the expedition (RA. Carlsbergfondets DANA-ekspeditioner. 1917–1973. 7a).

part of a clear plan to drum up interest, not least among leading politicians and business people, who might be interested in investing in Schmidt's next adventure. This is evident from the interviews Schmidt gave to the press in July 1930 upon his arrival back in Copenhagen after the circumnavigation. While there was plenty to talk about from the voyage itself, Schmidt made sure to frame his story in the context that now, having brought back all these wonders from around the World, it was time to establish a combined marine research centre with a museum and an aquarium for the public to view.[7]

It was no just Schmidt who dreamed of a new marine laboratory. In 1928, the Danish hydrographer and pioneer from the early days of ICES, Martin Knudsen contacted the International Education Board of the Rockefeller Foundation with an application for funds to establish a 'house of oceanography' in Denmark.[8] Nothing came out of this however, but Schmidt corresponded several times about this issue in the early 1930s. In the spring of 1931 Schmidt once more attempted to stimulate national interest by referring to a possible race with the Norwegians. Schmidt feared that if Denmark did not receive the support from the Rockefeller Foundation for a marine science research institute any time soon, then the Norwegians would get there first.[9]

Schmidt once more entered into negotiations with Lauder W. Jones from the Rockefeller Foundation, but he was as unsuccessful this time as he had been with his approach in the mid-1920s concerning a circumnavigation. The last known correspondence between Schmidt and Lauder W. Jones is a letter from 2 February 1933.[10]

This happened to be one of the last letters Schmidt would read in his life. On the 21 February 1933, Schmidt died. His heart condition had been a problem for more than a decade, and the frequent attacks of bronchitis during the 1920s were probably detrimental to his overall health. Then in February 1933 he was hit by influenza, which his body was not able to withstand.[11] By this point Ingeborg and Johannes Schmidt had lived in their spacious apartment adjoined with the Carlsberg Laboratorium for 23 years. But the apartment came with the job. The widowed Ingeborg Schmidt had to move out to allow Øjvind Winge and his family to move in. Schmidt died fairly young, at the age of 56 and Ingeborg outlived him by more than twenty years before she died in 1956.[12]

During Schmidt's final summers, in 1931 and 1932, Ingeborg and Johannes Schmidt visited Clermont-Ferrand in central France for spa treatments. Schmidt continued keeping a diary, a project he seems to have begun during the circumnavigation, 1928–30. From the entries in Clermont-Ferrand it is clear that he suffered from high blood pressure, something which was not easily treated before World War II. No doubt this would have affected his general well-being and ability to work hard, yet the diary also contained daily entries as to the exact geographical position of the research vessel *Dana*, when she was traversing the North Atlantic in 1931, and indeed the scientific outcome of Schmidt's oeuvre

7 KB. Johannes Schmidt 'Papirer'. Ut.374.IV.3. 'Avisklip', *Social Demokraten*, 1 July 1930.

8 KB. Johannes Schmidt 'Papirer'. Ut.374.VII.2. Copy of letter from M. Knudsen to Rockefeller Foundation 12 October 1928.

9 KB. Johannes Schmidt 'Papirer'. Ut.374.VI.4. Blandet Korrespondance 1912–1928. J. Schmidt to Johs Dalhoff 28 May 1931.

10 KB. Johannes Schmidt 'Papirer'. Ut.374.I.5. L. W. Jones to J. Schmidt, 2 February 1933.

11 Russell ES. Johannes Schmidt (1877–1933). *Journal du Conseil*. 1933;8(2):145–160.

12 Wolff, T. ,Schmidt, Ernst Johannes. In: Cedergreen Bech, S. ed. *Dansk Biografisk Leksikon*, Copenhagen, Gyldendal; 1983:13,148–149

as a scientist was still very prolific in his last years, and not least in the years to come thereafter.

Indo-Pacific eel

When the expedition plan for the circumnavigation was presented in the autumn of 1927, the mysteries of the freshwater eel were high on the list of items to be investigated. As it turned out Schmidt and his team of researchers managed to track the breeding grounds of all four species of freshwater eel living in the western part of the Indonesian Archipelago. As the *Dana* pushed into the Indian Ocean to the west of the Sunda Islands off Sumatra in the autumn of 1929, the first eel larvae ever caught in the Indian Ocean landed on the deck of the *Dana*. 'Will many more appear after this one?' Schmidt asked his diary on the 9 September. In the two evening stations the same day, a total of 12 larvae ranging from 37–53 millimetres were found in the nets. Clearly moved by the event, Schmidt hurried to wire a telegram to Copenhagen about these latest findings.[13]

The strategy for tracking the breeding grounds of the Indian Ocean eel was the same as the one used in the Atlantic Ocean. Cruising up and down along the West coast of Sumatra, Schmidt notes how the larvae become older and larger, as the ship approached the island of Nias just North of the equator. Then they travelled back South towards the town of Padang on the West coast of Sumatra.[14]

Following an intensive October voyage in the eastern parts of the Indian Ocean between Cocos Islands and Sumatra, the collections of eel larvae was piling up. Thus on the 20 October, Schmidt recorded in his diary how they had collected so many eel larvae of various sizes and sub-types that the puzzle of the breeding places of the tropical eel could now to be considered solved. 'The greatest day of the entire expedition', Schmidt notes, was celebrated with drinks before dinner for the officers and scientists, while all men on board received a Carlsberg beer for dinner.[15]

While Schmidt's own diary is filled with notes on eel sampling, the multitude of other creatures found in the deep blue sea were rarely noted. However, there was one exception in November 1929 when the expedition hauled on board 2 'toads' – a pet name for the Danish word for anglerfish, 'tudsefisk'. Schmidt notes that they were caught in the 3-metre

13 RA. *Schmidt, Ernst Johannes, biolog*: Materiale vedr. ekspeditioner og rejser (1876–1933) 2: m.m. 9 September 1929.

14 RA. *Schmidt, Ernst Johannes, biolog*: Materiale vedr. ekspeditioner og rejser (1876–1933) 2: m.m. 20 September 1929.

15 RA. *Schmidt, Ernst Johannes, biolog*: Materiale vedr. ekspeditioner og rejser (1876–1933) 2: m.m. 20 October 1929.

ring net on 3500 metres of wire. One of them seemed to be a new species, 'a female with a parasitic male attached both of them pretty and black.'[16]

Once the expedition had left the Indian Ocean and was heading up the Atlantic Ocean Schmidt was able to take stock of the main accomplishments within eel research in a report to the expedition committee: '...this expedition has caught and identified larvae of 6 of the 10 species of freshwater eel in the Pacific Oceans, while 5 of the 6 species of eel in the Indian Ocean have now been identified with regards to their larval state.' Before *Dana*'s visit to the Indo-Pacific area only 1 species was identified in the larval state, and that was discovered by me in an American collection of Leptocephali from Dutch East India.'[17]

As we learned in chapter 14 the fate of the Japanese eel remained an unsolved mystery for the Danish expedition in 1928–30 as a typhoon prevented its vessel from entering Japanese waters. The subsequent quest for charting the life history of the Japanese eel is a testimony of the historical significance of the many discoveries of breeding grounds conducted from the *Dana*. While Schmidt during his life-time charted a whole string of different eels from around the world, the discovery of the breeding grounds for the Japanese eel came about more than sixty years after Schmidt had left the Indo-Pacific. The fact that the breeding place of the Japanese eel was not discovered for so long after Schmidt's investigations, is a clear testimony of how much was accomplished by one dedicated researcher.[18] It was in 1992 that a team of Japanese scientists finally published their work on the breeding grounds of the Japanese eel, which are located in the area of the West Mariana Ridge to the West of the island of Guam in the Western Pacific Ocean.[19]

Hydrography and the Carlsberg Ridge

One of the more spectacular results from the circumnavigation came out of the hydrographical measurements, which included several hundred stations taken across the great oceans of the World. From the brief excursion into the Gulf of Panama in 1922 the *Dana* expedition had discovered a large area of anoxic water as the water samplers were lowered to a depth of around 400–500 metres. From this it might have been deduced that the nutrient contents in the Pacific Ocean would differ from those in the Atlantic Ocean. As the R/V *Dana* made its way across the Pacific Ocean the samples taken indicated that indeed a vast part of the Pacific

16 RA. *Schmidt, Ernst Johannes, biolog*: Materiale vedr. ekspeditioner og rejser (1876–1933) 2: m.m. 18 November 1929.
17 KB. Johannes Schmidt 'Papirer'. Ut.374.VI.1. 'Dana-Expeditionen, 1928–30', J. Schmidt to Komiteen for Dana-Expeditionen, 14 February 1930.
18 Tsukamoto K. Discovery of the spawning area for Japanese eel. *Nature*. 1992;356(6372):789–789–791.
19 Tsukamoto K. Discovery of the spawning area for Japanese eel. *Nature*. 1992;356(6372):789–789–791.

20 Wolff T. *Danske ekspeditioner på verdenshavene: Dansk havforskning gennem 200 år*. Copenhagen: Rhodos; 1967. p. 251–253; Thomsen H. Oxygen in the tropical pacific. *Nature*. 1931;127(3204):489.

Ocean contained very little dissolved oxygen. Dissolved oxygen is a fundamental building block for a marine ecosystem as it supports the growth of plankton, which is essential for the structure and functioning of the rest of any marine ecosystem. Yet, in the tropical parts of the Pacific Ocean the oxygen level was as little as 0.5% of what had been observed in nutrient rich waters. Having steamed for 7000 kilometres from Panama to the Marquesas Islands, the saturation had increased only to 11% of a nutrient rich area of the sea, and only as the *Dana* went past Tahiti did the levels increase. This observation not only provided information as to the state of the marine ecosystem, but also filled another blank spot on the map of the living oceans. The reason for the lack of nutrients is the dominant ocean current in the Eastern half of the Pacific Ocean. The surface water is rich in nutrients, but it is also lighter and therefore does not mix with the heavier anoxic water further below.[20]

Another ambitious part of the original expedition plan had been the desire to compare the 'Urnæring', or the ocean biomass. While the measurement of salinity, temperature, and oxygen levels was a standard procedure for most of the expeditions covered in this volume, the circumnavigation also included measurements of nutrients such as phosphate and nitrate. The presence of these chemical components

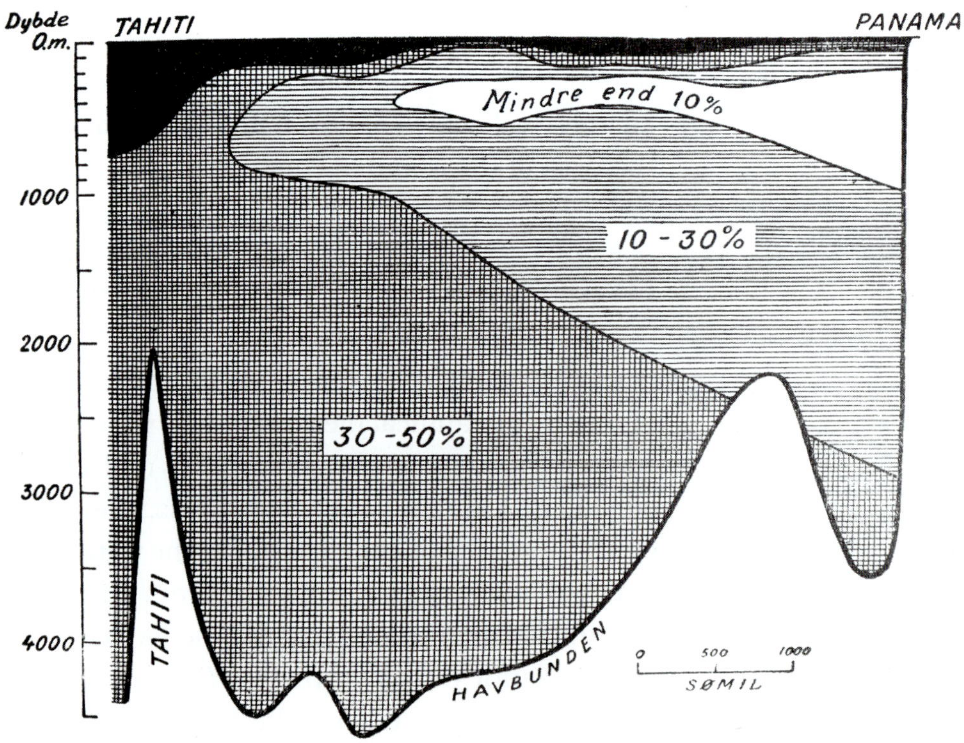

16.2 Levels of dissolved oxygen in the Eastern Pacific Ocean *Samples taken by the R/V Dana indicated that a vast part of the Pacific Ocean contained very little dissolved oxygen. Dissolved oxygen is a fundamental building block for a marine ecosystem, but in the tropical parts of the Pacific Ocean the oxygen level was as little as 0.5% of what had been observed in nutrient rich waters (Winge Ø. & Vedel Taaning Å. 1947).*

is vital for the growth of plankton, the foundation for ocean life. With regards to the great oceans it was found that frequently the surface water had a low content of nutrients, whereas great quantities of phosphate and nitrate were found in deeper layers in the water column. Towards the end of the circumnavigation the Mediterranean was the scene of an interesting deviance from this pattern. The Mediterranean was found to have a low content of nutrients – also in the deep layers of this sea. This has to do with the distinctive way in which ocean water flows in and out

16.3 Helge Thomsen with a water sampler
Helge Thomsen at work with the water sampler. Thomsen was a young student of natural history when he was hired to join the circumnavigation. He was in charge of hydrographical measurements for the entire trip and contributed with thousands of measurements of salinity and nitrogen in the deep sea (*RA. Carlsbergfondets DANA-ekspeditioner. 1917–1973. 7C*).

of the Strait of Gibraltar. The deeper layers of waters are heavier and flow out of the Strait of Gibraltar, to be replaced by surface water of lower nutrient content. This gives a relative deficit of nutrients in the Mediterranean Sea, which in turn explains why the quantity of marine animal life in the Mediterranean is less than in the great oceans. It is also the reason why the Mediterranean is a characteristic clear, azure blue.[21]

As we saw above, Schmidt's main motivation for going back to the Mediterranean in 1930 was to search for more eel larvae, and not least to show off in front of the Italians. As it turned out however, probably the greatest legacy of this leg of the circumnavigation was the abovementioned hydrographical measurements. In some ways this mirrored the first Mediterranean expedition with the *Thor*, when the most notable result came from J. N. Nielsen's hydrographical work.

When looking at a modern atlas there is a section of the North-western part of the Indian Ocean between the Seychelles and Sri Lanka, with the name 'The Carlsberg Ridge.' Carlsberg is probably the only private enterprise anywhere in the World that has a submarine geographical feature bearing its name. This is closely linked to the accomplishments of Schmidt's circumnavigation, but this time it was the fruitful combination of results from the echo sounder and the hydrographical samples, which prompted the breakthrough. Over the course of the two years of expedition the Atlas echo-sounder made 8042 soundings into the water column, but on their way to the Seychelles the soundings indicated that they passed an area between 65th and 70th Eastern Length, where the sea bottom was much higher than the depths of 4–5000 metres, which they encountered elsewhere. From temperature measurements on both sides of this bulge in the echo-sounding chart, it was possible to see that the temperature to the West of the ridge was dropping all the way to the sea floor 3500 metres below. On the eastern side of the ridge however, the temperature started to increase again at c. 3,500 metres below sea level. This difference in temperature suggested that the waters below this depth were entirely separated from each other. This, in combination with a couple of older soundings facilitated the conclusion that an entire new ridge had been found. The extent and direction of the Carlsberg Ridge was later charted by an Egyptian-English expedition.[22]

When the discovery was first reported to the leading journal *Nature*, Swedish marine scientist, Hans Petterson included a map of the Indian Ocean on which the submarine ridge was labelled 'The Johannes Schmidt Ridge,'[23] This was then subsequently changed to Carlsberg Ridge.

21 Thomsen H. Vandundersøgelser paa de store have. In: Winge Ø, Vedel Taaning Å, eds. *Naturforskeren Johannes Schmidt. Hans Liv og Ekspeditioner*. Copenhagen: Gyldendalske Boghandel Nordisk Forlag; 1947:167–173.

22 Thomsen H. Vandundersøgelser paa de store have. In: Winge Ø, Vedel Taaning Å, eds. *Naturforskeren Johannes Schmidt. Hans Liv og Ekspeditioner*. Copenhagen: Gyldendalske Boghandel Nordisk Forlag; 1947:171.

23 Petterson H. The 'Johannes Schmidt' Ridge in the Indian Ocean. *Nature*. 1934;134. (7 July 1934).

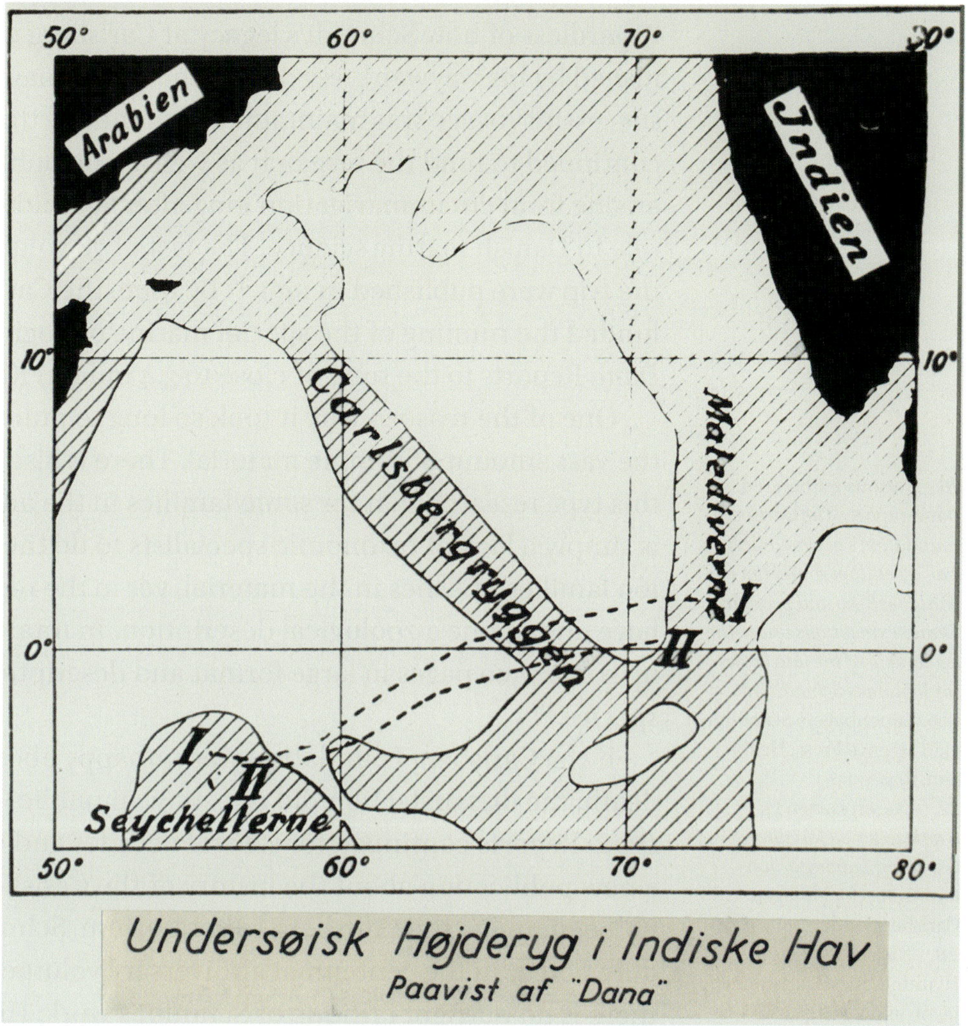

16.4 The Carlsberg Ridge
Thanks to the echo-soundings on board the Dana there is a section of the North-western part of the Indian Ocean between the Seychelles and Sri Lanka, with the name 'The Carlsberg Ridge.' Carlsberg is probably the only private enterprise anywhere in the World that has a submarine geographical feature bearing its name (*Carlsberg Archives*).

The *Dana* Reports and Carlsberg legacy

The mysteries of the Indo-Pacific eel addressed by Schmidt and the crew on board the *Dana* had top priority, and much of what came out of the eel explorations was sorted and analysed while the ship was at sea. Similarly the results of the hydrographical measurements were recorded along the route. Nonetheless, there were some research results from the circumnavigation that had to wait for closer inspection on a desk-top at a zoological museum, in Denmark or abroad. The material from for instance the 255 specimens of anglerfish, the wonder creatures living all of their lives at great depth, took years, even decades, to investigate.[24] Today there are still publications being produced based on the collection of anglerfishes from the *Dana* expeditions.[25]

24 Wolff T. *Danske ekspeditioner på verdenshavene: Dansk havforskning gennem 200 år*. Copenhagen: Rhodos; 1967. p. 188.
25 Pietsch TW. *Oceanic Anglerfishes. Extraordinary diversity in the deep sea.* Berkeley & Los Angeles: University of California Press; 2009.

Regardless of how Schmidt's legacy at Carlsberg might have been tarnished by his waywardness during the circumnavigation, the façade of a successful voyage was maintained, and indeed the Carlsberg Foundation continued to fund the research as well as the publication of the material arising from circumnavigation long after Schmidt had ceased to play a role in Danish marine science. The last of 78 volumes of research from the trip were published in 1991.[26] By then, the Carlsberg Foundation had funded the running of the special marine biological laboratory and the *Dana* Reports to the tune of close to 3.4 million Kroner.[27]

One of the reasons why it took so long to publish all the results was the vast amount of sample material. There is also an inherent risk in this type research that for some families in the animal kingdom there is simply a lack of taxonomic specialists to do the work. There are some 160 families of fishes in the material, yet in the reports only 44 of these have undergone a zoological description. In total however, this makes up some 7500 pages in large format and descriptions of around 350 new species.[28]

If the Carlsberg Foundation was unhappy about Schmidt after his death, the dissatisfaction did not show in public. The money for the *Dana* Reports continued for several decades and whenever there was a new publication about the history of the Carlsberg Foundation and its 'Section A' – the Carlsberg Laboratorium, Schmidt's efforts were duly noted. In the centennial anniversary volume about the laboratory, there is no outright critique of Schmidt's work. His long-term assistant Wilhelm Ege contributed to this volume with an article on Schmidt's expeditions, attached to the volume at the very end, after the otherwise chronological narrative spun around the various directors.[29] In 1976 the professor of economic history, Kristof Glamann who was also Chairman of the Board of Directors from 1969–1993, published a book on the history of the Carlsberg Foundation and its various branches. Here, Schmidt's effort at the Carlsberg Laboratory received the following discreet characterization: 'One of the directors of the Carlsberg Laboratory ended up in the Sargasso Sea. This was not particularly relevant to the brewery, but many others found joy in this.'[30]

Thus it was not until 2011, when the outgoing Chairman of the Carlsberg Foundation, Povl Krogsgaard-Larsen had his memoires published, that a critical light was cast on the career of Johannes Schmidt within the context of the over-all functioning of the Carlsberg Laboratorium, as discussed in chapter 5.[31] More recently Peder Olesen Larsen's *Iværksæt-*

26 Various authors, ed. Dana-report : *The Carlsberg Foundation's oceanographical expedition round the World 1928–30 and previous 'Dana'-expeditions under the leadership of the late Professor Johannes Schmidt. vol. 1–16*. Copenhagen: Scandinavian Science Press; Høst & Søn; 1934–1991.

27 Olesen Larsen P. *Iværksætter og forvaltere. Carlsberg Laboratorium 1876–2014*. Copenhagen: Carlsbergfondet; 2015. p. 91.

28 Nielsen, Khristian Hvidtfelt, Ries, Christopher Jacob and Østergaard, Kristian B., 'Danske Videnskabelige ekspeditioner i perioden', in: Hvidtfelt Nielsen, Kristian and Nielsen, Henry (eds.), *Viden uden Grænser 1920–1970, Dansk Naturvidenskabs Historie vol. 4*, Aarhus: Aarhus University Press; 2006. p. 290.

29 Ege, Vilhelm, Johannes Schmidt 1877–1933, In: Holter H, Max Møller K, eds. The Carlsberg Laboratory 1876–1976. Copenhagen: Rhodos; 1976:195–209.

30 Glamann K. Carlsbergfondet. Copenhagen: Rhodos; 1976: 110.

31 Bjerager A. *Bryggerens Lærling. Povl Krogsgaard-Larsens erindringer*. Gylling, Denmark: Gyldendal; 2011.

ter og forvaltere. Carlsberg Laboratorium 1876–2014 evaluated the deeds of Johannes Schmidt and his Board of Trustees in the context of how they contributed to the vision of the laboratory as outlined by I. C. Jacobsen in 1876.[32]

But despite these recent criticisms regarding Schmidt's contribution to the Carlsberg Foundation, it is hard to ignore the fact that Schmidt was internationally acclaimed in scientific circles, as is evidenced by the many official decorations he received.

Medals of impact

All the above chapters have in one way or another dealt with Schmidt's scientific production, which testifies to the fact his research has, on the whole, been widely considered and appreciated, despite the recent critiques. There is also another area in which Schmidt excelled and that was in medals and decorations for his endeavours. Schmidt goes down as one of the all-time most decorated marine scientists. During his own lifetime, especially in the 1920s, Johannes Schmidt was the recipient of a whole string of honours both in Denmark and internationally. From 1918 onwards he was elected to the Danish Royal Academy of Sciences and Letters. During the next 15 years, he became corresponding member or received a honourable membership from 29 different learned societies (see Table 16.1).[33]

In addition, Schmidt received no less than 15 official decorations from Denmark, Iceland, France, Italy, Norway, Spain, Tunisia, Morocco, Turkey, Greece, The Netherlands and what was then the Kingdom of Siam.[34] Realizing the importance of these kinds of recognition Schmidt also wrote several letters of recommendation to the Danish government as a way of promoting his colleagues abroad. Thus in 1924, Schmidt successfully recommended Bjarni Sæmundsson as well as the Faroese fisheries liaison, F. Øster for Danish knighthoods.[35] In 1922, Schmidt succeeded in putting forward the French geographer Charles Rabot for the official Danish distinction of Commander of the Order of the Dannebrog. The following year, Rabot was instrumental in having Schmidt appointed Commander of the French Legion d'Honneur. In letters the two commanders thanked each other for the favours they had bestowed on each other.[36]

32 Olesen Larsen P. *Iværksætter og forvaltere. Carlsberg Laboratorium 1876–2014*. Copenhagen: Carlsbergfondet; 2015.

33 The table is based on information from: Vedel Taaning Å. Hovedlinierne trækkes op. In: Winge Ø, Vedel Taaning Å, eds. *Naturforskeren Johannes Schmidt. Hans Liv og Ekspeditioner*. Copenhagen: Gyldendalske Boghandel Nordisk Forlag; 1947:184–185.

34 Anonymous, 'Schmidt, Johannes' in: *Kraks Blaa Bog, Fem tusinde nulevende Mænd og Kvinders Levnedsforløb indtil Aar 1930*, Copenhagen: O. C. Olsen & Co.; 1932: 897–898; Vedel Taaning Å. Hovedlinierne trækkes op. In: Winge Ø, Vedel Taaning Å, eds. *Naturforskeren Johannes Schmidt. Hans Liv og Ekspeditioner*. Copenhagen: Gyldendalske Boghandel Nordisk Forlag; 1947:185.

35 KB. Johannes Schmidt 'Papirer'. Ut.374.V.2. Privat og Blandet Korr. 1912–1927. J. Schmidt to Kammerherre A. Krieger, 18 April 1924.

36 KB. Johannes Schmidt 'Papirer'. Ut.374.I.4. J. Schmidt to C. Rabot, 2 November 1922; KB. Johannes Schmidt 'Papirer'. Ut.374.I.4. J. Schmidt to C. Rabot, 23 November 1923.

16.5 Schmidt's honourable positions and scholarly distinctions

Year of election	Position/ Distinction	Society	City	Country
1918	Fellow	The Royal Academy of Sciences and Letters	Copenhagen	Denmark
1921	Corresponding Fellow	Zoological Society of London	London	Great Britain
1921	Honourable Fellow	The Challenger Society	London	Great Britain
1922	Honourable Fellow	Nederlandsche Dierkundige Vereeniging	Den Helder	The Netherlands
1922	Medal of Honour	City of Le Havre	Le Havre	France
1923	Council member	Institut Océanographique	Paris	France
1923	Honourable Fellow	Société Zoologique de France	Paris	France
1923	Honourable Fellow	Société Centrale d'Aquiculture et de Pêche	Paris	France
1923	Honourable Fellow	Die Gesellschaft für Erdkunde zu Berlin	Berlin	Germany
1923	Honourable Doctor	University of Liverpool	Liverpool	Great Britain
1923	The Weldon Memorial Prize	University of Oxford	Oxford	Great Britain
1924	Corresponding Fellow	Académie de Sciences	Paris	France
1924	Foreign Fellow	Linnean Society of London	London	Great Britain
1925	Foreign Fellow	Masarykova Akademie	Prague	Czechoslovakia
1925	Honourable Fellow	Deutscher Fischerei-Verein	Berlin	Germany
1925	Honourable Fellow	Hamburgische Haupt Fischerei-Verein	Hamburg	Germany
1926	Foreign Fellow	Zoological Society of London	London	Great Britain
1926	Corresponding Fellow	Société d'êtudes Ocêaniennes	Paris	France
1926	Honourable Fellow	Fiskifélag Íslands	Reykjavik	Iceland
1926	Honourable Fellow	Washington Academy of Sciences	Washington D.C.	United States
1926	Honourable Fellow	New Zealand Institute (now The Royal Society of New Zealand)	Wellington	New Zealand
1927	Foreign Fellow	Det Norske Videnskaps-Akademi	Oslo	Norway
1927	Honourable Fellow	Royal Society of Edinburgh	Edinburgh	Great Britain
1930	Agassiz Medal	U.S. National Academy of Sciences	Washington D.C.	United States
1930	Galathea Medaljen	Det Kongelige Danske Geografiske Selskab	Copenhagen	Denmark
1930	Darwin Medal	Royal Society of London	London	Great Britain
1931	Honourable Fellow	Náttùrufrædisfjelag	Reykjavik	Iceland
1931	Médaille Geoffrey Saint-Hilaire	Société Nationale d'acclimatation de France	Paris	France
1932	Foreign Fellow	Societas Scientiarum Fennica	Helsinki	Finland

One of the most prestigious awards Schmidt received was the Agassiz Medal for Oceanography granted by the National Academy of Sciences in the USA. This medal recognized all the major achievements 'his work in the conduct of numerous oceanographic expeditions, his investigation on the life-history of eels, an outstanding investigation in marine biology of recent years, and for his publications in the genetics and variation of various fishes.'[37]

Finally, Schmidt will have been particularly honoured when he received the Darwin Medal from the Royal Society in London in 1930; the very same learned society which had previously bestowed this honour on Giovanni Batista Grassi. Schmidt felt that Grassi had had a hand in preventing Schmidt's 1912 article on the Atlantic origin of the herring from being published. In 1922 Schmidt got his revenge when they published his Sargasso Sea theory, but the crowning achievement no doubt was the reception of the Darwin Medal for an outstanding contribution in biology.

The continental drift – and eels

Schmidt's revelation that the breeding places of the freshwater eel were in the Sargasso Sea quickly received global attention, even in the field of geology. In 1915, German geologist and Arctic explorer, Alfred Wegener had published his theory of continental drift. According to Wegener's theory the continents on planet Earth were all on the move, resting on tectonic plates, which were drifting on top of a soft interior of the globe. Thus initially the Americas and Eurasia had been one large continent, but a rift between two plates had gradually slid them apart thereby creating the Atlantic Ocean in between. Today this is a well-recognized theory, but when Wegener first presented it, he faced much opposition.[38]

Over the years Wegener amassed more evidence to support his theory, and in the 4th radically expanded version of *Die Entstehung der Kontinente und Ozeane* from 1929, Schmidt's theory on Sargasso Sea as the breeding place of all Atlantic eel became part of the support for Wegener's theory. Here Wegener cited three different authors all of whom had spotted the possible connection between the eel's lengthy wanderings and the widening of the Atlantic Basin between 1924 and 1928. According to Wegener's own recollection he had himself discussed the connection between the eel in the continental drift with Johannes Schmidt in 1922.[39] Schmidt also recalled having met Wegener in Copenhagen. Although when Schmidt was asked by his friend Charles Rabot to contribute a piece on Wegener's book to the French journal, *La Nature*. Schmidt replied that he thought other colleagues in Denmark

37 KB. Johannes Schmidt 'Papirer'. Ut.374.II.1. D. White to C. Brun, 28 March 1930.
38 Vogel, Andreas, Einleitung zum Nachdruck, in: Vogel, Andreas ed., Wegener A. *Die Entstehung der Kontinente und Ozeane*. reprint of 1st and 4th edition. Braunschweig: Verlag Vieweg; 1980. 1–7.
39 Vogel, Andreas ed., *Wegener A. Die Entstehung der Kontinente und Ozeane*. reprint of 1st and 4th edition. Braunschweig: Verlag Vieweg; 1980. 107.

were more suited for the task.[40] Wegener was in Copenhagen on several occasions, and travelled to Greenland on Danish led expeditions. He also lectured in Copenhagen on his theory of the drifting continents in 1918.[41] In 1922, Wegener gave three lectures in Copenhagen for the Danish Society for Natural Sciences, and it is most likely that his personal communication with Schmidt stems from this occasion. The published transcript of his 1922 lecture in Copenhagen does not, however, contain information about far-flung fish migrations.[42] It is also possible that Schmidt met Wegener prior to 1922, although there is no mention of such a meeting in Schmidt's correspondence. There is little doubt, however, that they will have known about each other's research, as they will have been part of some of the same Arctic exploration circles in Copenhagen, although earlier editions of Wegener's book on the drift theory made no mention of Schmidt's theory.[43] The following year, January 1923 Schmidt published his Sargasso Sea theory in German and English.[44] News of this publication spread quickly and later the same year, the British marine scientist Thomas Wemyss Fulton had spotted the connection between Wegener's theory and Schmidt's theory, which he published as a letter to the top journal *Nature* that same year.[45]

40 KB. Johannes Schmidt 'Papirer'. Ut.374.I.4. J. Schmidt to C. Rabot, 5 December 1922; KB. Johannes Schmidt 'Papirer'. Ut.374.I.4. C. Rabot to J. Schmidt, 9 December 1922.

41 Rud M. *Grønlandsforskeren Alfred Wegener og de drivende kontinenter*. 2nd ed. Copenhagen: Vandkunsten; 2007: 89–98.

42 Wegener A. *Tre Foredrag holdte i Danmarks Naturvidenskabelige Samfund 1922*. Copenhagen: Danmarks Naturvidenskabelige Samfund; 1923.

43 Wegener A. Die Entstehung der Kontinente und Ozeane. 2nd ed. Braunschweig: Die Wissenschaft. Sammlung von Eizeldarstellungen aus den Gebieten der Naturwissenschaft und der Technik; 1920; Wegener A. Die Theorie der Kontinentalverschiebungen. *Zeitschrift der Gesellschaft für Erdkunde zu Berlin*. 1921(3/4):7–130; Vogel, Andreas ed., Wegener A. Die Entstehung der Kontinente und Ozeane. reprint of 1st and 4th edition. Braunschweig: Verlag Vieweg; 1980; Wegener A. Kontinentforskydningsteorien og dens betydning for de systematiske og de eksakte naturvidenskaber. *Naturens Verden*. 1923;7:193–217.

44 Schmidt JS. Die laichplätze des Flußaals. *Internationale Revue der gesamten Hydrobiologie und Hydrographie*. 1923;11(1-2):1–40. Schmidt J. The breeding places of the eel. *Philosophical Transactions of the Royal Society of London.Series B, Containing Papers of a Biological Character.* 1923;211:179–208; Schmidt J. Breeding places and migrations of the eel. *Nature*. 1923;111:51–54.

The Galathea II Expedition and beyond

Today, the *Carlsberg Foundation's Oceanographic Expedition Round the World* is forgotten about outside narrow groups of marine scientists and commentators of marine science history. In public memory this final demonstration of Johannes Schmidt's skills as a research organizer has fallen under the shadow of the so-called Galathea II Expedition 1950–52. The Galathea II Expedition was in many ways a continuation of Schmidt's efforts. First of all, this was another circumnavigation planned in Denmark, this time backed primarily by the Danish state and through donations from Danish expatriates. Secondly, the aim was to supplement Schmidt's research during the 1950–52 expedition with further investigations of pelagic marine life. As a third and perhaps the most important point, two veterans of Schmidt's circumnavigation headed the Galathea II Expedition: navy officer Svend Greve and

45 Fulton TW. The life-cycle of the eel in relation to Wegener's hypothesis. *Nature*. 1923 (March 17):359–360.

marine scientist Anton Bruun. Bruun was still in his twenties in 1928 and Greve was a Navy Lieutenant, and they were both still at the helm twenty years on. Greve took charge of the M/S *Galathea*, while Bruun was in charge of the scientific part of the expedition.[46]

Four reasons deserve mentioning when trying to explain why *Galathea II* is still remembered, and the *Dana II* expedition is not. First of all the times were different, 1952 is more recent, and whereas Schmidt had pioneered the use of mass media, public relations had become a mainstay of running such an expedition by 1950. Secondly, on board the *Galathea* there were three people, including the Danish writer Hakon Mielche, who had been given the task of handling and feeding the media interest.[47] On board the R/V *Dana* Schmidt had relied on his own skills as a communicator as well as the news reports written by members of the scientific staff Aage Vedel Taaning and Poul Jespersen. Thirdly, the *Galathea II* was also a much bigger vessel than R/V *Dana*. It was almost 300 feet, and there were more than 100 people on board during parts or all of the expedition. Another reason the Galathea II Expedition is still remembered is because a few years into the 21st century, the Danish newspaper Jyllands-Posten and the Danish state decided to initiate and carry out a sequel, the Galathea III Expedition.

Nonetheless, in substantial ways the *Galathea II* built on the successes and experiences of Schmidt's many travels from 1899 to 1933. There are also even more recent research initiatives that might be considered a legacy of the efforts of Johannes Schmidt and his colleagues more than a century ago. In the 1960s Danish marine scientists, and in particular a married couple called Jan and Inge Boëtius, began breeding experiments with eel in large tanks. The eels were induced with hormones, which caused the eggs of the female eel to ripen, and the specimens eventually spawned their eggs.[48] The hope was that one day it might be possible to breed eel in a controlled environment in the same way as salmon (*Salmo salar*) and several other fish species are bred through aqua-culture. If successful, this approach could help to lower the negative impact of over-fishing on stock size. Research into the challenges of eel farming is still on going.[49]

In addition to these experiments Jan Boëtius continued to work into the 1980s on the data collected on the *Margrethe* and *Dana* expeditions from the 1910s and 20s.[50]

Since then, several scientists have been working on other aspects of the life history of the eel. Today the population of eel in the Atlantic

46 Wolff T. *Danske ekspeditioner på verdenshavene: Dansk havforskning gennem 200 år*. Copenhagen: Rhodos; 1967.P. 251–253.

47 Wolff T. *Danske ekspeditioner på verdenshavene: Dansk havforskning gennem 200 år*. Copenhagen: Rhodos; 1967. P. 251–253.

48 Boëtius I, Boëtius J. Studies in the european eel, anguilla anguilla [L.]: *Experimantal induction of the male sexual cycle, its relation to temperature and other factors*. Andr. Fred. Høst & Søn; 1967; Boëtius J, Boëtius I, Hemmingsen AM, Bruun AF, Møller-Christensen E. Studies of ovarial growth induced by hormone injections in the european and american eel (*anguilla anguilla* L. and *anguilla rostrata* LeSueur). *Meddelelser fra Danmarks Fiskeri-og Havundersøgelser*. 1962;3(7/12):183–198.

49 Sørensen SR, Butts IAE, Munk P, Tomkiewicz J. Effects of salinity and sea salt type on egg activation, fertilization, buoyancy and early embryology of european eel, *anguilla anguilla*. *Zygote*. 2015:1–18.

50 Boetius J, Harding EF. A re-examination of Johannes Schmidt's atlantic eel investigations. *Dana: A journal of fisheries and marine research*. 1985; 4: 129–162.

Ocean is in dire straits. The entire stock of eel is estimated at a small fraction of the number of eel in the ocean just a few decades ago. There are several theories as to why this is the case, these include overfishing in European and American waters, destruction of habitats such as wetlands being drained, hydropower installations and similar barriers for upstream migration, and the accidental introduction into Atlantic waters of parasites and diseases.[51] These environmental challenges prompted the most recent Danish led deep-sea expedition to the Sargasso Sea. In 2014 the state research vessel, appropriately named R/V *Dana IV* went to the Sargasso Sea to try and catch an eel in the act of breeding, this time the leader of the expedition was Peter Munk of the Danish fisheries research centre, DTU-AQUA.[52]

Conclusion

While completely forgotten in the larger public memory of today, it should be emphasized that Johannes Schmidt was in fact a well-known public figure in his later years, often referred to as Eel-Schmidt, at least within Denmark. Around the World he was famous among fisheries scientists, physical oceanographers, and geneticists. As a side effect of his fame, and a testimony of his impact as a scientist, Schmidt received no less than 15 official decorations from states and governments, and 30 different memberships and medals from learned societies around the World.

This was closely related to the results from 35 years of research as a botanist, a geneticist, and a marine scientist, but Schmidt also knew how to approach the press, and in particular later in his life he made a substantial effort to popularize his research. This was partly associated with his quest to create a more secure institutional set up for marine science in Denmark. Following the fall out with Drachmann in particular, Schmidt will have been aware that the Carlsberg Laboratorium could no longer exist as a quasi zoological museum, once he was gone as Director of the Physiological Department. Right until he fell ill and died in 1933 Schmidt was actively pursuing American funding from the Rockefeller Foundation to secure the future of deep-sea marine science in Denmark.

While Schmidt never succeeded in attracting the necessary funds from the USA, the Carlsberg Foundation eventually secured Schmidt's own zoological collection from the *Dana* expeditions, which was stored

51 Rindom S, Tomkiewicz J, Munk P, et al. Eels in culture, fisheries and science in denmark. In: Tsukamoto K, Kuroki M, eds. *Eels and humans*. Japan: Springer Science+Business Media; 2014:41–60.

52 Reeh L. Dansk åleekspedition 2014 er retur efter fem ugers togt i sargassohavet. 2014. http://www.aqua.dtu.dk/Nyheder/2014/05/140506_Dansk-aaleekspedition-retur

in the attic of the Carlsberg Laboratorium. This was made possible through more than 50 years of sponsorship dedicated to working up the material from the *Dana* expeditions, chiefly the circumnavigation.

There was much to show and tell. Schmidt, with his spectacular results from the trawls and water samplers of the *Dana*, paved the way for the life history of most of the eel populations in the Indo-Pacific area to be written, something he had hoped for when in 1909 he published his first volume on the bio-geography of the World's eel populations. The deployment of the echo-sounder was equally important and led to the discovery of what is now The Carlsberg Ridge. The discovery of large anoxic areas in the Eastern Pacific Ocean was another related discovery that arose directly from the expedition.

Finally, the reading of Schmidt's many letters and articles have revealed a direct linkage between two of the most spectacular natural history theories of the 20th century, that is Schmidt's Sargasso Sea theory for the Atlantic eel, and Alfred Wegener's theory of the drift of continents. Both men seem to have developed their thinking independently of each other, but they definitely knew about each other, and they possibly met in Copenhagen during World War One, when Wegener lectured in Copenhagen, and they definitely talked to each other in 1922. The link between the two theories – the gradual separation of the Americas from Europe is the long-term reason why the Atlantic eel travel so far from their tropical birth-place to temperate Europe. Millions of years ago the same trip was much shorter. Marine scientist Fulton saw this link as early as 1923 right after Schmidt published his grand theory. Wegener then used the Sargasso Sea theory in support of his own theory as he published a revised and extended version of his ideas on continental drift in 1929.

As indicated in several previous chapters, much of Schmidt's work is still being cited. That is one tangible way of measuring scientific impact. Schmidt's legacy as a research entrepreneur and an expedition leader also deserve a mention. In particular there is a clear link between the circumnavigation from 1928–1930 and the circumnavigation of the *Galathea II* in 1950–52, as the leaders of the latter expedition, Anton Bruun and Svend Greve had also taken part in Schmidt's tour of the World twenty years earlier. Finally, the legacy lives on in Danish marine science, where explorations of the life history of the Atlantic eel and its birthplace somewhere in the Sargasso Sea are a much prioritized research field, even in the 21st century.

17 A fortunate man

The sum of a career or one's fortune can be expressed as an equation consisting of talent, chance, and hard work. Johannes Schmidt clearly had a talent for science as evident from his achievements and impact discussed throughout the preceding chapters, but in no smaller amount did Schmidt possess a talent for hard work. In Chapter 1 of this book, the story was told of how Schmidt was an early achiever, writing and publishing solid volumes on bacteriology and botany in his early twenties. He seemed never to be shy of hard work, whether he was operating in the field in Siam or on board the Biological Station in the summer months, or during the winter when he was writing up his research adding to a rapidly growing list of publications while also teaching at Copenhagen University. Through the subsequent chapters this pattern continued, as he was relentlessly pursuing opportunities for more work, be it for the Danish commission for marine science, his own expeditions to the Mediterranean, or when he took it upon himself to hold the dual positions of Director of the Physiological Department of the Carlsberg Laboratorium and paid member of the marine science commission. Within ICES Schmidt first got his own eel committee in 1912, which continued after World War One. Then from 1923, he was kept busy by the chairmanship of the North-Western Area Committee, while in 1927 he replaced Drechsel as Danish delegate to ICES and then from 1932 Vice President.

In Chapter 3, where the main story revolved around Schmidt's relationship with Petersen, Schmidt demonstrated a very conscious determination, as he notified Petersen that to make the most of his working life in his 30s, 'when a scientist has the greatest capacity for labour…' he would not be interested in working for Petersen. As it turned out, Schmidt worked hard beyond his 30s engaging with no less than 26 oceangoing explorations between 1898–1933. In fact, as his health started to slowly deteriorate as a consequence of his heart condition and frequent attacks of bronchitis, Schmidt stepped up the amount of travelling including the two trips around the World in 1925–26 and

1928–30. During the latter circumnavigation it is evident that the other crewmembers were affected by the demands and pace set by the expedition leader. Chapter 14 for instance, presented evidence from both Schmidt's own diary and from Bruun's diary that the search for eel larvae West of Sumatra in the autumn of 1929 was an extraordinary physical and mental tour de force for everyone on board.

In addition to hard work and talent, Schmidt was also in possession of good fortune. There is a double meaning to the English word fortune, which does not translate well into for instance Danish, Schmidt's native language. First of all fortune can be synonymous with chance, luck or the amount of success or failure of someone. Secondly, the word fortune can mean a large sum of money or assets. In both ways however, Schmidt was fortunate in his career as scientist. He had good strokes of luck, and he possessed and gathered the means to pursue his luck.

'One must *always* take the chance, when it is there, because most often it will never come back.'[1] This is how Anton Bruun quoted a frequent *bon mot* of Johannes Schmidt, and there is a definite recurrent theme in Schmidt's career, which revolves around handling risks. As stated in the introductory chapter, this is not foremost a psychological or existential portrait of a famous person of the past. Nonetheless, when looking at Schmidt's actions as a scientist and research politician he appears to have been someone, who in his professional career frequently was willing to assume responsibility, willing to take risks on his own behalf and on behalf of others, and someone literally willing to risk the wrath of his superiors go beyond the charted course of an expedition and far out into the sea to take a chance as he saw it. In *The Prince*, Nicolo Machiavelli's 16th century analysis of politics and power, good fortune is ranked high as a virtue of the great prince, who is the one willing to take great risks and he who knows how to adapt to changing conditions.[2] This was true for Schmidt, if we look upon him as someone setting out to conquer the World of marine science, and literally conquering specimens and territory from the natural World and foreign states.

In 1907 and 1908 Schmidt was at a watershed in his career. As we saw in Chapters 2 and 3, Schmidt had become tremendously successful while working as C. G. Joh. Petersen's assistant, but his decision to openly defy his boss while still a junior researcher was a successful choice in terms of adapting to the changing conditions, where Petersen was trying to undermine an independent Danish marine science com-

1 In Danish it reads: 'Man skal altid tage chancen, når den er der, for oftest faar man den aldrig igen.'

2 Machiavelli N. *Fyrsten*. 2. revised transl. from Italian ed. Hasselager: Helikon; 2006.

mission, while the rest of the group of leading Danish marine scientists were trying to undermine Petersen. The gamble paid off for Schmidt. When the commission was reconstructed the young Schmidt assumed much of the responsibility for fisheries science that was taken from Petersen.

Any oceangoing expedition is an operation full of risk, and it is worth mentioning that there were never any casualties on board Schmidt's cruises. Conducting a winter expedition in the Mediterranean, as Schmidt did in 1908–09 was unusual. However, when looking at the fate of the ageing M/S *Margrethe* with its hogging keel it was a particular high-risk operation to send her across the North Atlantic Ocean partly crossing against the wind (Chapter 7). Ove Paulsen warned Schmidt that the vessel was in a poor shape and the fact that she first lost her flying jib off Bermuda, before finally being torn to pieces on a Caribbean coral reef showed that a bigger disaster could have struck easily. But the gamble paid off, as the boxes with eel specimen were saved from the shipwreck.

In 1920, the M/S *Dana* was in dire straits during the first of the Atlantic expeditions after the war, but this appears to have been a result of poor ship building, as she was a brand new vessel. With regards to the *Dana* expeditions of the 1920s, Schmidt's risk assessments were not so much centred around the actual physical risk of jeopardizing the lives of his men at sea, rather more of an act of audacity on a managerial level. Drechsel was in charge of the expedition in Copenhagen, and as we saw in Chapter 11 it was an act of open defiance that caused Schmidt to change the expedition plan more or less as he himself saw fit, in the midst of the actual expedition. Again, Schmidt's gamble paid off in the sense that the trip to the Gulf of Panama reaped spectacular amounts of new information on marine life, and in the long run he managed to maintain a good working relationship with Drechsel.

Drechsel was also only informed of the events taking place in Iceland in the summer of 1924 (Chapter 12). Schmidt had a busy internationally approved work plan to follow on board the R/V *Dana*, where Aage Vedel Taaning was a trusted research assistant. Yet, in the face of what Schmidt and Drechsel found a grave injustice, namely the Danish government acknowledging the Norwegian presence in parts of Greenland in 1924, he installed Aage Vedel Taaning as his very own instrument of foreign policy. At the symbolical level this gamble allowed Schmidt to demonstrate the importance of Danish marine science to the leading

Danish, Icelandic, and Faroese politicians and decision makers. From a scientific point of view, the patriotic deed of marking cod in Greenland eventually paid off with the most unexpected result that the large cod appearing off shore in the Davis Strait in fact breed as far away as Iceland.

Nonetheless, the Carlsberg Foundation's Circumnavigation around the World was probably the most daring attempt to seize an opportunity as Schmidt deviated from the charted course. When Schmidt explained his philosophy of taking chances to the young Bruun, he was in the midst of solving large sections of the riddles surrounding the breeding places of eels in the Indian Ocean. As we saw in Chapter 15, Schmidt also spent these busy months in the autumn of 1929 debating with Drachmann and Ostenfeld (both representing the Carlsberg Foundation) to what extent he could do as he pleased, and to what extent he should show prudence with the Foundation's money and not least the timing of the return of the *Dana* to Copenhagen. In a rare outburst of direct criticism, Ostenfeld warned Schmidt that he could lose his job, if he did not in any way comply with the wishes of the Carlsberg Foundation's Board of Directors. This was a high-risk gamble, with the Board of Directors trying to get rid of him and attempting to lay him off at the Carlsberg Laboratorium. Again the gamble proved successful in the end. Schmidt did not lose his job over the fight with Drachmann. From a scientific point of view, there was little gain from entering the Mediterranean with regards to the eel, but the hydrographic sampling was highly successful (Chapter 16).

Looking back, Schmidt's return to Denmark in 1930 would have been easier if he had not been so stubborn and provocative towards Drachmann. There may be an irony in the fact that late in his career Schmidt was showing some of the same signs of self-wilfulness as Petersen had exhibited 20 years earlier, when Petersen was the uncrowned king of Danish marine science.

The manager

In the 1920s Schmidt emerged as a prince of Danish marine science in his own right. According to Machiavelli the intelligence of a ruler can be determined by looking at the men he surrounds himself with.[3] The men working for Schmidt eventually had splendid careers in their own

3 Machiavelli N. *Fyrsten*. 2. revised transl. from Italian ed. Hasselager: Helikon; 2006.

right. As we saw in Chapter 5, Øjvind Winge became a founding father of yeast genetics. Anton Bruun succeeded Schmidt in the 1940s and 1950s as the next leader of Danish deep-sea expeditions, after learning the ropes in the 1920s (Chapters 12–16). Poul Jespersen was a trusted assistant of Schmidt who in effect led large parts of the *Margrethe* expedition and the *Dana* expeditions to the North Atlantic, as well as the Circumnavigation whenever Schmidt was not working from the ship himself (Chapter 7, 11 & 14). Jespersen ended his career as Head of plankton studies at the Danish marine science commission.[4]

Career-wise, Aage Vedel Taaning was even more successful than Jespersen. Vedel Taaning learned the ropes during the *Dana* expeditions, and then worked more independently under Schmidt's guidance as part of the fisheries investigations in the North-Western Area in the 1920s and 30s. Vedel Taaning replaced Schmidt as the leader of the international programme in the North-Western Area in 1935. In 1951 the two independent marine science institutions in Denmark merged into one. The marine science commission under the auspices of ICES merged with C. G. Joh. Petersen's Biologisk Station to become the *Danmarks Fiskeri- og Havundersøgelser*, the foundation of what is now DTU-AQUA. Vedel Taaning also emulated the success of Schmidt internationally. From 1952–57 Vedel Taaning was Vice President of ICES, and subsequently he became president.[5] Many of the scientists mentioned above worked loyally for Schmidt for many years. The same goes for N. C. Andersen, the medical doctor who took care of hundreds of hydrographical samples, when helping out on the *Dana* expeditions in the 1920s.

Finally, Captain Georg Hansen deserves a special mention, as he was the captain of the ship on virtually all of Schmidt's expeditions from 1908 onwards. Hansen had the triple competence of being able to captain his ship, steaming and steering in concordance with the deployment of complicated fishing gear for deep sea pelagic trawling, while he also developed a pronounced ability to help with sorting out eel larvae, when they emerged in the trawls. Hansen even worked on his own, searching for eel larvae from the coast of Sicily in 1910–11 (Chapter 4), while on the second replacement trip of the M/S *Dana* in 1921, Hansen was in charge of the pelagic trawling (Chapter 10). Hansen was also a trusted companion of Schmidt in matters of the crew. Schmidt's diary indicates that Hansen's advice was sought in several instances when dealing with matters concerning the personnel during the circumnavigation. The importance of Hansen's presence is indicated by the fact

4 Wolff, Torben, Jespersen, Poul Christian, in: Engelstoft P, Dahl S, eds. *Dansk Biografisk Leksikon*. 3rd ed. Copenhagen: Gyldendal; 1980.
5 Wolff, Torben, Vedel Taaning, Aage, in: Engelstoft P, Dahl S, eds. Dansk Biografisk Leksikon. 3rd ed. Copenhagen: Gyldendal; 1980.

that when he fell ill before and during the circumnavigation he was brought back on board as soon as possible, even it meant long travels to get to the ship (Chapter 10, 11, 14 and 15).

During Schmidt's lifetime the men who worked for him rarely aired their opinion on what they thought of their leader. Bruun's diaries (Chapter 11, 14–15) and Knud Hensch Stephensen's diary (Chapter 10) were both written up as private notes, but neither of these men spent many lines on commenting about Schmidt's leadership. The careers of the younger scientists in particular, depended to a large extent on the possibility of joining one of more or Schmidt's expeditions. Ship time is highly prized, even today, and for many marine scientists working in the field, the acquisition of ship time is the real bottleneck for the advancement of much research. In the case of Schmidt's expeditions the destinations were often previously uncharted waters of the World's oceans, and the attraction will have been enormous for the young and ambitious. This also means that the scientists present on board the *S/S Thor*, the M/S *Margrethe*, the M/S *Dana,* and the R/V *Dana* would put up with endless working days which would eventually further their own careers. Criticising Johannes Schmidt would have to wait until after he died in 1933.

Sometime in the mid-1940s, a group of Johannes Schmidt's former friends, colleagues, and employees, all from Denmark, published a volume to commemorate the famous scientist. Aage Vedel Taaning and Øjvind Winge edited a collection of 19 articles on aspects of Johannes Schmidt's professional life and work.[6] Published in 1947, many of Schmidt's peers had passed away. Ostenfeld died in 1931, Sæmundsson in 1940, while Drechsel passed away in 1927. Still, some of the veterans from the early days of the North Atlantic cruises such as Ove Paulsen and Martin Knudsen were able to contribute, and Theodor Mortensen who explored Koh Chang Island in Thailand was able to provide an account of his time with Schmidt. The other contributors were Schmidt's younger colleagues, who in several cases had profited tremendously from Schmidt's fundraising abilities for the success of their own careers. The volume is a gem of anecdotes about the great man, and there are plenty of commentaries on his genius as a scientist and science entrepreneur. Yet, perhaps surprisingly, pleasant remarks about Schmidt's personal character are rare. Paulsen's introduction to Schmidt's family background and his youth is one exception, where Schmidt is attributed with a sense of humour as well as a fondness for plain people – and dogs![7]

6 Winge Ø, Vedel Taaning Å, eds. *Naturforskeren Johannes Schmidt. Hans Liv og Ekspeditioner*. Copenhagen: Gyldendalske Boghandel Nordisk Forlag; 1947.

7 Paulsen O. Slægt og ungdom. In: Winge Ø, Vedel Taaning Å, eds. *Naturforskeren Johannes Schmidt. Hans Liv og Ekspeditioner.* Copenhagen: Gyldendalske Boghandel Nordisk Forlag; 1947:9–11.

The impression left by Schmidt's collaborators is that of a man, who in the later parts of his career, that is from c. 1920 onwards, became quite openly self-conscious and perhaps a bit aloft in the face of his younger associates and assistants. In spite of being one of the editors, Øjvind Winge probably painted the most negative image of Schmidt's personal drive. Winge remembered that prior to the *Dana* expeditions Schmidt had given the appearance that he was utterly disinterested in becoming a popular and famous character in the public arena. Yet, following the great public interest from around the world, once the breeding places of the eel had been recognized in 1922, Schmidt developed a taste for being a celebrity. Thus, Schmidt told Winge that in reality he gained greater satisfaction from producing results for the joy of the larger public, than from presenting research accessible only to a very limited scientific audience.[8]

Schmidt's preference for popularizing and entertaining a broader audience at the expense of continuing his learned discussion with colleagues in the republic of letters, was not just something negative, it could be interpreted as a flaw in his character, and this is how it was used by Drachmann. When Schmidt came back from his two-year voyage around the World, Drachmann distanced himself from Schmidt's efforts as a popularizer of marine science. For sure, he was mainly frustrated with Schmidt for having exceeded his budget, but in his official critic, Schmidt was blamed for having created too much fuss and festivity around the homecoming of the expedition. Drachmann told him that, 'It would have been more dignified, if the reception had been more modest'.[9] From today's point of view, it appears strange to have criticized a successful scientist for effectively communicating his research to the outer world. Schmidt, however, seems to have been punished, also posthumously, for having stepped out of the role he was assigned as foremost a representative of an important chain in a line of succession of scientific lineage. The scientific work culture of early and mid-20th century Denmark, and Northwest Europe in general for that matter, was marked by an ideal of working hard, leading a slightly ascetic life, where modesty, endurance, and patriarchal respect were guiding virtues.

Schmidt certainly was a very hard working man, and during his 26 oceangoing expeditions he endured countless instances of rough weather, technical and financial problems, not to mention the sheer physical challenge of frequent long distance travels for a man, who suffered from high blood pressure in an age when effective treatment

8 Winge Ø. Andre Strenge anslaas – paa Carlsberg Laboratoriet. In: Winge Ø, Vedel Taaning Å, eds. *Naturforskeren Johannes Schmidt. Hans Liv og Ekspeditioner*. Copenhagen: Gyldendalske Boghandel Nordisk Forlag; 1947:78.

9 CFD. Vedr. DANA II, 1828–1930. 64/1930. Memorandom of meeting with J. Schmidt 8 July, by A. B. Drachmann, 8 September 1930.

had yet to be found. Modesty, however, does not seem to have weighed heavily on Schmidt's shoulders, neither as a promising young scientist, nor later in his life, when so many decorations and honours were bestowed upon him. This was expressed as a flaw in his reputation by contemporaries and in later commentary of Carlsberg origin. However, when assessing the background for Schmidt's many successes, there is no question his unbending self-confidence combined with an unyielding assertive presence in his work life was a huge personal asset.

In summary, Johannes Schmidt certainly was well situated for his chosen line of career as a scientist in early 20th century Denmark, and his family connections proved to be valuable assets on several occasions. From then on though, what carried him through to international scientific merit and world fame, was his unquestioned talent for organizing science and people alike, as well as a determination and energy, which was much later characterized as 'pathologically ambitious.'[10] In any case, Schmidt's sense of strategy and his determination to fulfil his ambitions were well developed from an early age on. Thereto his networking skills must have been impeccable, and he certainly knew how to handle people from a very early age. Talent, hard work, and fortune – in the sense of seeking opportunities were all characteristics, but then there was the second definition of fortune, the possession of assets and wealth.

The fundraiser

As a scientist Johannes Schmidt was one of the greatest Danish fundraisers in his own life-time, matched only in scale by the likes of Niels Bohr and August Krogh. Schmidt's fortune came from several supply chains, in this section they are accounted for as seven different types of sources; i) his marriage, ii) the Biologisk Station, iii) the Danish commission for Marine Science under the auspice of ICES, iv) Carlsberg Laboratorium, the Carlsberg Foundation, vi) Direct government and navy support, and finally vii) the private corporate funds from a number of different companies.

Much of it was ad hoc soft money for a specific expedition, although a steady supply of funds from his two main employers, the Carlsberg Laboratorium from 1910 and the Danish government via his membership of the Danish marine science commission from 1909 provided the main source of income for his growing marine science empire from the

10 Fort T. *The book of eels*. Great Britain: Harper Collins Publishers; 2002, p. 129.

1910s onwards. The earliest of Schmidt's permanent fortunes however, came through the institution of marriage, when he wedded Ingeborg in 1903. This brought him into the mercantile upper class of Danish society.

The family fortune played a significant role at several crucial stages in Schmidt's life. As we saw above, it is perfectly normal for a young natural scientist working in the field to be highly dependent on the chances of getting to work alongside the leading man in a specific research field. However, when C. G. Joh. Petersen offered Schmidt a position as the second in command of a restructured commission the young and aspiring scientist had the nerve to reject the invitation (Chapter 3). Schmidt's position in this case was aided by the fact that he was not totally dependent on his salary from working for Petersen at the Biologisk Station. He had other means. The benefits these offered were also apparent in how Schmidt was able to handle the expenses of his two Mediterranean expeditions. In the case of Schmidt spending much more than he had been granted for the 1908–09 expedition the Carlsberg Foundation appears to have accepted the offer from Schmidt's mother-in-law to cover the excess spending. When later the same year Schmidt's application for a further expedition was first turned down by the Board of Directors, he was able to draw once more on his private means, chipping in the considerable amount of 10,000 Danish Kroner – or c. three times his annual salary, when he first started as Director of the Carlsberg Foundation (Chapter 4). Then in the 1920s, Ingeborg Schmidt travelled on her own budget in both instances, when she and her husband sailed around the World (Chapter 13–15). This is worth mentioning as Ingeborg Schmidt had an important representational role in Schmidt's global adventures.

Working for the Biological Station and the University of Copenhagen as a teaching assistant until 1908, Schmidt had a decent annual income, roughly double that of a skilled craftsman in Copenhagen at the same time (Chapter 2–3). However, when Schmidt became Director of the Carlsberg Laboratory from 1 October 1910 until he passed away in 1933, his annual income rose significantly. The income and other assets at the Carlsberg Laboratorium have not been accounted for in previous accounts of Schmidt's life and career. To a large extent however, Schmidt's position at the laboratory became his fall back position, when times were rough for many other marine scientists during World War One. Schmidt instead resorted to path breaking research in genetics together with Øjvind Winge (Chapter 5).

A 1910 book by the American, Charles Kofoid, presented an overview of biological stations in Europe. There were many such stations around Europe at the time, but most of them were very small typically employing one or two science educated researchers.[11] As the Carlsberg Laboratorium became totally immersed in marine science during the 1910s and 1920s, this privately funded Copenhagen based research centre easily ranked among similar biological stations in Europe. Commenting on Schmidt's laboratory in 1947, Aage Vedel Taaning, emphasized the great practical benefits of having your own independent laboratory and your own assistants. Taaning found that Schmidt's working area was so detached from that of many other Danish biologists working in the field that he had contact with very few of them on a day-to-day basis. His own work took up all of his time, but this one-sided-ness was his strength in his day-to-day work in the laboratory as well as in the eye of the public, and the way Taaning saw it, Schmidt found his colleagues outside Denmark rather than in his own country.[12] As is clear from the above, various publications have touched upon how Schmidt made use of his various positions and sources of finance. This is however, the first quantitative assessment of the impact of possessing this fortune. In total, Schmidt disposed of 1,640,000 Danish Kroner from the Carlsberg Laboratorium (Chapter 6). This amount is significantly more than what he disposed of as a regular member of the Danish science commission.

When in 1902 Denmark decided to join the ranks of the founding nations of the International Council of the Exploration of the Sea (ICES), Schmidt was the right man at the right time. While already working for Petersen, the new internationally approved research programme entailed a significant rise in government funding for marine science in Denmark (Chapter 2). The Danish branch of ICES, so to speak, that is the Danish commission for marine science, came to a cross roads when Schmidt tackled Petersen head on in 1908 (Chapter 3). Initially it was not a foregone conclusion that ICES would continue to exist beyond its initial five year programme. As an institution ICES eventually gained more permanent status, not least after World War One, when Drechsel played a significant role in its continuation (Chapter 8). The Danish marine science commission became an institutional and fiscal setting Schmidt and his colleagues could depend on, when dreaming up new research topics and not least the continuation of fisheries investigations year after year, searching for change and variation in fish stocks and environmental forcing(Chapter 2–3 and 12).

11 Kofoid, C. A., *The Biological Stations of Europe*, Washington D.C.: United States Bureau of Education, 1910.

12 Vedel Taaning Å. Fiskeriundersøgelser i nordlige have efter krigen. In: Winge Ø, Vedel Taaning Å, eds. Naturforskeren Johannes Schmidt. Hans Liv og Ekspeditioner. Copenhagen: Gyldendalske Boghandel Nordisk Forlag; 1947:130.

Through the Danish marine science commission Schmidt earned a solid extra salary, and he also disposed of an annual lump sum to pay his research assistants, while the publication of results was a significant part of the expenses. Although this manuscript does not present the precise amount funnelled from government via the commission to Schmidt throughout his career, the figure is significant, and a conservative estimate would be at least 300,000 Danish Kroner over his career, not counting the value of disposing partly of the *S/S Thor* and the R/V *Dana* vessels for annual routine research, nor the specific funding for the 26 ocean expeditions.

The vast majority of Schmidt's funds have been accounted in the previous chapters. The Carlsberg Foundation was Denmark's foremost research foundation from its beginning in 1876 until sometime after the Second World War, when the Danish government started to develop schemes for setting up actual public sources of research funding. The Rask-Ørsted Fondet from the 1920s is counted as the first government initiative of this kind, but the amount of money they donated was a mere fraction of what the Carlsberg Foundation was capable of supplying.[13] Schmidt profited greatly from the Carlsberg Foundation right from the start of his career until half a century after he had passed away. Figure 17.1 shows the high frequency with which Schmidt was a recipient of grants from the Carlsberg Foundation.[14] Schmidt was the lead applicant in most cases, and he is credited in the annals of the Carlsberg Foundation as the principal investigator of close to 865,000 Danish Kroner. The 600,000 Kroner he raised for the circumnavigation bearing the name of the main sponsor is the single largest amount. The second largest grant however, came as a consequence of the *Thor* expeditions to the Mediterranean. While the foundation was generous with sponsoring the actual research cruise, the initial grant was only a tenth of the total amount. This has to do with the policy of the foundation to fund the publication of research results arising from previous grants. When the publication of results from the *Thor* expedition ended in the 1920s more than 115,000 Kroner had gone this way.

Several other colleagues profited from taking part in Schmidt's research projects and expeditions, starting with Theodor Mortensen on the Siam expedition. Over the years close to 73,000 Kroner was donated to research projects, where the main applicant referred to the fact that he was researching into material uncovered through one of Schmidt's expeditions. This pattern in fact continued long after Schmidt had died

13 Olesen Larsen, P. *Stadier på forskningens vej. Dansk forskningspolitik i går og i dag*. Odense: University of Southern Denmark Press; 2010.

14 The amounts mentioned in this section are found in Spang-Hanssen, E. *Carlsbergfondets Understøttelser 1876–1936*. Copenhagen: Bianco Lunos Bogtrykkeri A/S, 1937.

in 1933. From then on and until the final *Dana Report* was published in 1991, an additional 3.4 million Kroner was donated for the advancement of results coming out of the collections from the circumnavigation 1928–30 (Chapter 16).

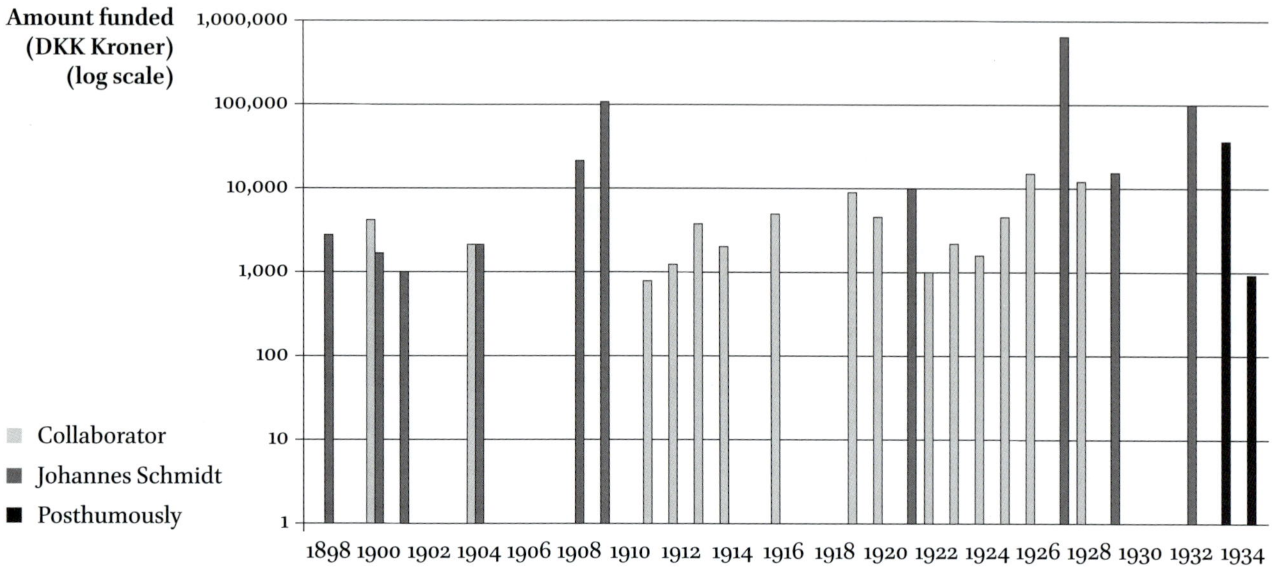

17.1 Donations by Carlsberg Foundation for Schmidt and collaborators

This brings us to the one sponsor, which supported every single expedition, and which should not be overlooked, namely the Danish government. As mentioned above, the government was the long-term sponsor of Denmark's involvement in ICES through the Danish marine science commission, but Schmidt was not fronting the actual commission. This was first Petersen, then Drechsel. However, in terms of extraordinary subsidies for Schmidt's many expeditions the government support was crucial. When in Siam, Schmidt represented the State of Denmark, and when he later embarked on his Mediterranean expeditions, he was borrowing a government vessel, the *S/S Thor*. During and after World War One, the *Thor* was no longer an option for deep sea research, but the government's decision to buy the *S/S John Quilliam* and convert her into R/V *Dana* was a direct effect of Schmidt and Drechsel's relentless fundraising effort from 1914 until 1921. Therefore, it is fair to argue that the government lump sum of 600,000 Danish Kroner in support of acquiring the *Dana*, should be counted as, at least partly the result of Schmidt's work as a fundraiser (Chapter 11).

The Kingdom of Denmark also supported Schmidt's unsuccessful fundraising tour to the USA in 1926, where Schmidt received a considerable amount of diplomatic and secretarial support from the Danish Legation in Washington D. C (Chapter 13). In a similar way, the various diplomatic representations around the World functioned as important local contact points, as the *Dana* made its way around the World, calling at ports in very diverse countries and different political and cultural environments. During the circumnavigation, there was a much bigger need for personnel than during the normal routine service of conducting fisheries investigations. Schmidt was able to borrow navy personnel, paid for by the Danish navy for the years of sailing around the World (Chapter 13).

Source of funding for Johannes Schmidt oceanographic expeditions and subsequent research, 1899-1932

Schmidt present on board	Year	Vessel	Destination	DK government	East Asiatic Company	Other private sponsors	Carlsberg Foundation	Carlsberg Laboratory	Family fortune
×	1899-1900	*Siam* & *Cathay*	Thailand	×	×		×		
×	1903	*Thor*	Iceland, Faroe Islands	×					
×	1904	*Thor*	Iceland, Faroe Islands	×					
×	1905	*Thor*	Iceland, Faroe Islands	×					
×	1906	*Thor*	Iceland, Faroe Islands	×					
×	1907	*Thor*	Iceland, Faroe Islands	×					
×	1908	*Thor*	Iceland, Faroe Islands	×					
	1908-1910	*Tjalfe*	Greenland, fisheries exploration	×					
×	1908-1909	*Thor*	Mediterranean winter cruise	×			×		×
×	1910	*Thor*	Mediterranean summer cruise	×			×		×
	1911	*Thor*	Iceland, Faroe Islands	×					
	1912	Rented vessel	Faroe Islands	×					
	1909-1911	various	Ships of opportunity, Mediterranean	×	×	×			
	1911-1915	various	Ships of opportunity, Atlantic Ocean	×	×	×			
×	1913	*Margrethe*	Atlantic Ocean	×		×		×	
×	1920	*Dana I*	Atlantic Ocean	×	×	×		×	
	1921	*Dana I*	Sargasso Sea	×	×	×		×	
×	1921-1922	*Dana II*	Atlantic Ocean, Panama Gulf	×		×	×	×	
	1923	*Dana II*	Iceland, Faroe Islands	×				×	
×	1924	*Dana II*	Greenland, Iceland, Faroe Islands	×				×	
×	1925	*Dana II*	Greenland, Iceland, Faroe Islands	×				×	
o	1925-1926	various	Circumnavigation					×	×
	1926	*Dana II*	Greenland, Iceland, Faroe Islands	×				×	
	1927	*Dana II*	Greenland, Iceland, Faroe Islands	×				×	
×	1928-1930	*Dana II*	Circumnavigation	×	×		×	×	
	1931	*Dana II*	Iceland, Faroe Islands	×				×	
	1932	*Dana II*	Iceland, Faroe Islands	×				×	

17.2 Sources of funding for Schmidt's expeditions, 1899–1932

This book is also the first publication to focus on the origin of Schmidt's non-governmental support for his expeditions. Most importantly, Schmidt developed an excellent relationship with the largest corporate firm in Denmark, the East Asiatic Company (EAC). Schmidt first benefitted from his connections to EAC in 1998, as they carried him and Theodor Mortensen to Siam at a discounted rate. For many years, at least from 1909 onwards, the EAC collected samples for Schmidt and his other Danish colleagues, when their modern freight vessels made their way through the Mediterranean en route to the Far East, or when they crossed the Atlantic Ocean back and forth between the Americas and Europe. This is a very early instance of a now well-established research method, the use of ships of opportunity. Another private firm lending a hand in this way was the United Danish Steamship Companies, DFDS.

Perhaps, the most generous offer from EAC was the one made in late 1916, at the time that Drechsel and Schmidt succeeded in raising sufficient funds to match their Atlantic ambitions. H. N. Andersen, the founder and owner of EAC built a brand new motor ship, which was constructed in such a way to make room for scientific research and more cabins than would normally be needed on an ocean freighter. Not only did this give Schmidt and Drechsel a head start, when marine scientists were able to roam the waters of the Atlantic Ocean after the end of World War One. The prospect of being able to use the M/S *Dana* was exploited as leverage by Drechsel to lure the Great Powers, especially The UK back into ICES after World War One (Chapter 10). The UK might very well have joined ICES regardless, but the attempt to use the availability of the *Dana* in this way speaks of the importance, and indeed great comparative advantage of controlling a research vessel, even if only for a few months. The generosity continued in 1920 when the first trip with the *Dana* was cut short due to leaks in the hull. While Drechsel and Schmidt were keen to demand compensation, the EAC appears to have conceded to the wishes of the marine scientists rather quickly, resulting in the second *Dana* expedition in 1921.

During the circumnavigation, the support from EAC continued as they assisted with shipping all the samples and specimens back to Copenhagen from ports around the World, courtesy of the company (Chapter 13–15). It is very difficult to put an exact figure to the value of this kind of assistance, but definitely it was highly welcomed asset, probably crucial in several instances. There was significant symbolic value in having the support of the largest corporate firm in Denmark at the time, headed by perhaps the most influential industrialist, H. N. Andersen. In the Danish World of business, only Emil Glückstadt, owner of the largest Danish bank at the time, Den Danske Landmandsbank, rivalled Andersen in terms of personal wealth. Glückstadt also supported Schmidt when on behalf of his bank he donated the single largest cash sum from a private company, namely the 150,000 Kroner for an Atlantic deep sea expedition. The ship owner Andreas Erlandsen and the attorney Zeuthen provided 100,000 Kroner and 10,000 Kroner respectively for the same cause (Chapter 8 and 9).

Drechsel was at least as important a figure as Schmidt, when addressing the private fortunes of the Copenhagen business world. This is also how one should interpret the interest from the fish exporter, Mikkelsen Vendsyssel, who lent his motor ship, the *Margrethe* to the Danish marine science commission in 1913 free of charge. Mikkelsen Vendsyssel was an old business acquaintance of Drechsel from when the two men tried to make a fortune with steamship trawling in the North Atlantic in the 1890s (Chapter 7).

Figure 17.2 charts all of Schmidt's oceangoing expeditions from 1899–1932. Wilhelm Ege originally compiled the list in 1933, but the version in this manuscript includes additional columns that use '×' to indicate the sponsors who supported specific expeditions. The column with 'DK Government' shows that it backed each and every expedition, although this also included both funding through the marine science commission and direct support, buying a vessel, adding personnel, or the diplomatic services offered by government ministries and foreign legations.

An accidental institutionalization

When looked upon as individual expeditions, there is a remarkable breadth in sponsorship, but when looked upon as a whole, there are several recurrent features, adding structure to the funding sources. In fact, there was such a steady flow of money for Schmidt's marine science adventures that when seen together, the main supporters formed an unprecedented *accidental* institutionalization of marine research around Johannes Schmidt.

From the beginning of 1902 onwards, the Danish marine science commission gradually became a permanent research institution. This has been accounted for in previous assessments of international marine science in the first half of the 20th century, for instance in the works on the history of ICES.[15] What is novel in this book however is the evidence presented for the range and significance of the role played by the Carlsberg Foundation. There was the strategic asset in the family fortune that could be used in times of trouble. There was the Carlsberg Laboratorium, which provided space and equipment for research, and finally either Schmidt or one of his collaborators were almost annual beneficiaries of the research endowments of the Carlsberg Foundation. This triple entry into the financial assets streaming out of the beer taps in Copenhagen was a vital aspect of Schmidt's fortune in every sense of this word.

Schmidt was the key actor and the most important 'spider in the web', in this accidental institutionalization. It was accidental in the sense, that it was not planned to play out in this way by any large interest group, and the way in which Schmidt's empire of marine science was dismantled within a few weeks following his death further testifies to the fact that at least the laboratory part of Schmidt's realm disappeared as a marine science institution within a year from February 1933 (Chapter 6). On the other hand, the fact that his superiors were not able to fire him in 1930 testifies to the solidity of the arrangement.

The role played by the Carlsberg Foundation itself however, solidified Danish deep-sea marine research for another half century through the support of the publication of the *Dana Reports* series. This is a longer time span than the duration of many *de jure* institutions, hence it is no exaggeration to claim that the Carlsberg Foundation to a large extent has played the role of providing *de facto* institutional support for Danish marine science expeditions ever since the Board of Directors first lent support to Schmidt in December of 1898 and until the *Dana Reports* ceased to be published in 1991.

To make all this happen however, Schmidt was never alone. There is a considerable added value to pointing to Schmidt's core network as quite fundamental for the functioning of his web of marine science. Schmidt was the key actor in a network of like-minded people with a deep and insightful interest in natural history. In addition to his skills and hard work, Schmidt's entry into this network is what catapulted him into the cockpit of Danish based international marine science in the first place. Schmidt's core network over the course of his career consisted of five people. One of them was his wife Ingeborg, as pointed out above. This section highlights the importance of the other four crucial

15 Rozwadowski, H. *The Sea Knows No Boundaries* (ICES/University of Washington Press, 2002); Went AEJ. *Seventy years agrowing. A history of the international council for the exploration of the sea, 1902–1972*. Copenhagen: International Council for the Exploration of the Sea; 1972.

business partners of Schmidt; Eugen Warming, Carl. H. Ostenfeld, Christian F. Drechsel, and C. G. Joh. Petersen.

Eugen Warming

Schmidt probably had a good grounding in natural science before entering university, through his upbringing close to his maternal uncle. In any case, Professor Eugen Warming spotted his talent at an early stage, and Warming became Schmidt's most important champion for academic promotion in the following years. Having witnessed Schmidt's top marks from university and early publications, Warming probably had a role to play in putting Schmidt's name forward when it came to finding a botanist for the Siam expedition. Permanent academic positions were hard to find around the turn of the 20th century, and Warming tried to keep Schmidt 'warm', as he hired him to teach as an assistant at the university from 1901–1908 (Chapter 1 and 2).

With regards to Schmidt's first Mediterranean expedition it is not clear at what stage Warming and Schmidt started talking to each other about it, but there is no reason to doubt that the issue of support from the Carlsberg Foundation was discussed with Warming prior to Schmidt's actual application. Warming's significance in the promotion of Schmidt's research plans are much more obvious in the case of the Carlsberg Foundation's decision to fund Schmidt's second trip to the Mediterranean. This was a split decision, where at first Jørgensen and Christiansen had turned down Schmidt's application. First of all, they may have wondered, why the man they had just hired as Emil Chr. Hansen's replacement at the Carlsberg Laboratorium, was going to start his career in beer research with an expensive summer cruise to the Mediterranean. Secondly, they guessed – correctly, that the subsequent costs of publishing the findings from this expedition would be a large expense for the foundation (Chapter 4). Warming's decision to launch a – successful – campaign among other professors in Copenhagen to have his fellows in the Board of Directors revoke the decision not to fund Schmidt, is highly unusual and probably unique in the past 140 years, as the Carlsberg Foundation usually donates research grants on a competitive basis.

Warming proved to be the strong arm in the Board of Directors and also in the Board of Trustees of the Carlsberg Laboratory, where Warming was the one who promoted Schmidt to the position as Director of the Physiological Department. Warming was not alone however, as Schmidt was well connected through his marriage to the adjunct members of the Board of Trustees, Rudolph Koefoed and Carl Jacobsen.

From 1914–1921 Warming was the chairman of the Board of Trustees and he will have approved all the important research in Schmidt's laboratory from hops to genetic studies on a whole string of animals, principally fish. Schmidt and Winge's research in genetic variation proved to be ground breaking (Chapter 5). Quite quickly these studies were picked up around the World earning Schmidt the 1923 Weldon Memorial Prize for his bio statistical studies (Chapter 6). For previous commentators of the Carlsberg Laboratory it has been a long-standing puzzle, why Schmidt, a rising star in fisheries science, was hired in the first place, and not least why his actions were approved of for almost two decades.

This book proposes that Schmidt's work, gradually distancing him from the original intentions of creating the laboratory, was much appreciated, perhaps even encouraged by Eugen Warming. Warming was a leading man in ecology himself, and he was someone passionately involved in the debate with Wilhelm Johannsen over the hereditary potential of environmental impacts. Schmidt explicitly addressed this issue in his hereditary studies (Chapter 6).

From the correspondence of the Board of Trustees it is clear that the members took an active interest in the work at the laboratory. Warming on the other was also very protective towards Schmidt's colleague, Albert Klöcker who had an unfortunate run as Director Extraordinaire trying to continue and complete with Emil Chr. Hansen's studies on yeast. Klöcker became vulnerable to critique, as his diabetes may have hampered his working capacity, as well as his constant shortage of money due to a poorly managed private household. As soon as Warming had stepped down from office, the Board of Trustees axed Klöcker. Later when the Board of Trustees was without Warming, it could have done the same to Schmidt, but it did not (Chapter 6).

Carl H. Ostenfeld

This seems to be clearly correlated with the fact that Warming was not the only one with strong sympathies for Schmidt. When Carl H. Ostenfeld took up the seat from Warming both on the Board of Directors and in the Board of Trustees, a windfall of free research time opened for Schmidt. Ostenfeld was a friend of Schmidt's since the days of the *Thor* expeditions in the 1900s, and from 1909 until Ostenfeld's death in 1931, they both held seats in the marine science commission. Ostenfeld was a colleague and overseer respectively in all three institutions funding most of Schmidt's research for more than two decades, and as their correspondence clearly reveals, he was someone Schmidt could trust. This is what makes Ostenfeld so important, which became evident in at least four key situations.

First of all, Ostenfeld was one of the instigators of the revolt against Petersen in 1908. Prior to the revolt Ostenfeld had tried, in vain, to push for the inclusion of Schmidt into the commission then dominated by Petersen (Chapter 3).

The second time Ostenfeld aligned with Schmidt in times of trouble was in relation to the sale of the Danish West Indies to the USA. Within the society De Danske Atlanterhavsøer, a lobby organization for the Danish Atlantic possessions, Ostenfeld led a successful rebellion against the inactive president of the society, Niels Neergaard, with the support of Schmidt, Ove Paulsen and two other natural scientists with stakes in the continuation of a Danish presence in the North Atlantic realm. The advocates for the Danish West Indies possessions lost the referendum, but the marine scientists nonetheless managed to secure themselves a favourable position in the aftermath of this debacle. While Schmidt and Drechsel secured the funds for a new Atlantic deep sea expedition Ostenfeld operated at the organizational level, as he became the inaugural President of the replacement society for Danish interests abroad, the *Dansk Samvirke* (Chapter 9).

As a third concrete instance, where Ostenfeld aligned with Schmidt, the Great Atlantic expedition with the R/V *Dana* deserves mentioning. Both men wanted badly to pay a visit to the Pacific Ocean, although this was not part of the expedition plan, which had been negotiated with Drechsel just a few days before the expedition went to sea. The costly price of entering the new Panama Canal was an obstacle though. Nonetheless, in terms of sticking to the agreed plan there was little Drechsel could do from Copenhagen, except from going along with the proposal to the Carlsberg Foundation to support the passing of the Panama Canal with 10,000 Kroner. Ostenfeld took part in this decision as a newly appointed board member. Probably quite unheard of, Ostenfeld came to profit directly from this grant, as he immediately boarded a ship to the West Indies to team up with Schmidt and the rest of the expedition. Eventually, some of the most important results of the entire expedition came out of their weeklong visit to the Pacific Ocean (Chapter 11).

Finally, Ostenfeld and Schmidt were the two chief architects behind the Carlsberg Foundation's Oceanographic Expedition Round the World. Schmidt was the main applicant and Ostenfeld worked his fellow members of the Board of Directors to secure their backing, when the actual research plan came about. Moreover, the board decided to manage the expedition themselves placing Ostenfeld at the helm of the expedition committee. In hindsight, this was not a good solution. Ostenfeld and Schmidt were too close to each other for Ostenfeld to act as an effective controller of what Schmidt was doing on the other side of the World. In addition to their long standing scientific collaboration, Ostenfeld's daughter Gertrud had accompanied the Schmidt couple around the World in 1925–26, a trip paid for by Ostenfeld, as he managed to secure an advance payment of his honorarium from the Carlsberg Foundation. Finally, Ostenfeld paid too little interest in the accounts associated with the circumnavigation in 1928–30. this seems to have been the result of unfortunate circumstances and lack of oversight. First he was travelling in North America for a long period, and when he came home, he was ill (Chapter 13 and 15).

As it turned out Drachmann lost patience with the existing arrangements and acted to replace Ostenfeld as chairman of the expedition committee. Schmidt failed to acknowledge the importance of an anchoring point in Copenhagen, taking care of manageing the expedition, while he was out at sea. Ostenfeld clearly did not have his eye on the ball sufficiently. The committee did not meet regularly, and the Carlsberg Foundation was not geared to efficient accounting of such a large project – until then, the single largest in the history of the foundation. By the summer of 1929, no one had an overview of the exact expenses of the circumnavigation so far. Drachmann became fearful that the costs would exceed the financial capacity of the foundation, while on the other side of the planet Schmidt felt an encroachment on his ability to manoeuvre freely as the designated leader of the expedition. Two determined personalities, Drachmann and Schmidt clashed to the extent that Schmidt came close to paying the price of losing his job (Chapter 15).

When the Board of Directors and the Board of Trustees finally became tired of Schmidt, they were not able to fire him, but the timing is interesting. Warming was long gone in the fall of 1930, and Ostenfeld was seriously ill. When Ostenfeld died the following year, Schmidt for the first time ever was without a trusted confidant in the Carlsberg Foundation. Warming, Ostenfeld, Koefoed, and Carl Jacobsen had all passed away, and Schmidt's initial network had shrunk. He was still however, married into the brewery's history, and a mighty man himself. Perhaps in the early 1930s, Schmidt also missed his collaboration with the deceased Christian F. Drechsel.

Christian F. Drechsel

During the earlier major expeditions, Schmidt had profited tremendously from Drechsel's bureaucratic strong-arm, and if Drechsel had still been alive and active by the summer of 1928, Schmidt and Ostenfeld would have acted wisely to place this veteran expedition manager in the cockpit of the circumnavigation. Drechsel was not a marine scientist by training, and today practitioners of marine science have largely forgotten about Drechsel. This is unlike Schmidt, Warming, and Ostenfeld, all of whom enjoy an afterlife with current citations of their scholarly work. Drechsel, however, was no less important for the significant success of marine science in Denmark for almost half a century.

Drechsel and C. G. Joh. Petersen formed a good team for almost a quarter of a century, before the two men separated in 1908. This means that when Drechsel took over the chairmanship of the Danish marine science commission he was already a veteran manager of fisheries research and someone the younger generation, such as Schmidt, could rely on and respect. The appointment of Drechsel as secretary general of ICES is another testimony of his status in the organizational ranks of international marine science, a position he was able to hold on to for another quarter of a century providing stability in the Danish commission as well as ICES during World War One (Chapter 2–4 & 12). Drechsel became a particularly important asset for Schmidt, when both men attempted to raise funds for deep-sea expeditions to the Atlantic Ocean. Here, Drechsel was able to add another dimension to Schmidt's network – contacts with the 'gold fish', the corporate elite of bankers and ship owners in Copenhagen. This was a unique asset of Drechsel. It was Drechsel who had a prior business contact with Mikkelsen Vendsyssel who lent out the M/S *Margrethe* (Chapter 7). Drechsel was also the one who led the negotiations with the East Asiatic Company concerning the construction of the M/S *Dana*, and it was also Drechsel who went to England to buy the R/V *Dana* on behalf of the Danish government (Chapter 10–11).

While Drechsel's fundraising effort is obvious, he also contributed to Schmidt's success in a less tangible way. Drechsel managed the accounts and expenses, procuring replacement fishing gear etc. from his desk in Copenhagen, while Schmidt and Ostenfeld for that matter could concentrate on their research. In Drechsel, Schmidt had met his match with regards to determination and willpower. Drechsel needed the deep-sea expedition to flag an early success for the reconstruction of ICES after World War One, while Schmidt dreamed about launching a circumnavigation to find the Indo-Pacific eel in addition to being first to chart the life history of the Atlantic eel. In the end, both men had their way (Chapter 11).

While it is a counterfactual hypothesis, there seems to be a strong indication that if Drechsel had been on board, when planning and executing the circumnavigation 1928–30, the whole operation would have been smoother. In any case it is striking that the lack of financial and logistical control during that particular expedition was a far cry from the care and attention marking Schmidt's other deep sea expeditions, as well as the more routinely conducted fisheries expeditions in Faroese, Icelandic, and Greenland waters.

C. G. Joh Petersen

Finally, Schmidt's first marine science champion and mentor within marine science, C. G. Joh. Petersen should be remembered for both his direct and indirect contribution to Danish deep-sea expeditions. If Drechsel was the one holding the purse strings, Petersen had a major influence on Schmidt's career in terms of his scientific merit from the time he first entered the towboat of the Biologisk Station until long after they had parted in 1908. This influence has been somewhat over looked in the literature to date.

Petersen was the most important Danish marine scientist during the last decades of the 19th century, and he was among the more important agents behind setting up ICES in 1902. Today, Petersen is remembered for pioneering habitat studies, but he had an equally successful run as an eel researcher, perhaps the most important in Europe around the turn of the 20th century. As early as 1890 Petersen used length distributions of fish samples to determine year-classes. He discovered that the yellow eel and the silver eel are the same species, but in different life stages, and he proposed that the eel was unique in the sense that it did not spawn in shallow inshore waters or in river systems. Rather the breeding place of the eel was to be found somewhere outside the areas where other natural scientists had been looking (Chapter 2).

This book has introduced the idea that it was in fact Petersen who asked Schmidt to search for eel in the North Atlantic, when he first sent Schmidt on a mission with the *S/S Thor* in 1903. Petersen therefore deserves at least some of the credit normally attributed to Schmidt for having found the first eel larvae in 1904. The close working relationship between Petersen and Schmidt until 1908 has been largely left out from subsequent scholarly literature. This book has tried to present evidence that Schmidt ceased to refer to Petersen's role in the *Thor* expeditions and the discovery of the eel larvae after their fallout in 1908. It is often the winners who write history, and Schmidt and his peers came out on top after their fight with Petersen (Chapter 3).

Petersen's legacy is, however, significant and includes various technological advances that, perhaps, were a bit underrated by Schmidt's generation, although they are recognized in the literature of the history of fisheries science. Petersen was a magnificent inventor of gear. He is widely credited for having constructed the 'Petersen bottom sampler', the 'Petersen young-fish trawl', and he was one of the first in the World to conduct marking experiments with numbered metal pieces. All of this innovation and technology was passed on to the next generation of marine scientists (Chapter 2).

Technological leaps

Deep-sea marine research is a very expensive enterprise, not least because it is demanding in terms of equipment. This is why the quest for funding was so central for much of Schmidt's work. The acquisition of the *S/S Thor* was a giant leap for marine science, and it was bought and equipped as a direct consequence of Denmark's international agreements within ICES. The *Thor* was able to sail for 12 days on one shipload of coal. If the weather was rough, the range was less. This means that from the shores of Iceland and the Faroe Islands the *Thor* had a significant geographical reach, unprecedented when compared to the sailing vessels used by the previous generation of marine scientists, so in many geographical areas Schmidt and his colleagues had the advantage of being the first to conduct their investigations. They were true explorers not knowing always what to expect when lowering their fishing gear and water samplers into the depths of the sea.

This is the technological background underpinning Schmidt's important conclusions as to the migration routes of many commercially important fish species in the North Atlantic. This included challenging and gradually replacing the paradigm of the so-called polar migration theory with the more contemporary migration triangle theory. Other notable achievements that benefitted from having access to state of the art methods and gear included the marking experiments with sea turtles in the Caribbean – the world's oldest experiments of this kind (Chapter 7).

As the search for Atlantic eel larvae led Schmidt further and further out into the Atlantic Ocean he needed another type of vessel, capable of operating more freely in open waters for several weeks at a time. For all its weaknesses the M/S *Margrethe* met that purpose, as she was a combined motor ship and sailing vessel, and the motor schooner M/S *Dana* had a similar advantage of a significant geographical reach (Chapter 7 and 10). In this way Schmidt was literally pushing the frontiers of eel research.

With the advent of the R/V *Dana* another physical frontier was pushed, this time downwards into the water column. With 10,000 meters of steel wire available on board, the research team of the Great Atlantic expedition was able to sample for larvae, plankton as well as pelagic fishes at a greater depth than anyone had attempted in tropical waters before them. As a consequence a tremendous number of new species were discovered (Chapter 11).

Finally the circumnavigation prompted another landmark for marine research technology, as Schmidt was the first to take an echo-sounder around the World, which led to the discovery of the Carlsberg Ridge (Chapter 16). Petersen's young-fish trawl for pelagic surveys, invented before 1900 in Danish waters and perfected by Schmidt's generation probing the Atlantic Ocean in the first three decades of the 20th century was also used round the world, on some occasions as far as possible from the Danish shores. There was however, one important aspect of Petersen's work, which set him apart from Schmidt and his colleagues, and that was politics.

Cosmopolitan conservatives

From 1814 onwards Denmark was a small and weak state. Following the Napoleonic Wars, known for good reason in Denmark as the 'Englandskrigen', and the subsequent Congress of Vienna, sovereign control over Norway was transferred from Copenhagen to Stockholm, and in 1864 another significant chunk of the former Danish Atlantic Empire was lost, as Schleswig-Holstein was conquered by the King of Prussia to be incorporated in the emerging great power of Germany. Denmark upheld a policy of neutrality in the face of being caught in the middle between the UK and Germany.

This was a successful policy during World War One, where Denmark stayed neutral, and even prospered from trading with both sides in the war. No one in their right mind disagreed with the desirability of not being squashed like Belgium in between the Great Powers, but this does not mean that everyone agreed as to how Danish foreign policy was to be enacted. In his biography of the Danish diplomat Henrik Kaufmann, the Danish historian, Bo Lidegaard presents the argument that Danish foreign policy formation before World War Two was divided between two sets of worldviews.[16]

First of all, there was a *national* take on foreign policy mainly represented by politicians and civil servants leaning towards the liberal party (Venstre), the social liberal party (Det radikale Venstre), and the social democratic workers' party (Socialdemokratiet). Between them, these three parties were in power throughout the period from 1901 until 1982, and for half a century the foreign policy focus was dictated by an overarching understanding that, what were most important for a small state like Denmark were socially progressive reforms to gradually improve the livelihood of the greater part of society. An educated, responsible and well-fed population was seen as the most important bulwark against the perils of totalitarianism, represented so prolifically in the 1920s and 30s by communism, fascism, and nazism.[17] Denmark's foreign policy doctrine of neutrality, however, did eventually shift in 1948 with the decision to enter the US led NATO alliance.

Several Danish diplomats including Henrik Kaufmann put forward an alternative vision for Danish foreign policy, dubbed *cosmopolitan* by Lidegaard. Many of these diplomats belonged to the conservative party (Det Konservative Folkeparti) and they looked at Denmark from the outside, as one of many states in the World. From their point of view

16 Lidegaard B. *I Kongens navn. Henrik Kauffmann i dansk diplomati 1919–1958.* Copenhagen: Samlerens Forlag; 1996.

17 Lidegaard B. *I Kongens navn. Henrik Kauffmann i dansk diplomati 1919–1958.* Copenhagen: Samlerens Forlag; 1996: 567–572.

how Denmark positioned itself in relation to other sovereign states was vital for its own interests. While there was no clear policy agenda for the development of Danish society through political reforms, there was a strong wish to uphold Denmark's position as a sovereign state. Therefore, Denmark should mark its territory and thereby earn the respect of other sovereign states, large and small alike.[18]

Some Danish industrialists during the first era of capitalist globalization (c. 1870–1914) belonged to this group of cosmopolitans, who often shared a similar ethos, or political take on the World.[19] Many of these men were very nationalistic in the sense that they valued culturally conservative ideals, but they were also internationalists, or cosmopolitan in the sense that they believed in the value of preserving a distinct Danish presence abroad, whether that be in the world of business, the world of politics, or the world of science.

Recently, Danish historian of science Kristian Hvidtfelt Nielsen proposed that the Galathea II Expedition from 1950–52 exhibited a similar cosmopolitanism, tied-in to Denmark's desire to promote itself on the World stage. Hvidtfelt Nielsen further argues that this expedition was the first attempt to use deep-sea marine research as a means to promote Danish values and civilization abroad.[20] There is no doubt that the Galathea II Expedition was 'a first' in the sense that this expedition included a dedicated team of professional communicators on board. This book however, argues that the cosmopolitan trait in the political motivation for sending the *Galathea II* round the world was a continuation of similar efforts tied into Denmark's waning North Atlantic Empire.

The cosmopolitan line of thinking and Danish deep-sea research extends back, at least to the early days of ICES around the turn of the 20th century. Once more some of the key players were Schmidt, Drechsel, Ostenfeld, and Warming – and perhaps more reluctantly, Petersen. All five men were eager travellers, and they all had stakes in research areas far from Danish waters. In the case of Drechsel, he even had a personal economic interest in Iceland based trawl fisheries dating back to the 1890s (Chapter 2).

Similar to the aristocratic diplomats at various Danish legations abroad, the marine scientists were globetrotters. Warming had travelled to Latin America and the Danish West Indies at a very early stage, Schmidt had been to Siam, and Ostenfeld went to Australia for research. Their view of the World and their view of Denmark's position in the World came to the forefront, whenever a conflict arose. Thus in the

18 Lidegaard B. *I Kongens navn. Henrik Kauffmann i dansk diplomati 1919–1958*. Copenhagen: Samlerens Forlag; 1996: 567–572.

19 Boje P. *Danmark og multinationale virksomheder før 1950*. Odense: Odense University Press; 2000; On this wave of globalization see: McNeill JR, McNeill WH. *The human web: A bird's-eye view of world history*. WW Norton & Company; 2003.

20 Nielsen KH. Postcolonial partnerships: Deep sea research, media coverage and (inter)national narratives on the Galathea deep sea expedition from 1950 to 1952. *The British Journal for the History of Science*. 2010;43(1):75–98.

chapters above there are four clear cases, where these *cosmopolitan conservatives* showed their colours.

The first clear case of this was the fight with Petersen in 1908. While personal disagreement and considerations on how to promote one's career certainly played a role, there was a clear political divide, where Petersen was happy with having discovered the breeding places and migration patterns of the most important species. The job was solved, as he stated in 1907. Petersen genuinely believed that further research would mainly benefit the Great Powers, who would exploit the fisheries on and outside of the continental shelf of Europe. Petersen would much rather focus his interests on serving the social and economic interests of fishermen and fishing communities in what he saw as Denmark proper, not counting in the far flung possessions in the North Atlantic. In short, Petersen was the marine science equivalent of the social liberal national approach to foreign policy (Chapter 3).

Drechsel, Ostenfeld, and Schmidt had a completely opposite take on Denmark's presence, which became evident when Schmidt stated outright that there were not only scientific reasons for staying in the North Atlantic Ocean, but also substantial political reasons, in particular the reinforcement of Denmark's interests in Iceland. They were the conservative cosmopolitans, and they found a meeting place for promoting the idea that Denmark still had a noteworthy North Atlantic empire, when they all joined the society, De Danske Atlanterhavsøer.

The second turning point when the marine scientists showed their political colours was in relation to the sale of the Danish West Indies. The political uprising in the fall of 1916 led to the formation of an informal maritime Danish North Atlantic alliance consisting of the marine scientists, influential merchant capitalists, ship owners, and leading bankers. They shared a belief in the importance of upholding a Danish presence in the Atlantic realm for financial, scientific, and not least patriotic reasons. Next to the business people, Danish marine scientists made up the one group that would stand to lose the most, if Denmark's Atlantic Empire began to shrink. Schmidt's expert witness testimony to the government hearing reads like a consistent culturally conservative position, where Schmidt saw the West Indies colony as integral to Danish society (Chapter 9).

Although the evidence is circumstantial, this book proposes that the main reason, why Schmidt succeeded in raising such a vast sum of money from the corporate World in the fall of 1916 is a result of the abovementioned political alliance. *Cui Bono*, one could ask, who benefitted? Considering the fact that the East Asiatic Company, the leading Danish banks, and the group of marine scientists all had assets in and around the Virgin Islands *after* the sale of the islands to the USA, they had a mutual interest in securing the presence of Danish deep sea research in Caribbean waters – and not least in the, relatively speaking, nearby Sargasso Sea.

The third instance of political cosmopolitanism highlighted in the above is the race for Greenland waters in 1924, when Schmidt initiated his very own foreign policy initiative by way of improvisation, while in Iceland. Sending Taaning to Greenland to mark cod was also a demarcation of Danish territorial presence. Notably, Drechsel blamed the national wing of the liberal party for having caved in to the Norwegian claim to a special status in Greenland (Chapter 12).

Finally, the circumnavigation was a showcase in its own right for Danishness, as Schmidt brought along films about Denmark courtesy of the Ministry of Foreign Affairs, while his assistants wrote newspaper columns along the way. Nonetheless, it is worth noting that Schmidt spoke differently when addressing an audience far away from any territorial claim of Denmark's, than he did, when speaking about the North Atlantic realm. This quote from a lecture in Sydney serves to show that Denmark has a place among nations like any other sovereign state: 'When we visit and meet with the scientific workers of other countries, we are not regarded as foreigners, but as Fellow Scientists. Science recognizes no Boundaries, no Class, and no Creed, all working to a common end, to increase knowledge and to uplift humanity' (Chapter 14).

The idea of not recognizing any boundaries is the opposite of what he spoke about with regards to the Greenland waters in 1924, which were to be reserved for the Danes. It is also the opposite of what he said about promoting fisheries in the Danish West Indies in 1916, and it is a far cry from the territorial exclusivity he had stood up for in relation to the North Atlantic fisheries research in 1908. This is not however, because of a change of mind during Schmidt's 30 years at the helm of Danish deep-sea marine research. Instead, Schmidt merely expressed a scientist's soft diplomacy version of Hugo Grotius' distinction between Mare Clausum (the Danish North Atlantic) and Mare Liberum (the rest of the World).

In public or in letters Schmidt rarely stated which political party he voted for, but probably he belonged to the conservative party. This also fitted with his 1920 membership card from the society known as the *Sønderjydsk Forening for København og Omegn*, which was deposited in his private archive.[21] This society was one of several patriotic societies, which emerged in Denmark following the outcome of the 1919 Treaty of Versailles. Here the Danish government recommended that a referendum was held in the Duchies of Schleswig-Holstein as to whether or not parts of this area would become part of the Danish state. Covering roughly the southern half of the Jutland Peninsula, Schleswig-Holstein had been part of the realm of the Danish state until 1864, when Prussia incorporated the territory into what soon became The German Empire. By 1919 there was still a sizeable ethnic Danish population in Schleswig-Holstein. A referendum was held in February of 1920, when a majority in the northern part of Schleswig voted for a return to Denmark.[22] The patriotic societies such as the one above worked actively in support of ethnic Danes entrapped in Germany due to the results of the referendum. Some of these societies even advocated that the referendum should be ignored, so the Danish state could have a larger chunk, of what is still today Northern Germany. As mentioned above Vagn Jacobsen of the Carlsberg Brewery as well as Eugen Warming were fellow advocates of an activist and offensive territorial policy on this issue.

In summary, cosmopolitan conservatism was a common denominator for Schmidt and his core network consisting of Drechsel, Ostenfeld, and Warming. Petersen did not fit this formula, and this was a significant factor in his fallout with Schmidt and others. Furthermore, this cosmopolitan world-view was an intricate part of Schmidt's oeuvre as a scientist. It was not merely a discourse he would turn on, whenever this was politically opportune, although frequently it did turn out to be politically opportune for him to play the nationalist card. As described above, Norway and Italy took turns playing the antagonist when Schmidt was setting the stage for yet another trip to the Mediterranean or the Northern North Atlantic Ocean. Nonetheless, the diaries of Bruun and Hensch Stephensen both painted an image of Schmidt as a deeply convinced nationalist patriot. When Schmidt would not speak German to his close collaborator Professor Ehrenbaum, while he was on board the *Dana* in 1921, Schmidt was not putting on a show for a fundraising scheme. He was just being rude (Chapter 11).

As a curiosity Schmidt actually met his fellow cosmopolitan Dane, Henrik Kaufmann in the Dutch East Indies in 1929 (Chapter 14). There is good reason to believe that Schmidt liked Kaufmann. At any rate,

21 KB. Johannes Schmidt 'Papirer'. Ut.374.V.2. Privat og Blandet Korr. 1912–1927. Membership card for 1920.

22 Lammers, Karl Christian, '1914–1920 Fra borgfred til Påskekrise', in Søren Mørch (ed.) *Danmarks Historie, 7*, 117–120.

Schmidt included a portrait image of Kaufmann in his popular book on the circumnavigation, indicating that the encounter with Kaufmann was a noteworthy event, even though he was neither a marine scientist, nor an exotic species or indigenous tribe, which were the typical motives for this particular publication.[1]

An informal maritime empire

The specific character of ocean explorations such as the huge costs of equipping vessels, manning them often for years of travels and handling the findings of explorations makes the concept of Big Science just as relevant for marine science prior to WWII, as the concept is today for areas such as particle research and space exploration.

In Denmark, a small power by European standards, big science took off at the start of the 20th century when Danish marine science took to the deep sea. The personification of this was the Danish marine scientist Johannes Schmidt, the foremost fundraiser in Danish marine science in the first third of the 20th century. He was particularly apt at raising funds for deep-sea explorations in the Atlantic, and crowned his achievements by successfully launching the world's first circumnavigation with a distinct biogeographic purpose in 1928–30. The scientific legacy of these explorations amounts to discovering the breeding places of the Atlantic eel and later eels in the Pacific and Indian Oceans. Perhaps less spectacular but similarly ground-breaking were the marking experiments conducted over several decades to map the migration patterns of cod around Iceland.

These expeditions, launched in Denmark unfolded as interplay between scientific strategies, colonial desires and changing natural resource availability in the Atlantic Ocean. In the cold waters around Iceland and Greenland there was a clear incentive to enhance knowledge on the emerging fisheries for cod and halibut, both for direct fisheries gains and also in order to keep Norwegian geopolitical and fisheries resource driven interests at bay. In the Danish West Indies, the picture is less clear with regards the prospects of deep-sea explorations creating immediate economic gains for local fisheries. Here, the overall Danish effort in marine science should rather be seen as a way of unfolding an informal colonial empire through ocean sciences in a time when

23 Schmidt J, ed. *Dana's Togt Omkring Jorden 1928–1930*. Copenhagen: Gyldendalske Boghandel; 1932

the Danish political colonial presence was folding. Drawing on Brian Payne's concept of environmental territorialization, this was a sustained act of *science based environmental territorialization*. Paraphrasing Donald Worster: 'individuals as well as societies are shaped by natural as well as cultural influences.'

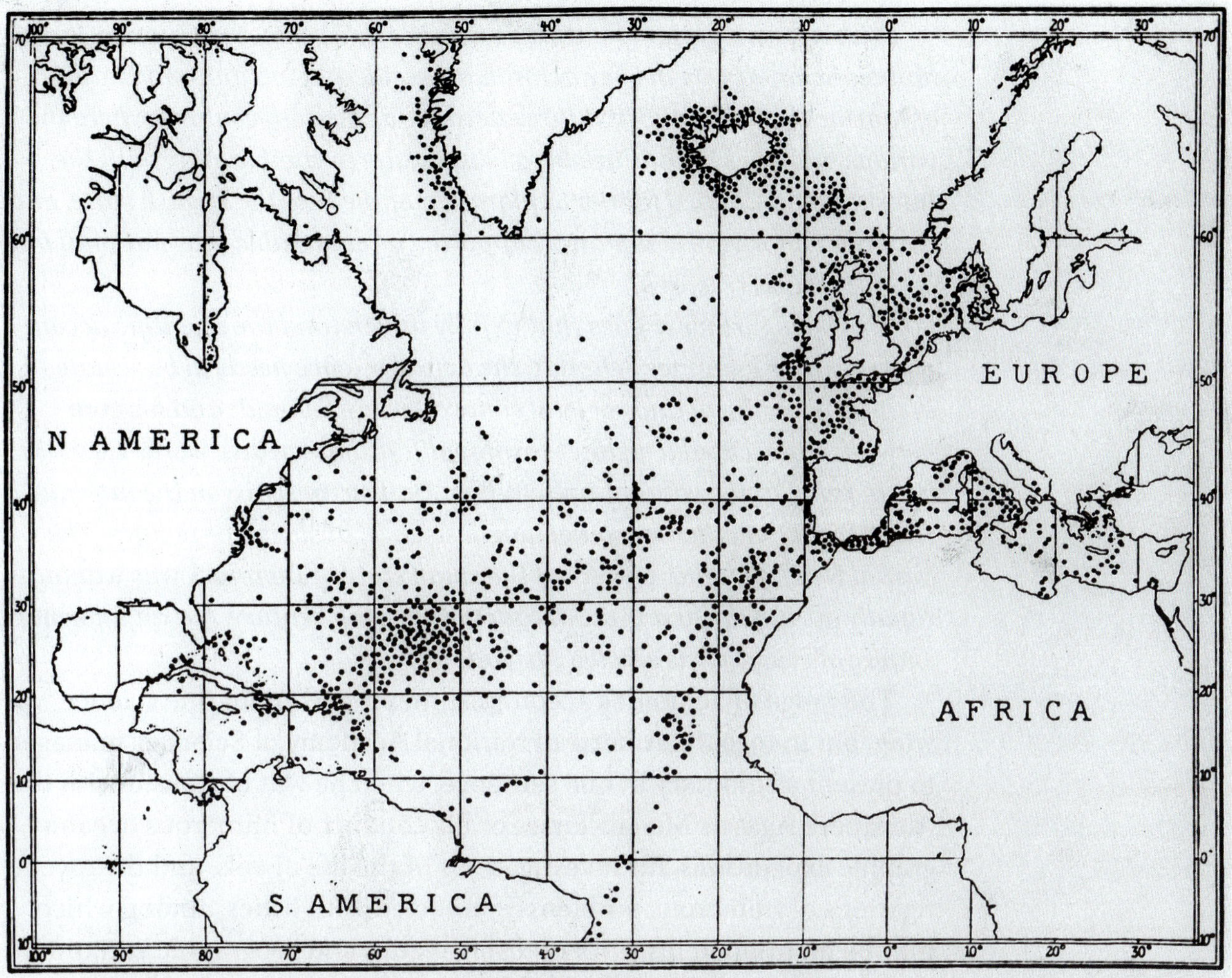

Udvalg af de danske Havundersøgelses-„Stationer" i Atlanterhavet fra 1903 til 1925. Det dér udførte Arbejde udgjorde Grundlaget for Dana-Ekspeditionens Undersøgelser paa Jordomsejlingen.

17.3 Danish stations in the Atlantic Ocean 1903–1925
This map of the Atlantic Ocean shows the presence of Danish marine research in between 1903–1925. The message from the Danish marine science commission is clear: A small nation contributes on a large scale (*Carlsberg Archives*).

Conclusion

The main results coming out of this investigation into the life and career of Johannes Schmidt have been presented in the concluding sections of each of the previous sixteen chapters. In this final chapter these results have been summarized and framed in relation to the over-all argument in this book which is the following:

Within a core group of natural scientists and science managers, Johannes Schmidt was the key actor in an accidental institutionalization of marine science in the first three decades of the 20th century, where the various branches of the Carlsberg Foundation in combination with the International Council for the Exploration of the Sea, the Danish state, and a number of private companies supported a remarkable international hub for World class research.

This book further argues that to fully understand the scientific accomplishments of Johannes Schmidt, the actual science needs to be situated in the right political and societal context, where Schmidt and his core network represented a distinct cosmopolitan conservative world view with a firm belief in the value of upholding a Danish presence on the international scene, whether in connection to science, business, or politics. The formal North Atlantic Empire of the small state of Denmark was waning rapidly during the first three decades of the 20th century, but an informal empire of science was erected instead.

The range of Schmidt's accomplishments within marine science was wide, but in 1930 the American National Academy of Sciences managed to present a summary in one sentence, when he was bestowed with the Alexander Agassiz Medal: 'First For his conduct of numerous oceanographic expeditions, his investigations of the life of eels, and the investigations of numerous problems connected with fishes, among which may be mentioned his work on fish genetics and geographic variation.'

Finally, he was also a fortunate man. He was fortunate in his choice of spouse, and he was fortunate with his choice of mentors, and also with the scientific positions he obtained chiefly within the Danish marine science commission and the Carlsberg Laboratorium. In addition he shrewdly exploited the chances arising over a scientific career spanning 35 years of scientific publications. While failure was part of his long quest for success, he was largely successful over the long run. Late in life he was fortunate enough not to be fired from his job.

As his Scottish colleague D'Arcy W. Thompson told Schmidt in a letter from 1924; 'You are a very happy man, if I may say so. You see your work growing in your hands from year to year. And how splendidly it has grown, and is still growing, there are not many who know better than – Your faithful friend.'[24]

24 KB. Johannes Schmidt 'Papirer'. Ut.374.V.4. Blandede papirer og breve fra 1923–26. D'Arcy W. Thompson to J. Schmidt, 16 January 1924.

Summary

1 Formative years

The opening chapter sets the scene for the protagonist, explaining how the young Schmidt was brought up in a stimulating scientific environment in the house of his maternal uncle, Johan Kjeldahl, acclaimed head of the Chemical Department of the Carlsberg Laboratory in Copenhagen. Ties with Carlsberg were further tightened in 1903 as Schmidt married the daughter of the CEO of Carlsberg Breweries, controlled by the Carlsberg Foundation. One of the board members was Eugen Warming the world renowned founder of ecology as a scientific discipline. Warming was also Schmidt's professor at Copenhagen University and very influential in supporting Schmidt's first expedition, to Thailand in 1899–1900, which laid the foundation for his doctoral thesis on mangrove forests defended in 1903. By the age of 26, Schmidt had also published a couple of books in bacteriology, but he had yet to find his career path as a scientist.

2 The legend of the unintentional eel larva

This came about in 1903 when Denmark's then most famous marine scientist, C. G. Joh. Petersen hired the young Schmidt. Petersen invited Schmidt to lead a new Danish expedition that would investigate the commercially important fish species of the waters around the Faroes and Iceland for the next five years. Schmidt had developed an interest in marine science as an assistant on Petersen's floating Biologisk Stations in 1902. The International Council for the Exploration of the Sea (ICES) was also set up in 1903, and Petersen was made chair of the Danish component, the Kommissionen for Havundersøgelser.

In 1904, Schmidt made his most famous discovery, the first catch of eel larvae outside European coastal waters. This is well known in history of marine science literature, since the breeding place of the freshwater eel had been a mystery since the time of Aristotle. In the literature Schmidt is often portrayed as the great genius who found the larva almost by accident. This chapter also argues that Schmidt's initial successes as a marine scientist were a result of being the right person at the right time and place. With the research vessel, S/S *Thor*, and custom made fishing gear Danish marine scientists could suddenly access the deep sea in ways never explored before. The reality was that is was his boss, Petersen, who had asked Schmidt to search for eel larvae. Petersen, who is more famous for pioneering marine habitat mapping, ought to also be heralded as Europe's leading eel researcher around the turn of the 20th century. The reason why this is not the case is the leitmotiv of the following chapter.

3 A marine science coup d'état

This chapter highlights the importance of agency in the sustainment of research institutions. As the first five-year-term of ICES came to an end in 1908 Denmark's position was at a watershed. Petersen found that the North Atlantic cruises had fulfilled their purpose and tried to cut most of the government funding when applying for their successor. He also dreamt of merging his own Biologisk Station with the Danish Kommissionen for Havundersøgelser into one strong research unit. These moves alienated Petersen from his collaborators, chiefly fisheries consultant, Chr. Drechsel, plankton biologist C. H. Ostenfeld, and hydrographer Martin Knudsen, as the jobs of their assistants became endangered. Schmidt was caught in the middle but clearly sided with the rebels, who came out on top. Petersen was ousted from the commission and lost his seat among Danish delegates to ICES. After the coup, Schmidt and his colleagues more or less stopped referring to Petersen's work on the eel in their research papers and commemorative manuscripts. Therefore, this part of Petersen's world leading research has generally been absent from previous historical documentation.

4 The Mediterranean expeditions (1908–1910)

This chapter connects for the first time, Johannes Schmidt's two Mediterranean expeditions with two of his foremost sponsors, the Danish government and the Carlsberg Foundation. A third sponsor was his mother-in-law. The chapter further argues that Schmidt was interested in the eels in the Mediterranean for much longer than previously believed and continued with his efforts there until 1911. Although evidence for a Mediterranean breeding ground was never found. Instead other considerable advances, especially in hydrography, were made. For instance the observation that the salinity and temperature of intermediate layers of water dictate the flow in and out of the Strait of Gibraltar. This is one reason Schmidt was able to raise funds for a second trip. However, what has not been known before is that the Carlsberg Foundation had initially declined Schmidt's application for a second trip. The decision was eventually revoked after Schmidt and Eugen Warming mobilized a string of important professors to write letters of support. The Board of Directors of Carlsberg had good reason to hesitate, since many of the board members were also on the Board of Trustees of the Carlsberg Laboratory, which had just hired Johannes Schmidt to be the new head of the laboratory's Physiological Department.

5 Director at Carlsberg laboratory (1910–1933)

Surprisingly, Schmidt continued his work in marine science during his 23 years as director of a laboratory, whose raison d'être was to promote beer related research. Previous commentators have assumed that Schmidt's marriage was the prime mover behind his promotion to this job, but this study argues that Eugen Warming once more was the driving force behind the promotion, and following Warming's retirement Schmidt's old comrade of arms, Ostenfeld joined the Board of Trustees. Next to his modest research programme centred on hops, Schmidt engaged in ground breaking genetic studies. Using large-scale laboratory and *in situ* investigations Schmidt identified hereditary aspects of quantifiable characteristics such as the number of vertebrae in fish and chicken, which won him the 1923 Weldon Memorial Prize.

6 Carlsberg appreciation

This chapter explains for the first time how Schmidt managed to turn the Carlsberg Laboratory into a formidable marine biology laboratory. The high annual income received by the laboratory from Carlsberg made the brewing organization a prime supporter of marine science. Surprisingly, the Board of Trustees as well as the CEO of the Carlsberg Brewery seemed completely satisfied with Schmidt's work throughout the 1910s and 20s, since they took no action against the laboratory leader, unlike Schmidt's colleague Alberg Klöcker, who was axed unceremoniously once the board became displeased with the management of his laboratory.

7 Atlantic adventures

Schmidt's search for eel drove his research further and further out into the Atlantic, beyond reach of the steam engine of the S/S *Thor*. Through the aid of ships of convenience the sampling for eel larvae continued, but a real breakthrough came in 1913, when Schmidt borrowed the M/S *Margrethe*. This chapter tells the dramatic story of how the ship wrecked just before reaching port in St. Thomas. Luckily the eel samples were rescued, and the surviving bottles of larvae enabled Schmidt to deduce that the breeding ground of the freshwater eel was somewhere in the Sargasso Sea. The rest of the expedition was rescheduled as coastal surveys around the Danish West Indies, which resulted in hitherto undisclosed fisheries investigations and tagging experiments on the resident sea turtles, some of the first in the world. One reason for the lack of information about this part of the expedition is the outbreak of WWI in 1914, and the 1916 US acquisition of the Danish West Indies.

8 Fundraising for Atlantic deep sea expedition

Following the dramatic voyage of the *Margrethe*, this chapter tells the story of how Schmidt and Chr. Drechsel were eagerly searching for funds to make another expedition into the Atlantic, even as early as April 1914. The well-connected Drechsel was both general secretary of ICES and chairman of the Danish Kommissionen for Havundersøgelser and the former port director in Copenhagen, but their efforts were cut short by the outbreak of World War One. This was detrimental to Atlantic marine science not least because of the immanent dangers from submarine warfare, but also the breakdown of much pre-war international scientific collaboration. Then suddenly, in November 1916, funding was secured for a Danish Schmidt-led expedition for when the war was over. The main sponsor was the East Asiatic Company, which also promised to build a research vessel. Meanwhile the largest Danish bank, Landmandsbanken was the second most important sponsor, with ship owner Andreas Erlandsen and attorney Zeuthen taking third and fourth places, respectively. The following chapter is devoted to teasing out, why these private companies suddenly had an interest in sponsoring Schmidt's research projects.

9 Resistance from the sea

Following the sale of the Danish West Indies, marine scientists had every reason to doubt that the Danish government would sustain a Danish presence of marine science in the Caribbean. Moreover, the war seemed to be dragging on with the USA still a few months away from declaring war on Germany. However, what looked somber from the outside may in fact have been a crucial turning point for the scientific dreams of Schmidt, Drechsel, Ostenfeld, and the rest of the prominent Danish marine scientists at the time. The turning point was that the events surrounding the sale of the Danish Caribbean possessions had brought to the table an informal maritime Danish Atlantic alliance consisting of the marine scientists, merchant capitalists, ship owners, and leading bankers. They shared a belief in importance of upholding a Danish presence in the Atlantic realm for financial, scientific, and not least patriotic reasons. Through agitation and diverse political moves centred on the society, De Danske Atlanterhavsøer, the marine scientists flagged their will to defend the mutual interests of the informal alliance. This meant that the marine scientists could now turn to their friends in the private sector for assistance in sustaining Danish marine science. Drawing on previously unexploited archives, this is the first analysis of Danish foreign policy in relation to marine science and the issue of the Danish West Indies.

10 The Atlantic expeditions with the M/S *Dana*

This chapter is the first in-depth exploration of the planning and execution phase of two major Atlantic deep sea explorations undertaken with Schmidt at the helm, as well as one replacement expedition. The first and second expeditions were in 1920 and again for a few months in 1921 with the M/S *Dana*, on loan from the East Asiatic Company, which had been built partly to serve the interests of the marine scientists. From an international comparative perspective the promises of Danish deep-sea investigations were used as leverage by Chr. Drechsel in his attempt to re-assemble international support for ICES after WWI. Schmidt on the other hand was interested on two fronts. First of all, he wanted to solve the riddle of the breeding place of the Atlantic eel once and for all, secondly, he dreamed about launching an expedition beyond the Atlantic Ocean into the Pacific, possibly all around the World.

11 The great Atlantic expedition with the R/V *Dana*

In 1921–22, following the government acquisition of the research vessel, *Dana II*, Schmidt and Drechsel were the masterminds behind the launch of their largest expedition so far. The Danish explorations flew the ICES flag on what became path breaking research cruises. Using custom made deep sea pelagic nets for fish and plankton the riddle of the breeding place of the Atlantic freshwater eels was finally settled, while dozens of new species of anglerfish and invertebrate saw the light of the day for the first time. *Dana II* became the first research vessel to cross the Panama Canal and conduct investigations on both sides of the Americas with the same gear. Schmidt became world famous as a consequence.

12 Marking cod and territory in the North Atlantic

This chapter connects for the first time, Schmidt as a laboratory researcher and the main man in international fisheries science in the 1920s, where the resurging international collaboration of marine science was greatly aided by the resourceful Carlsberg Laboratory. The pioneering racial studies of the 1910s revolutionized the study of North Atlantic commercial fish species such as cod and plaice, and the research headed by Schmidt laid the foundation for future conservation management. While international collaboration was at the forefront, the research presented here, reveals an undercurrent of political animosity between two of the leading fisheries nations, Norway and Denmark. This animosity was spearheaded by the Norwegian claim to jurisdiction over parts of Greenland. Schmidt was conscious of the symbolic value, or soft diplomacy, of re-directing parts of the Danish marine research effort towards Greenland.

13 Organizing the Carlsberg circumnavigation of the world

This chapter traces the origin of Schmidt's global research aspirations. Contrary to published wisdom, Schmidt dreamt of making a circumnavigation from at least 1916 onwards. Following his stardom from 1922 onwards, and his great academic freedom at the Carlsberg Laboratory he now pursued this dream. Schmidt's 1925–26 trip to Tahiti and the USA is described in the literature, and he managed to publish several articles on the life history of eels in the Pacific. Yet, the fact that especially the American leg of the trip was a distinct fundraising tour is completely new. Backed by the Danish ministry of foreign affairs Schmidt met with influential American funding organizations, but did not have any immediate success. In 1927, Schmidt turned to the Carlsberg Foundation once more to find support for his adventures. This time the long-awaited circumnavigation was on the table, and the Carlsberg Foundation put at Schmidt's disposal what, at the time, was its largest single donation. Once more, it was Schmidt's enduring interest in the eel, which had already made him world famous, that drove the science programme of the expedition. Next to the Carlsberg Foundation, the East Asiatic Company and the Danish government backed the expedition which included frequent presentations of promotional films about Denmark. The chapter follows closely the organizational and planning aspects of the circumnavigation that was to commence in June of 1928.

14 Around the world in two years

This chapter tells the story of how the Circumnavigation became a display of publicity around the world. Schmidt's mastery of connecting PR work and promotion of Denmark, with the promotion and communication of his own research is not mentioned in the literature. The diaries kept by Schmidt and Anton Bruun provide a clear picture of 'life on board' the research vessel, and some of the values that dictated the way in which Schmidt ran his ship. This also gives a rare insight into the prevailing colonial attitude in the worldviews of the Danish explorers. In spite of careful planning the weather and the availability of coal came to play a major restraining role for the expedition, which never reached two of the four designated hotspot areas (Japan and Solomon Island Archipelago) highlighted on the original expedition plan.

15 Stretching the expedition, stretching ties to Carlsberg

It has never been known, either at the time or later, that one year into the expedition Schmidt had a substantial fight with the Carlsberg Foundation over the duration, expenses, and travel plan for the circumnavigation. Schmidt's idea in the middle of the journey, to extend the trip by three months, and add significantly to the cruise plan has never before been reported. Schmidt's falling out with the Board of Directors of the Carlsberg Foundation took him to the brink of losing his job at the Carlsberg Laboratory. However, Schmidt had become an uncontrollable force in the 'Carlsberg family', a sort of cuckoo in a warbler's nest.

16 Measures of impact and legacy

Research wise, the circumnavigation, 1928–1930 became another example of how Schmidt and his generation of Danish marine scientists excelled through the use of the most sophisticated technology available on the ocean territories, where they sailed. Thus for the first time an echo sounding machine made it around the world and resulted in the discovery of what is now the Carlsberg Ridge in the Indian Ocean. In the Pacific the discovery of huge anoxic areas with low primary productivity was another major result, while the breeding grounds of most of the World's eel populations (but not the Japanese) were charted along the way.

This chapter also analyses Schmidt's impact and legacy as a Scientist and expedition leader. His research proved inspirational for many subsequent areas as diverse as Wegener's theory of the drift of continents and the sustained Danish tradition of eel research, where the latter is still ongoing. While almost forgotten in the larger public memory of today, it should be emphasized that in his day Johannes Schmidt was a celebrity, world famous among marine scientists and a well-known figure in Denmark in the 1920s and early 1930s.

17 A fortunate man

The final chapter synthesizes Schmidt's professional life and works of science. Within a core group of natural scientists and science managers, Johannes Schmidt was the key actor in an accidental institutionalization of marine science in the first three decades of the 20th century, where the various branches of the Carlsberg Foundation in combination with the International Council for the Exploration of the Sea, the Danish state, and a number of private companies supported a remarkable international hub for World class research. This book further argues that to fully understand the scientific accomplishments of Johannes Schmidt, the actual science needs to be situated in the correct political and societal contexts. Schmidt and his core network represented a distinct cosmopolitan conservative world-view with a firm belief in the value of upholding a Danish presence on the international scene, whether in connection to science, business, or politics. The former North Atlantic Empire of the small state of Denmark was waning rapidly during the first three decades of the 20th century, but an informal empire of science was erected instead.

Archival sources

Carlsberg Archives, Copenhagen

Carlsberg Arkiv, Valby(CA):
Carlsberg Archives (CA), Carlsberg Laboratorium (CL), boxes no. E06-E10 covering 1909–1935. 'Korrespondance, regnskaber og beretninger'.

Box E06

CA. *Carlsbergfondet, Carlsberg Laboratorium.* E06.'1908–1913'.Korrespondance, regnskaber og beretninger.

CA. Carlsbergfondet, Carlsberg Laboratorium. E06.'1909'. J. Schmidt to Board of Trustees, 22 September 1909.

CA. *Carlsbergfondet, Carlsberg Laboratorium.* E06.'1909'. S. M. Jørgensen to J. Schmidt, 2 October 1909.

CA. *Carlsbergfondet, Carlsberg Laboratorium.* E06.'1911'. J. Schmidt to Board of Trustees, 23 January 1911.

CA. *Carlsbergfondet, Carlsberg Laboratorium.* E06.'1911'. J. Schmidt to Board of Trustees, 13 May 1911.

CA. *Carlsbergfondet, Carlsberg Laboratorium.* E06.'1911'. J. Schmidt to Board of Trustees, 10 June 1911.

CA. Carlsbergfondet, Carlsberg Laboratorium. E06.'1911'. J. Schmidt to Board of Trustees, 27 September 1911.

CA. *Carlsbergfondet, Carlsberg Laboratorium.* E06.'1912'. J. Schmidt to Board of Trustees, 30 September 1912.

CA. Carlsbergfondet, Carlsberg Laboratorium. E06.'1912'. R. Koefoed to Board of Trustees, 10 October 1912.

CA. Carlsbergfondet, Carlsberg Laboratorium. E06.'1912'. S. M. Jørgensen to A. Klöcker, 27 October 1912.

CA. Carlsbergfondet, Carlsberg Laboratorium. E06.'1912'. A. Klöcker to Board of Trustees, 31 October 1912.

CA. Carlsbergfondet, Carlsberg Laboratorium. E06.'1912'. J. Schmidt to Board of Trustees, 19 November 1912.

CA. Carlsbergfondet, Carlsberg Laboratorium. E06.'1912'. S. M. Jørgensen, 'Administrative Bestemmelser vedrørende Carlsberg Laboratoriet.' 1912.

CA. Carlsbergfondet, Carlsberg Laboratorium. E06.'1913'. J. Schmidt to S. M. Jørgensen, 22 May 1913.

CA. Carlsbergfondet, Carlsberg Laboratorium. E06.'1913'. S. M. Jørgensen to R. Koefoed, 28 May 1913.

CA. Carlsbergfondet, Carlsberg Laboratorium. E06.'1913'. S. M. Jørgensen to A. Klöcker, 31 May 1913.

CA. Carlsbergfondet, Carlsberg Laboratorium. E06.'1913'. R. Koefoed to Board of Trustees, 2 June 1913.

CA. *Carlsbergfondet, Carlsberg Laboratorium*. E06.1913. Letters from J. Schmidt to Board of Trustess 15, 16 and 23 September 1913.

Box E07

CA. *Carlsbergfondet, Carlsberg Laboratorium*. E07.1914. Letter from J. Schmidt to Board of Trustees, 24 December 1913.

CA. *Carlsbergfondet, Carlsberg Laboratorium*. E07.'1915'. J. Schmidt to Board of Trustees, 9 August 1915.

CA. *Carlsbergfondet, Carlsberg Laboratorium*. E07.'1915'. E. Warming to Board of Trustees, 15 August 1915.

CA. *Carlsbergfondet, Carlsberg Laboratorium*. E07.'1915'. J. Schmidt to Board of Trustees, 30 September 1915.

CA. Carlsbergfondet, Carlsberg Laboratorium. E07. '1919'. A. Klöcker to E. Warming, 9 September 1919.

CA. Carlsbergfondet, Carlsberg Laboratorium. E07. '1919'. A. Klöcker to E. Warming, 11 October 1919.

Box E08

CA. Carlsbergfondet, Carlsberg Laboratorium. E08. 1920. No. 6. Letter from J. Schmidt to Board of Trustees, 14 January 1920.

CA. Carlsbergfondet, Carlsberg Laboratorium. E08. 1920. No. 21. Letter from E. Warming to Board of Trustees, 10 June 1920.

CA. Carlsbergfondet, Carlsberg Laboratorium. E08. 1920. No. 29. Letter from J. Schmidt to Board of Trustees, 7 July 1920.

CA. Carlsbergfondet, Carlsberg Laboratorium. E08. 1921. Letter from E. Warming to Board of Trustees, 28 March 1921.

CA. Carlsbergfondet, Carlsberg Laboratorium. E08. 1921. Letter from E. Warming to Board of Trustees, 2 April 1921.
CA. Carlsbergfondet, Carlsberg Laboratorium. E08. 1921. Letter from J. Schmidt to E. Warming, 3 April 1921.
CA. Carlsbergfondet, Carlsberg Laboratorium. E08. 1921. Letter from E. Warming to Board of Trustees, early September 1921.
CA. Carlsbergfondet, Carlsberg Laboratorium. E08.1922. No. 5. Letter from J. Hjelmslev to V. Henriques, 1 April 1922.
CA. Carlsbergfondet, Carlsberg Laboratorium. E08.1922. No. 5. Letter from C. Ostenfeld to V. Henriques, April 1922.
CA. *Carlsbergfondet, Carlsberg Laboratorium*. E08.1922. No. 5. Letter from V. Henriques to Board of Directors, 5 April 1922.
CA. *Carlsbergfondet, Carlsberg Laboratorium*. E08.1922. No. 28. Letter from V. Henriques to Board of Trustees, 6 October 1922.
CA. *Carlsbergfondet, Carlsberg Laboratorium*. E08.1923. No. 10. Letter from V. Henriques to Board of Trustees, 14 April 1923.
CA. *Carlsbergfondet, Carlsberg Laboratorium*. E08.1923. Letters from J. Schmidt to V. Henriques, 18 April, 10 May & 25 May 1923.
CA. *Carlsbergfondet, Carlsberg Laboratorium*. E08.1923. No. 23. Letter from V. Henriques to Board of Trustees, 14 September 1923.
CA. *Carlsbergfondet, Carlsberg Laboratorium*. E08.1923. No. 23. Letter from J. Schmidt to V. Henriques, 16 October 1923.
CA. *Carlsbergfondet, Carlsberg Laboratorium*. E08.1923. No. 36. Letter from J. Schmidt to Board of Trustees, 16 October 1923.
CA. Carlsbergfondet, Carlsberg Laboratorium. E08. 1924. Letter from J. Schmidt to V. Henriques, 10 April 1924
CA. Carlsbergfondet, Carlsberg Laboratorium. E08. 1924. no. 11. Letter from J. Schmidt to Board of Trustees, 22 April 1924.
CA. Carlsbergfondet, Carlsberg Laboratorium. E08.1924. No. 21. Letter from J. Schmidt to V. Henriques, 21 August 1924.
CA. *Carlsbergfondet, Carlsberg Laboratorium*. E08.1924. Letter from J. Schmidt to V. Henriques, 7 July 1924.
CA. *Carlsbergfondet, Carlsberg Laboratorium*. E08.1924. No. 27. Letters from J. Schmidt to V. Henriques, 21 September 1924.
CA. *Carlsbergfondet, Carlsberg Laboratorium*. E08.1924. No. 26. Letter from J. Schmidt to Board of Trustees, 24 September 1924.
CA. *Carlsbergfondet, Carlsberg Laboratorium*. E08.1924. No. 33. Letters from J. Schmidt to V. Henriques, 17 October 1924.

CA. *Carlsbergfondet, Carlsberg Laboratorium.* E08.1925. Letter from J. Schmidt to V. Henriques, 1 May 1925.
CA. *Carlsbergfondet, Carlsberg Laboratorium.* E08.1925. No. 14. Letter from J. Schmidt to Board of Trustees, 18 May 1925.
CA. *Carlsbergfondet, Carlsberg Laboratorium.* E08.1925. No. 28. Letter from J. Schmidt to Board of Trustees, 30 September 1925.
CA. Carlsbergfondet, Carlsberg Laboratorium. E08. 1926. Letter from J. Schmidt to V. Henriques, 10 November 1926.
CA. Carlsbergfondet, Carlsberg Laboratorium. E08. 1926. Letter from J. Schmidt to V. Henriques, 22 November 1926.
CA. Carlsbergfondet, Carlsberg Laboratorium. E08. 1926. Letter from V. Henriques to Board of Trustees, 26 November 1926.
CA. *Carlsbergfondet, Carlsberg Laboratorium.* E08.1926. no. 3. Letter from Ministry of Education, Nina Bang to the Carlsberg Foundation's Board of Directors, 8 January 1926.
CA. *Carlsbergfondet, Carlsberg Laboratorium.* E08.1926. no. 28. Letter from J. Schmidt to Board of Trustees, 30 September 1926.
CA. *Carlsbergfondet, Carlsberg Laboratorium.* E08.1926. Letter from J. Schmidt to V. Henriques, 5 November 1926.

Box E09

CA. Carlsbergfondet, Carlsberg Laboratorium. E09.1927. Letter from J. Schmidt to Board of Trustees, 30 September 1927.
CA. *Carlsbergfondet, Carlsberg Laboratorium.* E09.1927. Letter from J. Schmidt to V. Henriques, 29 December 1927.
CA. Carlsbergfondet, Carlsberg Laboratorium. E09.1928. Postcard from J. Schmidt to V. Henriques, 1 July 1928.
CA. Carlsbergfondet, Carlsberg Laboratorium. E09.1928. Letter from J. Schmidt to Board of Trustees, 30 August 1928.
CA. Carlsbergfondet, Carlsberg Laboratorium. E09.1928. Letter from J. Schmidt to Board of Trustees, 30 September 1928.
CA. Carlsbergfondet, Carlsberg Laboratorium. E09.1928. Letter from J. Schmidt to Board of Trustees, 17 October 1928.
CA. *Carlsbergfondet, Carlsberg Laboratorium.* E09. 1929. Letter from V. Ege to V. Henriques, Annual Report 1929.
CA. *Carlsbergfondet, Carlsberg Laboratorium.* E09. 1930. Letter from J. Schmidt to V. Henriques, 19 October 1930.

Box E10
CA. *Carlsbergfondet, Carlsberg Laboratorium.* E10.'1933'. No. 10. S. P. L. Sørensen to Board of Trustees, 9 March 1933.
CA. Carlsbergfondet, Carlsberg Laboratorium. E10.1933. no. 19. P. Boysen Jensen to Valdemar Henriques, 11 April 1933.
CA. Carlsbergfondet, Carlsberg Laboratorium. E10.1933. no. 20. Valdemar Henriques to P. Boysen Jensen, 12 April 1933.
CA. Carlsbergfondet, Carlsberg Laboratorium. E10.1933. no. 21. Øjvind Winge to P. Boysen Jensen, 15 April 1933.
CA. *Carlsbergfondet, Carlsberg Laboratorium.* E10.1933. no. 26. Ø. Winge to Board of Trustees, 15 May 1933.
CA. *Carlsbergfondet, Carlsberg Laboratorium.* E10.1933. no. 27. Circular Board of Trustees, 7 June 1933.
CA. *Carlsbergfondet, Carlsberg Laboratorium.* E10.1933. no. 31. Ø. Winge to Board of Trustees, 13 June 1933.
CA. *Carlsbergfondet, Carlsberg Laboratorium.* E10.1933. no. 30. Board of Trustees to Board of Directors, 15 June 1933.

Carlsberg Archives (CA), Carlsberg Laboratorium (CL), no. E17, 'Forhandlingsprotokol, Gærfysiologisk Underafdeling, Regnskabsbog, Bestyrelsen, 1876–1953.'
CA. Carlsberg Laboratorium, E17. Forhandlingsprotokol I, 18 September 1909, p. 248–255.
CA. Carlsberg Laboratorium, E17. Forhandlingsprotokol II, 108th meeting, 17 March 1933, p. 29–32.
CA. Carlsberg Laboratorium, E17. Forhandlingsprotokol II, 109th meeting, 24 March 1933, p. 33.

CA. *Bladudklip, Direktøren,* 1 June 1909–13 September 1910, *Dannebrog,* 28 August 1909.
CA. *Bladudklip, Direktøren,* 1 June 1909–13 September 1910, 'Carlsbergs ny Gærfysiolog', *Politiken,* 8 October 1909.
CA. *Bladudklip, Direktøren,* 1 June 1909–13 September 1910, 'Carlsberg-Laboratoriet', *Børsen,* 10 October 1910.

CA. *Carlsbergfondet, Carlsberg Laboratorium,* E31. Endelige budgetter og noter 1914–1972.

Carlsberg Foundation, Copenhagen

Carlsberg Foundation, Board of Directors (CFD):

CFD. *Carlsbergfondets Direktions Protokol*, vol. 1–3, 7 April 1909.
CFD. *Carlsbergfondets Direktions Protokol*, vol. 5, 25 February 1926.
CFD. *Carlsbergfondets Direktions Protokol*, vol. 6, 7 September 1927.
CFD. *Carlsbergfondets Direktions Protokol*, vol. 6, 12 October 1927.
CFD. *Carlsbergfondets Direktions Protokol*, vol. 6, 23 November 1927.
CFD. *Carlsbergfondets Direktions Protokol*, vol. 6, 13 February 1928.
CFD. *Carlsbergfondets Direktions Protokol*, vol. 6, 3 September 1929.
CFD. *Carlsbergfondets Direktions Protokol*, vol. 6, 1 October 1929.
CFD. *Carlsbergfondets Direktions Protokol*, vol. 6, 18 October 1929.

CFD. 1898–99. No. 47. Application from J. Schmidt, 24 December 1898.
CFD. 1906–07. no. 28.
CFD. 1908–09. no. 17. Application from J. Schmidt, 20 October 1908.
CFD. 1909–10. no. 227d. Application from J. Schmidt, 23 November 1909.
CFD. 1909–10. no. 227d. J. Schmidt to E. Warming, 28 December 1909.
CFD. 1909–10. no. 227d. M. Knudsen to Board of Directors, 31 December 1909.
CFD. 1909–10. no. 227d. H. Jungersen to Board of Directors, 3 January 1910.
CFD. 1909–10. no. 227d. J. Schmidt to E. Warming, 4 January 1910.
CFD. 1909–10. no. 227d. E. Warming to Board of Directors, 4 January 1910.
CFD. 1909–10. no. 227d. J. Schmidt to K. Erslev, 30 January 1910.
CFD. 1909–10. no. 227d. J. Schmidt: Cruise report concerning 'Den Danske oceanografiske Expedition til Middelhavet 1908–1909', p. 5.
CFD. 1912–13, no. 209. Application from J. Schmidt, 12 November 1912.
CFD. 1912–13, no. 209. J. Schmidt to E. Warming, 14 November 1912.
CFD. 1921–22. No. 73. Application from C. F. Drechsel, 3 December 1921.

CFD. *DANA*. BS3 1. Dana-Expeditionen: Gratialer, 8 July 1930.
CFD. *DANA*. BS3 1. Accounts for S/S 'DANA' Expeditionen, 1/5–1928–31/3–1931.
CFD. *DANA*. BS3 1. Regnskab for Bearbejdelsen af 'Dana-Expeditionen', 24 October 1933.

CFD. *DANA*. BS3 2. Jens With Skibsproviantering, Copenhagen, 12 June 1928.

CFD. *DANA*. BS3 2. Het Manado Veem Scheepsleveranties, 5 July 1929.

CFD. *DANA*. BS3 2. Paul Lanier Ship Chandler, Mahé Seychelles, 13 December 1929.

CFD. Vedr. DANA II, 1828–1930. 103/1928. Minutes, DANA Committee, 1st meeting, 9 May 1928.

CFD. Vedr. DANA II, 1828–1930. 103/1928. Expedition Plan, DANA Committee, 1st meeting, 9 May 1928.

CFD. Vedr. DANA II, 1828–1930. 103/1928. Letter from J. Schmidt to Carlsbergfondets Kvæstur, 5 August 1929.

CFD. Vedr. DANA II, 1828–1930. 103/1928. Letter from C. H. Ostenfeld to A. B. Drachmann, 11 August 1929.

CFD. Vedr. DANA II, 1828–1930. 103/1928. Letter from J. Schmidt to G. C. Amdrup, 20 August 1929.

CFD. Vedr. DANA II, 1828–1930. 103/1928. Letter from J. Schmidt to G. C. Amdrup, 26 August 1929.

CFD. Vedr. DANA II, 1828–1930. 103/1928. Letter from A. Drachmann to J. Schmidt, 9 October 1929.

CFD. Vedr. DANA II, 1828–1930. 103/1928. Letter from G. C. Amdrup to A. Drachmann, 10 October 1929.

CFD. Vedr. DANA II, 1828–1930. 103/1928. Telegram from J. Schmidt to G. C. Amdrup, 15 October 1929

CFD. Vedr. DANA II, 1828–1930. 103/1928. A. B. Drachmann to Komiteen for Carlsbergfondets Dana-Expedition, 8 November 1929.

CFD. Vedr. DANA II, 1828–1930. 103/1928. M. Knudsen to A. B. Drachmann, 9 November 1929.

CFD. Vedr. DANA II, 1828–1930. 103/1928. A. B. Drachmann to M. Knudsen, 11 November 1929

CFD. Vedr. DANA II, 1828–1930. 103/1928. M. Knudsen to A. B. Drachmann, 13 November 1929.

CFD. Vedr. DANA II, 1828–1930. 103/1928. Letter from Gotfred Johansen to Carlsbergfondets Direktion, 17 September 1929.

CFD. Vedr. DANA II, 1828–1930. 103/1928. J. Schmidt to A. B. Drachmann, 17 November 1929.

CFD. Vedr. DANA II, 1828–1930. 103/1928. Memorandum A. B. Drachmann, 20 November 1929.

CFD. Vedr. DANA II, 1828–1930. 103/1928. A. B. Drachmann to J. Schmidt, 3 December 1929.

CFD. Vedr. DANA II, 1828–1930. 103/1928. J. Schmidt to Komiteen for Dana-Expeditionen, 31 December 1929.

CFD. Vedr. DANA II, 1828–1930. 103/1928. A. B. Drachmann to J. Schmidt, 4 January 1930.

CFD. Vedr. DANA II, 1828–1930. 103/1928. J. Schmidt to Komiteen for Dana-Expeditionen, 25 January 1930.

CFD. *Vedr. DANA* II, *1828–1930*. 64/1930. Memorandom of meeting with J. Schmidt 8 July, by A. B. Drachmann, 8 September 1930.

Archives of the Danish Film Institute

The Carlsberg Foundation's oceanographical expedition round the World 1928–30. 20 minutes, 1931.

Rigsarkivet (Public Records Office), Copenhagen

RA. Folketælling 1890, Staden København, Udenbys Vester, Vesterbrogade 125 Matrikel 43. E.1. (p. 293).

RA. Folketælling 1890, Staden København, Udenbys Klædebo, Østerbrogade 58 Matrikel 136. E. (p. 114).

RA. Kirkebøger, Københavns Amt, Valby Sogn, 1892 V. – 1905 V. 17. p. 120. 17. September 1903.

Rigsarkivet(RA): Anton Fr Bruun: Sagligt ordnet korrespondance, rapporter mm, 'Forhandlingsprotokoller 1917–1973; Dana-ekspeditionerne, 1924–1930', no 6768

RA. *Anton Fr. Bruun. Danaekspeditionerne 1924–1930*. Dagbog fra 'Dana' Expeditionen 1924.

RA. *Anton Fr. Bruun. Danaekspeditionerne 1924–1930*. Diary from 1927

RA. *Anton Fr. Bruun. Danaekspeditionerne 1924–1930*. Privat Dagbog, 7–14 October 1929.

RA. *Anton Fr. Bruun. Danaekspeditionerne 1924–1930*. Avisklip, *Nationaltidende,* 14 June 1928.

RA. *Anton Fr. Bruun. Danaekspeditionerne 1924–1930*. Avisklip, *Nakskov Tidende,* 11 January 1929.

RA. *Anton Fr. Bruun. Danaekspeditionerne 1924–1930*. Avisklip, *Fyns Stifttidende*, 3 October 1929.
RA. *Anton Fr. Bruun. Danaekspeditionerne 1924–1930*. Avisklip, *Fyns Venstreblad*, 2 November 1929.
RA. *Anton Fr. Bruun. Danaekspeditionerne 1924–1930*. Avisklip, *Politiken*, 16 February 1930.
RA. *Anton Fr. Bruun. Danaekspeditionerne 1924–1930*. Avisklip, *Berlingske Aftenavis*, 30 March 1930.
RA. *Anton Fr. Bruun. Danaekspeditionerne 1924–1930*. Avisklip, *Il Mattino*, 27 May 1930.
RA. *Anton Fr. Bruun. Danaekspeditionerne 1924–1930*. Avisklip, *Berlingske Tidende*, 26 June 1930.

Rigsarkivet(RA), Carlsbergfondets DANA-ekspeditioner 1917–1973

RA. *Carlsbergfondets DANA-ekspeditioner. 1917–1973*. 1. Gæstebog, 1922–1940.
RA. *Carlsbergfondets DANA-ekspeditioner. 1917–1973*. 1. Forhandlingsprotokoller. Regnskab 'Dana'-Ekspeditionen.
RA. *Carlsbergfondets DANA-ekspeditioner. 1917–1973*. 1. Forhandlingsprotokoller. 5th meeting of the Board of the 'Dana'-Ekspeditionen, 20 November 1920.
RA. *Carlsbergfondets DANA-ekspeditioner. 1917–1973*. 2–4. J. Schmidt to C. Schmiegelow, 11 August 1916.
RA. *Carlsbergfondets DANA-ekspeditioner. 1917–1973*. 2–4. J. Schmidt to C. Schmiegelow, 10 December 1918.
RA. *Carlsbergfondets DANA-ekspeditioner. 1917–1973*. 5. Photographs.
RA. *Carlsbergfondets DANA-ekspeditioner. 1917–1973*. 7a-7c. Photographs.

Rigsarkivet(RA), Dansk Biologisk Station

RA: *Dansk Biologisk Station, Johannes Schmidts embedsarkiv*: Korrespondance om åleforskning (1908–1933) E28: 1908–1913: B. Sæmundsson to C. F. Drechsel, 10 January 1903.
RA: *Dansk Biologisk Station, Johannes Schmidts embedsarkiv*: Korrespondance om åleforskning (1908–1933) E28: 1908–1913: H. Jakobsen to J. Schmidt, 1 October 1909.
RA: *Dansk Biologisk Station, Johannes Schmidts embedsarkiv*: Korrespondance om åleforskning (1908–1933) E28: 1908–1913: Schmiegelow to J. Schmidt, 7 September 1909 & J. Schmidt to Schmiegelow, 15 November 1909.

RA: *Dansk Biologisk Station, Johannes Schmidts embedsarkiv*: Korrespondance om åleforskning (1908–1933) E28: 1908–1913: G. B. Grassi to J. Schmidt, 3 November 1909 & 10 January 1910.

RA: *Dansk Biologisk Station, Johannes Schmidts embedsarkiv*: Korrespondance om åleforskning (1908–1933) E28: 1908–1913: Agreement between J. Schmidt and Kommissionen for Havundersøgelser.

RA: *Dansk Biologisk Station, Johannes Schmidts embedsarkiv*: Korrespondance om åleforskning (1908–1933) E28: 1908–1913: J. Hjort to J. Schmidt, 31 January 1910.

RA: *Dansk Biologisk Station, Johannes Schmidts embedsarkiv*: Korrespondance om åleforskning (1908–1933) E28: 1908–1913: E. J. Allen to J. Schmidt, 21 February 1910.

RA: *Dansk Biologisk Station, Johannes Schmidts embedsarkiv*: Korrespondance om åleforskning (1908–1933) E28: 1908–1913: J. Schmidt to E. J. Allen, 24 February 1910.

RA: *Dansk Biologisk Station, Johannes Schmidts embedsarkiv*: Korrespondance om åleforskning (1908–1933) E28: 1908–1913: 5 November 1910.

RA: *Dansk Biologisk Station, Johannes Schmidts embedsarkiv*: Korrespondance om åleforskning (1908–1933) E28: 1908–1913. Letters from G. Hansen to J. Schmidt, 28 March and 15 April 1913.

RA: *Dansk Biologisk Station, Johannes Schmidts embedsarkiv*: Korrespondance om åleforskning (1908–1933) E28: 1908–1913. 'Instruks for Undersøgelsern i Atlanterhavet' from J. Schmidt to G. Hansen & P. Jespersen, 23 June 1913.

RA. *Dansk Biologisk Station, Johannes Schmidts embedsarkiv*: Fiskeriundersøgelser og skildpaddemærkning i Dansk Vestindien (1914–1916) E30: Letter from J. Schmidt to Gouvernementet for de dansk-vestindiske Øer, 16 February 1914.

RA. *Dansk Biologisk Station, Johannes Schmidts embedsarkiv*: Fiskeriundersøgelser og skildpaddemærkning i Dansk Vestindien (1914–1916) E30: Memorandum from Th. Mortensen to Gouvernementet for de dansk-vestindiske Øer, 4 April 1906.

RA. *Dansk Biologisk Station, Johannes Schmidts embedsarkiv*: Fiskeriundersøgelser og skildpaddemærkning i Dansk Vestindien (1914–1916) E30: 'Beretning om Fiskeriet i Dansk Vestindien (Kladde)'.

Rigsarkivet (RA), De Danske Atlanterhavsøer

RA.10154. *De Danske Atlanterhavsøer*. 1902–1917 Forhandlingsprotokol for foreningen, bind 1–4.

RA.10154. *De Danske Atlanterhavsøer*. 1904–1916. Forhandlingsprotokoller for sektionen i Dansk Vestindien, bind 5.
RA.10154. *De Danske Atlanterhavsøer.* 1902–1919. Medlemsprotokoller, Bind 9.
RA.10154. *De Danske Atlanterhavsøer.* 1903–1919. Kopibøger med registre, bind 18.

Rigsarkivet(RA), Kommissionen for Havundersøgelser

RA. *Komm. Danmarks Fiskeri- og Havundersøgelser*: Forhandlingsprotokoller (1900–1960) N1.: 1st meeting, 16 January 1900.
RA. *Komm. Danmarks Fiskeri- og Havundersøgelser*: Forhandlingsprotokoller (1900–1960) N1.: 4th meeting, 5 December 1900.
RA. *Komm. Danmarks Fiskeri- og Havundersøgelser*: Forhandlingsprotokoller (1900–1960) N1.: 10th meeting, 5 July 1902.
RA. *Komm. Danmarks Fiskeri- og Havundersøgelser:* Forhandlingsprotokoller (1900–1960) N1.: 19th meeting, 18 February 1903.
RA. *Komm. Danmarks Fiskeri- og Havundersøgelser:* Forhandlingsprotokoller (1900–1960) N1.: 20th meeting 18 April 1903.
RA. *Komm. Danmarks Fiskeri- og Havundersøgelser:* Forhandlingsprotokoller (1900–1960) N1.: 21st meeting 24 April 1903.
RA. *Komm. Danmarks Fiskeri- og Havundersøgelser:* Forhandlingsprotokoller (1900–1960) N1.: 24th meeting 24 November 1903.
RA. *Komm. Danmarks Fiskeri- og Havundersøgelser:* Forhandlingsprotokoller (1900–1960) N1.: 27th meeting, 22 April 1904.
RA. *Komm. Danmarks Fiskeri- og Havundersøgelser:* Forhandlingsprotokoller (1900–1960) N1.: 29th meeting, 14 February 1905.
RA. Komm. Danmarks Fiskeri- og Havundersøgelser: Forhandlingsprotokoller (1900–1960) N1.: 30th meeting, 17 November 1905.
RA. Komm. Danmarks Fiskeri- og Havundersøgelser: Forhandlingsprotokoller (1900–1960) N1.: 31st meeting, 4 January 1906.
RA. Komm. Danmarks Fiskeri- og Havundersøgelser: Forhandlingsprotokoller (1900–1960) N1.: 32nd meeting, 26 January 1906.
RA. Komm. Danmarks Fiskeri- og Havundersøgelser: Forhandlingsprotokoller (1900–1960) N1.: 33rd meeting, 18 April 1906.

RA. Komm. Danmarks Fiskeri- og Havundersøgelser: Forhandlingsprotokoller (1900–1960) N2: 38th meeting, 11 November 1907.
RA. Komm. Danmarks Fiskeri- og Havundersøgelser: Forhandlingsprotokoller (1900–1960) N2: 39th meeting, 21 January 1908.

RA. Komm. Danmarks Fiskeri- og Havundersøgelser: Forhandlingsprotokoller (1900–1960) N2: 40th meeting, 5 February 1908.
RA. Komm. Danmarks Fiskeri- og Havundersøgelser: Forhandlingsprotokoller (1900–1960) N2: 41st meeting, 22 April 1908.
RA. Komm. Danmarks Fiskeri- og Havundersøgelser: Forhandlingsprotokoller (1900–1960) N2: 42nd meeting, 23 July 1908.
RA. *Komm. Danmarks Fiskeri- og Havundersøgelser:* Forhandlingsprotokoller (1900–1960) N2: 43rd meeting, 16 September 1908.
RA. *Komm. Danmarks Fiskeri- og Havundersøgelser:* Forhandlingsprotokoller (1900–1960) N2: 44th meeting, 5 January 1909.
RA. *Komm. Danmarks Fiskeri- og Havundersøgelser:* Forhandlingsprotokoller (1900–1960) N2: 45th meeting, 27 February 1909.
RA. *Komm. Danmarks Fiskeri- og Havundersøgelser:* Forhandlingsprotokoller (1900–1960) N2: 46th meeting, 23 March 1909.
RA. *Komm. Danmarks Fiskeri- og Havundersøgelser:* Forhandlingsprotokoller (1900–1960) N2: 49th – 60th meetings 1909.
RA. *Komm. Danmarks Fiskeri- og Havundersøgelser:* Forhandlingsprotokoller (1900–1960) N2: 61st meeting, 15 February 1910.

RA. *Komm. Danmarks Fiskeri- og Havundersøgelser*: Forhandlingsprotokoller (1900–1960) N3: 69th meeting, 6 February 1911.
RA. *Komm. Danmarks Fiskeri- og Havundersøgelser*: Forhandlingsprotokoller (1900–1960) N3: 72nd meeting, 4 May 1911.
RA. *Komm. Danmarks Fiskeri- og Havundersøgelser*: Forhandlingsprotokoller (1900–1960) N3: 77th meeting, 29 January 1912.
RA. *Komm. Danmarks Fiskeri- og Havundersøgelser*: Forhandlingsprotokoller (1900–1960) N3: 85th meeting, 26 November 1912.
RA. *Komm. Danmarks Fiskeri- og Havundersøgelser*: Forhandlingsprotokoller (1900–1960) N3: 86th meeting, 5 December 1912.

RA. *Komm. Danmarks Fiskeri- og Havundersøgelser*: Forhandlingsprotokoller (1900–1960) N4: 96th meeting, 20 October 1913.
RA. *Komm. Danmarks Fiskeri- og Havundersøgelser*: Forhandlingsprotokoller (1900–1960) N4: 97th meeting, 13 November 1913.
RA. *Komm. Danmarks Fiskeri- og Havundersøgelser*: Forhandlingsprotokoller (1900–1960) N4: 98th meeting, 20 November 1913.
RA. *Komm. Danmarks Fiskeri- og Havundersøgelser*: Forhandlingsprotokoller (1900–1960) N4: 99th meeting, 27 December 1913.

RA. *Komm. Danmarks Fiskeri- og Havundersøgelser*: Forhandlingsprotokoller (1900–1960) N4: 102nd meeting, 17 March 1914.
RA. *Komm. Danmarks Fiskeri- og Havundersøgelser:* Forhandlingsprotokoller (1900–1960) N4: 103rd meeting, 26 March 1914.
RA. *Komm. Danmarks Fiskeri- og Havundersøgelser*: Forhandlingsprotokoller (1900–1960) N4: 105th meeting, 2 June 1914.
RA. *Komm. Danmarks Fiskeri- og Havundersøgelser*: Forhandlingsprotokoller (1900–1960) N4: 106th meeting, 21 September 1914.
RA. *Komm. Danmarks Fiskeri- og Havundersøgelser*: Forhandlingsprotokoller (1900–1960) N4: 108th meeting, 23 November 1914.
RA. Komm. Danmarks Fiskeri- og Havundersøgelser: Forhandlingsprotokoller (1900–1960) N4: 109th meeting, 18 January 1915.
RA. *Komm. Danmarks Fiskeri- og Havundersøgelser:* Forhandlingsprotokoller (1900–1960) N4: 113th meeting, 7 January 1916.
RA. *Komm. Danmarks Fiskeri- og Havundersøgelser*: Forhandlingsprotokoller (1900–1960) N4: 115th meeting, 15 April 1916.
RA. *Komm. Danmarks Fiskeri- og Havundersøgelser*: Forhandlingsprotokoller (1900–1960) N4: 117th meeting, 7 November 1916.

RA. Komm. Danmarks Fiskeri- og Havundersøgelser: A 22 Dana-ekspeditionerne 1917–1924, J. Schmidt to C. F. Drechsel, 21 August 1916.
RA. Komm. Danmarks Fiskeri- og Havundersøgelser: A 22 Dana-ekspeditionerne 1917–1924, J. Schmidt to C. F. Drechsel, 31 August 1916.
RA. Komm. Danmarks Fiskeri- og Havundersøgelser: A 22 Dana-ekspeditionerne 1917–1924, J. Schmidt to C. F. Drechsel, 3 September 1916.
RA. Komm. Danmarks Fiskeri- og Havundersøgelser: A 22 Dana-ekspeditionerne 1917–1924, Den Danske Landmandsbank to C. F. Drechsel, 30 December 1916.
RA. Komm. Danmarks Fiskeri- og Havundersøgelser: A 22 Dana-ekspeditionerne 1917–1924, J. Schmidt to C. F. Drechsel, 30 January 1917.
RA. Komm. Danmarks Fiskeri- og Havundersøgelser: A 22 Dana-ekspeditionerne 1917–1924, C. F. Drechsel to J. Schmidt, 17 May 1917.
RA. Komm. Danmarks Fiskeri- og Havundersøgelser: A 22 Dana-ekspeditionerne 1917–1924, J. Schmidt to C. F. Drechsel, 27 July 1917.
RA. Komm. Danmarks Fiskeri- og Havundersøgelser: A 22 Dana-ekspeditionerne 1917–1924, J. Schmidt to C. F. Drechsel, 8 December 1918.
RA. Komm. Danmarks Fiskeri- og Havundersøgelser: A 22 Dana-ekspeditionerne 1917–1924, C. F. Drechsel to H. N. Andersen, 10 December 1918.

RA. Komm. Danmarks Fiskeri- og Havundersøgelser: A 22 Dana-ekspeditionerne 1917–1924, H. N. Andersen to C. F. Drechsel, 14 December 1918.

RA. Komm. Danmarks Fiskeri- og Havundersøgelser: A 22 Dana-ekspeditionerne 1917–1924, C. Schmiegelow to J. Schmidt, 19 December 1918.

RA. Komm. Danmarks Fiskeri- og Havundersøgelser: A 22 Dana-ekspeditionerne 1917–1924, C. F. Drechsel to C. Schmiegelow, 21 January 1919.

RA. Komm. Danmarks Fiskeri- og Havundersøgelser: A 22 Dana-ekspeditionerne 1917–1924, C. Schmiegelow to C. F. Drechsel, 7 February 1919.

RA. Komm. Danmarks Fiskeri- og Havundersøgelser: A 22 Dana-ekspeditionerne 1917–1924, Kjøbenhavns Handelsbank to C. F. Drechsel, 30 May 1919.

RA. Komm. Danmarks Fiskeri- og Havundersøgelser: A 22 Dana-ekspeditionerne 1917–1924, J. Schmidt to C. F. Drechsel, 31 May 1919.

RA. Komm. Danmarks Fiskeri- og Havundersøgelser: A 22 Dana-ekspeditionerne 1917–1924, J. Schmidt to C. F. Drechsel, 1 January 1920.

RA. Komm. Danmarks Fiskeri- og Havundersøgelser: A 22 Dana-ekspeditionerne 1917–1924, C. F. Drechsel to J. Schmidt, 7 January 1920 & 9 January 1920.

RA. Komm. Danmarks Fiskeri- og Havundersøgelser: A 22 Dana-ekspeditionerne 1917–1924, J. Schmidt to C. F. Drechsel, 15 May 1920.

RA. Komm. Danmarks Fiskeri- og Havundersøgelser: A 22 Dana-ekspeditionerne 1917–1924, J. Schmidt to C. F. Drechsel, 22 May 1920.

RA. Komm. Danmarks Fiskeri- og Havundersøgelser: A 22 Dana-ekspeditionerne 1917–1924, C. F. Drechsel to J. Schmidt, 4 June 1920.

RA. Komm. Danmarks Fiskeri- og Havundersøgelser: A 22 Dana-ekspeditionerne 1917–1924, J. Schmidt to C. F. Drechsel, 9 July 1920.

RA. Komm. Danmarks Fiskeri- og Havundersøgelser: A 22 Dana-ekspeditionerne 1917–1924, J. Schmidt to C. F. Drechsel, 6 November 1920.

RA. Komm. Danmarks Fiskeri- og Havundersøgelser: A 22 Dana-ekspeditionerne 1917–1924, Nørgaard to C. F. Drechsel, 15 December 1920.

RA. Komm. Danmarks Fiskeri- og Havundersøgelser: A 22 Dana-ekspeditionerne 1917–1924, Revideret Koncept.

RA. Komm. Danmarks Fiskeri- og Havundersøgelser: A 22 Dana-ekspeditionerne 1917–1924, Instructions for Havundersøgelses Ekspeditionen 'DANA', 24 August 1921.

RA. Komm. Danmarks Fiskeri- og Havundersøgelser: A 23 Dana-ekspeditionerne 1917–1924, C. F. Wandel to C. F. Drechsel, 4 October 1916.

RA. Komm. Danmarks Fiskeri- og Havundersøgelser: A 23 Dana-ekspeditionerne 1917–1924, C. F. Drechsel to C. Schmiegelow, 3 November 1916.

RA. Komm. Danmarks Fiskeri- og Havundersøgelser: A 23 Dana-ekspeditionerne 1917–1924, C. F. Drechsel to L. Zeuthen, 22 January 1917.

RA. Komm. Danmarks Fiskeri- og Havundersøgelser: A 23 Dana-ekspeditionerne 1917–1924, L. Zeuthen to C. F. Drechsel, 7 February 1917.

RA. Komm. Danmarks Fiskeri- og Havundersøgelser: A 23 Dana-ekspeditionerne 1917–1924, C. Schmiegelow to C. F. Drechsel, 9 February 1917.

RA. Komm. Danmarks Fiskeri- og Havundersøgelser: A 23 Dana-ekspeditionerne 1917–1924, C. F. Drechsel to C. Schmiegelow, 1 March 1917.

RA. *Komm. Danmarks Fiskeri- og Havundersøgelser:* A 23 Dana-ekspeditionerne 1917–1924, Store Nordiske to C. F. Drechsel, 9 March 1917.

RA. *Komm. Danmarks Fiskeri- og Havundersøgelser*: A 23 Dana-ekspeditionerne 1917–1924, E. Warming to C. F. Drechsel, 28 July 1920.

Rigsarkivet(RA): Paulsen, Ove botaniker, og hustru: Dagbog mm (1874–1947)

Rigsarkivet(RA): *Paulsen, Ove botaniker, og hustru: Dagbog m.m. (1874–1947)*. No. 5. Diary 18 November 1897.

Rigsarkivet(RA): *Paulsen, Ove botaniker, og hustru: Dagbog m.m. (1874–1947)*. No. 5. Diary 8 December 1900.

Rigsarkivet(RA), Schmidt, Ernst Johannes, biolog

RA. *Schmidt, Ernst Johannes, biolog*: Materiale vedr. ekspeditioner og rejser (1876–1933) 1: m.m.

RA. *Schmidt, Ernst Johannes, biolog*: Materiale vedr. ekspeditioner og rejser (1876–1933) 1: Letter from J. Schmidt to the Board of Directors of the Carlsberg Foundation, 6 April 1909.

RA. *Schmidt, Ernst Johannes, biolog*: Materiale vedr. ekspeditioner og rejser (1876–1933) 1: Blade fra min Dagbog fra en Rejse til Tahiti. Marts – April 1926.

RA. *Schmidt, Ernst Johannes, biolog*: Materiale vedr. ekspeditioner og rejser (1876–1933) 2: m.m.

Rigsarkivet(RA): Stephensen, Knud Hensch: 'Dagbog fra Dana-ekspeditionen til Det karibiske Hav 1920–1922'

RA. *Stephensen, Knud Hensch*: 'Dagbog fra Dana-ekspeditionen til Det karibiske Hav 1920–1922'. 3 December 1921.

Royal Library, Copenhagen

Det Kongelige Bibliotek, Centre for Manuscripts and Book Studies(KB)

Det kongelige Bibliotek, Centre for Manuscripts and Books, (CMB). *Johannes Schmidt*: 'Papirer', Ut. 374. (hereafter abbreviated 'KB. Johannes Schmidt 'Papirer'. Ut. 374').

Box I.2

KB. Johannes Schmidt 'Papirer'. Ut.374.I.2. J. Schmidt to C. H. Ostenfeld, 20 March 1922.

KB. Johannes Schmidt 'Papirer'. Ut.374.I.2. J. Schmidt to H. Boesgård, 26 May 1922

KB. Johannes Schmidt 'Papirer'. Ut.374.I.2. H. Boesgård to J. Schmidt, 12 June 1922.

Box I.3

KB. Johannes Schmidt 'Papirer'. Ut.374.I.3. C. F. Drechsel to J. Schmidt, 1 September 1921.

KB. Johannes Schmidt 'Papirer'. Ut.374.I.3. J. Schmidt to C. F. Drechsel, 12 September 1921.

KB. Johannes Schmidt 'Papirer'. Ut.374.I.3. C. F. Drechsel to J. Schmidt, 29 September 1921.

KB. Johannes Schmidt 'Papirer'. Ut.374.I.3. C. F. Drechsel to J. Schmidt, 28 October 1921.

KB. Johannes Schmidt 'Papirer'. Ut.374.I.3. J. C. Wiley to C. F. Drechsel, 10 November 1921.

KB. Johannes Schmidt 'Papirer'. Ut.374.I.3. Minutes 160th meeting of Kommissionen for Havundersøgelser, 18 November 1921.

KB. Johannes Schmidt 'Papirer'. Ut.374.I.3. C. F. Drechsel to J. Schmidt, 17 December 1921.

KB. Johannes Schmidt 'Papirer'. Ut.374.I.3. J. Schmidt to C. F. Drechsel, 14 December 1921.

KB. Johannes Schmidt 'Papirer'. Ut.374.I.3. J. Schmidt to C. F. Drechsel, 19 December 1921.

KB. Johannes Schmidt 'Papirer'. Ut.374.I.3. C. F. Drechsel to J. Schmidt, 5 January 1922.

KB. Johannes Schmidt 'Papirer'. Ut.374.I.3. C. F. Drechsel to J. Schmidt, 20 February 1922.

KB. Johannes Schmidt 'Papirer'. Ut.374.1.3. J. Schmidt to C. F. Drechsel, 2 January 1922.
KB. Johannes Schmidt 'Papirer'. Ut.374.1.3. J. Schmidt to C. F. Drechsel, 20 January 1922.
KB. Johannes Schmidt 'Papirer'. Ut.374.1.3. J. Schmidt to C. F. Drechsel, 5 February 1922.
KB. Johannes Schmidt 'Papirer'. Ut.374.1.3. J. Schmidt to C. F. Drechsel, 16 June 1922.
KB. Johannes Schmidt 'Papirer'. Ut.374.1.3. '1921–1924.' Minutes 11th meeting of the Dana Expedition Committee, 26 August 1922.

Box 1.4

KB. Johannes Schmidt 'Papirer'. Ut.374.1.4. J. Schmidt to Kammerherre Bernhoft, 19 July 1922.
KB. Johannes Schmidt 'Papirer'. Ut.374.1.4. '1922–1925'. K. Erslev to J. Schmidt, 27 July 1922.
KB. Johannes Schmidt 'Papirer'. Ut.374.1.4. '1922–1925'. J. Schmidt to Board of Trustees, 27 July 1922.
KB. Johannes Schmidt 'Papirer'. Ut.374.1.4. J. Schmidt to C. Rabot, 2 November 1922.
KB. Johannes Schmidt 'Papirer'. Ut.374.1.4. J. Schmidt to Captain Laub, 15 November 1922.
KB. Johannes Schmidt 'Papirer'. Ut.374.1.4. J. Schmidt to C. Rabot, 22 November 1922.
KB. Johannes Schmidt 'Papirer'. Ut.374.1.4. J. Schmidt to C. Rabot, 23 November 1922.
KB. Johannes Schmidt 'Papirer'. Ut.374.1.4. J. Schmidt to C. Rabot, 5 December 1922.
KB. Johannes Schmidt 'Papirer'. Ut.374.1.4. J. Schmidt to S. Madsen, 6 December 1922
KB. Johannes Schmidt 'Papirer'. Ut.374.1.4. S. Madsen to J. Schmidt, 8 December 1922.
KB. Johannes Schmidt 'Papirer'. Ut.374.1.4. C. Rabot to J. Schmidt, 9 December 1922.
KB. Johannes Schmidt 'Papirer'. Ut.374.1.4. Various letters from 1922–1923.
KB. Johannes Schmidt 'Papirer'. Ut.374.1.4. University of Oxford to J. Schmidt, November 1923.

Box 1.5

KB. Johannes Schmidt 'Papirer'. Ut.374.1.5. B. Helland-Hansen to J. Schmidt, 20 January 1924.

KB. Johannes Schmidt 'Papirer'. Ut.374.1.5. C. F. Drechsel to J. Hjort, 22 January 1924.

KB. Johannes Schmidt 'Papirer'. Ut.374.1.5. J. Schmidt to B. Helland-Hansen, 23 January 1924.

KB. Johannes Schmidt 'Papirer'. Ut.374.1.5. J. Schmidt to B. Helland-Hansen, 25 January 1924.

KB. Johannes Schmidt 'Papirer'. Ut.374.1.5. J. Hjort to C. F. Drechsel, 29 January 1924.

KB. Johannes Schmidt 'Papirer'. Ut.374.1.5. B. Helland-Hansen to J. Schmidt, 30 January 1924.

KB. Johannes Schmidt 'Papirer'. Ut.374.1.5. J. Schmidt to B. Helland-Hansen, 1 February 1924.

KB. Johannes Schmidt 'Papirer'. Ut.374.1.5. J. Schmidt to C. F. Drechsel, 26 March 1924.

KB. Johannes Schmidt 'Papirer'. Ut.374.1.5. J. Schmidt to B. Helland-Hansen, 4 April 1924.

KB. Johannes Schmidt 'Papirer'. Ut.374.1.5. J. Schmidt to B. Helland-Hansen, 10 April 1924.

KB. Johannes Schmidt 'Papirer'. Ut.374.1.5. J. Schmidt to B. Helland-Hansen, 29 April 1924.

KB. Johannes Schmidt 'Papirer'. Ut.374.1.5. J. Schmidt to B. Helland-Hansen, 1 May 1924.

KB. Johannes Schmidt 'Papirer'. Ut.374.1.5. Axel Topsøe-Jensen to C. F. Drechsel, 22 May 1924.

KB. Johannes Schmidt 'Papirer'. Ut.374.1.5. C. F. Drechsel to J. Schmidt, 31 May 1924.

KB. Johannes Schmidt 'Papirer'. Ut.374.1.5. J. Schmidt to C. F. Drechsel, 13 June 1924.

KB. Johannes Schmidt 'Papirer'. Ut.374.1.5. C. F. Drechsel to J. Schmidt, 13 June 1924.

KB. Johannes Schmidt 'Papirer'. Ut.374.1.5. J. Schmidt to C. F. Drechsel, 18 June 1924.

KB. Johannes Schmidt 'Papirer'. Ut.374.1.5. P. Jespersen to J. Schmidt, 21 June 1924.

KB. Johannes Schmidt 'Papirer'. Ut.374.1.5. A. V. Taaning to J. Schmidt, 2 July 1924.

KB. Johannes Schmidt 'Papirer'. Ut.374.1.5. C. F. Drechsel to J. Schmidt, 7 July 1924.
KB. Johannes Schmidt 'Papirer'. Ut.374.1.5. J. Schmidt to Count C. Moltke, 14 July 1924.
KB. Johannes Schmidt 'Papirer'. Ut.374.1.5. News clip from Berlingske Tidende, 26 July 1924.
KB. Johannes Schmidt 'Papirer'. Ut.374.1.5. J. Schmidt to C. F. Drechsel, 2 August 1924.
KB. Johannes Schmidt 'Papirer'. Ut.374.1.5. J. Schmidt to C. F. Drechsel, 14 August 1924.
KB. Johannes Schmidt 'Papirer'. Ut.374.1.5. J. Schmidt to B. Helland-Hansen, 3 September 1924.
KB. Johannes Schmidt 'Papirer'. Ut.374.1.5. L. W. Jones to J. Schmidt, 2 February 1933.

Box 1.6

KB. Johannes Schmidt 'Papirer'. Ut.374.1.6. 'Korrespondance 1925-(32)'. P. C. Paulsen to J. Schmidt, 30 May 1923.

Box 1.7.

KB. Johannes Schmidt 'Papirer'. Ut.374.1.7. 'Korrespondance 1927–1929'. C. H. Ostenfeld to J. Schmidt, 5 August 1927.
KB. Johannes Schmidt 'Papirer'. Ut.374.1.7. 'Korrespondance 1927–1929'. C. H. Ostenfeld to J. Schmidt, 1 September 1927.
KB. Johannes Schmidt 'Papirer'. Ut.374.1.7. 'Korrespondance 1927–1929'. G. Schott to J. Schmidt, 6 December 1927.
KB. Johannes Schmidt 'Papirer'. Ut.374.1.7. 'Korrespondance 1927–1929'. C. H. Ostenfeld to C. Schmiegelow, 20 February 1928.
KB. Johannes Schmidt 'Papirer'. Ut.374.1.7. 'Korrespondance 1927–1929'. Department of Foreign Affairs to J. Schmidt, 16 April 1928.
KB. Johannes Schmidt 'Papirer'. Ut.374.1.7. 'Korrespondance 1927–1929'. J. Schmidt to Rechnitzer, 12 May 1928.
KB. Johannes Schmidt 'Papirer'. Ut.374.1.7. 'Korrespondance 1927–1929'. Rechnitzer to J. Schmidt, 12 May 1928.
KB. Johannes Schmidt 'Papirer'. Ut.374.1.7. 'Korrespondance 1927–1929'. J. Schmidt to Press Office of the Department of Foreign Affairs, 30 May 1928.
KB. Johannes Schmidt 'Papirer'. Ut.374.1.7. 'Korrespondance 1927–1929'. P. Jespersen to J. Schmidt, 18 November 1928.

KB. Johannes Schmidt 'Papirer'. Ut.374.I.7. 'Korrespondance 1927–1929'. C. H. Ostenfeld to J. Schmidt, 25 November 1928.
KB. Johannes Schmidt 'Papirer'. Ut.374.I.7. 'Korrespondance 1927–1929'. E. Reventlow to J. Schmidt, 9 April 1929.
KB. Johannes Schmidt 'Papirer'. Ut.374.I.7. 'Korrespondance 1927–1929'., J. Schmidt to E. Reventlow, 24 June 1929.
KB. Johannes Schmidt 'Papirer'. Ut.374.I.7. 'Korrespondance 1927–1929'. J. Schmidt to Rechnitzer, 4 October 1929.
KB. Johannes Schmidt 'Papirer'. Ut.374.I.7. 'Korrespondance 1927–1929'. Danish Research Exploration Party. Tentative Programme, February 1929.

Box I.8

KB. Johannes Schmidt 'Papirer'. Ut.374.I.8. 'Korrespondance 1927–1929'. Minutes, 4th meeting of Komiteen for Dana-Expeditionen, 22 October 1929.
KB. Johannes Schmidt 'Papirer'. Ut.374.I.8. 'Korrespondance 1927–1929'. A. B. Drachmann to J. Schmidt, 25 November 1929.
KB. Johannes Schmidt 'Papirer'. Ut.374.I.8. 'Korrespondance 1927–1929'. C. H. Ostenfeld to J. Schmidt, 30 April 1930.
KB. Johannes Schmidt 'Papirer'. Ut.374.I.8. 'Korrespondance 1927–1929'. J. Schmidt to Rechnitzer, 6 May 1930.
KB. Johannes Schmidt 'Papirer'. Ut.374.I.8. 'Korrespondance 1927–1929'. J. Schmidt to F. V. Mortensen, 8 May 1930.
KB. Johannes Schmidt 'Papirer'. Ut.374.I.8. 'Korrespondance 1927–1929'. C. H. Ostenfeld to J. Schmidt, 14 May 1930.

Box II.1

KB. Johannes Schmidt 'Papirer'. Ut.374.II.1. D. White to C. Brun, 28 March 1930.

Box II.4

KB. Johannes Schmidt 'Papirer'. Ut.374.II.4. A. Bøving to J. Schmidt, 7 March 1925
KB. Johannes Schmidt 'Papirer'. Ut.374.II.4. Wickliffe Rose to S. Sørensen, 14 April 1925.
KB. Johannes Schmidt 'Papirer'. Ut.374.II.4. C. Brun to Wickliffe Rose, 13 January 1926.

KB. Johannes Schmidt 'Papirer'. Ut.374.II.4. Wickliffe Rose to C. Brun, 22 January 1926.

KB. Johannes Schmidt 'Papirer'. Ut.374.II.4. C. Brun to J. Schmidt, 8 May 1926.

KB. Johannes Schmidt 'Papirer'. Ut.374.II.4. C. Brun to J. Schmidt, 22 July 1926.

KB. Johannes Schmidt 'Papirer'. Ut.374.II.4. Wickliffe Rose to C. Brun, 8 January 1927.

KB. Johannes Schmidt 'Papirer'. Ut.374.II.4. J. Schmidt to C. Brun, 28 February 1927.

KB. Johannes Schmidt 'Papirer'. Ut.374.II.4. Austin H. Clark to C. Brun, 16 July 1927.

KB. Johannes Schmidt 'Papirer'. Ut.374.II.4. C. Brun to J. Schmidt, 21 July 1927.

KB. Johannes Schmidt 'Papirer'. Ut.374.II.4. J. Schmidt to C. Brun, 16 October 1927.

Box III

KB. Johannes Schmidt 'Papirer'. Ut.374.III. 'Forskellige Breve fra Johannes Schmidt til Bjarni Sæmundsson'. Letter from J. Schmidt to B. Sæmundsson, 10 April 1903.

KB. Johannes Schmidt 'Papirer'. Ut.374.III. 'Forskellige Breve fra Johannes Schmidt til Bjarni Sæmundsson'. Letter from J. Schmidt to B. Sæmundsson, 16 October 1903.

KB. Johannes Schmidt 'Papirer'. Ut.374.III. 'Forskellige Breve fra Johannes Schmidt til Bjarni Sæmundsson'. Letter from J. Schmidt to B. Sæmundsson, no date, but from context it is surely spring of 1906.

Box IV.3

KB. Johannes Schmidt 'Papirer'. Ut.374.IV.3. 'Avisklip', *Scoial Demokraten*, 1 July 1930.

KB. Johannes Schmidt 'Papirer'. Ut.374.IV.3. 'Avisklip', *Dansk Fiskeritidende*, 9 July 1930.

Box V.1

KB. Johannes Schmidt 'Papirer'. Ut.374.V.1. Privat, 'April 1908'. Copy of letter from C. F. Drechsel, M. Knudsen and C. H. OStenfeld to the Ministry of Agriculture, 23 September 1908.

KB. Johannes Schmidt 'Papirer'. Ut.374.v.1. Privat, 'April 1908'. Copy of Letter from C. G. J. Petersen to Ministry of Agriculture, received by M. Knudsen 25 July 1908.

KB. Johannes Schmidt 'Papirer'. Ut.374.v.1. Privat, 'April 1908'. Copy of Letter from J. N. Nielsen to M. Knudsen 28 April 1908.

KB. Johannes Schmidt 'Papirer'. Ut.374.v.1. Privat, 'April 1908'. Letters from C. F. Drechsel to J. Schmidt, 19 June 1916.

KB. Johannes Schmidt 'Papirer'. Ut.374.v.1. Privat, 'April 1908'. Draft memorandum by C. F. Drechsel, June 1916.

KB. Johannes Schmidt 'Papirer'. Ut.374.v.1. Privat, 'April 1908'. Report by J. Schmidt, undated.

KB. Johannes Schmidt 'Papirer'. Ut.374.v.1. Privat, 'April 1908'. Report by A. C. Johansen, 18 May 1908.

KB. Johannes Schmidt 'Papirer'. Ut.374.v.1. Privat, 'Instrukser og Breve, ca. 1902–1908'. Letter from E. Warming to J. Schmidt, 2 May 1902.

KB. Johannes Schmidt 'Papirer'. Ut.374.v.1. Privat, 'Instrukser og Breve, ca. 1902–1908', Copy of Instructions from C. G. J. Petersen to J. Schmidt, 23 April 1903.

KB. Johannes Schmidt 'Papirer'. Ut.374.v.1. Privat, 'Instrukser og Breve, ca. 1902–1908'. Instructions from C. G. J. Petersen to J. Schmidt, 9 March 1904.

KB. Johannes Schmidt 'Papirer'. Ut.374.v.1. Privat, 'Instrukser og Breve, ca. 1902–1908'. Letter from C. G. J. Petersen to J. Schmidt, 28 June 1904.

KB. Johannes Schmidt 'Papirer'. Ut.374.v.1. Privat, 'Instrukser og Breve, ca. 1902–1908'. News clip from undated newspaper, July 1904.

KB. Johannes Schmidt 'Papirer'. Ut.374.v.1. Privat, 'Instrukser og Breve, ca. 1902–1908'. News clip from Politiken, July 1904.

KB. Johannes Schmidt 'Papirer'. Ut.374.v.1. Privat, 'Instrukser og Breve, ca. 1902–1908'. Sørensen in Politiken, July 1904.

KB. Johannes Schmidt 'Papirer'. Ut.374.v.1. Privat, 'Instrukser og Breve, ca. 1902–1908'. William Sørensen in *Politiken*, 6 August 1904.

KB. Johannes Schmidt 'Papirer'. Ut.374.v.1. Privat, 'Instrukser og Breve, ca. 1902–1908'. Letter from C. G. J. Petersen to J. Schmidt, 4 July 1905

KB. Johannes Schmidt 'Papirer'. Ut.374.v.1. Privat, 'Instrukser og Breve, ca. 1902–1908'. Letter from C. G. J. Petersen to J. Schmidt, 9 July 1905.

KB. Johannes Schmidt 'Papirer'. Ut.374.v.1. Privat, 'Instrukser og Breve, ca. 1902–1908'. Letter from C. G. J. Petersen to J. Schmidt, 24 July 1905.

KB. Johannes Schmidt 'Papirer'. Ut.374.V.1. Privat, 'Instrukser og Breve, ca. 1902–1908'. Letter from C. G. J. Petersen to J. Schmidt, 4 August 1906.

KB. Johannes Schmidt 'Papirer'. Ut.374.V.1. Privat, 'Instrukser og Breve, ca. 1902–1908'. Letter from J. Hjort to J. Schmidt, 21 May 1907.

KB. Johannes Schmidt 'Papirer'. Ut.374.V.1. Privat, 'Correspondance 1980–1916'. Letter from J. E. V. Boas to J. Schmidt, 1 May 1907.

KB. Johannes Schmidt 'Papirer'. Ut.374.V.1. Privat, 'Instrukser og Breve, ca. 1902–1908'. Letter from C. G. J. Petersen to J. Schmidt, 29 May 1907.

KB. Johannes Schmidt 'Papirer'. Ut.374.V.1. Privat, 'Instrukser og Breve, ca. 1902–1908'. Memo from J. Schmidt no date.

KB. Johannes Schmidt 'Papirer'. Ut.374.V.1. Privat, 'Instrukser og Breve, ca. 1902–1908'. Note by J. Schmidt, spring of 1908.

KB. Johannes Schmidt 'Papirer'. Ut.374.V.1. Privat, 'Instrukser og Breve, ca. 1902–1908'. Letter from C. G. J. Petersen to J. Schmidt, 9 March 1908.

KB. Johannes Schmidt 'Papirer'. Ut.374.V.1. Privat, 'Instrukser og Breve, ca. 1902–1908'. Letter from J. Schmidt to C. G. J. Petersen, 9 March 1908. (letterhead says '1907', but it is clear from the rest of the correspondence that the correct year is 1908).

KB. Johannes Schmidt 'Papirer'. Ut.374.V.1. Privat, 'Instrukser og Breve, ca. 1902–1908'. Letter from C. G. J. Petersen to J. Schmidt, 11 March 1908.

KB. Johannes Schmidt 'Papirer'. Ut.374.V.1. Privat, 'Instrukser og Breve, ca. 1902–1908'. Draft letter from J. Schmidt to C. G. J. Petersen, 12 March 1908.

KB. Johannes Schmidt 'Papirer'. Ut.374.V.1. Privat, 'Instrukser og Breve, ca. 1902–1908'. Letter from C. G. J. Petersen to J. Schmidt, 13 March 1908.

KB. Johannes Schmidt 'Papirer'. Ut.374.V.1. Privat, 'Instrukser og Breve, ca. 1902–1908'. Letters from C. G. J. Petersen to J. Schmidt, 27 March and 24 April 1908.

KB. Johannes Schmidt 'Papirer'. Ut.374.V.1. Privat Korrespondance 1902–1916. Letters from G. Hansen to J. Schmidt, 26 & 29 March 1911.

KB. Johannes Schmidt 'Papirer'. Ut.374.V.1. Privat Korrespondance 1902–1916. Letter from *Staatliche Fischereidirektion*, Hamburg to J. Schmidt, 26 April 1913.

KB. *Johannes Schmidt 'Papirer'*. Ut.374.V.1. Privat Korrespondance 1902–1916. 'Privat, 1913'. Letter from J. Schmidt to H. N. Andersen, 30 June 1913.

KB. Johannes Schmidt 'Papirer'. Ut.374.V.1. Privat Korrespondance 1902–1916. Letter from J. Schmidt to Kaptain Kall, 26 December 1911.

KB. Johannes Schmidt 'Papirer'. Ut.374.V.1. Privat Korrespondance 1902–1916. Letter from Commandør F. Gotschalk to J. Schmidt, 24 November 1912.

KB. Johannes Schmidt 'Papirer'. Ut.374.V.1. Privat Korrespondance 1902–1916. 'Privat kuvert især breve 1912–15'. Letter from O. Paulsen to J. Schmidt, 2 June 1913.

KB. Johannes Schmidt 'Papirer'. Ut.374.V.1. Privat Korrespondance 1902–1916. 'Privat kuvert især breve 1912–15'. Letter from J. Schmidt to A/S Motorfabrikken 'Dan', 19 May 1913

KB. Johannes Schmidt 'Papirer'. Ut.374.V.1. Privat Korrespondance 1902–1916. 'Privat kuvert især breve 1912–15'. Letter from G. Hansen to J. Schmidt 7 June 1913.

KB. Johannes Schmidt 'Papirer'. Ut.374.V.1. Privat 1908–1916. Breve fra Skrivebordet. Letter from O. Paulsen to J. Schmidt 15 June 1916.

KB. Johannes Schmidt 'Papirer'. Ut.374.V.1. Privat 1908–1916. Breve fra Skrivebordet. Letter from C. F. Drechsel to J. Schmidt 20 June 1916.

KB. Johannes Schmidt 'Papirer'. Ut.374.V.1. Privat 1908–1916. Breve fra Skrivebordet. Letter from C. F. Drechsel to J. Schmidt 28 June 1916.

KB. Johannes Schmidt 'Papirer'. Ut.374.V.1. Privat 1908–1916. Breve fra Skrivebordet. Letter from A d. Richelieu to J. Schmidt 30 June 1916.

Box V.2

KB. Johannes Schmidt 'Papirer'. Ut.374.V.2. Privat og Blandet Korr. 1912–1927. Letter from A d. Richelieu to J. Schmidt, 24 November 1916.

KB. Johannes Schmidt 'Papirer'. Ut.374.V.2. Privat og Blandet Korr. 1912–1927. Letter from A d. Richelieu to J. Schmidt, 11 February 1919.

KB. Johannes Schmidt 'Papirer'. Ut.374.V.2. Privat og Blandet Korr. 1912–1927. Letter from A d. Richelieu to J. Schmidt 5 March 1919.

KB. Johannes Schmidt 'Papirer'. Ut.374.V.2. 'Privat og Blandet korr. 1912–1927'. P. C. Paulsen to J. Schmidt, 27 January 1920.

KB. Johannes Schmidt 'Papirer'. Ut.374.V.2. Privat og Blandet Korr. 1912–1927. Membership card for 1920.

KB. Johannes Schmidt 'Papirer'. Ut.374.V.2. 'Privat og Blandet korr. 1912–1927'. P. C. Paulsen to J. Schmidt, 2 January 1923.

KB. Johannes Schmidt 'Papirer'. Ut.374.V.2. Privat og Blandet Korr. 1912–1927. J. Schmidt to F. Dalgas, 29 March 1923.

KB. Johannes Schmidt 'Papirer'. Ut.374.V.2. Privat og Blandet Korr. 1912–1927. J. Schmidt to Kammerherre A. Krieger, 18 April 1924.

KB. Johannes Schmidt 'Papirer'. Ut.374.V.2. Privat og Blandet Korr. 1912–1927. Press release by J. Schmidt 22 April 1924.

Box V.3

KB. Johannes Schmidt 'Papirer'. Ut.374.V.3. Kommandør Drechsels Papirer fra DANA-Ekspeditionen 1920. C. F. Drechsel to DANA-Committee 11 June 1919.

KB. Johannes Schmidt 'Papirer'. Ut.374.V.3. Kommandør Drechsels Papirer fra DANA-Ekspeditionen 1920. C. F. Drechsel to DANA-Committee 17 June 1919.

KB. Johannes Schmidt 'Papirer'. Ut.374.V.3. Kommandør Drechsels Papirer fra DANA-Ekspeditionen 1920. C. F. Drechsel to DANA-Committee 1 November 1919.

KB. Johannes Schmidt 'Papirer'. Ut.374.V.3. Kommandør Drechsels Papirer fra DANA-Ekspeditionen 1920. Circular from C. F. Drechsel to DANA Committee 3 December 1921.

KB. Johannes Schmidt 'Papirer'. Ut.374.V.3. Kommandør Drechsels Papirer fra DANA-Ekspeditionen 1920. Travel account from C. H. Ostenfeld to DANA Committee 5 April 1922.

Box V.4

KB. Johannes Schmidt 'Papirer'. Ut.374.V.4. Blandede papirer og breve fra 1923–26. D'Arcy W. Thompson to J. Schmidt, 16 January 1924.

Box V.6

KB. Johannes Schmidt 'Papirer'. Ut.374.V.6. Statsradiofonien to J. Schmidt, February 1931.

Box VI.1

KB. Johannes Schmidt 'Papirer'. Ut.374.VI.1. 'Dana-Expeditionen, 1928–30', J. Schmidt to C. H. Ostenfeld, 25 August 1927.

KB. Johannes Schmidt 'Papirer'. Ut.374.VI.1. 'Dana-Expeditionen, 1928–30', Minutes, 2nd meeting of the Komiteen for Dana Expeditionen, 8 June 1928

KB. Johannes Schmidt 'Papirer'. Ut.374.VI.1. 'Dana-Expeditionen, 1928–30', Minutes, 3rd meeting of the Komiteen for Dana Expeditionen, 14 July 1928

KB. Johannes Schmidt 'Papirer'. Ut.374.VI.1. 'Dana-Expeditionen, 1928–30', C. Ostenfeld to J. Schmidt, 2 June 1929.

KB. Johannes Schmidt 'Papirer'. Ut.374.VI.1. 'Dana-Expeditionen, 1928–30', J. Schmidt to A. Drachmann, 3 September 1929

KB. Johannes Schmidt 'Papirer'. Ut.374.VI.1. 'Dana-Expeditionen, 1928–30', J. Schmidt Memorandum, 4 September 1929.

KB. Johannes Schmidt 'Papirer'. Ut.374. VI.1. 'Dana-Expeditionen, 1928–30', C. Ostenfeld to J. Schmidt, 1 November 1929.

KB. Johannes Schmidt 'Papirer'. Ut.374.VI.1. 'Dana-Expeditionen, 1928–30', J. Schmidt to Komiteen for Dana-Expeditionen, 30 November 1929.

KB. Johannes Schmidt 'Papirer'. Ut.374.VI.1. 'Dana-Expeditionen, 1928–30', J. Schmidt to Komiteen for Dana-Expeditionen, 1 December 1929.

KB. Johannes Schmidt 'Papirer'. Ut.374.VI.1. 'Dana-Expeditionen, 1928–30', Circulair Drachmann to Komiteen for Dana-Expeditionen, 10 January 1930.

KB. Johannes Schmidt 'Papirer'. Ut.374.VI.1. 'Dana-Expeditionen, 1928–30', J. Schmidt to Komiteen for Dana Expeditionen, 6 February 1930.

KB. Johannes Schmidt 'Papirer'. Ut.374.VI.1. 'Dana-Expeditionen, 1928–30', J. Schmidt to C. Ostenfeld, 7 February 1930.

KB. Johannes Schmidt 'Papirer'. Ut.374.VI.1. 'Dana-Expeditionen, 1928–30', J. Schmidt to Komiteen for Dana-Expeditionen, 8 February 1930.

KB. Johannes Schmidt 'Papirer'. Ut.374.VI.1. 'Dana-Expeditionen, 1928–30', J. Schmidt to Komiteen for Dana-Expeditionen, 14 February 1930.

KB. Johannes Schmidt 'Papirer'. Ut.374.VI.1. 'Dana-Expeditionen, 1928–30', J. Schmidt to C. H. Ostenfeld, 16 April 1930.

KB. Johannes Schmidt 'Papirer'. Ut.374.VI.1. 'Dana-Expeditionen, 1928–30', J. Schmidt to C. H. Ostenfeld, 17 April 1930.

KB. Johannes Schmidt 'Papirer'. Ut.374.VI.1. 'Dana-Expeditionen, 1928–30', J. Schmidt to C. H. Ostenfeld, 21 May 1930.

KB. Johannes Schmidt 'Papirer'. Ut.374.VI.1. 'Dana-Expeditionen, 1928–30', J. Schmidt to Komiteen for DANA-Expeditionen, 9 June 1930.

Box VI.2

KB. Johannes Schmidt 'Papirer'. Ut.374.VI.2. 'Manuskripter'. Eksamens-opgave 1898.

KB. Johannes Schmidt 'Papirer'. Ut.374.VI.2. J. Schmidt to G. Tate Regan, 30 June 1922.

Box VI.3

KB. Johannes Schmidt 'Papirer'. Ut.374.VI.3. Preliminary Programme, BAAS Liverpool Meeting 1923.

KB. Johannes Schmidt 'Papirer'. Ut.374. 'Taler og Foredrag'. Speech, Town Hall, Wellington J. Schmidt, 15 January 1929.

KB. Johannes Schmidt 'Papirer'. Ut.374. 'Taler og Foredrag'. Speech, Royal Society of New South Wales, Sydney J. Schmidt, 5 February 1929.

KB. Johannes Schmidt 'Papirer'. Ut.374. 'Taler og Foredrag'. Speech, Den kongelige Skydebane, Copenhagen, J. Schmidt, 25 September 1930.

Box VI.4

KB. Johannes Schmidt 'Papirer'. Ut.374.VI.4. Blandet Korrespondance 1912–1928. Letter from C. F. Drechsel to J. Schmidt 6 April 1914.

KB. Johannes Schmidt 'Papirer'. Ut.374.VI.4. Blandet Korrespondance 1912–1928. Draft letter from J. Schmidt to A d. Richelieu 20 March 1919.

KB. Johannes Schmidt 'Papirer'. Ut.374.VI.4. Blandet Korrespondance 1912–1928. J. Schmidt to Johs Dalhoff 28 May 1931.

Box VII.2

KB. Johannes Schmidt 'Papirer'. Ut.374.VII.2. Copy of letter from M. Knudsen to Rockefeller Foundation 12 October 1928.

KB. Johannes Schmidt 'Papirer'. Ut.374.VII.2. Ministry of Seafaring and Fisheries to J. Schmidt 13 December 1930.

The Archives of the Museum of Natural History

Various photos

Published sources

Adler A. Drachmann, Anders Bjørn. In: Engelstoft P, Dahl S, eds. *Dansk Biografisk Leksikon*. 3rd ed. Copenhagen: Gyldendal; 1980:28–30.

Allen RC. *The British industrial revolution in global perspective*. Cambridge University Press Cambridge; 2009.

Andersen HN. *Tilbageblik og udvikling*. Copenhagen: Østasiatisk Kompagni; 1952.

Andersen, V. *Danmark-Filmen*, Copenhagen, 1925.

Anonymous, ed. *Betænkning angiven af den i henhold til lov nr. 294 af 30. september 1916 nedsatte rigsdagskommission angaaende de Dansk Vestindiske øer.* Copenhagen: J. H. Schultz; 1916.

Anonymous, ed. *Betænkning over forholdene paa de Dansk-Vestindiske øer afgiven af den allerhøjeste resolution af 18. november 1902 anordnede Vestindiske kommission.*Copenhagen: J. H. Schultz; 1903

Anonomous. *Carlsbergfondet Fundats*, 2013. http://www.carlsbergfondet.dk/~/media/pdf-filer/Fundats/CF%20Fundats%202013.ashx

Anonymous, De 'Dana', in: *Bataviaasch Nieuwsblad*, 11 November, 1929: 2;

Anonymous De 'Dana', in: *Het nieuws van den dag voor Nederlandsch-Indië*, 11 November, 1929: 2.

Anonymous. De danmark-film. *De Sumatra Post*. 12 November 1929:3.

Anonymous. Deense Diepzee-Expeditie, *De Sumatra Post*, 22 January 1929.

Anonymous, *Fiskeriberetning*, Ministry of Agriculture (for the years 1888–1910).

Anonymous, 'Foreningens Generalforsamling 1916', in: *Atlanten*, vol. 4, (1915–1917), p. 449–472.

Anonymous. *Forhandlingerne om overdragelsen af de Dansk-Vestindiske øer til de Amerikanske Forenede Stater 1915–1916*. Copenhagen: Udenrigsministeriet; 1916.

Anonymous. Gunstige resultaten der 'Dana'-expeditie. *Soerabaiasch Handelsblad*. 22 February 1930:2.

Anonymous, Het expeditie-vaartuig 'Dana', in: *De Indische Courant*, 1929, 11 November: 1

Anonymous, 'Medlemsfortegnelse', in: *Atlanten*, vol. 1, 1904–1906. p. 56

Anonymous, Messina earthquake and tsunami of 1908. In: *Encyclopaedia Britannica. Encyclopaedia Britannica Online Academic Edition* ed. Encyclopædia Britannica Inc.; 2013; accessed 21 November. <http://www.britannica.com/EBchecked/topic/1483443/Messina-earthquake-and-tsunami-of-1908>.

Anonymous, Marine Life. A Floating Museum. The *Dana* in Port. Queer Species of Fish. *The Sydney Morning Herald*, 6 February 1929 p 18.

Anonymous, Mostly personal. People of the Week, in: *Western Mail*, Thursday 28 February 1929 p 4

Anonymous, Oceanographische expeditie, in: *De Indische Courant*, 26 June 1928: 3.

Anonymous, Ocean Study. The *Dana* Expedition. Professor Schmidt Arrives, in: *Sydney Morning Herald*, 23 January 1929: 15.

Anonymous, ed. *Sandheden om Dansk Vestindien. Dansk Vestindisk samfunds skrifter 1.* Copenhagen: H. Hagerups Forlag; 1916.

Anonymous, 'Schmidt, Johannes' in: *Kraks Blaa Bog, Fem tusinde nulevende Mænd og Kvinders Levnedsforløb indtil Aar 1932*, Copenhagen: O. C. Olsen & Co.; 1930: 897–898

Anonymous, Solving Ocean Mysteries – Scientific Expedition, in: *Kalgoorlie Miner*, Western Australia, 23 May 1928: 3.

Anonymous, Studying the eel. Professor Schmidt's Lecture, in: Sydney Morning Herald, 7 February 1929: 17.

Anonymous. *Statistiske meddelelser*. 1909;4. series, vol. 40, issue 1–3. p. 10.

Anonymous. *Vendsyssel packing co. A/S 1898–1948 blade af selskabets historie*. Denmark: Vendsyssel Packing Co. A/S; 1948.

Anonymous, 'Eventyret om Aalen og *Dana*'s Mænds Gerning paa det 11 Maaneders Togt', in: *Vore Herrer*, 7. årgang (16), 17 August 1922.

Aoyama J. Life history and evolution of migration in catadromous eels (genus Anguilla). *Aqua-BioScience Monographs*. 2009;2(1):1–42.

Barfoed N. *Manden bag helten*, Copenhagen: Gyldendal, 2011.

Bjerager A. *Bryggerens Lærling. Povl Krogsgaard-Larsens erindringer.* Gylling, Denmark: Gyldendal; 2011

Bjorndal KA, Bolten AB, Chaloupka MY. Green turtle somatic growth model: Evidence for density dependence. *Ecol Appl*. 2000;10(1):269–282

Blegvad H. Christian Frederik Drechsel, *Den Store Danske Encyclopædi*, Copenhagen: Gyldendal, http://www.denstoredanske.dk/Dansk_Biografisk_Leksikon/Forsvar_og_politi/S%C3%B8officer/C._F._Drechsel (accessed, 2 January 2014).

Blegvad H. *Dansk biologisk station gennem 50 år, 1889–1939*. Copenhagen: C. A. Reitzels Forlag; 1943.

Blegvad H. *Fiskeriet i Danmark*. København: Selskabet til udgivelse af kulturskrifter; 1946. vol. 1.

Blom I. Kampen om Eirik Raudes Land. Pressgruppepolitikk i Grønlandsspørgsmålet 1921–1931. Oslo: Gyldendal Norsk Forlag; 1973.

Boesgaard H. De sidste Omdrejninger. In: Schmidt J, ed. *Dana's Togt omkring Jorden*. Copenhagen: Gyldendal; 1932:352–364.

Boëtius I, Boëtius J. Studies in the european eel, *anguilla anguilla* [L.]: *Experimantal induction of the male sexual cycle, its relation to temperature and other factors*. Andr. Fred. Høst & Søn; 1967

Boëtius J, Boëtius I, Hemmingsen AM, Bruun AF, Møller-Christensen E. Studies of ovarial growth induced by hormone injections in the European and American eel (*anguilla anguilla* L. and *anguilla rostrata* LeSueur). *Meddelelser fra Danmarks Fiskeri-og Havundersøgelser*. 1962;3(7/12):183–198.

Boetius J, Harding EF. A re-examination of Johannes Schmidt's Atlantic eel investigations. *Dana: A journal of fisheries and marine research*. 1985; 4: 129–162.

Boje P. *Danmark og multinationale virksomheder før 1950*, Odense: Odense University Press; 2000.

Boussoulengas A. 80 years of oceanographic research in the Aegean Sea. In: Lenz W, Deacon M, eds. *Ocean sciences: Their history and relation to man. Proceedings of the 4th international congress on the history of oceanography*. Hamburg: Deutsches Hydrographisches Institut; 1990:245–249.

Braudel F. *On history.* University of Chicago Press; 1982.

Bravo M, Sörlin S. Narrative and practice – an introduction. In: Bravo M, Sörlin S, eds. *Narrating the Arctic: A cultural history of Nordic scientific prac*tices. Science History Publications/USA; 2002:3–32.

Bruun AF. Bjarni Sæmundsson 1867–1940. *Dansk Fiskeritidende*. 1941:139.

Bruun AF. Jordomsejlingen med S/S 'Dana' (II) 1928–30. In: Winge Ø, Vedel Taaning Å, eds. Naturforskeren Johannes Schmidt. Hans Liv og Ekspeditioner. Copenhagen: Gyldendalske Boghandel Nordisk Forlag; 1947:143–158.

Bruun AF. Middelhavstogternes resultater. In: Winge Ø, Vedel Taaning Å, eds. *Naturforskeren Johannes Schmidt. Hans Liv og Ekspeditioner*. Copenhagen: Gyldendalske Boghandel Nordisk Forlag; 1947:52–56

Bruun AF. The Life and Work of Professor Johannes Schmidt. *Estratto dalla Rivista di Biologia*. 1934;16(1):3–22

Bruun AF. Til New Zealand og Tahiti for at fange Aal. In: Winge Ø, Vedel Taaning Å, eds. *Naturforskeren Johannes Schmidt. Hans Liv og Ekspeditioner*. Copenhagen: Gyldendalske Boghandel Nordisk forlag; 1947: 132–136.

Capanna E. Grassi versus Ross: Who solved the riddle of malaria? *International Microbiology*. 2006;9(1):69–74.

Carpine-Lancre J. Oceanographic Sovereigns: Prince Albert I of Monaco and King Carlos I of Portugal. In: Deacon M, Rice T, Summerhayes C, eds. *Understanding the Oceans. A century of exploration*. London and New York: UCL Press; 2001:56–68.

Christensen C. Eug. Warming. En Levnedsskildring. In: Kolderup Rosenvinge L, ed. *Eugenius Warming 3. november 1841–2. april 1924*. Copenhagen; 1924:7–30

Christensen C. 'Weis, Frederik Anton', in: Cedergreen Bech, Svend (eds.) *Dansk Biografisk Leksikon*, Copenhagen: Gyldendal; 1984 vol. 15: 376.

Coesel PFM, Alfinito S. *Staurastrum multinodulosum* (desmidiaceae), a little-known plankter from a crater-lake in Lazio (Italy). *Algological Studies*. 2006;122(1):1–6.

Debora, 'Paa Aalejagt Til Tahiti', *Tidens Kvinder*, 30 September 1926, 6–7.

Drechsel CF, Knudsen M, Ostenfeld CH, Petersen CGJ. *De Internationale Havundersøgelser 1902–1907*. Copenhagen: Kommissionen for Havundersøgelser / C. A. Reitzel; 1908.

Drechsel CF. *Island og Fiskeriet*, Copenhagen: J. H. Schultz' Bogtrykkeri, 1906.

Drechsel CF. 'Nogle Bemærkninger om Fiskeriet ved Island og Færøerne og dets Fremtid', in: *Atlanten*, vol. 1, 1904–1906. p. 109–117.

Drechsel CF. *Oversigt over vore Saltvandsfiskerier i Nordsøen og Farvandene indenfor Skagen*. 1890. Dansk Fiskerimuseum; 1988, first pub. 1890

Ege V. Johannes Schmidt 1877–1933, In: Holter H, Max Møller K, eds. *The Carlsberg Laboratory 1876–1976*. Copenhagen: Rhodos; 1976:195–209.

Eggers-Lura A. *Admiralen, Kongen & Kaptajnen. Danskere i elefanternes rige*. Copenhagen: Høst & Søn; 1998.

Ellemose S. *Kompagniet. – H. N. Andersens ØK 1884–2007*, Gylling: Jyllands-Postes Forlag, 2007.

Ernst PG. *Det store eventyr om H. N. Andersen og Ø. K.* Copenhagen: Frederik E. Pedersens forlag; 1937.

Fenchel T, Bendtsen J, Sand-Jensen K. *Naturen i Danmark: Havet*. Copenhagen: Gyldendal; 2006.

Finley C. *All the fish in the sea. Maximum Sustainable Yield and the Failure of Fisheries Management*, (The University of Chicago Press, 2011).

Fort T. *The book of eels.* Great Britain: Harper Collins Publishers; 2002.

Frazer NB, Ladner RC. A growth curve for green sea turtles, *Chelonia mydas*, in the US Virgin Islands, 1913–14. *Copeia.* 1986(3):798–802.

Fort T. *The Book of Eels*. Great Britain: Harper Collins Publishers; 2002.

Fulton TW. The life-cycle of the eel in relation to Wegener's hypothesis. *Nature.* 1923(March 17):359–360.

Garstang W. *Larval forms and zoological verses.*; Basil Blackwell, 1951, 1985 (reprinted 1985, Chicago: University of Chicago Press).

Gillis JR. 'Filling the Blue Hole in Environmental History', http://johnrgillis.com/works/bluehole2.pdf

Gillis JR. *The human shore: Seacoasts in history*. Chicago & London: The University of Chicago Press; 2012

Glamann K. *Beer and Brewing in Pre-Industrial Denmark*. Odense: University of Southern Denmark Press; 2005.

Glamann K. *Bryggeren I. C. Jacobsen på Carlsberg*. 2011 Special edition ed. Copenhagen: Gyldendal; 1990:

Glamann K. *Carlsbergfondet.* Copenhagen: Rhodos; 1976.

Glamann K. *The Story of Emil Chr. Hansen.* Copenhagen: Carlsberg Foundation; 2009.

Goodrich HB. Mendelian inheritance in fish. *The Quarterly Review of Biology* 1929;4(1):83–99

Grassi GB. The Reproduction and Metamorphosis of the Common Eel (*Anguilla vulgaris*). *Proceedings of the Royal Society of London*. 1896;60 (359–367):260–271.

Greve S, Nielsen E. Marquesas og Tahiti. In: Schmidt J, ed. *Dana's Togt omkring Jorden*. Copenhagen: Gyldendal; 1932:72–92.

Gøbel E. *A Guide to the Sources for the History of the Danish West Indies (U.S. Virgin Islands), 1671–1917*, (Odense: University of Southern Denmark Press, 2002).

Gøbel E. 'Sankt Thomas havn i det 19. og tidlige 20. århundrede', in: *Handels- og Søfartsmuseets Årbog*, 60, (2001), p. 23–64.

Hamblin J. *Oceanographers and the Cold War. Disciples of Marine Science*, (Seattle & London: University of Washington Press, 2005).

Hagemann GA. *Danmark og dets Vestindiske øer*. Copenhagen: J. H. Schultz; 1916.

Hansen G. *Red Vestindien! et forslag i den 11. time.* Copenhagen: H. Meyers Bogtrykkeri; 1916.

Hátún H, Payne M, Beaugrand G, et al. Large bio-geographical shifts in the north-eastern Atlantic Ocean: From the subpolar gyre, via plankton, to blue whiting and pilot whales. *Prog Oceanogr*. 2009;80(3):149–162.

Heinberg Å. *H. N. Andersen. Stifteren af Det Østasiatiske Kompagni*, Copenhagen: De Unges Forlag, 1952.

Holm, P. (1996). Technology transfer and social setting: The experience of Danish steam trawlers in the north sea and off iceland, 1879–1903. In Holm P, Janzen O & Thór J (Eds.), *Northern seas yearbook 1996* (pp. 113–157). Esbjerg: Fiskeri- og Søfartsmuseets Forlag.

Holter H. An outline of the history of the two departments. In: Holter H, Max Møller K, eds. *The Carlsberg Laboratory 1876–1976*. Copenhagen: Rhodos; 1976:24–49.

Hornby O. *Kolonierne i Vestindien*. Copenhagen: Politikens Forlag; 1980.

Hubbard J. *A Science on the Scales: The rise of Canadian Atlantic fisheries biology, 1898–1939*, (Toronto: University of Toronto Press, 2005)

Hubbs CL. Death Notice: Anton Frederik Bruun. *Copeia*. 1962;1962(2):481–482

Jakobsson J. Ocean travellers. ICES *Marine Science Symposia*. 2002;215:330–342.

Janicki C. Giovanni Battista Grassi. *Naturwissenschaften*. 1926;14(13):261–269.

Jansen CR. 'Schmiegelow, Chr.' in: Cedergreen Bech, S. ed. *Dansk Biografisk Leksikon*, Copenhagen, Gyldendal; 1983:13, 160–161.

Jensen AS & Hansen PM. Investigations on the Greenland Cod (*Gadus callarias*), in: Schmidt, J. (ed.) *Scientific report of the North-Western Area Committee for 1928–1929, Rapports et Procès-Verbaux des Réunions,* vol. 72, (1931), Pp. 1–41.

Jespersen KJV. *Rytterkongen – et portræt af Christian 10.* Copenhagen: Gyldendal; 2007.

Jespersen P. Danske handelsskibes og skonnerten *Margrethe*'s indsats. In: Winge Ø, Vedel Taaning Å, eds. *Naturforskeren Johannes Schmidt. Hans Liv og Ekspeditioner*. Copenhagen: Gyldendalske Boghandel Nordisk Forlag; 1947:57–69.

Jespersen P. Togter med M/S 'Dana' (I) 1920–21. In: Winge Ø, Vedel Taaning Å, eds. *Naturforskeren Johannes Schmidt. Hans liv og ekspeditioner*. Copenhagen: Gyldendalske Boghandel Nordisk Forlag; 1947:80.

Jespersen P. Vest paa til Panama. In: Schmidt J, ed. *Dana's Togt omkring Jorden 1928–1930*. Copenhagen: Gyldendal; 1932:39–56.

Jespersen P, Vedel Taaning Å. Den store atlantiske ekspedition med S/S 'Dana' (II) 1921–22. In: Winge Ø, Vedel Taaning Å, eds. *Naturforskeren Johannes Schmidt. Hans liv og ekspeditioner*. Copenhagen: Gyldendalske Boghandel Nordisk Forlag; 1947: 89–107.

Johannsen W. 'Johan Kjeldahl 1849–1900', In: Holter, H, Max Møller, K, eds. *The Carlsberg Laboratory 1876–1976*, Copenhagen: Rhodos; 1976. p. 52–62.

Johansen AP. *Foreningen De Danske Atlanterhavsøer og Foreningen Dansk Samvirke: Privatarkiv nr. 10154 og 10279*, Copenhagen: 1974.

Jølle HD. *Innledning til en historiefaglig biografi tillegg til Nansen. Opdageren.* [Ph.D.]. Tromsø: Universitetet i Tromsø; 2013.

Kaarsted T. *Admiralen. Andreas de Richelieu – forretningsmand og politiker i Siam og Danmark.* Odense: Odense Universitetsforlag; 1990.

Kjær B. A Danish National Institution: The Carlsberg Foundation. *Scandinavian Review*. 1993;81(2):87–93.

Kjørup S. Billeder, kilder og gode idéer. *Personalhistorisk Tidsskrift*. 1999(2):259–266.

Knudsen M. De internationale havundersøgelser. In: Winge Ø, Vedel Taaning Å, eds. *Naturforskeren Johannes Schmidt. Hans liv og ekspeditioner*. Copenhagen: Gyldendalske Boghandel Nordisk Forlag; 1947:108–112

Knudsen H. *Videnskabens mand. Fysiologen, formidleren og forskningsaktivisten Poul Brandt Rehberg*. Aarhus: Aarhus Universitetsforlag; 2010.

Koch CH. En ny videnskabshistorisk genre? *Personalhistorisk Tidsskrift*. 1999(2):280–286.

Kondrup JJ. Biografen og den biograferede. *Personalhistorisk Tidsskrift*. 1999(1999: 2):252–258.

Kofoid CA. *The Biological Stations of Europe*, Washington DC: United States Bureau of Education, 1910.

Kraft OB. *Ung mand undervejs – erindringer fra årene 1912–1926*. Copenhagen: Nyt Nordisk Forlag Arnold Busck; 1958.

Lammers KC. '1914–1920 Fra borgfred til Påskekrise', in Søren Mørch (ed.) *Danmarks Historie, 7*, 82–138.

Lange O. *Jorden er ikke større… – H. N. Andersen, ØK og storpolitikken 1914–1937*. Viborg: Gyldendal; 1988.

Larsen P. 'Niels Nielsen', in Holter, H, Max Møller, K, eds. *The Carlsberg Laboratory 1876–1976*, Copenhagen: Rhodos; 1976. p. 231–238.

Lehmann J. (ed.). *For eller Imod. Udtalelser med Henblik paa det eventuelle Salg af de Danske Vestindiske Øer*. Copenhagen: Lehmann & Stage; 1916

Lidegaard B. *I kongens navn. Henrik Kauffmann i Dansk diplomati 1919–1958*. Copenhagen: Samlerens Forlag; 1996.

Lidegaard B. *Overleveren. 1914–1945*, Carsten Due-Nielsen, Ole Feldbæk & Nikolaj Petersen (eds.): *Dansk Udenrigspolitiks Historie, vol. 4*, Copenhagen: Gyldendal Leksikon, 2003.

Ljungman AV. 1879. Bidrag till lösningen af Frågan om de stora sillfiskenas sekulära periodicitet. *Nordisk Tidsskrift for Fiskeri* 5, 257–268.

Læssøe Müller P. 'Rothe, Carl', in: Povl Engelstoft & Svend Dahl (eds.), *Dansk Biografisk Leksikon*, vol. 20, Copenhagen: J. H. Schultz Forlag; 1941, p. 219–220.

Machiavelli N. *Fyrsten*. 2. revised transl. from Italian ed. Hasselager: Helikon; 2006.

Mayor AG. Detecting ocean currents by observing their hydrogen-ion concentration. *Proc Am Philos Soc*. 1919;58(2):150–160.

McClenachan L. Social conflict, overfishing and disease in the Florida sponge fishery, 1849–1939. In: Starkey DJ, Holm P, Barnard M, eds. *Oceans past: Management insights from the history of marine animal populations*. Earthscan; 2008:25–46.

McNeill JR, McNeill WH. *The human web: A bird's-eye view of world history*. WW Norton & Company; 2003.

Meylan AB. Status of the hawksbill turtle (*eretmochelys imbricata*) in the Caribbean region. *Chelonian Conservation and Biology*. 1999;3(2):177–184.

Mills E. *Biological Oceanography – An Early history, 1870–1960*, (Toronto: University of Toronto Press, (2nd ed.) 2012).

Morcos S, Smed J, Svansson A. The eel and Mediterranean oceanography – A Danish contribution to ocean science. In: Morcos S, ed. *Ocean sciences bridging the millenia: A spectrum of historical accounts*. Beijing: UNESCO; 2004:197–207.

Mortensen MT. *Pensionatets kulturhistorie. fællesskab, flirt og frikadeller*. Aarhus: Aarhus Universitetsforlag; 2015.

Mortensen T. Børgesen F. 'En biologisk station i Dansk Vestindien', in: *Atlanten*. 1904;1(6):1–15.

Mortensen T. Siam-Ekspeditionen, in: Winge Ø, Vedel Taaning Å, eds. *Naturforskeren Johannes Schmidt. Hans Liv og Ekspeditioner*. Copenhagen: Gyldendalske Boghandel Nordisk Forlag; 1947: 12–16.

Mortensen T. Fra de dansk-vestindiske øer. *Geografisk Tidsskrift*. 1907;19:73–94.

Mou G. Lidt om *Dana*'s Radiostation. In: Schmidt J, ed. *Dana's Togt omkring Jorden 1928–1930*. 1932:365–368.

Munch P. *Erindringer 1914–1918 – Under Den første Verdenskrig*. Copenhagen: Nyt Nordisk Forlag Arnold Busck, 1961

Møller F. 'Foreningens Stiftelse og første Virksomhed. Uddrag af Foreningens Forhandlingsprotokol': in: *Atlanten*, vol. 1, 1904–1906. p. 23–32.

Møller Hansen M, Sparholt H. Den europæiske ål. http://www.galathea3.dk/dk/Menu/Forskning/Den+europ%C3%A6iske+%C3%A5l/Projektbeskrivelse. Accessed 04/19, 2013.

Nielsen H & Nielsen K. Danmark og Nobelprisen gennem 100 år, in: *Aktuel Naturvidenskab*, 2001, 1: 33–35.

Nielsen KH. Postcolonial partnerships: Deep sea research, media coverage and (inter)national narratives on the *Galathea* deep sea expedition from 1950 to 1952. *The British Journal for the History of Science*. 2010;43(1):75–98.

Nielsen KH, Harbsmeier M, Ries CJ. Studying Scientists and Scholars in the Field', in: Nielsen KH, Harbsmeier M, Ries CJ (eds.): *Scientists and Scholars in the Field*, (Aarhus: Aarhus University Press, 2012)

Nielsen KH, Ries CJ, Østergaard KB. 'Danske Videnskabelige ekspeditioner I perioden', in: Nielsen KH, Nielsen H (eds.), *Viden uden Grænser 1920–1970, Dansk Naturvidenskabs Historie vol. 4*, Aarhus: Aarhus University Press; 2006. p. 283–302.

Nørregaard G. *Vore gamle tropekolonier, vol. 4. dansk vestindien 1880–1917. reformforsøg og forhandlinger.* Copenhagen: Fremad; 1966.

Olesen Larsen P. Anderledes – og tak for det. *Carlsbergfondet Årsskrift*. 2011;2011:58–77.

Olesen Larsen P. *Iværksætter og forvaltere. Carlsberg Laboratorium 1876–2014*. Copenhagen: Carlsbergfondet; 2015.

Olesen Larsen P. *Stadier på forskningens vej. Dansk forskningspolitik i går og i dag*. Odense: University of Southern Denmark Press; 2010.

Ostenfeld CH. 'Dana'-ekspeditionen. *Danmarksposten*. 1922;3(8):125–127.

Ostenfeld CH. Nygaard G. On the phytoplankton of the Gatun Lake, Panama Canal. *Dansk Botanisk Arkiv*. 1925;4:1–16.

Ostermann H. 'Ryder, C. H.', in: Povl Engelstoft & Svend Dahl (eds.), *Dansk Biografisk Leksikon*, vol. 12, Copenhagen: J. H. Schultz Forlag; 1937. p. 485–486.

Paulsen O. Havundersøgelserne i de nordlige Have begynder. In: Winge Ø, Vedel Taaning Å, eds. *Naturforskeren Johannes Schmidt. Hans Liv og Ekspeditioner*. Copenhagen: Gyldendalske Boghandel Nordisk Forlag; 1947: 17–25.

Paulsen O. I europæiske Farvande. In: Schmidt J, ed. *Dana's Togt omkring Jorden*. Copenhagen: Gyldendal; 1932:19–38.

Paulsen O. Middelhavstogterne med 'Thor'. In: Winge Ø, Vedel Taaning Å, eds. *Naturforskeren Johannes Schmidt. Hans Liv og Ekspeditioner*. Copenhagen: Gyldendalske Boghandel Nordisk Forlag; 1947: p. 45–51.

Paulsen O. Slægt og ungdom. In: Winge Ø, Vedel Taaning Å, eds. *Naturforskeren Johannes Schmidt. Hans Liv og Ekspeditioner.* Copenhagen: Gyldendalske Boghandel Nordisk Forlag; 1947:9–11.

Payne BJ. *Fishing a Borderless Sea: Environmental Territorialism in the North Atlantic 1818–1910*, East Lansing: Michigan State University Press. 2010.

Pedersen J. Hvilken kamp for at undslippe (kritik mod Thomas Söderqvist). Historisk Tidsskrift. 1999(99):480–491.

Pedersen J. *The Carlsberg Foundation*. Copenhagen: Bianco Luno; 1956.

Petersen CGJ. *Aalen. Hvad man ved om den samt anvisning paa at forøge udbyttet af Aalesfiskeriet*. Copenhagen: Det Schubotheske Forlag, 1904.

Petersen CGJ. *Det videnskabelige udbytte af Kanonbaaden 'Hauchs' Togter i de danske Farvande indenfor Skagen, i Aarene 1883–86*. Copenhagen: Høst & Søns Forlag; 1893.

Petersen CGJ. Fiskenes biologiske Forhold i Holbæk Fjord. *Beretning fra Den Danske Biologiske Station*. 1890–91;1:123–183.

Petersen CGJ. Hvor og under hvilke Forhold kunne Rødspættens Æg udvikle sig til Unger indenfor Skagen? *Fra Den Danske biologiske Station*. 1904;12 (1902–03):3–23.

Petersen CGJ. Larval Eels (leptocephalus brevirostris) of the Atlantic Coast of Europe. *Meddelelser fra Kommissionen for Havundersøgelser: serie Fiskeri*. 1905;1(5):3–5.

Petersen CGJ. *Om Biologisk Stations Virksomhed fra 1889–1914. Efterskrift 1918*. Copenhagen: Centraltrykkeriet; 1919.

Petersen CGJ. *Om de skalbærende Molluskers Udbredningsforhold i De Danske Have indenfor Skagen*. Copenhagen: Andr. Fred. Høst & Søns Forlag, 1888.

Petersen CGJ. The Common Eel (*Anguilla vulgaris* T.) gets a particular Breeding-dress before its Migration to the Sea. *Report from the Dansk Biologisk Station*. 1896;5.

Petersen CGJ. What is overfishing? *Journal of the Marine Biological Association*. 1903;6:587–594

Petersen J. *The Carlsberg Foundation*, Copenhagen: Bianco Luno; 1956.

Petterson H. The 'Johannes Schmidt' Ridge in the Indian Ocean. *Nature*. 1934;134. (7 July 1934).

Pietsch TW. *Oceanic anglerfishes. extraordinary diversity in the deep sea*. Berkeley & Los Angeles: University of California Press; 2009.

Pontoppidan E. *Det første Forsøg paa Norges naturlige Historie vol. 1–2*. Copenhagen: 1752–53.

Possing B. *Viljens styrke: Nathalie zahle: En biografi om dannelse, køn og magtfuldkommenhed*. Gyldendal; 1992

Poulsen EM. *Kommissionen for Danmarks Fiskeri- og Havundersøgelser 1902–1952*. Copenhagen: Kommissionen for Danmarks Fiskeri- og Havundersøgelser / C. A. Reitzel; 1952.

Poulsen B. 'Fisheries', In: Kotzé L & Morse S (eds.) *The Berkshire Encyclopedia of Sustainability vol. 9. Afro-Eurasia: Assessing Sustainability*, Great Barrington, MA: Berkshire Publishing Group, (2012), 136–140.

Poulsen B. Marin miljøhistorie – samfundsrelevant tværvidenskab. In: Enevoldsen T, Jelsøe E, eds. *Tværvidenskab i teori og praksis*. Copenhagen: Hans Reitzel; 2012:180–200.

Prytz S. *Warming – botaniker og rejsende*. Lynge: Bogan's Forlag A/S; 1984

Rancke-Madsen E, Koefoed RG. In: Cedergreen Bech S, ed. *Dansk Biografisk Leksikon*, vol. 8. 3rd ed. Copenhagen: Gyldendal; 1981:146–147.

Reeh L. Dansk åleekspedition 2014 er retur efter fem ugers togt i sargassohavet. 2014. http://www.aqua.dtu.dk/Nyheder/2014/05/140506_Dansk-aaleekspedition-retur

Richelieu A. Landstingsmand admiral A. de Richelieus tale i landstinget. In: Anonymous, ed. *Sandheden om dansk vestindien. dansk vestindisk samfunds skrifter 1*. Copenhagen: H. Hagerups Forlag; 1916:23–29.

Ries C. Tropesejlads. In: Schmidt J, ed. *Dana's Togt omkring Jorden 1928–1930*. Copenhagen: Gyldendal; 1932:139–152.

Rindom S, Tomkiewicz J, Munk P, et al. Eels in culture, fisheries and science in denmark. In: Tsukamoto K, Kuroki M, eds. *Eels and Humans*. Japan: Springer Science+Business Media; 2014:41–60

Rozwadowski, H., *Fathoming the Ocean. The Discovery and Exploration of the Deep Sea*, (Cambridge, MA & London: The Belknap Press of the Harvard University Press, 2005).

Rozwadowski H. *The sea knows no boundaries. A century of marine science under* ICES. Seattle and London: ICES/University of Washington Press; 2002.

Rud M. *Grønlandsforskeren Alfred Wegener og de drivende kontinenter*. 2nd ed. Copenhagen: Vandkunsten; 2007.

Russell ES. Johannes Schmidt (1877–1933). *Journal du Conseil.* 1933;8(2):145–160.

Sand-Jensen K. Naturhistorie. In: Kjærgaard PC, ed. *Lys over landet 1850–1920. dansk naturvidenskabs historie 3*. Aarhus: Aarhus Universitetsforlag:141–192.

Sandbeck T. *Danske havforskningsskibe gennem 250 år*. Stenstrup: Skib Forlag; 2007

Sander F. 'Jacobsen, Vagn' in: Engelstoft P & Dahl S (eds.), *Dansk Biografisk Leksikon*, vol. 11, Copenhagen: J. H. Schultz Forlag; 1937. p. 336–338

Schiönning H. Albert Klöcker 1862–1923, in: Holter H, Møller KM, eds. *The Carlsberg Laboratory 1876–1976*. Copenhagen: Rhodos; 1976:191–194.

Schmidt I. *Blade af Carlsbergs historie. Kaptajn S. A. van der Aa Kühle og hans liv på Carlsberg*. Copenhagen: Bianco Lunos Bogtrykkeri; 1956.

Schmidt J. 'Aale-Spørgsmålet i det Indo-Pacifiske Omraade. Vil Dansk Samvirke's Medlemmer Bidrage til dets Løsning?', in: *Danmarksposten*, (1923), 12: 215–216.

Schmidt J. *Bidrag til Kundskab om Skuddene hos den gamle Verdens Mangrovetræer*. Copenhagen: Det nordiske Forlag; 1903.

Schmidt J. Breeding places and migrations of the eel. *Nature*. 1923;111:51–54.

Schmidt J. Can different clones be characterised by the number of marginal teeth in the leaves. *Comptes-rendus des travaux du Laboratoire Carlsberg*. 1918;14(2):1–24.

Schmidt J. Contributions to the Life-History of the Eel (*Anguilla vulgaris*, flem.). *Rapport et Proces-Verbaux des Reunions*. 1906;5:137–274.

Schmidt J. (ed.) *Dana's togt omkring jorden 1928–1930*. Copenhagen: Gyldendalske Boghandel; 1932.

Schmidt J. *Danske Undersøgelser i Atlanterhavet og Middelhavet over Ferskvandsaalens Biologi*. Copenhagen: C. A. Reitzel; 1912

Schmidt J. Die Laichplätze des Flußaals. *Internationale Revue der gesamten Hydrobiologie und Hydrographie*. 1923;11(1-2):1–40.

Schmidt J. (ed.) *Flora of Koh Chang, Contributions to the knowledge of the vegetations in the Gulf of Siam, 1900–1916*, (Reprint from Botanisk Tidsskrift). Copenhagen: Bianco Luno, 1916.

Schmidt J. Indledning in: Schmidt J (ed.) *Dana's Togt Omkring Jorden 1928–1930*. Copenhagen: Gyldendalske Boghandel; 1932, p. 9–18.

Schmidt J. (ed.) *Investigations on Hops* (*Humulus lupulus L.*) *1913–1918*, i-xi, (reprint from *Comptes-rendus des travaux du Laboratoire Carlsberg* vol. 10–14), Copenhagen, 1918.

Schmidt J. Live Specimens of Spirula, Nature 110, 788–790 (9 December 1922).

Schmidt J. Marking experiments with turtles in the Danish Eest Indies. *Meddelelser fra Kommissionen for Havundersøgelser: serie Fiskeri*. 1916;5(1).

Schmidt J. On the amount of lupulin in plants raised by crossing. *Comptes-rendus des travaux du Laboratoire Carlsberg*. 1915;11:165–183.

Schmidt J. On the aroma of hops. *Comptes-rendus des travaux du Laboratoire Carlsberg*. 1915;11:149–163.

Schmidt J. On the aroma of in plants raised by crossing. *Comptes-rendus des travaux du Laboratoire Carlsberg*. 1917;11:330–332.

Schmidt J. 'On the contents of Oxygen in the ocean on both sides of Panama', *Science*, 1925; 61, 1588, 592–593.

Schmidt J. On the distribution of the fresh-water eels (anguilla) throughout the world: I. atlantic ocean and adjacent regions, a biogeographical investigation. *Meddelelser fra Kommissionen for Havundersøgelser: serie Fiskeri.* 1909;3(7).

Schmidt J. On the distribution of the freshwater eels throughout the world. part II. indo-pacific region. *Det Kongelige Danske Videnskabernes Selskabs Skrifter*. 1925;8(10(4)):329–382.

Schmidt J. On the early larval stages of the fresh-water eels (*anguilla*) and some other North Atlantic murænoids. Meddelelser fra Kommissionen for Havundersøgelser: serie Fiskeri. 1916;5(4):1–20.

Schmidt J. On the flowering time of plants raised by crossing. *Comptes-rendus des travaux du Laboratoire Carlsberg*. 1915;11:188–198.

Schmidt J. Racial Investigations: 10. The Atlantic Cod (*Gadus callarias* L.) and local Races of the same. *C R Trav Lab Carlsberg*. 1930;18(6): 1–71.

Schmidt J. Racial studies in fishes. I. Statistical investigations with *Zoarces viviparus* L. *Journal of Genetics.* 1918;7:105–119.

Schmidt J. Racial studies in fishes. II. Experimental investigations with *Lebistes reticulatus* (Peters) Regan. *Journal of Genetics*. 1919;8(3):147–153.

Schmidt J. Racial studies in fishes. III. Diallel crossings with trout (*salmo trutta* L.). *Journal of Genetics.* 1919;9(1):61–67.

Schmidt J. Racial studies in fishes. IV. Experimental investigations with *Zoarces viviparus* L. *Journal of Genetics.* 1920;10(3):179–191.

Schmidt J. Diallel crossings with the domestic fowl. *Journal of Genetics.* 1922;12(3):241–245.

Schmidt J. (ed.) Racial Investigations, 1917–1930, i-x, (reprint from *Comptes-rendus des travaux du Laboratoire Carlsberg*, vol. 13–18). Copenhagen, 1931.

Schmidt J. (ed.), *Report on the Danish oceanographical expeditions 1908—1910 to the Mediterranean and adjacent seas — vol 1: Introduction, Hydrography, Deposits of the Sea Bottom*. Copenhagen: 1912.

Schmidt J. (ed.), *Report on the Danish oceanographical expeditions 1908—1910 to the Mediterranean and adjacent seas.* Reports no. 1–10. Copenhagen: 1912–.

Schmidt J. (ed.) *Scientific reports of the north western area committee 1924–29*. 3 reports. ICES, *Rapport et Proces-Verbaux des Reunions*, vol. 39, 57, 72; 1929.

Schmidt J. The breeding places of the eel. *Philosophical Transactions of the Royal Society of London.Series B, Containing Papers of a Biological Character*. 1923;211:179–208.

Schmidt J. (ed.) *The Carlsberg Foundation's oceanographical Expedition round the World 1928–30*. Report no. 2. Copenhagen: ; 1932.

Schmidt J. (ed.) *The Danish 'Dana'-Expeditions 1920–22 in the North Atlantic and Gulf of Panama : Leader: Johs. Schmidt : Oceanographical reports ed. by the 'Dana'-committee* : No. 1–8. 2 vol. Copenhagen: ; 1926–31.

Schmidt J. The distribution of the pelagic fry and the spawning regions of the gadoids in the North Atlantic from Iceland to Spain. *Rapports et procès-verbaux des Réunions du Conseil International pour l'Exploration de la Mer.* 1909;10(4):150.

Schmidt J. The freshwater eels of New Zealand. *Transactions and Proceedings of the Royal Society of New Zealand.* 1928:379–388.

Schmidt J. The transatlantic migration of the eel-larvae. *Nature.* 1924; 113:12.

Schmidt J. The growth in length of hop stems and its diurnal periodicity. *Comptes-rendus des travaux du Laboratoire Carlsberg.* 1913;10:233–251.

Schmidt J. The occurrence of the wild hop in Denmark. *Comptes-rendus des travaux du Laboratoire Carlsberg.* 1917;11:314–329.

Schmidt J. The rotational movement of hop stems and its diurnal periodicity. *Comptes-rendus des travaux du Laboratoire Carlsberg.* 1913; 10:267–283.

Smith TD. *Scaling fisheries: The science of measuring the effects of fishing, 1855–1955.* 2nd revised 2007 ed. New York: Cambridge University Press; 1994

Schwach V. *Havet, fisken og Vitenskapen. Fra fiskeriundersøgelser til havforskningsinstitutt 1860–2000,* Oslo: Havforskningsinstituttet; 2000.

Schwach V. *Til havs med vitenskapen. Fiskerirettet havforskning 1860–1970.* [Dr. Philos.]. Oslo: Universitetet i Oslo; 2011.

Schwerdtner-Máñez, K. & Poulsen, B. (eds.) *Perspectives on Oceans Past: A Handbook of Marine Environmental History,* (Springer Science + Business Media B. V., 2016).

Schwerdtner Mánez K, Holm P, Blight L, Coll M, MacDiarmid A, Ojaveer H, Poulsen B, Tull M. *The future of the oceans past: Towards a global marine historical research initiative.* I: PLoS One, Vol. 9, Nr. 7, e101466, (2014).

Secor DH. Historical roots of the migration triangle, ICES *Marine Science Symposia.* 2002;215:323–329.

Sinclair M. Marine biodiversity: ICES, Johannes Schmidt, and the Carlsberg connection. ICES *Insight.* 2010(47):4–9.

Sinclair M, Solemdal P. The development of 'population thinking' in fisheries biology between 1878 and 1930. *Aquatic Living Resources.* 1988;1(3):189–213.

Sjøqvist V. *Erik Scavenius: Danmarks udenrigsminister under to verdenskrige. Statsminister 1942–1945.* Copenhagen: Gyldendal, 1973.

Smed J. 'Ekkolod og Asdic', in: *Naturens Verden,* 32, (1948), p. 177–183.

Smed J. Martin Knudsen – the oceanographer, in: ICES *Marine Science Symposia*, 215: 124–131. 2002.

Spang-Hanssen E. *Carlsbergfondets Understøttelser 1876–1936*. Copenhagen: Bianco Lunos Bogtrykkeri A/S; 1937.

Spärck R, 'Jespersen, P.', in: Engelstoft P & Dahl S (eds.), *Dansk Biografisk Leksikon, vol. 11*, Copenhagen: J. H. Schultz Forlag; 1937. p. 594–595.

Strubberg AC. Nyttefiskenes Livshistorie studeres i Nordatlanten og sydpaa til Biscaya. In: *Naturforskeren Johannes Schmidt. Hans Liv og Ekspeditioner*. Copenhagen: Gyldendalske Boghandel Nordisk Forlag; 1947:26–36.

Szybalski W. My Road to Øjvind Winge, the Father of Yeast Genetics, *Genetics* (1 May, 2001) vol. 158 no. 1, 1–6.

Sæmundsson B. *Oversigt over Islands Fiske med Oplysninger om deres Forekomst, vigtigste biologiske Forhold og økonomiske Betydning*. Copenhagen: Kommissionen for Havundersøgelser, C. A. Reitzel; 1908. p. 106–107.

Söderqvist T. *Hvilken kamp for at undslippe. En biografi om immunologen og nobelpristageren Niels Kaj Jerne*. Copenhagen: Borgens Forlag; 1998.

Söderqvist T. Hvilken kamp for at undslippe. Svar til Jan Pedersen. *Historisk Tidsskrift*. 1999;99(2);

Sørensen SR, Butts IAE, Munk P, Tomkiewicz J. Effects of salinity and sea salt type on egg activation, fertilization, buoyancy and early embryology of european eel, *Anguilla anguilla*. *Zygote*. 2015:1–18.

Tarbensen K. *For alle danmarks fiskere – Danmarks Fiskeriforening 125 år, 1887–2012*. Taulov: Danmarks fiskeriforening; 2012.

Thamdrup B, Bristow LA, Canfield DE, Dalsgaard T, De Brabandere L, Revsbech NP. 'Om iltsvind, bakterier, og det marine kvælstofkredsløb', paper presented at 17. Dansk Havforskermøde, Roskilde Universitet, 2013.

Thomsen H. Oxygen in the tropical Pacific. *Nature*. 1931;127(3204):489–489.

Thomsen H. Togtets længste stræk. In: Schmidt J, ed. *Dana's togt omkring jorden 1928–1930*. Copenhagen: Gyldendal; 1932:57–71.

Thomsen H. Vandundersøgelser paa de store have. In: Winge Ø, Vedel Taaning Å, eds. *Naturforskeren Johannes Schmidt. Hans Liv og Ekspeditioner*. Copenhagen: Gyldendalske Boghandel Nordisk Forlag; 1947:167–173.

Tsukamoto K. Discovery of the spawning area for Japanese eel. *Nature*. 1992;356(6372):789–789–791.

Toft P. Kriser og nedgang. den økonomiske udvikling i Dansk Vestindien 1875–1917. *Erhvervshistorisk Årbog*. 1982;32:146–191.

Tovborg Jensen A. The men behind the scene (The Board of Trustees), in: Holter H, Møller KM (eds.) *The Carlsberg Laboratory 1876–1976*. Copenhagen: Rhodos; 1976: 259–281.

Van Ginneken VJT, Maes GE. The European eel (*Anguilla anguilla*, Linnaeus), its lifecycle, evolution and reproduction: A literature review. *Rev Fish Biol Fish*. 2005;15(4):367–398.

Various authors, ed. *Dana*-report : The Carlsberg Foundation's oceanographical expedition round the World 1928–30 and previous 'Dana'-expeditions under the leadership of the late Professor Johannes Schmidt. vol. 1–16. Copenhagen: Scandinavian Science Press; Høst & Søn; 1934–1991.

Vedel Taaning Å. Aaleundersøgelserne begynder for alvor. In: Winge Ø, Vedel Taaning Å, eds. *Naturforskeren Johannes Schmidt. Hans Liv og Ekspeditioner*. Copenhagen: Gyldendalske Boghandel Nordisk Forlag; 1947:37–44.

Vedel Taaning Å. Fiskeriundersøgelser i nordlige have efter krigen. In: Winge Ø, Vedel Taaning Å, eds. *Naturforskeren Johannes Schmidt. Hans Liv og Ekspeditioner*. Copenhagen: Gyldendalske Boghandel Nordisk Forlag; 1947:113–131.

Vedel Taaning Å. Hovedlinierne trækkes op. In: Winge Ø, Vedel Taaning Å (eds.) *Naturforskeren Johannes Schmidt. Hans Liv og Ekspeditioner*. Copenhagen: Gyldendalske Boghandel Nordisk Forlag; 1947:174–187.

Vendsyssel K. M. *Danmarks fiskehandels- og havfiskeriforening – spredte meddelelser om fiskeri og fiskehandel i anledning af foreningens 40 aars jubilæum den 18. september 1921*. Aalborg: Danmarks Fiskehandels- og Havfiskeriforening; 1921.

Wegener A. *Die Entstehung der Kontinente und Ozeane*. reprint of 1st and 4th edition. Braunschweig: Verlag Vieweg; 1980.

Wegener A. *Die Entstehung der Kontinente und Ozeane*. 4th ed. Braunschweig: Die Wissenschaft. Sammlung von Eizeldarstellungen aus den Gebieten der Naturwissenschaft und der Technik; 1929.

Wegener A. *Die Entstehung der Kontinente und Ozeane*. 2nd ed. Braunschweig: Die Wissenschaft. Sammlung von Eizeldarstellungen aus den Gebieten der Naturwissenschaft und der Technik; 1920.

Wegener A. Die Theorie der Kontinentalverschiebungen. *Zeitschrift der Gesellschaft für Erdkunde zu Berlin*. 1921(3/4):7–130.

Wegener A. Kontinentforskydningsteorien og dens betydning for de systematiske og de eksakte naturvidenskaber. *Naturens Verden*. 1923;7:193–217.

Wegener A. *Tre Foredrag holdte i Danmarks Naturvidenskabelige Samfund 1922*. Copenhagen: Danmarks Naturvidenskabelige Samfund; 1923.

Went AEJ. *Seventy years agrowing. A history of the international council for the exploration of the sea, 1902–1972*. Copenhagen: International Council for the Exploration of the Sea; 1972.

Westergaard M. 'Øjvind Winge 1886–1964', in: Holter H, Max Møller K, eds. *The Carlsberg Laboratory 1876–1976*. Copenhagen: Rhodos; 1976:
Winge Ø. Andre strenge anslaas – paa Carlsberg Laboratoriet. In: Winge Ø, Vedel Taaning Å (eds.) Naturforskeren Johannes Schmidt. Hans liv og ekspeditioner. Copenhagen: Gyldendalske Boghandel Nordisk Forlag; 1947:70–79.
Winge Ø. The location of eighteen genes in *lebistes reticulatus*. *Journal of Genetics*. 1927;18(1):1–43.
Winge Ø. Crossing-over between the X-and the Y-chromosome in Lebistes. *Journal of Genetics*. 1923;13(2):201–217.
Winge Ø. A peculiar mode of inheritance and its cytological explanation. *Journal of Genetics*. 1922;12(2):137–144.
Winge Ø. One-sided masculine and sex-linked inheritance in *lebistes reticulatus*. *Journal of Genetics*. 1922;12(2):145–162.
Winge Ø, Vedel Taaning Å, eds. *Naturforskeren Johannes Schmidt. Hans Liv og Ekspeditioner*. Copenhagen: Gyldendalske Boghandel Nordisk Forlag; 1947.
Wolff T. *Danske ekspeditioner på verdenshavene: Dansk havforskning gennem 200 år*. Copenhagen: Rhodos; 1967.
Wolff T, Jespersen PC. In: Engelstoft P, Dahl S, eds. *Dansk Biografisk Leksikon*. 3rd ed. Copenhagen: Gyldendal; 1980.
Wolff T, Jungersen HFE. in: Cedergreen Bech, S. ed. *Dansk Biografisk Leksikon*, Copenhagen, Gyldendal; 1981:7, 487–488.
Wolff T, Schmidt EJ. In: Cedergreen Bech, S. ed. *Dansk Biografisk Leksikon*, Copenhagen, Gyldendal; 1983:13,148–149
Wolff T, Vedel Taaning Å. in: Engelstoft P, Dahl S, eds. *Dansk Biografisk Leksikon*. 3rd ed. Copenhagen: Gyldendal; 1980.
Worm Begtrup J. *Dansk-Vestindiske Nationalbank 1904–1935 – En undersøgelse af bankens oprettelse, drift og lukning, herunder dens ret til at trykke penge med Dansk værdi på et Amerikansk ejet territorium*. Unpublished MA-thesis, University of Copenhagen, 2008.
Worm Begtrup J. 'Den Dansk-Vestindiske Nationalbank 1904–1935', in: *Erhvervshistorisk Årbog*, 59, (2010) p. 183–211.
Worster D. Living in nature: Biography and Environmental History. In: Myllyntaus T, ed. *Thinking through the Environment. Green Approaches to Global History*. Cambridge: White Horse Press; 2011:28–39.

Index

A

B

C

D

E

F

G

H

I

J

K

L

M

N

S

T

U

V

W

X

Y

Z